Practical Collider Physics

Practical Collider Physics

Andy Buckley
School of Physics and Astronomy, University of Glasgow, Glasgow, UK

Christopher White
School of Physical and Chemical Sciences, Queen Mary University of London, London, UK

Martin White
School of Physical Sciences, University of Adelaide, Adelaide, Australia

IOP Publishing, Bristol, UK

ISBN   978-0-7503-2444-1 (ebook)
ISBN   978-0-7503-2442-7 (print)
ISBN   978-0-7503-2445-8 (myPrint)
ISBN   978-0-7503-2443-4 (mobi)

DOI   10.1088/978-0-7503-2444-1

Version: 20211201

IOP ebooks

British Library Cataloguing-in-Publication Data: A catalogue record for this book is available from the British Library.

Published by IOP Publishing, wholly owned by The Institute of Physics, London

IOP Publishing, Temple Circus, Temple Way, Bristol, BS1 6HG, UK

US Office: IOP Publishing, Inc., 190 North Independence Mall West, Suite 601, Philadelphia, PA 19106, USA

Supplementary information to accompany this book is available at (https://www.adelaide.edu.au/pcp).

Contents

Preface

The life of a particle physicist is, in many ways, a charmed one. Students entering the field rapidly discover an amazing world of international collaboration and jet-setting, plus fascinating physics problems that are at the frontier of human understanding. Furthermore, particle physics is extraordinarily broad in the range of subjects and techniques it encompasses: from quantum field theory to statistics, to computational methods and 'big data' processing, to detector development, operations, and engineering. This makes the field exciting for students with a wide range of computational and mathematical backgrounds.

Nevertheless, there is one time of year that every supervisor of graduate students dreads, and that is the arrival of new students. This has nothing to do with the students themselves, and everything to do with the fact that much of what they need to know is not written down anywhere. Despite a mountain of excellent books on quantum field theory, statistics, and experimental particle physics, there are no books that summarise the many pieces of received wisdom that are necessary to actually *do* something with hadron collider data. For the supervisors, this entails having the same conversations every year in an increasingly weary tone. For the poor arriving students, this makes the journey into particle physics considerably more stressful than it has to be, as they are left to piece together the toolkit they need to get them through their day job, whilst suffering from the false impression that everyone else is too clever for them.

The purpose of this book is to summarise the practical methods and tools of high-energy collider physics in a convenient form, covering the minimum amount of material that one needs to know about both theoretical and experimental methods for hadron colliders. Although the book is divided into a mostly theoretical part, followed by a mostly experimental part, we strongly encourage theorists and experimentalists to read both parts to gain a complete understanding of how hadron collider physics works in practice. It is impossible to properly interpret Large Hadron Collider (LHC) data without a great deal of experimental knowledge, and neither is it possible for experimentalists to properly complete their work without knowing a reasonable amount of theory. In essence, our book digests the normal supervisor–student mentoring conversations into a convenient form, so that students can hit the ground running as they start a graduate project in the field. As a reflection of this role, we have also tried to make this a *practical* manual, focused on concepts and strategies needed to make progress on concrete research tasks, and—as far as we can manage—with a minimum of formality. For supervisors, we hope that the book will provide a handy reference for those senior moments when they cannot quite remember how something works.

The first part of the book mostly covers theoretical topics. Chapter 1 provides a brief, non-technical review of most of the concepts that we treat in detail later in the book, including the particles of the Standard Model, the different types of particle collider, relativistic kinematics and various concepts that are important to understand collider data. Chapter 2 covers a reasonable minimum of the topics in quantum field theory that must be understood to have a solid understanding of LHC data, including the gauge principle, the Standard Model, renormalisation and parton distribution functions. Clearly this will be insufficient for anyone seeking a deep knowledge of quantum field theory, and the aim instead is to provide a solid grounding that can be supplemented by the many excellent existing treatments of the subject. In Chapter 3, we describe the physics that connects the more fundamental concepts in quantum field theory to the data that is actually taken and observed in experiments, including concepts such as resummation, parton showers, hadronisation, the underlying event and jet algorithms. Much of this chapter summarises the physics behind Monte Carlo event generators. In Chapter 4, we provide a brief introduction to the sorts of theories that might supercede the Standard Model, the hunt for which is the primary reason to build particle colliders in the first place. Chapter 5 presents the statistical theory that is needed to make sense of any of the experimental observations at the LHC.

The second part of the book is mostly focussed on experimental techniques. We first describe the physics of particle detectors in Chapter 6, in addition to the techniques by which particles are reconstructed. Whilst experimentalists will be reading the chapter to understand their apparatus, theorists should read it to understand what factors control and limit the scope of what we can say about particle theories. In Chapter 7, we complete the picture by describing the computing and data processing environments at the LHC experiments. We then embark on a journey to explore, in detail, how data at the LHC is analysed to test the Standard Model and search for new particles. Topics common to all data analyses are described in Chapter 8, followed by a review of different direct particle search techniques in Chapters 9 and 10. High precision measurements are detailed in Chapter 11, and the reinterpretation of LHC analyses is considered in Chapter 12. Finally, in Chapter 13 we conclude the book with a quick review of what to look out for in the next decade of particle physics which, with delays in funding and building schedules, will probably remain current for a good few years as the book ages!

Our thanks to the many colleagues who have provided input to this process, either directly through draft-review and feedback, by authoring the many insightful papers that introduced and refined the concepts to come, or by serving as our own mentors along our journeys in high-energy physics thus far. Chapters 1 to 4 are largely based on a course that Chris taught at Queen Mary University of London, with some additional material from Andy and Martin. We are grateful to the students for their insightful comments and general enthusiasm, and owe particular thanks to Nadia Bahjat-Abbas for her many useful suggestions regarding the notes and exercises. Chapter 5 was primarily written by Andy, and refined by Martin for a physics data-analysis and modelling course taught at the University of Adelaide. We thank the Australian cohort of students for their skill in weeding out confusions and typos. Part II consists of extra original material by Martin (most of Chapters 6, 9, 10, 12 and 13) and Andy (most of Chapters 7, 8 and 11), but with additional material from all authors in each case. We suspect that a two-semester course could cover the whole book, or that a single-semester course could contain highlights from Parts I and II as required. Martin is acknowledged for devising and kickstarting the project, organising the writing schedule and leading the administrative work. Chris and Martin wish to thank Andy for his amazing additional work on the typesetting of the draft, plus his wonderful hand-drawn illustrations. We further thank Jan Conrad, Christoph Englert, Chris Gutschow, Thomas Kabelitz, Zach Marshall, Josh McFayden, Knut Morå, and Chris Pollard for their excellent feedback on the draft, and give special thanks to Jack Araz for his illustrative implementation of top-quark reconstruction performance.

Finally, we express our immense gratitude to our partners—Jo, Michael and Catherine—and our children Alec, Edith, Toby, Alfie and Henry. Their enduring love, support, and sanity were vital as we completed the book, large parts of which were written during the bewildering events of 2020–21. Although this book has been declared boring by a six year old, we hope that both commencing graduate students and seasoned practitioners will find much to enjoy in it.

Author biographies

Andy Buckley

Andy Buckley is a Senior Lecturer in particle physics at the University of Glasgow, Scotland. He began his career in collider physics as a PhD student at the University of Cambridge, developing Cerenkov-ring reconstruction and CP-violation data-analyses for the LHCb experiment. Later he worked on MC event generation, MC tuning, and data-preservation as a postdoctoral researcher at the Institute for Particle Physics Phenomenology in Durham, UK, before resuming experimental physics work on the ATLAS experiment at University College London, the University of Edinburgh, CERN, and finally Glasgow. His work on ATLAS has included responsibility for MC event modelling, from detector simulation to event generation; measurements of soft-QCD, b-jets and top-quark physics; and searches for the Higgs boson and new-physics $H \rightarrow b\bar{b}$ decay channels. Outside ATLAS, he is leader of the Rivet and LHAPDF projects, a work-package leader in the MCnet research & training network, and active in BSM phenomenology studies via the Contur, TopFitter, and GAMBIT collaborations.

Christopher White

Christopher White is a Reader in Theoretical Physics at Queen Mary University of London. He obtained his PhD at the University of Cambridge, specialising on high-energy corrections to the structure of the proton. He then moved to Nikhef, Amsterdam (the Dutch National Centre for Nuclear and High Energy Physics), where he broadened his research into hadron collider physics, including Monte Carlo simulation, the description of low momentum ('soft') radiation in QCD, and various aspects of top-quark and Higgs physics. Following further appointments at the Universities of Durham and Glasgow, Chris moved to his current position, where he has continued his collider-physics research alongside more formal work looking at relationships between theories like QCD, and (quantum) gravity.

Martin White

Martin White is a Professor in particle astrophysics, and Deputy Dean of Research in the Faculty of Sciences at the University of Adelaide, Australia. He obtained a PhD in high energy physics at the University of Cambridge, working as a member of the ATLAS experiment with interests in silicon-detector physics, supersymmetry searches, and supersymmetry phenomenology. He then moved to the University of Melbourne before arriving at Adelaide in 2013 to start a particle astrophysics group. Within the ATLAS collaboration, he spent several years performing tests of the silicon tracker before and after installation, whilst also searching for supersymmetric particles. Most recently, he developed a new theoretical framework for modelling resonance searches in the diphoton final state. Outside of ATLAS, he is Deputy Leader of the GAMBIT collaboration, a team of 70 international researchers who perform global statistical fits of new physics models, and is also a project leader in the DarkMachines collaboration, which seeks to find new applications of machine learning techniques in dark-matter research. He has broad interests in particle astrophysics phenomenology and data-science, including the development of new data-analysis techniques for collider and dark-matter search experiments, and studies of a wide range of Standard Model extensions. He is the author of the Physics World Discovery book *What's Next for Particle Physics?*

Part I

Theory and methods for hadron colliders

IOP Publishing

Practical Collider Physics

Andy Buckley, Christopher White and Martin White

Chapter 1

Introduction

The nature and fate of our universe have fascinated humankind for millenia, and we remain fascinated by the big questions underlying our existence: why is there something rather than nothing? How did the Universe get here, and how will it end? What are the basic building blocks of Nature, that operate at the smallest possible distance scales? Nowadays, such questions fall within the remit of **fundamental physics**. In particular, **particle physics** deals with the basic constituents of matter and their fundamental interactions, and forms part of the larger framework of **high-energy physics**, which may include other exotic objects such as strings or branes. These subjects overlap with astrophysics and cosmology, which describe the Universe at its largest scales, since we know that the Universe expanded outwards from a finite time in the past (the Big Bang). As we turn the clock back, the Universe gets smaller and hotter, such that high-energy physics processes become relevant at early times, whose imprints can be measured today.

Thousands of years of scientific research have culminated in our current understanding: that the Universe contains **matter**, that is acted on by **forces**. All forces that we observe are believed to be a consequence of four **fundamental forces**: electromagnetism, the strong nuclear force (that holds protons and neutrons together), the weak nuclear force (responsible for certain kinds of nuclear decay), and gravity. For the first three forces, we know how to include the effects of special relativity and quantum mechanics. The consistent theory that combines these two ideas is **quantum field theory** (QFT). For gravity, our best description is **general relativity** (GR). It is not yet known whether or not a sensible quantum theory of this force exists, but there are hints that it is needed to describe extreme regions of the Universe where classical GR breaks down, such as the centre of black holes, or the Big Bang itself. In particle physics, we can safely neglect gravity for the most part, as it will be much too weak to be noticed.

The matter particles can be divided into two types, First, there are the leptons, that feel the electromagnetic and weak forces. These are the electron e^-, together

Table 1.1. The matter particle content of the Standard Model, where the mass units are explained in the text. All particles have spin 1/2.

		Quarks			Leptons	
		Mass/GeV	EM charge		Mass/GeV	EM charge
1st generation	$\begin{pmatrix} u \\ d \end{pmatrix}$	0.002 0.005	$+\frac{2}{3}$ $-\frac{1}{3}$	$\begin{pmatrix} e^- \\ \nu_e \end{pmatrix}$	0.0005 ?	-1 0
2nd generation	$\begin{pmatrix} c \\ s \end{pmatrix}$	1.3 0.1	$+\frac{2}{3}$ $-\frac{1}{3}$	$\begin{pmatrix} \mu^- \\ \nu_\mu \end{pmatrix}$	0.106 ?	-1 0
3rd generation	$\begin{pmatrix} t \\ b \end{pmatrix}$	173 4.2	$+\frac{2}{3}$ $-\frac{1}{3}$	$\begin{pmatrix} \tau^- \\ \nu_\tau \end{pmatrix}$	1.78 ?	-1 0

with its heavier partners (the muon μ^- and tauon τ^-). Associated with each of these is a corresponding neutrino ν_i, whose name charmingly means 'little neutral one' in Italian. Next, there are the quarks, which feel the electromagnetic, strong and weak forces. These have odd historical names: up (u), down (d), charm (c), strange (s), top (t) and bottom (b). We can arrange all of these particles into three generations, where the particles in higher generations have exactly the same quantum numbers as the lower generations, apart from the mass. We summarise all masses and charges for the quarks and leptons in table 1.1. Each of them also has a corresponding **anti-particle**: this has the same mass, but all other quantum numbers (e.g. charge) reversed.

As we will see later on, quarks carry a type of charge (not related to electro-magnetic charge) called **colour**. Furthermore, free quarks are never observed—they are instead confined into composite particles called **hadrons**, whose net colour charge is zero. We may further subdivide hadrons into **baryons**, which contain three quarks[1], and **mesons**, containing a quark/anti-quark pair. Examples of baryons include the proton $p(uud)$ and neutron $n(udd)$. For mesonic examples, you may have heard of the **pions** $\pi^+(u\bar{d})$, π^0(a quantum superposition of $u\bar{u}$ and $d\bar{d}$) and $\pi^-(d\bar{u})$.

The basic idea of QFT is that fields are described by equations that can have wave-like solutions. The quantum aspect tells us that these waves cannot have continuous energy, but instead travel in distinct particle-like 'quanta'. Put more simply, forces are themselves carried by particles[2]. When matter particles interact due to a given fundamental force, they do so by exchanging the appropriate force-carrying particle. The carrier of the electromagnetic force is the **photon**, usually written γ (n.b. 'gamma rays' are a certain kind of high-energy electromagnetic wave). The weak force turns out to be carried by three particles: the W^-, W^+ and Z^0 ('W and

---

[1] One can also have anti-baryons, containing three anti-quarks.
[2] Likewise, matter particles also arise as quanta of fields, which are no more or less fundamental than the fields corresponding to the forces.

Table 1.2. The force–particle content of the Standard Model, where all particles have spin 1.

Force	Carrier(s)	Mass/GeV	EM Charge
Electromagnetism	γ (photon)	0	0
Strong	g (gluon)	0	0
Weak	$W^{\pm}, Z^0$	80.4 (W bosons), 91.2 (Z boson)	+1, −1, 0

Z bosons'), where the superscript denotes the electromagnetic charge. The W^+ and W^- are mutual antiparticles, whereas the γ and Z^0 are their own antiparticle. The strong force is carried by the **gluon**[3]. Finally, if a quantum theory of gravity exists, we would call the appropriate force carrier the **graviton**. The masses of the force carriers are summarised in table 1.2. In particular, we see that the carriers of the weak force are massive, but all other force carriers are massless. In most quantum gravity theories, the graviton is also massless, a property which is heavily constrained by astrophysical data.

The theory that describes the various matter (anti-)particles and all forces (excluding gravity) is called the **Standard Model of Particle Physics**, or SM for short. It is a QFT, where all matter and force 'particles' are described by fields. As well as the above content, it needs one more thing. It turns out that the equations of the SM are mathematically inconsistent unless there is an additional field called the **Higgs field**. The field has a particle associated with it—the **Higgs boson** (H), which was discovered as recently as 2012. We will discuss these topics in much more detail in later Chapters. For now, however, we note that a number of open puzzles remain, including the following non-exhaustive list:

- Why are there three generations of particles?
- Why do they have their particular masses/charges, and not other values?
- Why is there more matter than antimatter in the Universe, when the SM implies these are created in equal amounts?
- What are the mysterious **dark matter** and **dark energy** that seem to be required from astrophysical measurements, and which are not in the SM?
- Why are there three separate forces? Might these *unify* into a single force at high energy, which might also include gravity?
- Is there a deeper explanation for the Higgs field/boson?

Many theories **Beyond the Standard Model** (BSM) have already been proposed to answer some of these questions, but to date there is no unambiguous signature of any of them. Our main way of testing such theories is to use particle accelerator, or **collider** experiments. The history of particle accelerators goes back many decades, and indeed such machines were crucial in establishing the Standard Model itself. The current flagship experiment in the world is the **Large Hadron Collider** (LHC)

---

[3] So called because it 'glues the proton together'. Yes really.

at CERN, near Geneva. It is the most complicated machine ever built, and contains a number of experiments looking at various aspects of high-energy physics. It has already discovered the Higgs boson and, at the end of its second run, has yet to take 96% of its projected data. At the time of writing, possible new facilities are being discussed around the world, with a view to a new facility coming online after the next decade or so. It is thus a good time to be a collider physicist!

The aim of this book is to examine the theory that is used at modern particle colliders, and to describe how modern experimental analyses are actually carried out. This is a vast subject, and thus we will restrict our attention to the two **general physics detector** (GPD) experiments at the LHC, namely ATLAS and CMS. We will not focus so much on heavy-ion physics experiments (e.g. ALICE), flavour physics (e.g. LHCb) or neutrino physics, which are very exciting facilities, but are distinct in many respects and are described in detail elsewhere.

1.1 Types of collider

Having introduced the state of the art in particle physics, let us now examine particle collider experiments in more detail. The most general definition we can give of a collider is a machine that accelerates charged particles, and smashes them into other particles. The beams are typically accelerated and focused using electric and magnetic fields, hence why the particles need to be charged. After the collision, the resulting debris is picked up by **detectors**, and their output analysed to try to ascertain what happened. There are two main types of experiment, illustrated schematically in Figure 1.1.

In a **fixed-target experiment** (Figure 1.1(a)), a beam is incident on a stationary target. In terms of the particles, this means that a moving particle of some type collides with a stationary particle (possibly of a different type). In a **colliding beam experiment**, both particles are moving, and usually such experiments are built so that the overall momentum of the colliding particles is zero (i.e. such that the lab frame corresponds to the **centre-of-mass** frame).

Examples of fixed-target experiments include Rutherford's original experiment that discovered the atomic nucleus (1909), a collider called the Bevatron (1950s), and the Stanford Linear Accelerator Complex (SLAC) experiments of the 1960s–1970s, that did much to unravel the structure of the proton. Examples of colliding beam experiments include HERA (Hamburg, 1990s), the Tevatron (Fermilab, near Chicago, 1990s–2000s), and the LHC (CERN, near Geneva, 2000s–present). Looking at these examples, we see that fixed target experiments were more common in the past, and colliding beam experiments more recently. This is partly because colliding beams are much more difficult from an engineering point of view, so that

(a)   (b)

Figure 1.1. (a) A fixed–target experiment; (b) a colliding beam experiment.

there is little incentive to build a colliding beam experiment unless you really have to. The great disadvantage of a fixed-target experiment, however, is that most of the energy in the initial state goes into kinetic energy in the final state, and thus is not available for making new particles. Thus, colliders designed to discover heavy new particles have to involve colliding beams, in which all of the energy (in principle) is available for new particle creation.

Colliders can be linear (e.g. SLAC, mentioned above), or circular (e.g. HERA, the Tevatron, the LHC). In the latter case, powerful magnets are used to deflect the beams: at the LHC, for example, the so-called 'dipole magnets' used for this purpose weigh a couple of tons each[4]! The advantage of circular colliders is that the particle can go round the circle multiple times, getting faster each time. Thus, one can get away with less powerful electric fields to accelerate them. The main disadvantage of circular colliders is that charged particles undergoing circular trajectories emit **synchotron radiation**, and lose energy at a rate

$$\left| \frac{\mathrm{d}E}{\mathrm{d}t} \right| \propto m^{-4} r^{-2},$$

where m is the mass of the particle, and r the radius of the circle. Because this effect involves a high power of inverse mass, the effect is much worse for light particles, such as electrons and positrons. For this reason, modern (anti)-proton colliders are circular, but future $e^{\pm}$ colliders are more likely to be linear (although circular $e^{\pm}$ collider proposals are currently on the table). Note that the above formula suggests we can mitigate the effects of synchotron radiation by having a very large radius. This is why the circumference of the LHC (which occupies the same tunnel as the former $e^{\pm}$ collider LEP) is 27 km!

In this book, we will not worry about how beams are produced and manipulated. This belongs to the field of **accelerator physics**, which would easily fill an entire tome by itself. Rather, we will only care about what particles each beam contains. Given that these particles will be moving very fast, we will need to describe them using the appropriate language of special relativity. This is the subject of the following Section.

1.2 Relativistic kinematics

In modern colliders, particles are moving very close to the speed of light, so that relativistic effects become mandatory (indeed, together with quantum mechanics, they are built into quantum field theory). The most convenient way to talk about special relativity is to use four-vectors, and we here list some basic properties so as to introduce notation for what follows. Recall that we can combine space and time into the four-dimensional vector

$$x^{\mu} = (ct, x, y, z), \tag{1.1}$$

---

[4] If you visit CERN, you can see a spare dipole magnet on the grass outside the main cafeteria. The size is such that you will often see visiting children (and sometimes grown adults) climbing on it.

in which we have adopted Cartesian coordinates for the spatial components. Here μ is an index that goes from 0 to 3, where 0 labels the time component, i.e.

$$x^0 = ct, \quad x^1 = x, \quad x^2 = y, \quad x^3 = z.$$

Given these components, we can also define the related components

$$x_\mu = (ct, -x, -y, -z). \tag{1.2}$$

That is, a four-vector with a *lower* index has its spatial components flipped with respect to the corresponding four-vector with an *upper* index. This is just a definition, but is convenient when we talk about combining four-vectors to make dot products etc. Note that the vector x^μ describes the location of an 'event' in four-dimensional **spacetime**, also known as 'Minkowski space'. That this is the right language to use essentially follows from the fact that Lorentz transformations in special relativity mix up space and time, so that it no longer makes sense to separate them, as we do in Newtonian physics.

Given the position four-vector, we can define the four-momentum of a particle

$$p^\mu \equiv m \frac{\mathrm{d}x^\mu}{\mathrm{d}\tau}, \tag{1.3}$$

where x^μ is the position of the particle in spacetime, m its *rest mass*, and τ its **proper time**, namely the time that the particle experiences when at rest. If the particle is moving, the proper time is dilated according to the formula:

$$t = \gamma\tau, \quad \gamma = \left(1 - \frac{v^2}{c^2}\right)^{-1/2}, \tag{1.4}$$

where $v = |v|$ is the magnitude of the three-velocity of the particle. This allows us to write

$$p^\mu = m\gamma \frac{\mathrm{d}x^\mu}{\mathrm{d}t} = (\gamma mc, \gamma mv), \tag{1.5}$$

which is usually written instead as

$$p^\mu = \left(\frac{E}{c}, p\right), \tag{1.6}$$

where the **relativistic energy and three-momentum** are given, respectively, by

$$E = \gamma mc^2, \quad p = \gamma mv. \tag{1.7}$$

We define other useful relativistic formulae in appendix A.1.

Given any two four-vectors, we may define a dot product for them:

$$a \cdot b = a^0 b^0 - a^1 b^1 - a^2 b^2 - a^3 b^3, \tag{1.8}$$

where on the right-hand side we have used the component notation. The definition of equation (1.2) (which generalises to any four-vector) allows us to write the dot product neatly as

$$a \cdot b = a^\mu b_\mu \equiv b^\mu a_\mu, \tag{1.9}$$

where we have adopted Einstein's famous **summation convention**, in which repeated indices are summed over (i.e. there is an implicit $\sum_{\mu=0}^{3}$ in equation (1.9)). Note that the dot product vector is *not* the same as the usual dot product for three-dimensional Euclidean space. It includes timelike and spacelike components for a start, and also has a relative minus sign between the timelike and spacelike terms. One may then prove that the dot product of any two four-vectors is invariant under Lorentz transformations. In particular, this means that any dot product of four-momenta can be evaluated in any Lorentz frame we like, as the answer will always be the same. Following convention, we will often write the dot product of[5] a four-vector with itself as

$$a^2 \equiv a \cdot a = (a^0)^2 - (a^1)^2 - (a^2)^2 - (a^3)^2. \tag{1.10}$$

From equation (1.6), we find (in a general frame where E, $|\boldsymbol{p}| \neq 0$)

$$p^2 = \frac{E^2}{c^2} - |\boldsymbol{p}|^2. \tag{1.11}$$

In the rest frame of a massive particle, however, we have

$$p^\mu = (mc, 0) \quad \Rightarrow \quad p^2 = m^2 c^2, \tag{1.12}$$

so that combining equations (1.11)) and (1.12) yields the **energy–momentum relation**

$$E^2 - c^2 |\boldsymbol{p}|^2 = m^2 c^4. \tag{1.13}$$

From now on, we will adopt **natural units**, which are ubiquitous throughout relativistic quantum field theory. This involves ignoring factors of $\hbar$ and c, which is often stated by saying that one sets $\hbar = c = \epsilon_0 = \mu_0 = 1$, where the latter two quantities are the permittivity and permeability of free space, respectively. This simplifies virtually all equations we will come across in this book. However, natural units can also be horribly confusing if you actually have to convert to numbers in SI units. The rule is: given any quantity in natural units, put in the right factors of $\hbar$ and c to get the dimensions right for the quantity you are converting to. One can often check which factors are needed by appealing to example equations relating the quantities of interest. For now, note that equation (1.13) simply becomes

$$E^2 - |\boldsymbol{p}|^2 = m^2. \tag{1.14}$$

Note that one might encounter numerical instabilities in the calculation of m when $|\boldsymbol{p}| \gg m$; we provide some helpful tips for avoiding this issue in appendix A.2.

---

[5] The notation here is unfortunately ambiguous, as a^2 looks like the square of a number. Whether or not we are talking about the square of a number or a four-vector will hopefully always be clear from the context.

Another quantity we will use throughout is the **invariant mass** of a set of particles with four-momenta $\{p_i, p_j, ...p_l\}$ (i.e. where indices label particle numbers):

$$s_{ij...l} = (p_i + p_j + ... + p_l)^2, \qquad (1.15)$$

e.g. for two particles we have

$$s_{ij} = (E_i + E_j)^2 - \boldsymbol{p}_i \cdot \boldsymbol{p}_j.$$

From equation (1.14), we see that mass and momentum both have dimensions of energy in natural units. Also, the formula for the de Broglie wavelength

$$\lambda = \frac{hc}{p}$$

implies that length has dimensions of (energy)$^{-1}$. We can thus choose dimensions of energy to express any quantity we encounter. It is conventional in high-energy physics to measure energy in **electron volts** (eV), where the definition of 1 eV is the energy gained by an electron moving through a potential difference of 1 volt. In SI units, 1eV amounts to 1.6×10^{-19} J. For particle physics, 1eV is a pathetic amount of energy[6]. For example, the mass of the electron (a very light particle in the scheme of things) is 0.5 MeV, and the mass of the proton is around 1 GeV. The current energy of an LHC beam is 6.5 TeV!

1.3 Events, cross-sections and luminosity

In a given collider experiment, what we generally care about is when the beam particles interact and produce something interesting (e.g. particles from a BSM theory, interesting SM particles that allow us to do a precision measurement etc). Each instance of this happening is called a **scattering event**, or just **event** for short. You will also hear people classify types of event by name, e.g. 'a top quark pair event' (meaning that a top quark and an anti-top quark were produced before decaying), a 'Higgs event', and so on. The job of a given detector collaboration (more specifically, a given subgroup of a detector collaboration) is to separate interesting events from boring ones, and also to tell the different types of interesting event apart. As the rest of this book will convince you, this is an extraordinarily complicated thing to try to do, and indeed the detectors filter out some boring stuff before data is even recorded.

Clearly the **event rate** (i.e. the number of events of a certain type per unit time) will depend on two things:

(i) The interactions of the individual beam particles.
(ii) How the beams are made up (e.g. how many particles each has in it, the area of the beams etc).

To make the first of these more formal, consider a purely classical situation in which you throw tennis balls at a target, as depicted in Figure 1.2(a).

---

[6] To give some context, the energy levels of an atom are typically separated by a few eV.

(a)   (b)

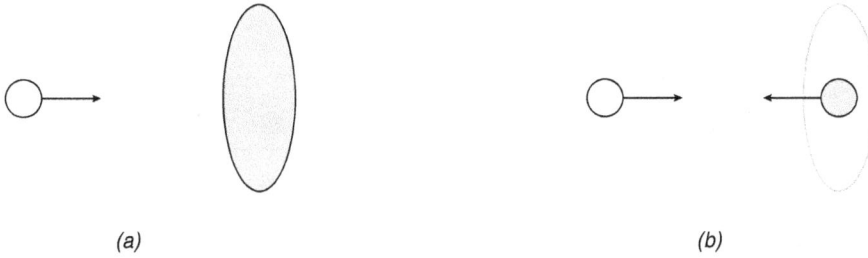

Figure 1.2. (a) Throwing a tennis ball at a classical target; (b) one quantum mechanical particle scattering on another.

The probability for the tennis ball to hit the target is clearly proportional to the cross-sectional area of the latter. We want to talk instead about the collision of two beams made of quantum mechanical 'particles'. The details of this are lot more complicated and fuzzy than the classical collisions we are familiar with from everyday life. However, we can simplify this in our minds by considering something like Figure 1.2(b), where we identify the particle on the left as being the incident particle, and the particle on the right as being the target. There is no well-defined sharp edge for the target particle as there is in the tennis ball example. Instead, there is some region of space such that, if the incident particle enters it, it will interact with the target particle and thus possibly be deflected. This will proceed via the exchange of force-carrying particles. Such a situation corresponds to an **elastic collision**, in which the two scattering particles remain intact. Another possibility is an **inelastic collision**, in which the particles may break up, or annihilate each other. In this case too, however, the incident particle must enter some region around the target particle in order to interact.

It follows from the above discussion that each target particle has an effective cross-sectional area for interaction. More formally, the scattering probability is related to some quantity σ, with dimensions of area (or (energy)$^{-2}$ in natural units). It is called the **scattering cross-section**, and depends on the properties of *both* colliding particles (e.g. the probability may depend on the charges of both particles, or their masses, etc). The rules of QFT tell us, at least in principle, how to calculate the cross-section for a given scattering process, and we will see how to do this in detail later in the book. For now, we can simply note that if N is the total number of scattering events (i.e. excluding events where the particles do not interact), the **event rate**, which is directly related to the interaction probability, must satisfy

$$\frac{\mathrm{d}N}{\mathrm{d}t} \propto \sigma \quad \Rightarrow \quad \frac{\mathrm{d}N}{\mathrm{d}t} = \mathcal{L}(t)\sigma, \tag{1.16}$$

which defines the **instantaneous luminosity** $\mathcal{L}(t)$. This depends on the parameters of the colliding beams, and thus can depend on time in general. Given that N is dimensionless, $\mathcal{L}$ has dimensions of (area)$^{-1}$(time)$^{-1}$. We see also that if $\mathcal{L}$ increases, then the event rate increases. Thus, $\mathcal{L}$ measures the 'brightness' of the beams in some sense. To quantify this further, consider a beam of particles of type a, incident on a

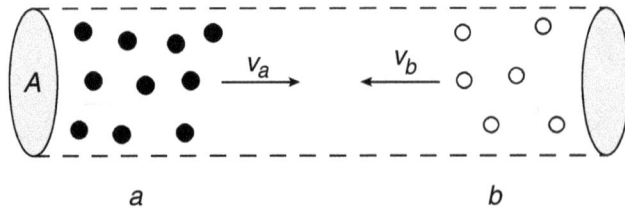

Figure 1.3. A beam of a-type particles incident on a target beam of b-type particles.

target beam of particles of type b, as depicted in Figure 1.3. Let each beam i have cross-sectional area A and uniform speed v_i, with n_i particles per unit volume. In time δt, each a-type particle moves a distance $v_a \delta t$ to the right, and each b-type particle moves a distance $v_b \delta t$ to the left. Thus, there is a volume

$$V = (v_a + v_b)A\delta t \tag{1.17}$$

in which the particles encounter each other, in time δt. The total number of b particles in this volume is

$$N_b = n_b V = n_b (v_a + v_b)A\delta t, \tag{1.18}$$

and the total probability of interaction is given by the fraction of the total cross-sectional area taken up by the cross-sections of each b-type particle:[7]

$$N_b \frac{\sigma}{A}.$$

This is the probability that a single a-type particle interacts, and thus the total interaction probability is given by multiplying this by the number of a-type particles in the volume V, N_a. Furthermore, the probability will be equal to the number of expected events in time δt. This follows from the fact that processes with a constant probability per unit time are described by a Poisson distribution, where the mean number of events per unit time is equal to the probability per unit time, as we discuss in more detail in Chapter 5. Provided we take δt small enough, we can treat the probability per unit time as constant. We thus have that the expected number of events in time δt is given by

$$\delta N = \frac{N_a N_b \, \sigma}{A} \tag{1.19}$$
$$= N_b \, n_a (v_a + v_b)\sigma\delta t,$$

where we have used equation (1.18). We can identify

$$\mathcal{F}_a = n_a(v_a + v_b) \tag{1.20}$$

---

[7] Here we are assuming that the beams are sufficiently diffuse that the cross-sections associated with each particle do not overlap, i.e. the particles scatter incoherently.

as the **flux** of incident particles (i.e. the number per unit time, per unit area). To see this, note that the number of a-type particles is given by

$$N_a = n_a V = n_a(v_a + v_b)A\delta t$$

and thus

$$\mathcal{F}_a = \frac{N_a}{A\delta t},$$

consistent with the above interpretation. Then, we can rearrange equation (1.19) to find that the cross-section is given by

$$\sigma = \frac{1}{N_b \mathcal{F}_a} \frac{\delta N}{\delta t}. \tag{1.21}$$

In words: the cross-section is the event rate per unit target particle, per unit flux of incident particle. This is a very useful interpretation, and also starts to tell us how we can calculate cross-sections in practice. Note that we considered the b-type particles as targets above. We could just as well have chosen the a-type particles, and thus interpreted the cross-section as being the event rate per unit a-type particle, per unit flux of b-type particle. The denominator factor in equation (1.21) is symmetric under interchanging a and b (as can be seen from the above derivation), and thus it does not matter which interpretation we choose.

From the definition of luminosity (equation (1.16)), we find that the number of events in some small time δt is

$$\delta N = \mathcal{L}\sigma\delta t,$$

and equation (1.19) then implies

$$\mathcal{L}dt = \frac{N_a N_b}{A}. \tag{1.22}$$

Thus, the luminosity is proportional to the number of particles in each beam, and is inversely proportional to the beam area. This makes sense given that we expected $\mathcal{L}$ to somehow measure the 'brightness' of each beam: a brighter beam has more particles in it, or has the particles concentrated in a smaller cross-sectional area.

The above discussion is clearly very schematic and simplified. However, it is a lot more correct than you might think, when applied to real beams. At the LHC, for example, protons collide in bunches with frequency f. If the beams are taken to have a Gaussian profile with width and height σ_x and σ_y respectively, then the luminosity is given by

$$\mathcal{L} = \frac{fN_1 N_2 N_b}{4\pi\sigma_x\sigma_y}, \tag{1.23}$$

where N_i is the number of protons per bunch in beam i, and N_b is the number of bunches. The denominator is a measure of the area of each beam, and thus equation (1.23) looks a lot like equation (1.22).

A given type of event i (e.g. top-pair, Higgs...) has its own associated cross-section σ_i, such that the total cross-section is given by the sum over each (mutually exclusive) event type:

$$\sigma = \sum_i \sigma_i. \tag{1.24}$$

It is perhaps not clear how to think about each σ_i as representing an area. However, we do not in fact have to do this: following equation (1.16), we can define each partial cross-Section in terms of the event rate for each process i, i.e.

$$\sigma_i = \frac{1}{\mathcal{L}} \frac{dN_i}{dt}. \tag{1.25}$$

Ultimately, the information content of this equation is that the event rate for process i factorises into a part which depends only on the nature of the colliding particles (the cross-section), and a part which depends on the structure of each beam (the luminosity). That the cross-section happens to have units of area is interesting, but we do not have to actually think about what this area means.

Given how complicated real beams are, the luminosity is never calculated from first principles. Rather, it can be measured experimentally. Theorists never have to worry about any of this: experimentalists usually present results for cross-sections directly, for comparison with theory. For bizarre historical reasons[8] the conventional unit of cross-section is the **barn**, where $1\,b \equiv 10^{-24}\,cm^2 = 10^{-28}\,m^2$. For most colliders, this is a stupendously large cross-section. For example, the total cross-section at the LHC (which includes the vast number of events in which the protons remain intact and don't do anything particularly interesting) is about 0.1b.

Luminosity gives us a convenient way to talk about how much data a collider has taken. From equation (1.16), we may define

$$N = \sigma L, \tag{1.26}$$

where

$$L = \int_0^T \mathcal{L} dt \tag{1.27}$$

is the **integrated luminosity**. This is a measure of how much potential data has been taken in time T, as the luminosity in some sense measures the possible number of particle collisions. Conventionally, L is quoted in units of **inverse femtobarns** (fb^{-1})—as we shall see, this odd formulation in which smaller SI prefixes correspond to larger amounts of integrated luminosity combines conveniently with the measurement of process cross-sections in (femto)barns to allow easy estimation of event counts.

---

[8] During the Manhattan project which created the first atomic bomb, American physicists chose the code word 'barn' to represent the approximate cross-sectional area presented by a typical nucleus, in order to obscure the fact that they were working on nuclear structure. A barn was considered to be a large target when using particle accelerators that had to hit nuclei.

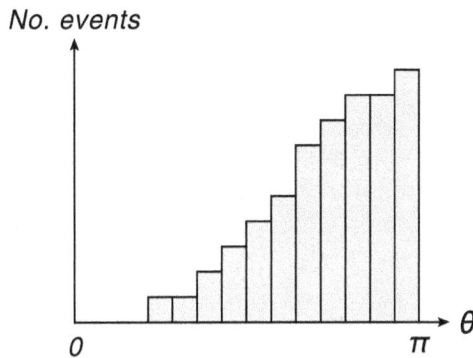

Figure 1.4. Schematic depiction of a differential cross-section, for the angle between two top quarks.

As an example, the Tevatron collider collected $\simeq 10$ fb^{-1} of data at its highest energy. If the cross-section for top pair production was 7.5 pb, we can ask how many top quark pairs were made. To find this, we first note that 7.5 pb $= 7.5 \times 10^3$ fb. Then the number of top pair events is

$$N_{t\bar{t}} = 7.5 \times 10^3 \times 10 = 75000.$$

1.4 Differential cross-sections

The cross-section is a single number for a given process, and tells us something about how likely this process is to occur. Often, however, we might want more information than this. For example, imagine we make a top/antitop quark pair. We may wish to know: what is the probability to have a given angle θ between the top quark and the antitop? We would measure this by collecting lots of $t\bar{t}$ events, and plotting a **histogram**[9], whose bins correspond to the different angle ranges we measure—see Figure 1.4 for a sketch. For infinitely many bins of vanishing width, this would converge to some smooth function $f(\theta)$. Different theories might predict different forms for this function, which thus gives us a huge amount of information for telling different theories apart.

Let us scale

$$f(\theta) \to \frac{f(\theta)}{L}$$

so that the integral of the function $f(\theta)$ becomes equal to the total cross-section:

$$\int_0^\pi \mathrm{d}\theta\, f(\theta) = \sigma_{t\bar{t}}. \tag{1.28}$$

---

[9] We will more formally introduce concepts from statistics and probability in Chapter 5.

The function $f(\theta)$ is then called the **differential cross-section**, and is given by

$$f(\theta) = \frac{\mathrm{d}\sigma_{t\bar{t}}}{\mathrm{d}\theta}. \tag{1.29}$$

Here we focussed on the angle between a pair of top quarks, but this clearly generalises. If we have a cross-section for some process, σ_i, and observable $\mathcal{O}$ (which was θ in the example above), then the differential cross-section for $\mathcal{O}$ is

$$\frac{\mathrm{d}\sigma_i}{\mathrm{d}\mathcal{O}}.$$

It represents the **distribution** of $\mathcal{O}$, and can be calculated in a given theory for comparison with data (after measuring lots of events). Differential cross-sections contain a large amount of information, as the kinematics of particle production can be very different in different (new) physics theories. Often, experiments present results for **normalised** differential cross-sections

$$\frac{1}{\sigma_i}\frac{\mathrm{d}\sigma_i}{\mathrm{d}\mathcal{O}},$$

which are defined such that the integral of the distribution is unity:

$$\int \mathrm{d}\mathcal{O}\frac{1}{\sigma_i}\frac{\mathrm{d}\sigma_i}{\mathrm{d}\mathcal{O}} = 1.$$

The reason this quantity is sometimes preferred is that many experimental uncertainties (and also theory uncertainties) can cancel in the ratio of the differential and total cross-sections.

1.5 Particle detectors

Events are measured by particle detectors. At a colliding beam experiment, these are typically cylindrical, surrounding the beam pipe. A rough theorist's view of a detector is shown (side on) in Figure 1.5. If we were to cut a slice through the detector, it would look something like Figure 1.6. It consists of different layers of material, each of which is designed to detect different kinds of particle. For example, the *electromagnetic calorimeter* (ECAL) detects photons and electrons/positrons, and the *hadronic calorimeter* detects hadrons, as its name suggests. Muons pass through the ECAL and HCAL, and are detected by the outer *muon detectors*.

The central *tracking detector* (typically made of many pixels and strips of silicon) is used for reconstructing the paths of charged particles, which ionise the strips and

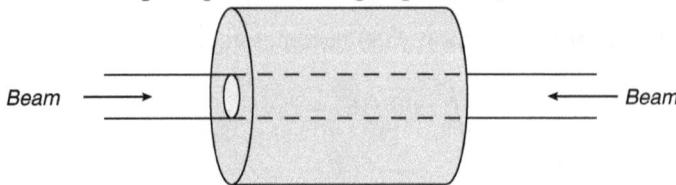

Figure 1.5. Schematic side-on view of a detector in a colliding beam experiment.

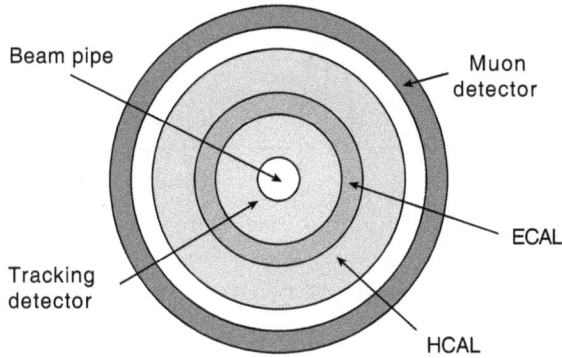

Figure 1.6. Schematic cross-sectional view of a detector in a colliding beam experiment.

thus give a pattern of dots which can be joined up to make a track. A magnetic field is applied, and the curvature of each particle track then provides information on the momentum of the particle. The calorimeters measure energy, and by comparing deposits in the calorimeters with tracks, we can unambiguously identify most particles. Note that almost all particles that pass through a given layer would leave some deposit there, even if the layer is not designed to measure them. Thus, a muon (as well as being seen in the tracker) would typically leave a small energy deposit in the ECAL and HCAL, even though it is in the inner tracker and muon detectors that are designed to measure the muon properties. The exceptions to this rule amongst Standard Model particles are the (anti)-neutrinos, which are so weakly interacting that they pass out of the detector completely without being seen. This gives rise to a missing four-momentum, with both energy and three-momentum components. The detector will also miss particles that pass down the beampipe, or through cracks/gaps in the detector. Experimentalists typically have to understand the geometry of their detector extremely well, to be able to correct for such effects where necessary.

How can we classify a given scattering event? Again from a highly simplified theorist's point of view, we can think of there being a list of particles (hadrons, leptons, photons etc), each with a four-momentum as measured by the detector, plus an additional four-vector representing the missing momentum. We need to choose a coordinate system in which to talk about the four-momenta, and it is conventional to choose this such that the incoming beams are in the $+z$ and $-z$ directions. The (x, y) plane is then transverse to the beam axis, as shown in Figure 1.7. It is then convenient to split the three-momentum of a particle $\boldsymbol{p} = (p_x, p_y, p_z)$ into a two-dimensional momentum transverse to the beam axis:

$$\boldsymbol{p}_T \equiv (p_x, p_y), \quad |\boldsymbol{p}_T| = \sqrt{p_x^2 + p_y^2}, \tag{1.30}$$

and the longitudinal component p_z. Note that the notation for the **transverse momentum** $\boldsymbol{p}_T$ is the same as if this were a three-vector rather than a two-vector. Furthermore, people often say 'transverse momentum' to mean the *magnitude* of this two-vector, rather than the vector itself. What is being talked about will usually be clear from the context.

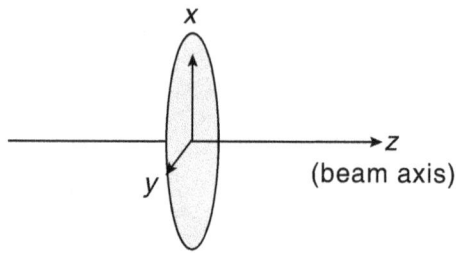

Figure 1.7. Coordinate system conventionally used for colliding beam experiments.

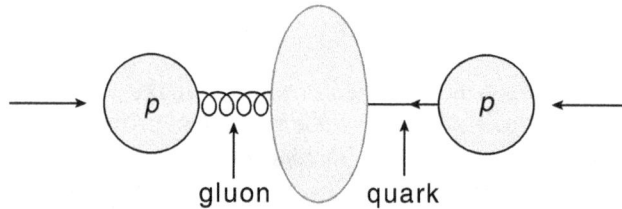

Figure 1.8. Collision of a gluon and quark, which emerge from two colliding protons.

In fact, p_z is not a convenient variable to use, particularly at hadron colliders (e.g. with pp or $p\bar{p}$ beams). To see why, note that (anti)-protons are composite particles, composed of (anti-)quarks and gluons. As we will see later on, in a hadronic scattering process, individual particles leave the proton and travel parallel to its original direction, before colliding (a schematic example is shown in Figure 1.8). In such a situation, we know the momenta of the incoming protons, as we have control over how the beams are designed. However, we do not know what the momenta of the incoming (anti-)quarks and gluons are, other than that they will be parallel with the incoming proton momenta (provided the protons are moving fast enough). It is therefore convenient to use variables which are invariant under Lorentz boosts (changes in velocity) in the z direction. This is already true for the transverse momentum p_T, which is why it is common to separate this off from the three-momentum. To see this, note that the four-momentum transforms under Lorentz boosts in the z direction according to

$$
\begin{pmatrix} E \\ p_x \\ p_y \\ p_z \end{pmatrix} \rightarrow \begin{pmatrix} E' \\ p'_x \\ p'_y \\ p'_z \end{pmatrix} = \begin{pmatrix} \gamma & 0 & 0 & -\beta\gamma \\ 0 & 1 & 0 & 0 \\ 0 & 0 & 1 & 0 \\ -\beta\gamma & 0 & 0 & \gamma \end{pmatrix} \begin{pmatrix} E \\ p_x \\ p_y \\ p_z \end{pmatrix}
$$

$$
= \begin{pmatrix} \gamma(E - \beta p_z) \\ p_x \\ p_y \\ \gamma(p_z - \beta E) \end{pmatrix},
$$

(1.31)

where $\gamma = (1 - v^2)^{-1/2}$, and $\beta = v$ in natural units, with v the velocity associated with the boost. We see that p_x and p_y do not change, and thus neither does p_{T}. However, p_z does change (it mixes with the energy E), and thus it is convenient to replace it by another variable. To this end, it is common to define the **rapidity**

$$y = \frac{1}{2} \ln \left(\frac{E + p_z}{E - p_z} \right). \tag{1.32}$$

Under a boost, this transforms as follows:

$$
\begin{aligned}
y \to y' &= \frac{1}{2} \ln \left(\frac{E' + p'_z}{E' - p'_z} \right) \\
&= \frac{1}{2} \ln \left(\frac{\gamma E - \beta \gamma p_z + \gamma p_z - \beta \gamma E}{\gamma E - \beta \gamma p_z - \gamma p_z + \beta \gamma E} \right) \\
&= \frac{1}{2} \ln \left(\frac{(1 - \beta)(E + p_z)}{(1 + \beta)(E - p_z)} \right),
\end{aligned}
$$

from which we find

$$y' = y + \frac{1}{2} \ln \left(\frac{1 - \beta}{1 + \beta} \right).$$

The second term is constant for a given boost, and it therefore follows that *differences* in rapidity are the same in all frames related by Lorentz boosts in the z direction. Upon encountering it for the first time, though, rapidity is a bit more abstract to think about than p_z, which is a component of momentum. What does rapidity actually mean?

To help visualise this, we have drawn a cylindrical particle detector in Figure 1.9, and written the value of rapidity associated with various directions. We see that rapidity essentially measures how *forwards* or *backwards* a particle is, where forwards means 'closer to the $+z$ direction'. The $+z$ axis corresponds to infinite rapidity, for example, and rapidity is zero for a particle that is produced centrally. We can make this idea more precise for particles that are highly energetic, i.e. with $E \gg m$, such that $E \simeq |p|$. Then one has

$$p_z \simeq E \cos \theta,$$

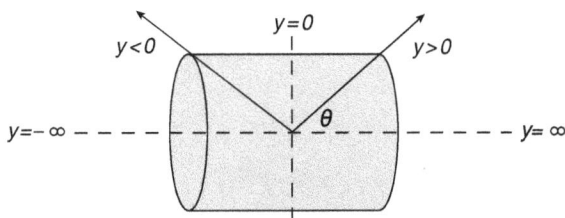

Figure 1.9. Various detector directions, and how they correspond to rapidity.

where θ is the angle between the particle direction and the beam axis, as shown in Figure 1.9. Equation (1.32) then becomes

$$y \simeq \frac{1}{2} \ln\left(\frac{E(1 + \cos\theta)}{E(1 - \cos\theta)}\right) = \frac{1}{2} \ln\left(\frac{1 + \cos\theta}{1 - \cos\theta}\right), \tag{1.33}$$

which we can simplify using the curious trigonometric identity

$$\tan^2\left(\frac{\theta}{2}\right) = \frac{1 - \cos\theta}{1 + \cos\theta}. \tag{1.34}$$

We thus obtain $y \simeq \eta$ if $E \gg m$, where

$$\eta = -\ln\left(\tan\frac{\theta}{2}\right) \tag{1.35}$$

is called the **pseudorapidity** (or pseudo-rapidity). Experimentalists tend to like this, as it is given in terms of the polar angle θ, that can be accurately measured. It also helps us to visualise rapidity a bit more easily, by backing up the claim above that higher $|y|$ corresponds to particles that are less centrally produced.

Very often we are only interested in which direction a particle went in, regardless of the magnitude of its momentum. Each direction is associated with a single point on a cylinder surrounding the beam pipe, as shown in Figure 1.10(a). We can fully specify this point using the rapidity y, and also the azimuthal angle ϕ around the detector, which is also labelled in the figure. Note that ϕ is in the transverse plane to the beam axis, so is itself invariant under Lorentz boosts in the z direction. A visual representation that is commonly used is to 'unwrap' the cylinder, to form a plane, whose coordinates are (y, ϕ). This is shown in Figure 1.10(b), where we have obviously condensed the infinite extent in the y direction! Each point in this plane represents a possible direction of a particle in the detector. Given two points, we can define a distance between them:

$$\Delta R = \sqrt{\Delta y^2 + \Delta\phi^2}, \tag{1.36}$$

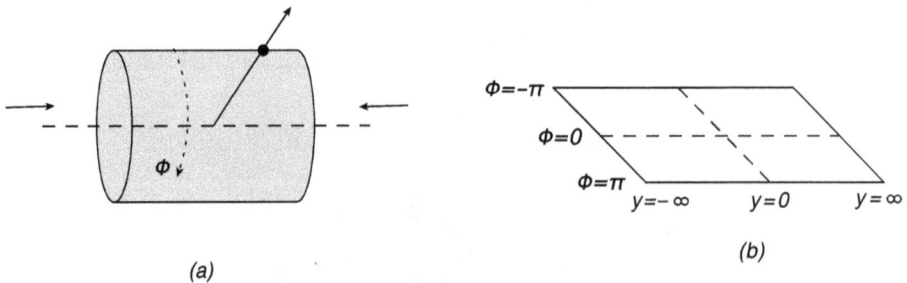

Figure 1.10. (a) Each particle direction cuts a cylindrical surface surrounding the beampipe at a single point; (b) we can unwrap this cylinder and use coordinates (y, ϕ) in the resulting plane.

which is dimensionless (i.e. both y and ϕ are dimensionless). This distance will be useful later in the book, particularly when we talk about jet physics. Note also that you will often see this formula with η replacing y.

Further reading

Excellent introductions to particle physics at an undergraduate level can be found in the following books:
- 'Modern Particle Physics', M Thomson, Cambridge University Press.
- 'An Introduction to the Standard Model of Particle Physics', D A Greenwood and W N Cottingham, Cambridge University Press.

Exercises

1.1 For four-vectors, we draw a distinction between upstairs and downstairs indices x^μ and x_μ. Why do we not usually bother with this for three-vectors in non-relativistic physics (e.g. with components x_i)?

1.2
   (a) Consider a fixed-target experiment, in which a beam of particles of mass m_1 and energy E_1 is incident on a stationary target of particles of mass m_2. Let p_1 and p_2 denote the four-momenta of the initial state particles.
      (i) If we want to make a new particle of M in the final state, explain why the invariant mass $s = (p_1 + p_2)^2$ satisfies

      $$s \geqslant M^2.$$

      (ii) By considering the four-momenta in the lab frame or otherwise, show that the invariant mass is given by

      $$s = m_1^2 + m_2^2 + 2E_1 m_2,$$

      and hence that the minimum beam energy to make the new particle is given by

      $$E_1 \geqslant \frac{M^2 - m_1^2 - m_2^2}{2m_2}.$$

      (iii) What is happening when M is much larger than m_2?
   (b) Now consider a colliding beam experiment, consisting of two beams of energy E (containing particles of equal mass) incident on each other in the centre-of-mass frame.
      (i) What is the minimum beam energy to make a particle of mass M?
      (ii) Is the fixed-target or colliding beam experiment more suitable for looking for new physics particles of high mass?

1.3 Name an advantage and a disadvantage of a circular collider, as opposed to a linear one.

1.4 Convert the following to SI units, from natural units:
    (i) The mass of the proton ($\simeq 0.938$ GeV).
    (ii) An area of 1 GeV^{-2}. What is this in barns?
    (iii) The energy of a proton in an LHC beam (6.5 TeV).

1.5 What are the dimensions of angular momentum in natural units?

1.6

    (a) The design luminosity of the LHC (which has since been surpassed!) was 1×10^{34} cm^{-2}s^{-1}. If it were to run for the whole of September, what would the corresponding integrated luminosity be in inverse femtobarns?

    (b) The main production mode for the Higgs boson has a cross-section of roughly 50 pb at the centre-of-mass energy 13 TeV. How many Higgs bosons would have been made in September, assuming the running conditions in part (a)?

1.7 Is it more important to build colliders with higher luminosity, or higher centre-of-mass energy?

1.8 For the following particle species, state whether you expect each one to leave a large energy deposit in the electromagnetic calorimeter or hadronic calorimeter, or to leave a track in the inner tracker or muon detector: (i) electron; (ii) proton; (iii) neutron; (iv) neutrino; (v) muon.

1.9 In the detector of a circular collider, a particle has a measured four-momentum $p^\mu = (10, 3, 4, 5)$ in GeV. Assuming the beam axis corresponds to the z-direction, answer the following:
    (i) What is the particle's mass?
    (ii) What is the magnitude of its transverse momentum?
    (iii) What is its rapidity?
    (iv) What is its pseudo-rapidity?

1.10 Heavy particles tend to have low magnitude of rapidity $|y|$ in the ATLAS and CMS detectors. Why?

IOP Publishing

Practical Collider Physics

Andy Buckley, Christopher White and Martin White

Chapter 2

Quantum field theory for hadron colliders

So far we have seen how to describe colliders, at least roughly. However, we have not described in detail what actually happens when the particles collide with each other. According to equations (1.16) and (1.25), this is given by the cross-section σ_i for a given process, and thus we need to know how to calculate it. This involves understanding the quantum field theory (QFT) for each force in nature as described by the SM, as well as any possible BSM interactions that may correspond to new physics. In particular, we need to know how to describe the **strong force** in great detail, for two main reasons:

1. The LHC collides protons, made of quarks and gluons. Thus, *any* scattering process involves the strong force somewhere.
2. Even at e$^\pm$ colliders, there is lots of quark and/or gluon radiation, given that the strong force, as its name suggests, is the strongest force in nature. Thus, radiation accompanying a given scattering process is typically dominated by quarks and gluons, and to a lesser extent by photons.

The part of the Standard Model (SM) describing quarks and gluon is **quantum chromodynamics** (QCD). It is a type of QFT called a **non-abelian gauge theory**, which refers to the abstract mathematical symmetry underlying its structure. We will build up to this complicated theory by first considering a simpler case, namely that of quantum electrodynamics (QED). This will introduce some of the ideas necessary for the more complicated QCD case, and we will then proceed to the latter theory, and examine its consequences in detail. After we have done so, we will briefly explain how similar ideas can be used in the rest of the SM, namely the combination of the electromagnetic and weak forces to make the electroweak theory.

Our aim in this chapter is to introduce the minimal working knowledge that is required for a good understanding of issues affecting contemporary experimental analyses. We do not claim to be a specialist resource on quantum field theory, for which we refer the reader to the further reading resources collected at the end of the chapter.

2.1 QED as a gauge theory

QED is the quantum version of electromagnetism, coupled to electrons. Its original version has now been superceded by the electroweak theory that forms a large part of the SM of particle physics, but its original incarnation provides a useful warm-up case for examining the strong and weak forces. You will have probably first learnt about electromagnetism (some years ago) through Maxwell's equations, which themselves distill hundreds of years of careful experiments. As it happens, we can derive the whole of electromagnetism from an abstract symmetry principle called **gauge symmetry**. Let us see how this works.

First, imagine that we have a theory containing free electrons and positrons only. This is described by the **Dirac Lagrangian**

$$\mathcal{L} = \bar{\psi}(i\gamma^\mu \partial_\mu - m)\psi. \tag{2.1}$$

Here ψ is a four component spinor field, whose four degrees of freedom describe the two spin degrees of freedom of the electron, and the two spin states of the positron. The matrix γ^μ is a four-by-four matrix in spinor space and satisfies the **Dirac algebra**[1]

$$\{\gamma^\mu, \gamma^\nu\} = 2\eta^{\mu\nu}. \tag{2.2}$$

The quantity $\bar{\psi} \equiv \psi^\dagger \gamma^0$ is a so-called **adjoint spinor**, where $\psi^\dagger$ denotes the Hermitian conjugate of ψ (i.e. the complex conjugate of the transpose). Finally, m is the mass of the electron (or positron). Given equation (2.1), we can see that it is invariant under the transformation

$$\psi \rightarrow e^{i\alpha}\psi, \tag{2.3}$$

where $\alpha \in \mathbb{R}$ is some constant number. To show this, note that equation (2.3) implies

$$\bar{\psi} \rightarrow e^{-i\alpha}\bar{\psi}, \tag{2.4}$$

so that

$$\mathcal{L} \rightarrow (e^{-i\alpha}\bar{\psi})(i\gamma^\mu \partial_\mu - m)(e^{i\alpha}\psi) = \mathcal{L},$$

where we have used the fact that we can cancel the factors of $e^{i\alpha}$ and its complex conjugate if α is constant.

In fact, we can provide a geometric interpretation of this symmetry, which also explains where it comes from. Given that $\psi(x)$ is a complex field, it has a phase at every point. This phase must lie in the range $(0, 2\pi)$, and we can represent the phase at a given point in spacetime by an arrow on the unit circle, as shown in Figure 2.1. We can then think of the phase of the entire field as having a set of arrows in

---

[1] In a slight abuse of notation, we follow convention in omitting an implicit (four-dimensional) identity matrix in spinor space on the right-hand side.

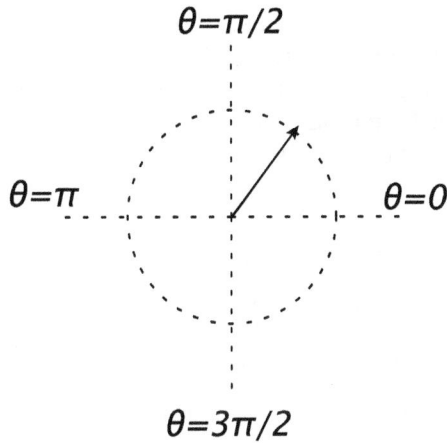

Figure 2.1. Representation of the phase of the electron field at a given point in spacetime, as an arrow on the unit circle.

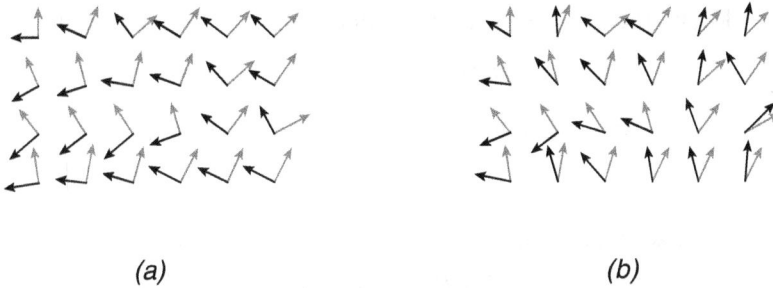

(a)

(b)

Figure 2.2. The red arrows correspond to the phase of the electron field throughout spacetime. The black arrows correspond to a gauge transformation that is (a) global; (b) local.

spacetime, with one at each point. An example is shown in Figure 2.2(a), where the red arrows might correspond to some set of phases at different points.

For constant α, the field of equation (2.3) is obtained from ψ by rotating all the arrows by an angle α, and we are obviously allowed to do this. That is, we are always free to redefine where on the unit circle the zero of phase is. This corresponds to rotating the unit circle, or alternatively to keeping the circle fixed, but rotating each arrow. This is called a **global gauge transformation**, where 'global' refers to the fact that we change the phase by the same amount *everywhere* in spacetime. The word 'gauge' is an old-fashioned term to do with calibrating measurements, which relates in this case to choosing where we set the zero of phase.

Global gauge symmetry is an interesting property by itself. But it turns out that QED has a much more remarkable symmetry than this. Namely, the theory is invariant under the transformation

$$\psi \rightarrow e^{i\alpha(x)}\psi \tag{2.5}$$

i.e. where we have now allowed the phase shift $\alpha(x)$ to be different at every point in spacetime. This corresponds to rotating the arrows at each point by *different* amounts, as shown in Figure 2.2(b). By analogy with the word 'global', this is called a **local gauge transformation**. Let us try to impose it on the electron field and see what happens. Again we have

$$\psi \to e^{i\alpha(x)}\psi \quad \Rightarrow \quad \bar{\psi} \to e^{-i\alpha(x)}\bar{\psi}.$$

The Dirac Lagrangian now transforms as

$$
\begin{aligned}
\mathcal{L} &\to e^{-i\alpha(x)}\bar{\psi}\Big[i\gamma^\mu \partial_\mu(e^{i\alpha(x)}\psi) - me^{i\alpha}\psi\Big] \\
&= e^{-i\alpha(x)}\bar{\psi}\Big[i\gamma^\mu e^{i\alpha(x)}\partial_\mu\psi - \gamma^\mu(\partial_\mu\alpha)e^{i\alpha}\psi - me^{i\alpha}\psi\Big] \\
&= \bar{\psi}(i\gamma^\mu\partial_\mu - m)\psi - (\partial_\mu\alpha)\bar{\psi}\gamma^\mu\psi \\
&= \mathcal{L} - (\partial_\mu\alpha)\bar{\psi}\gamma^\mu\psi.
\end{aligned}
\tag{2.6}
$$

We thus see that the electron theory by itself is not invariant under local gauge transformations. However, something interesting happens if we try to patch this up. If we look closely, the problem is that $\partial_\mu\psi$ does not transform particularly simply:

$$\partial_\mu\psi \to \partial_\mu(e^{i\alpha}\psi) = e^{i\alpha}[\partial_\mu\psi + i(\partial_\mu\alpha)\psi].$$

If we could instead find some modified derivative D_μ such that

$$D_\mu\psi \to e^{i\alpha(x)}D_\mu\psi \tag{2.7}$$

under a local gauge transformation, then the alternative Lagrangian

$$\mathcal{L}' = \bar{\psi}(i\gamma^\mu D_\mu - m)\psi \tag{2.8}$$

would be locally gauge invariant. Indeed such a D_μ exists, and is called the **covariant derivative**, where the word 'covariant' means 'transforms nicely under gauge transformations'. To find D_μ, we can try the form

$$D_\mu\psi = [\partial_\mu + ieA_\mu(x)]\psi, \tag{2.9}$$

where the constant $e \in \mathbb{R}$ and the factor of i is introduced for later convenience, and $A_\mu(x)$ is some function of spacetime. Then provided we can find an $A_\mu(x)$ such that equation (2.7) is satisfied, we will have achieved our task of constructing a gauge-invariant theory. To be fully general, we should include the possibility that A_μ itself can change under a gauge transformation, which for now we can simply write as

$$A_\mu(x) \to A'_\mu(x) \equiv A_\mu(x) + \delta A_\mu(x). \tag{2.10}$$

Then we find

$$
\begin{aligned}
D_\mu\psi &\to \partial_\mu(e^{i\alpha}\psi) + ieA'_\mu(e^{i\alpha}\psi) \\
&= e^{i\alpha}\Big[\partial_\mu\psi + i(\partial_\mu\alpha)\psi + ie(A_\mu + \delta A_\mu)\psi\Big] \\
&= e^{i\alpha}[D_\mu\psi + i(\partial_\mu\alpha + e\delta A_\mu)\psi].
\end{aligned}
$$

Thus, we can satisfy equation (2.7) provided we impose

$$\delta A_\mu = -\frac{1}{e}\partial_\mu\alpha, \quad \text{if} \quad \psi \to e^{i\alpha}\psi, \tag{2.11}$$

but where A_μ is otherwise arbitrary. It is certainly possible to find such an A_μ, and thus we have confirmed that we are able to make a locally gauge-invariant theory if we want to. However, a theory which has an almost completely arbitrary spacetime quantity A_μ in its Lagrangian is not much use—how are we supposed to interpret what A_μ is meant to be? To answer this, let us make the bold assertion that $A_\mu(x)$ is itself a physical field that actually exists in nature (in as much as ψ does). Then we could construct some equations of motion for A_μ, and *they* would tell us what A_μ is meant to be in any given situation! These equations of motion will not be very interesting unless we have a kinetic term for the field A_μ, involving derivatives. Furthermore, the kinetic term will have to be gauge invariant by itself, and it is then useful to define the quantity

$$F_{\mu\nu} = \partial_\mu A_\nu - \partial_\nu A_\mu. \tag{2.12}$$

This is indeed gauge invariant: equation (2.11) implies

$$\partial_\mu A_\nu \to \partial_\mu A_\nu - \frac{1}{e}\partial_\mu\partial_\nu\alpha,$$

and thus

$$F_{\mu\nu} \to \partial_\mu A_\nu - \frac{1}{e}\partial_\mu\partial_\nu\alpha - \partial_\nu A_\mu + \frac{1}{e}\partial_\nu\partial_\mu\alpha = F_{\mu\nu},$$

where the second equality follows from the fact that partial derivatives commute, so that the terms in α cancel. Typically, equations of motion are **second-order** in time and space derivatives. We thus need a kinetic Lagrangian term for A_μ that contains two derivatives, and is gauge invariant by itself (for which it helps that we have found $F_{\mu\nu}$). Indeed, up to some irrelevant caveats, it turns out that the only possibility is

$$\mathcal{L}_{\text{kin}} = -\frac{1}{4}F_{\mu\nu}F^{\mu\nu}, \tag{2.13}$$

where the overall numerical factor is conventional, and turns out to be particularly convenient. Our complete gauge-invariant Lagrangian, now involving the two physical fields $\psi(x)$ and $A_\mu(x)$, is

$$\mathcal{L}_{\text{QED}} = -\frac{1}{4}F_{\mu\nu}F^{\mu\nu} + \bar\psi[i\gamma^\mu D_\mu - m]\psi. \tag{2.14}$$

You may be wondering why we have labelled this as the Lagrangian of QED. This is because the field A_μ, introduced from symmetry principles alone, turns out to correspond to the electromagnetic field! To see this, let us write out equation (2.14) in full:

$$\mathcal{L}_{\text{QED}} = -\frac{1}{4}(\partial_\mu A_\nu - \partial_\nu A_\mu)(\partial^\mu A^\nu - \partial^\nu A^\mu) - \bar{\psi}[i\gamma^\mu \partial_\mu - m]\psi - A_\mu j^\mu, \qquad (2.15)$$

where we introduced

$$j^\mu = e\bar{\psi}\gamma^\mu \psi. \qquad (2.16)$$

The equation of motion for A_μ is given by the **Euler–Lagrange equation**

$$\partial_\alpha \frac{\partial \mathcal{L}}{\partial(\partial_\alpha A_\beta)} = \frac{\partial \mathcal{L}}{\partial A_\beta}. \qquad (2.17)$$

It can then be shown that

$$\frac{\partial \mathcal{L}}{\partial(\partial_\alpha A_\beta)} = -\partial^\alpha A^\beta + \partial^\beta A^\alpha = -F^{\alpha\beta}; \qquad \frac{\partial \mathcal{L}}{\partial A_\beta} = -j^\beta,$$

so that equation (2.17) becomes

$$\partial_\alpha F^{\alpha\beta} = j^\beta. \qquad (2.18)$$

Furthermore, the definition of the field strength tensor, equation (2.12), implies the so-called **Bianchi identity**

$$\partial_\alpha F_{\mu\nu} + \partial_\nu F_{\alpha\mu} + \partial_\mu F_{\nu\alpha} = 0. \qquad (2.19)$$

Equations (2.18) and (2.19) constitute the known forms of the Maxwell equations in relativistic notation. If they are new to you, then we can use them to derive the usual form (i.e. non-relativistic notation) by writing the components of the gauge field explicitly as

$$A^\mu = (\phi, A), \qquad (2.20)$$

where ϕ and A are the electrostatic and magnetic vector potential respectively, and we have used natural units. The physical electric and magnetic fields are given by

$$E = -\nabla \phi - \frac{\partial A}{\partial t}, \quad B = \nabla \times A. \qquad (2.21)$$

We can also define

$$j^\mu = (\rho, j), \qquad (2.22)$$

where we will interpret ρ and j in what follows. We can then find the components of $F^{\mu\nu}$. Firstly, the antisymmetry property $F^{\mu\nu} = -F^{\nu\mu}$ implies that the diagonal components

$$F^{00} = F^{ii} = 0,$$

where $i \in \{x, y, z\}$, and for once we do not use the summation convention. For the other components, recall that

$$\partial_\mu = \left(\frac{\partial}{\partial t}, \nabla\right) \quad \Rightarrow \quad \partial^\mu = \left(\frac{\partial}{\partial t}, -\nabla\right).$$

Then we have

$$F^{0i} = \partial^0 A^i - \partial^i A^0 = \left[\frac{\partial A}{\partial t} + \nabla\phi\right]_i = -E_i = -F^{i0}.$$

In words: the component F^{0i} represents the ith component of the electric field vector. Next, we have

$$F^{ij} = \partial^i A^j - \partial^j A^i = -F^{ji}.$$

As an example, we consider

$$F^{xy} = -\partial_x A_y + \partial_x A_y = -[\nabla \times A]_z = -B_z = -F^{yx}.$$

Similarly, one may verify that

$$F^{yz} = -F^{zy} = -B_x, \qquad F^{xz} = -F^{zx} = B_y.$$

Putting everything together, the field strength tensor has upstairs components

$$F^{\mu\nu} = \begin{pmatrix} 0 & -E_x & -E_y & -E_z \\ E_x & 0 & -B_z & B_y \\ E_y & B_z & 0 & -B_x \\ E_z & -B_y & B_x & 0 \end{pmatrix}. \tag{2.23}$$

Now let us examine the first of our supposed Maxwell equations, equation (2.18). We can split this into equations for each value of ν. The case $\nu = 0$ gives

$$\partial_\mu F^{\mu 0} = \partial_0 F^{00} + \partial_i F^{i0} = \nabla \cdot E = j^0 = \rho.$$

The case $\nu = x$ gives

$$\partial_\mu F^{\mu x} = \partial_0 F^{0x} + \partial_j F^{jx}$$
$$= -\frac{\partial E_x}{\partial t} + \partial_x F^{xx} + \partial_y F^{yx} + \partial_z F^{zx}$$
$$= -\frac{\partial E_x}{\partial t} + \partial_y B_z - \partial_z B_y$$
$$= \left[-\frac{\partial E}{\partial t} + \nabla \times B\right]_x$$
$$= j_x.$$

Note that in the final two lines, the subscript x refers to the component of a three-vector, and thus can be written upstairs or downstairs. Similarly, one finds

$$\partial_\mu F^{\mu i} = \left[-\frac{\partial E}{\partial t} + \nabla \times B\right]_i = j_i$$

for all $i \in \{x, y, z\}$. We have thus shown the following:

$$\nabla \cdot \mathbf{E} = \rho, \quad \nabla \times \mathbf{B} = \mathbf{j} + \frac{\partial \mathbf{E}}{\partial t}, \tag{2.24}$$

which you will hopefully recognise as two of Maxwell's equations. The remaining two come from the Bianchi identity, as you can explore in the exercises.

To summarise, we have seen that all of electromagnetism can be derived by requiring an abstract symmetry of the electron field ψ called **(local) gauge invariance**. This demanded that we introduce A_μ (the electromagnetic field), which is not surprising: local gauge invariance means we can change the phase of ψ independently at each spacetime point. The field A_μ allows us to compensate for this, making local gauge invariance possible.

This is all very well, but why should we want to impose gauge invariance in the first place? First of all, the mere fact that EM follows from gauge invariance means that we could simply regard gauge invariance as some fundamental principle in nature, without asking where it comes from. It certainly feels more 'fundamental' than positing the existence of all of Maxwell's equations themselves with no underlying cause. However, gauge invariance is also crucial in forming a consistent quantum theory of electromagnetism. It turns out that if one were to naïvely try to make a quantum theory, it would have mathematical problems (due to states of negative norm). Gauge invariance removes the problem, but this still does not tell us that it has to be there, or where it ultimately comes from. Indeed, some modern ideas suggest that gauge invariance is not absolutely fundamental, but instead emerges from some more underlying description of either QFT or a high energy replacement of it (e.g. string theory).

In any case, the success of gauge invariance in electromagnetism suggests that it could be useful for the other forces in nature. Indeed it is! Before moving on, though, note that if we considered a mass term for the photon in the QED Lagrangian, this would change under a gauge transformation as follows:

$$m^2 A_\mu A^\mu \rightarrow m^2 (A_\mu + \partial_\mu \alpha)(A^\mu + \partial^\mu \alpha).$$

Thus, mass terms for gauge fields are not gauge invariant! If we want to use gauge invariance to make a theory with massive force carriers (e.g. the W and Z bosons that carry the weak force), we have to be more clever. This is where the **Higgs boson** comes from, as we will see later. It is not a problem for the strong force, however, as gluons are massless.

2.2 Quarks and colour

For the electron field, we represented the phase by an arrow at each point in spacetime. (Local) gauge invariance meant that we could make arbitrary rotations of this arrow, which gave rise to the electromagnetic field A_μ. Quark fields also have such a phase, and thus couple to electromagnetism. However, we can also ask if there are other properties of the fields that can be represented by different arrows.

The above discussion suggests that, if this is true, we can introduce other types of gauge invariance, and thus more force fields!

Indeed, such an extra arrow exists for quarks. They carry a type of conserved charge called **colour**, not to be confused with electromagnetic charge. Whereas the latter has two types (+, −), colour charge has three types, conventionally called (r, g, b) (for 'red', 'green' and 'blue'). Note that these are just labels, and have nothing to do with actual colours. Any other arbitrary labels would have sufficed. Quarks have spin 1/2, and thus can be described by a Dirac spinor field, as for electrons. We can then write three such fields, one for each colour. It is then convenient to collect these into a single vector-like object

$$\psi_i = (\psi_r, \psi_g, \psi_b), \tag{2.25}$$

where $i \in \{r, g, b\}$ is a **colour index**. The vector we have formed lives in an abstract **colour space** at each point in spacetime, where an arrow in this space tells us how much redness, greenness and blueness the (vector) quark field has at that point (Figure 2.3).

The size or magnitude of the arrow then tells us the *overall* colour charge. This is sometimes called an **internal space** (i.e. internal to the quark field), to avoid confusion with actual spacetime. Similar to the phase of the electron field, we are always free to redefine what we mean by redness, greenness and blueness. In other words, the theory of quarks should be invariant under 'rotations' of the arrow in colour space, at all points in spacetime simultaneously. This is a **global gauge transformation**, directly analogous to the phase in QED. Any such rotation must act on the quark field as

$$\psi_i \rightarrow \psi'_i = U_{ij}\psi_j, \tag{2.26}$$

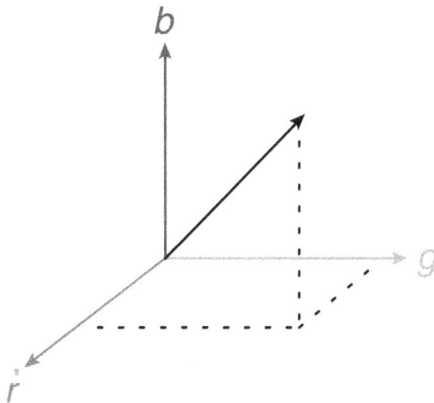

Figure 2.3. The quark field can be thought of as carrying different components, one for each colour. This leads to an abstract vector space at each point in spacetime, where the arrow specifies how much of each colour is present.

for some numbers $\{U_{ij}\}$. Let us now denote by Ψ the three-dimensional column vector whose components are given in equation (2.25). We can then write equation (2.26) as

$$\Psi' = \mathbf{U}\Psi, \qquad (2.27)$$

where $\mathbf{U}$ is a constant three-by-three matrix. Given $\psi_i \in \mathbb{C}$, $\mathbf{U}$ will also be complex. Furthermore, if colour is conserved, the size of the arrow cannot change under these transformations. We also want to exclude reflections, which correspond to inversions of one or more of the axes. Thus, we require

$$\det(\mathbf{U}) = 1. \qquad (2.28)$$

Following what we learned in QED, we can now demand *local* gauge invariance, by promoting $\mathbf{U} \rightarrow \mathbf{U}(x)$, corresponding to *different* rotations of the colour arrow at different spacetime points. To see how to do this in the present case, though, we need to learn a bit more about the mathematics underlying such symmetries. This is the subject of the following section.

2.3 Lie groups

The branch of mathematics underlying symmetry transformations is **group theory**, where a given set of transformations is described by a mathematical object called a **group**. A group G is any set of objects with a multiplication rule, satisfying the following properties:

(i) *Closure*. The product of any two elements in the group is also an element of the group:

$$ab \in G, \quad a, b \in G. \qquad (2.29)$$

(ii) *Associativity*. The multiplication rule is such that

$$(ab)c = a(bc) \qquad (2.30)$$

for all elements $a, b, c \in G$.

(iii) There is an *identity element* $e \in G$ such that

$$ae = ea = a \qquad (2.31)$$

for all elements $a \in G$.

(iv) Every element has an *inverse*, such that

$$aa^{-1} = a^{-1}a = e. \qquad (2.32)$$

Note that an element may be its own inverse—for example, this applies to the identity e itself.

These four properties are clearly satisfied by the phase rotations we encountered in QED. We can also see that they apply to the colour rotations acting on the quark field. First, note that we may write the product of two rotations acting on a quark field as

$$\mathbf{U_1 U_2 \Psi},$$

which means 'do rotation 2, then rotation 1'. The effect of this will be equivalent to some overall rotation

$$\mathbf{U_3 = U_1 U_2},$$

and the determinant of the latter is given by

$$\det(\mathbf{U_3}) = \det(\mathbf{U_1}) \det(\mathbf{U_2}) = 1.$$

That is, the product of two matrices of unit determinant itself has unit determinant, which fulfils the closure property of the group. For associativity, it is sufficient to note that matrix multiplication is itself associative. For property (iii), the identity element is simply the identity matrix $\mathbf{I}$. Finally, for property (iv), we can use the fact that all (complex) rotation matrices are invertible, where the inverse is given by the Hermitian conjugate,

$$\mathbf{U}^\dagger = (\mathbf{U}^\mathrm{T})^*.$$

To see this, we can first use the fact that (complex) rotations do not change the size of the colour arrow ($\Psi^\dagger \Psi$) by definition. Under a rotation this transforms as

$$\Psi^\dagger \Psi \rightarrow \Psi'^\dagger \Psi' = (\mathbf{U}\Psi)^\dagger (\mathbf{U}\Psi)$$
$$= \psi^\dagger (\mathbf{U}^\dagger \mathbf{U}) \Psi,$$

which then implies

$$\mathbf{U}^\dagger \mathbf{U} = \mathbf{I} \quad \Rightarrow \quad (\mathbf{U})^{-1} = \mathbf{U}^\dagger,$$

as required. Such matrices (i.e. whose inverse is the Hermitian conjugate) are called **unitary**, and we have now shown that these matrices fulfil all the properties necessary to form a group. To summarise, it is the group associated with 3×3 complex unitary matrices of unit determinant. This is conventionally called SU(3), where the 'S' stands for 'special' (unit determinant), the 'U' for 'unitary', and the number in the brackets denotes the dimension of the matrix. Likewise, we can also give a fancy name to the phase rotations of QED, by noting that any number of the type $e^{i\alpha}$ is a unitary 1×1 matrix, however daft it may feel to say so. Thus, the group associated with QED is U(1).

In general, groups can consist of **discrete** or **continuous** transformations, or both. For example, rotations about a single axis have a continuous parameter (an angle) associated with them, but reflections do not, and are therefore discrete. In the rotation example, the number of elements of the group is (continuously) infinite, but the number of parameters needed to describe all the group elements (i.e. an angle) is finite. Continuous groups are known as **Lie groups**, and the number of parameters needed to describe all the group elements is called the **dimension** of the group. It is not the number of group elements—this is infinite! Let us clarify this further with a couple of examples.

1. U(1) is the group of rotations about a single axis, and there is a single rotation angle associated with this, so that we have

$$\dim[U(1)] = 1.$$

2. SU(N) is the group associated with complex $N \times N$ matrices of unit determinant. A complex $N \times N$ matrix has $2N^2$ degrees of freedom (i.e. N^2 real, and N^2 imaginary parts). Not all of these are independent if the matrix is special and unitary: there is a set of N^2 equations

$$(U^\dagger)_{ij} U_{jk} = \delta_{ik}$$

associated with unitarity, and $\det[U] = 1$ imposes a further condition, leaving

$$2N^2 - N^2 - 1 = N^2 - 1$$

independent degrees of freedom. Thus,

$$\dim[SU(N)] = N^2 - 1.$$

For the particular group relevant for QCD, we find

$$\dim[SU(3)] = 8.$$

An important theorem says that *any* element of a Lie group G can be written in terms of matrices called **generators**, where the number of generators is equal to the dimension of the group (i.e. the number of independent degrees of freedom in each group element). Following physics conventions, we will label the set of generators $\{T^a\}$, where $a = 1, \dots, \dim(G)$. Then the expression relating group elements $\mathbf{U}$ to the generators is

$$\mathbf{U} = \exp[i\theta^a \mathbf{T}^a], \tag{2.33}$$

where we have used the summation convention. Equation (2.33) is called **Lie's theorem**, and the numbers $\{\theta^a\}$ are continuous parameters—one for each generator—such that a given set of values uniquely corresponds to a given group element. Note that the exponential of a matrix is defined by its Taylor expansion, where the zeroth-order term is understood to be the identity matrix, i.e.

$$\mathbf{U} = \mathbf{I} + \sum_{n=1}^{\infty} \frac{1}{n!} (i\theta^a \mathbf{T}^a)^n. \tag{2.34}$$

There is not a unique choice for the generators $\mathbf{T}^a$: all that matters is that they have the right degrees of freedom to be able to describe every group element (e.g. they must be linearly independent). Any change in the generators can be compensated by a change in the definition of the continuous parameters, and vice versa.

The above discussion is very abstract, so let us give a concrete example, and perhaps the simplest that we can consider. We saw above that the group U(1) has elements that can be represented by

$$e^{i\alpha} = \exp[i\theta^1 \mathbf{T}^1], \quad \theta^1 = \alpha, \quad \mathbf{T}^1 = 1.$$

On the right we see that each group element has precisely the form of equation (2.33), where there is a single generator (which we may take to be the number 1), and a single continuous parameter θ^1, which we may identify with the parameter α. A more complicated example is SU(3), where it follows from the above discussion that we expect eight generators. A popular choice is the set of so-called **Gell-Mann matrices**, and you will find them in the exercises. For actual calculations in QCD, we will never need the explicit form of the generators, which stands to reason: they are not unique, so the effect of making a particular choice must cancel out in any physical observable.

For any given Lie group G, the generators obey an important set of consistency relations. To see why, note that the group closure property

$$\mathbf{U}_1 \mathbf{U}_2 = \mathbf{U}_3,$$

together with Lie's theorem (equation (2.33)) implies

$$e^{i\theta_1^a \mathbf{T}^a} e^{i\theta_2^b \mathbf{T}^b} = e^{i\theta_3^c \mathbf{T}^c}, \tag{2.35}$$

where $\{\theta_i^a\}$ is the set of continuous parameters associated with the group element $\mathbf{U}_i$. On the left-hand side, we can rewrite the factors as a *single* exponential, provided we combine the exponents according to the **Baker–Campbell–Hausdorff formula**[2]

$$e^{\mathbf{X}} e^{\mathbf{Y}} = e^{\mathbf{X} + \mathbf{Y} + \frac{1}{2}[\mathbf{X}, \mathbf{Y}] + \cdots}, \tag{2.36}$$

where

$$[\mathbf{X}, \mathbf{Y}] \equiv \mathbf{X}\mathbf{Y} - \mathbf{Y}\mathbf{X}$$

is the **commutator** of $\mathbf{X}$ and $\mathbf{Y}$, and the neglected terms in equation (2.36) contain higher-order nested commutators. Applying equation (2.36) to equation (2.35), we find

$$\exp\left[i(\theta_1^a + \theta_2^a)\mathbf{T}^a - \frac{1}{2}\theta_1^a \theta_2^b [\mathbf{T}^a, \mathbf{T}^b] + \ldots\right] = \exp[i\theta_3^c \mathbf{T}^c]. \tag{2.37}$$

By taking the logarithm of both sides, the only way this equation can be true is if the commutator on the LHS is itself a superposition of generators, such that all higher order nested commutators also ultimately reduce to being proportional to a single generator. That is, we must have some relation of the form

$$[\mathbf{T}^a, \mathbf{T}^b] = i f^{abc} \mathbf{T}^c, \tag{2.38}$$

---

[2] If you are unaware of this result, you should find a description of it in any good mathematical methods textbook.

where the summation convention is implied as usual, and the set of numbers $\{f^{abc}\}$ (where all indices range from 1 to the dimension of the group) is such that the indices match up on both sides of the equation. Equation (2.38) is indeed true for the generators of *any* Lie group, and is called the **Lie algebra** of the group. The numbers $\{f^{abc}\}$ are called the **structure constants** of the group. Whilst the generators are non-unique, the structure constants are fixed for a given Lie group. That is, we may take a given Lie algebra or group as being *defined* by its structure constants[3], so that different structure constants define different Lie groups. From equation (2.38) and the fact that the commutator is antisymmetric, we see that

$$f^{abc} = -f^{bac}.$$

In fact, one may prove that f^{abc} is antisymmetric under interchange of any two of its indices. In full this implies

$$f^{abc} = f^{bca} = f^{cab} = -f^{bac} = -f^{acb} = -f^{cba}. \tag{2.39}$$

One of the reasons that the Lie algebra is useful is that it allows us to check whether a given choice of generator matrices is indeed an appropriate basis for a given Lie group. We can choose *any* set of matrices that satisfy the Lie algebra, and such a set is then called a **representation** of the group generators. There is clearly a large amount of freedom involved in choosing a representation, which then in turn implies a representation for the group elements themselves. For $SU(N)$, for example, we can use $N \times N$ complex matrices of unit determinant, which corresponds to the defining property of the group. However, this is not the only possibility. It is possible to find generator matrices that obey the Lie algebra of $SU(N)$, even though they are not $N \times N$, which appears confusing at first sight. However, the confusion is (hopefully) removed upon fully appreciating that a group is only defined by its Lie algebra. Then the defining property of $SU(N)$ is that its Lie algebra is the algebra associated with generators of $N \times N$ complex matrices of unit determinant. This *does not* mean that a given representation of the group always has $N \times N$ matrices, as some higher dimensional matrices may obey the same Lie algebra. Thus, the representations of $SU(N)$ have many possible dimensionalities, only one of which corresponds to the original defining property of the group.

Choosing an $N \times N$ representation (in line with the original group definition) has a special name: it is called the **fundamental representation**. For example, the fundamental representation of $SU(3)$ consists of 3×3 complex matrices of unit determinant. It is the representation that is relevant for quark fields, which live in a complex three-dimensional colour space. Another commonly used representation for a given Lie group G is the so-called **adjoint representation**, in which one uses matrices whose dimension is equal to the dimension of the group, $\dim(G)$. Indeed, one may show (see the exercises) that the matrices

$$(T^a)_{bc} = -if^{abc} \tag{2.40}$$

---

[3] Strictly speaking, there may be more than one Lie group associated with the same Lie algebra, a complication that will not bother us here.

obey the Lie algebra[4]. The adjoint representation acts on vectors whose dimension is that of the Lie group, and we will see such vectors later.

In any given representation R, the generators satisfy the following identities (not necessarily obvious!),

$$\text{tr}[\mathbf{T}^a\mathbf{T}^b] = T_R\delta^{ab}, \quad \sum_a \mathbf{T}^a\mathbf{T}^a = C_R\mathbf{I}, \tag{2.41}$$

where T_R is a non-unique constant that defines the normalisation of the generator matrices, and C_R is completely fixed for a given group G and representation R. The latter is called the **quadratic Casimir**. As an example, in the fundamental representation of $SU(N)$ we have

$$T_F = \frac{1}{2}, \quad C_F = \frac{N^2 - 1}{2N}, \tag{2.42}$$

where the first of these is a choice. In the adjoint representation of $SU(N)$ we have

$$T_A = C_A = N. \tag{2.43}$$

These definitions will be useful later when we calculate cross-sections involving quarks and gluons. Before moving on, let us comment on a useful point regarding notation. In some books on QFT and/or QCD, people use a different symbol to denote generators in the fundamental and adjoint representations. That is, they use $\mathbf{t}^a$ and $\mathbf{T}^a$ to denote a generator in the fundamental and adjoint representations, respectively. Given that this avoids a large amount of potential confusion in what is already a very complicated subject, we will adopt this practice here too.

2.4 QCD as a gauge theory

In Section 2.2, we showed that the quark field has an abstract arrow in colour space at each point in spacetime, and that the quark field is invariant under global gauge transformations, corresponding to complex rotations of this arrow. We then took a mathematical detour to see how to describe these transformations more formally. We can now use what we have learned to see how to make the theory invariant under *local* gauge transformations. First, let us write a Lagrangian for the quark field, which will be a generalisation of the Dirac Lagrangian for the electron field (equation (2.1)):

$$\mathcal{L}_{\text{quark}} = \bar{\Psi}(i\gamma^\mu\partial_\mu - m)\Psi. \tag{2.44}$$

Here we have assumed a single flavour of quark of mass m for now, and used vector notation to describe the three components of the field in colour space. As in QED, the key to making this Lagrangian locally gauge invariant is to replace the derivative ∂_μ with a suitable covariant derivative, such that the latter acting on the quark field transforms nicely under gauge transformations. In this case, the covariant derivative

---

[4] Note that for colour indices, index placement (upstairs or downstairs) is irrelevant, and we therefore follow existing conventions.

must be **matrix-valued** in colour space in general, given that it acts on a vector. So we are looking for some matrix $\mathbf{D}_\mu$, such that

$$\mathbf{D}_\mu\Psi \to \mathbf{U}\mathbf{D}_\mu\Psi, \quad \text{if} \quad \Psi \to \mathbf{U}\Psi. \tag{2.45}$$

That is, the covariant derivative acting on the quark field must transform *in the same way* as the quark field itself, directly analogous to the QED case. Note that $\mathbf{D}_\mu$ must be a 3×3 matrix if it acts on the quark field, so it is associated with the fundamental representation of SU(3).

In QED, the covariant derivative was given by equation (2.9), and studying this in more detail allows us to motivate a suitable *ansatz* for QCD. Firstly, the first term of equation (2.9) corresponds to an infinitesimal translation, in that for any function $f(x^\nu)$ we have

$$f(x^\nu + a^\nu) = f(x^\mu) + a^\mu\partial_\mu f(x^\nu) + \mathcal{O}(a^2).$$

The second term of equation (2.9) corresponds to an infinitesimal gauge transformation. To see this, note that we can consider the particular U(1) gauge transformation (which must correspond to multiplying by some number $e^{i\alpha}$)

$$e^{iea^\mu A_\mu} = \left[1 + iea^\mu A_\mu + \ldots\right].$$

Here we have expanded the gauge transformation assuming a^μ to be small, and we see that the second term involves the quantity ieA_μ, which enters the second term of equation (2.9). The latter thus indeed corresponds to some sort of infinitesimal gauge transformation, if we contract with a given four-vector.

Putting things together, we find that the covariant derivative should correspond to generating an infinitesimal translation, together with an infinitesimal gauge transformation. This tells us how to look for a suitable QCD covariant derivative. From equation (2.33), an infinitesimal gauge transformation in the fundamental representation has the form[5]

$$1 + i\theta^a\mathbf{t}^a + \mathcal{O}(\theta^a)^2,$$

which suggests a covariant derivative of the form

$$\mathbf{D}_\mu = \partial_\mu + ig_s\mathbf{A}_\mu, \quad \mathbf{A}_\mu = A_\mu^a\mathbf{t}^a, \tag{2.46}$$

where the factor of g_s is the analogue of e in QED, and is called the **strong coupling constant**—we will see why later. The first term in equation (2.46) generates a translation as desired, and contracting the second term with some four-vector a^μ corresponds to an infinitesimal gauge transformation with

$$\theta^a = g_s a^\mu A_\mu^a. \tag{2.47}$$

To see if the covariant derivative of equation (2.46) works, first note that equation (2.45) implies

---

[5] Here and in the following we have left identity matrices $\mathbf{I}$ implicit, as is done in many textbooks.

$$\mathbf{D}_\mu \Psi \to \mathbf{U} \mathbf{D}_\mu \mathbf{U}^{-1} \mathbf{U} \Psi$$

and hence

$$\mathbf{D}_\mu \to \mathbf{U} \mathbf{D}_\mu \mathbf{U}^{-1}. \tag{2.48}$$

One may show that this is satisfied provided the matrix-valued quantity $\mathbf{A}_\mu(x)$ behaves under a gauge transformation as

$$\mathbf{A}_\mu \to \mathbf{A}'_\mu = \mathbf{U} \mathbf{A}_\mu \mathbf{U}^{-1} + \frac{i}{g_s}(\partial_\mu \mathbf{U})\mathbf{U}^{-1}. \tag{2.49}$$

It is certainly possible to fulfil this condition, such that equation (2.48) then implies that if we promote the partial derivative in equation (2.44) to be a covariant derivative:

$$\mathcal{L}_{\text{quark}} = \bar{\Psi}(i\gamma^\mu \mathbf{D}_\mu - m)\Psi, \tag{2.50}$$

then this transforms as

$$\begin{aligned} \mathcal{L}_{\text{quark}} \to {} & \bar{\Psi}\mathbf{U}^{-1}(i\gamma^\mu \mathbf{U} \mathbf{D}_\mu \mathbf{U}^{-1} - m)\mathbf{U}\Psi \\ = {} & \bar{\Psi}(i\gamma^\mu \mathbf{D}_\mu - m)\Psi. \end{aligned} \tag{2.51}$$

Thus, we have found a quark Lagrangian that is invariant under local gauge transformations, provided the covariant derivative of equation (2.46) is such that equation (2.49) is true. In QED, we resolved the issue of how to find the quantity A_μ by requiring that it be a field that actually exists, whose equations of motion provide a solution for it. We can do exactly the same thing here, and say that

$$\mathbf{A}_\mu = A_\mu^a \mathbf{t}^a \tag{2.52}$$

is a field present in Nature: the **gluon**. A major difference with respect to QED is that the field has an extra index $a = 1, \ldots, \dim(G)$, where G is the Lie group behind our gauge transformations. This is commonly referred to as a 'gauge index'. In particular, $G = SU(3)$ for QCD, and thus $a = 1, \ldots, 8$. It is relatively straightforward to see why this must be the case. The reason the gauge field arises is that, if we rotate the colour arrow by different amounts at every point in spacetime, we can compensate for this by defining an additional quantity (the gauge field) at every point, such that the total theory is gauge invariant. A given local gauge transformation has $\dim(G)$ independent degrees of freedom from equation (2.33). Thus, the gauge field needs to have the same number of degrees of freedom in order to compensate for all possible local gauge transformations.

We can in fact view QED as a special case of the general framework discussed above, valid for any Lie group G. For QED, $G = U(1)$, which has a single generator $\mathbf{t}^1 = 1$. We can then write the gauge field as

$$\mathbf{A}_\mu = A_\mu^a \mathbf{t}^a = A_\mu^1 \equiv A_\mu$$

i.e. it has only a single component, so we can drop the superscript. Furthermore, the transformation of equation (2.49) reduces to

$$A'_\mu = A_\mu - \frac{1}{g_s}\partial_\mu\alpha, \quad \text{if} \quad \mathbf{U} = e^{i\alpha},$$

consistent with equation (2.11).

So far we have constructed a locally gauge invariant quark Lagrangian. To complete the theory, we need to add the kinetic term for the gluon field $\mathbf{A}_\mu$, where for QED we made use of the field strength tensor of equation (2.12), which is gauge invariant by itself. This will not be true in QCD, as $\mathbf{F}_{\mu\nu}$ becomes matrix-valued in colour space (given that the gauge field itself is). It turns out that we also need to generalise the definition of equation (2.12), to something that transforms more nicely under gauge transformations. To this end, consider the commutator of two QED covariant derivatives acting on an arbitrary electron field:

$$\begin{aligned}
[D_\mu, D_\nu]\Psi &= (\partial_\mu + ieA_\mu)(\partial_\nu + ieA_\nu)\Psi - (\mu \leftrightarrow \nu) \\
&= [\partial_\mu\partial_\nu + ieA_\mu\partial_\nu + ie(\partial_\mu A_\nu) + ieA_\nu\partial_\mu - e^2 A_\mu A_\nu]\Psi - (\mu \leftrightarrow \nu) \\
&= ie(\partial_\mu A_\nu - \partial_\nu A_\mu)\Psi \\
&= ieF_{\mu\nu}\Psi.
\end{aligned}$$

We therefore see that applying two covariant derivatives to the electron field, then reversing the order and taking the difference, is the same as multiplying the electron field with the electromagnetic field strength tensor (up to an overall factor). Furthermore, the fact that the electron field we are considering is arbitrary means that we can formally *define* the field strength tensor as

$$[D_\mu, D_\nu] = ieF_{\mu\nu}. \tag{2.53}$$

The reason this particular definition is so useful is that it is very straightforward to generalise to QCD. Upon doing this, one may consider the commutator of two QCD derivatives:

$$\begin{aligned}
[\mathbf{D}_\mu, \mathbf{D}_\nu] &= (\partial_\mu + ig_s\mathbf{A}_\mu)(\partial_\nu + ig_s\mathbf{A}_\nu) - (\mu \leftrightarrow \nu) \\
&= \left[\partial_\mu\partial_\nu + ig_s(\mathbf{A}_\mu\partial_\nu + \mathbf{A}_\nu\partial_\mu) + ig_s(\partial_\mu\mathbf{A}_\nu) - g_s^2\mathbf{A}_\mu\mathbf{A}_\nu\right] - (\mu \leftrightarrow \nu) \\
&= ig_s(\partial_\mu\mathbf{A}_\nu - \partial_\nu\mathbf{A}_\mu + ig_s[\mathbf{A}_\mu, \mathbf{A}_\nu]) \\
&= ig_s\mathbf{F}_{\mu\nu},
\end{aligned}$$

where the field strength tensor in this case is

$$\mathbf{F}_{\mu\nu} = \partial_\mu\mathbf{A}_\nu - \partial_\nu\mathbf{A}_\mu + ig_s[\mathbf{A}_\mu, \mathbf{A}_\nu]. \tag{2.54}$$

Using equation (2.52), we may rewrite this as

$$\begin{aligned}
\mathbf{F}_{\mu\nu} &= \mathbf{t}^a\left[\partial_\mu A_\nu^a - \partial_\nu A_\mu^a\right] + ig_s A_\mu^b A_\nu^c[\mathbf{t}^b, \mathbf{t}^c] \\
&= \mathbf{t}^a\left[\partial_\mu A_\nu^a - \partial_\nu A_\mu^a - g_s f^{abc} A_\mu^b A_\nu^c\right],
\end{aligned}$$

where we have used equation (2.38) in the second line. We may thus write

$$\mathbf{F}_{\mu\nu} = F_{\mu\nu}^a \mathbf{t}^a, \tag{2.55}$$

where the components are

$$F_{\mu\nu}^a = \partial_\mu A_\nu^a - \partial_\nu A_\mu^a - g_s f^{abc} A_\mu^b A_\nu^c. \tag{2.56}$$

Note that this reduces to the QED definition of equation (2.12) if $f^{abc} = 0$, which makes sense: as we discussed above, for the group U(1) there is only a single generator $\mathbf{t}^1 = 1$, and thus

$$[\mathbf{t}^a, \mathbf{t}^b] = 0$$

for all choices of a and b (both of which must be 1!). There are then no non-trivial structure constants, so that the final term in equation (2.56) vanishes. Lie groups where the elements all commute with each other (including the generators) are called **Abelian**. Conversely, if $f^{abc} \neq 0$ for any a, b or c, the group is called **non-Abelian**. Thus, QCD is an example of a **non-Abelian gauge theory**, and the form of the field strength tensor, equation (2.56), is our first hint that QCD is much more complicated than the **Abelian gauge theory** of QED!

Let us now see how to make a kinetic term out of the non-Abelian field strength tensor. One may show (see the exercises) that

$$\mathcal{L}_{\text{kin}} = -\frac{1}{2}\text{tr}[\mathbf{F}_{\mu\nu}\mathbf{F}^{\mu\nu}] \tag{2.57}$$

is gauge-invariant, where the trace takes place in colour space. Using equations (2.42) and (2.55), we may write this in terms of the component fields as follows:

$$\mathcal{L}_{\text{kin}} = -\frac{1}{4}F_{\mu\nu}^a F^{\mu\nu a}. \tag{2.58}$$

To construct the complete Lagrangian of QCD, we must add our kinetic term for the gluon to the quark Lagrangian, where we sum over six copies of the latter: one for each quark flavour out of (u, d, c, s, t, b). We then finally arrive at

$$\mathcal{L}_{\text{QCD}} = -\frac{1}{4}F_{\mu\nu}^a F^{\mu\nu a} + \sum_{\text{flavours } f} \bar{\Psi}_f(i\gamma^\mu \mathbf{D}_\mu - m_f)\Psi_f. \tag{2.59}$$

Some comments are in order:

- Once we say how may quark fields there are, and the fact that these are in the fundamental representation of a gauge group G, the Lagrangian is *completely fixed* by local gauge invariance!
- The Lagrangian of equation (2.59) may look simple, but do not be fooled! QCD is a spectacularly complicated theory, that nobody fully understands. Much of this lack of understanding relates to the fact that we cannot solve the quantum equations of the theory exactly in many situations of physical interest.

- Equation (2.59) is the pure QCD Lagrangian, including only the strong interactions of the quarks. The latter also have weak and electromagnetic interactions, which need to be added separately.

To analyse the theory in more detail, we can write out all terms in full to obtain

$$\mathcal{L}_{\text{QCD}} = -\frac{1}{4}\left(\partial_\mu A_\nu^a - \partial_\nu A_\mu^a\right)\left(\partial^\mu A^{\nu a} - \partial^\nu A^{\mu a}\right) + \sum_f \bar{\Psi}_f(i\gamma^\mu \partial_\mu - m_f)\Psi_f$$

$$+ g_s f^{abc} A_\mu^b A_\nu^c \partial^\mu A^{\nu a} - \frac{g_s^2}{4} f^{abc} f^{ade} A_\mu^b A_\nu^c A^{\mu d} A^{\nu e} \qquad (2.60)$$

$$- g_s \sum_f \bar{\Psi}_f \gamma^\mu \mathbf{t}^a \Psi_f A_\mu^a.$$

The quadratic terms look similar to the corresponding terms in QED, and there is also a term (in the third line) describing interactions of the gluon A_μ^a with the (anti-)quarks, similar to the coupling of the photon to electrons and positrons in QED. In QCD, however, this interaction is more complicated as it involves a colour generator. The physical meaning of this is that (anti-)quarks can change their colour by emitting/absorbing a gluon.

The second line in equation (2.60) contains cubic and quartic terms in the gluon field. Thus, the gauge field can interact with itself! This is consistent with what we said above: if (anti-)quarks can change colour by emitting a gluon, then gluons themselves must carry colour charge if colour is to be conserved overall. Then gluons can interact with other gluons. We can see that the gluon self-interaction terms all involve structure constants, and indeed vanish if $f^{abc} = 0$, which is the case in QED. This is why the photon does not interact with itself, and thus carries no electromagnetic charge. The gluon self-interactions may not look too innocuous here, but such terms are the main reason why non-Abelian gauge theories are much more complicated than Abelian ones.

2.5 Spontaneous symmetry breaking and the Higgs mechanism

We have so far seen that fundamental forces in nature can be associated with abstract local gauge symmetries. Requiring that our theories possess such symmetries involves introducing vector fields, which in the quantum theory give rise to force-carrying particles. The symmetries we considered so far were the U(1) symmetry of the original QED theory, and the SU(3) symmetry of QCD. In both of these contexts, the symmetry is exact. However, it may also be the case in nature that a local gauge symmetry appears, but is **broken**. Furthermore, we will see that this is intimately bound up with the issue of describing the non-zero mass of fundamental particles.

To illustrate the basic ideas, let us start by considering a theory containing a single complex scalar field ϕ, with Lagrangian

$$\mathcal{L} = (\partial^\mu \phi)^\dagger (\partial_\mu \phi) - V(|\phi|), \qquad (2.61)$$

where $V(|\phi|)$ is a potential energy. We have chosen that this depends only on the magnitude of the field, in which case equation (2.61) is invariant under the global gauge transformation

$$\phi \to e^{i\alpha}\phi \quad \Rightarrow \quad \phi^* \to e^{-i\alpha}\phi^*, \tag{2.62}$$

which may be compared with the similar gauge transformation which acts on fermions (equations (2.3) and (2.4)). Using the methods of Section 2.1, we may then promote this symmetry to a local gauge symmetry, by replacing the ordinary derivatives in equation (2.61) with the covariant derivative

$$D_\mu = \partial_\mu + igA_\mu,$$

where we use a different symbol g for the coupling constant, to emphasise the fact that we are in a different theory to QED. We must then also add the usual kinetic terms for the gauge field A_μ, upon which we arrive at the Lagrangian

$$\mathcal{L} = (D^\mu\phi)^\dagger (D_\mu\phi) - V(|\phi|) - \frac{1}{4}F^{\mu\nu}F_{\mu\nu}. \tag{2.63}$$

This describes the interaction of a charged scalar field with a photon-like particle. Note that the gauge transformations in equation (2.62) crucially relied upon the fact that the field was complex, which makes perfect sense: a complex field has two degrees of freedom, so can give rise to the positive and negative charge states that represent a scalar particle and its antiparticle.

So far we have left the form of the potential function $V(|\phi|)$ unspecified, but let us now consider the specific choice

$$V(|\phi|) = -\mu^2\phi^*\phi + \lambda(\phi^*\phi)^2, \tag{2.64}$$

for arbitrary parameters μ and λ. For a potential energy bounded from below, we require $\lambda > 0$. If the quadratic term were to have opposite sign (or $\mu^2 < 0$), this would act like a normal mass term for the scalar field, and the overall potential energy would have a minimum for $\phi = \phi^* = 0$. However, very different behaviour is obtained with the term as written (where we assume $\mu^2 > 0$). In that case, the minimum potential energy for the scalar field does not occur for $|\phi| = 0$, but for the **vacuum expectation value (VEV)**

$$|\phi| \equiv \frac{v}{\sqrt{2}} = \sqrt{\frac{\mu^2}{2\lambda}}, \tag{2.65}$$

where the constant v has been introduced for later convenience. This minimum can be seen in Figure 2.4(a), which shows the potential as a function of ϕ. One sees the presence of what look like two discrete minima on the plot, but it must be remembered that these are in fact continuously connected: the condition of equation (2.65) describes a circle in the (ϕ, ϕ^*) plane, such that the true form of the potential is that of Figure 2.4(b).

The vacuum of the theory will be associated with the minimum of the potential energy. However, unlike the previous cases of QED and QCD, we now have a

(a)   (b)

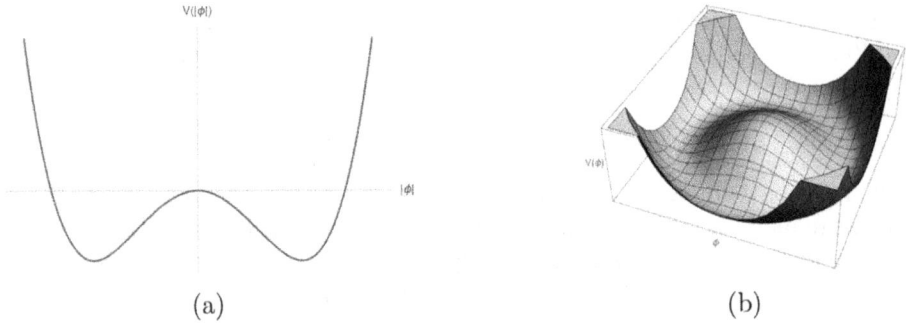

Figure 2.4. (a) The scalar potential of equation (2.64); (b) form of the potential in the (ϕ, ϕ^*) plane, showing the continuously connected circle of minima.

situation in which there is *no unique vacuum*. Instead, there are infinitely many possible vacua, corresponding to all of the points on the circle in the (ϕ, ϕ^*) plane such that $V(|\phi|)$ is minimised. To make this more precise, note that the general complex field ϕ that satisfies equation (2.65) can be written as

$$\phi = \phi_0 \, e^{i\Lambda} \tag{2.66}$$

where the arbitrary phase $\Lambda \in [0, 2\pi]$ labels all possible points on the circle. Note that different points on the circle—and thus different vacua—are related by the gauge transformations of equation (2.62). Choosing a vacuum state then amounts to fixing a particular value of Λ, after which gauge invariance is no longer manifest in the Lagrangian of the theory. The conventional terminology for this is that the gauge symmetry is **spontaneously broken**, although in practice the gauge invariance takes a different form: we are still allowed to gauge transform the fields, provided we also change the vacuum upon doing so.

Let us now examine the consequences of the above discussion, where without loss of generality we may choose $\Lambda = 0$ to simplify the algebra. We can then expand the complex field ϕ as follows:

$$\phi(x) = \frac{1}{\sqrt{2}}(v + \chi)e^{i\eta/v}, \quad \chi, \eta \in \mathbb{R} \tag{2.67}$$

Here both χ and η represent deviations from the modulus and phase of the vacuum field, respectively, where the inverse factor of v in the exponent keeps its argument dimensionless, and factors of $\sqrt{2}$ have been introduced so as to simplify later equations. We may substitute the expansion of equation (2.67) into equation (2.63), yielding

$$\begin{aligned}
\mathcal{L} = &-\frac{1}{4}F^{\mu\nu}F_{\mu\nu} - gvA^\mu\partial_\mu\eta + \frac{g^2v^2}{2}A^\mu A_\mu \\
&+ \frac{1}{2}\partial^\mu\chi \, \partial_\mu\chi - \lambda v^2\chi^2 + \frac{1}{2}\partial^\mu\eta \, \partial_\mu\eta + \dots ,
\end{aligned} \tag{2.68}$$

where the ellipsis denotes interaction terms (i.e. cubic and higher in the fields). In the first and second lines, we can recognise the usual kinetic term for the gauge field, together with kinetic terms for the two (real) scalar fields χ and η. In the second line, we see a mass term for the field χ, where $m_\chi^2 = 2\lambda v^2$ is the squared mass. There is no such term for the η field, indicating that the latter is massless. This turns out to be an example of a general result known as **Goldstone's theorem**: the spontaneous breaking of a continuous symmetry inevitably gives rise to massless scalar fields known as **Goldstone bosons**, where there is one such boson for every degree of freedom of the original symmetry that has been broken. Here, the choice of vacuum broke a rotational symmetry with one degree of freedom, thus there is only a single Goldstone boson.

There is a simple physical interpretation of Goldstone's theorem, which we can examine in the present case by looking at Figure 2.4(b). A given vacuum state corresponds to a point on the circle of minimum energy, and the two scalar fields χ and η correspond to perturbing the field in the radial and azimuthal directions, respectively. The first of these takes one away from the minimum by climbing a potential well that is quadratic to a first approximation: this is associated with the second term in the second line of equation (2.68), i.e. the mass term for the χ field. On the other hand, perturbations in the azimuthal direction take one around the circle, costing no energy. This explains the lack of a mass term for η, and hence the presence of a massless Goldstone boson.

As well as possible mass terms for the scalar fields, we also see a mass term for the photon (the third term in the first line of equation (2.68)). Comparing with the conventional mass term for a vector field, we find

$$\frac{1}{2} m_A A^\mu A_\mu \quad \Rightarrow \quad m_A = gv.$$

This mass is conceptually very different from the case discussed in Section 2.1, in which a mass term was added by hand. We saw that such a term was not gauge invariant, which is not the case here: we have started with a fully gauge-invariant theory of a scalar field, and broken the symmetry spontaneously. This generates the mass for the gauge field, as can be seen explicitly in equation (2.5) given that the mass is proportional to the vacuum expectation value of the scalar field. At high energies, the field will no longer be constrained to have minimum energy, and the manifest gauge symmetry will be restored.

We have not yet drawn attention to the second term on the right-hand side of equation (2.68), which is puzzling in that it constitutes a mixing between the gauge field and azimuthal perturbation η. This suggests we have not fully understood the physical particle spectrum of the theory, and we can verify this by counting degrees of freedom. Before the gauge symmetry is broken, we have a massless vector field, which has two polarisation states. Combined with the complex scalar ϕ, this gives four degrees of freedom in total. After the breaking, we have a massive vector field (three polarisation states), and two real scalars, giving five degrees of freedom. One of these must then be spurious, and can be gotten rid of. To do this, we can make the gauge transformation

$$\phi \rightarrow \phi \, e^{-i\eta/v},$$

after which the Lagrangian of equation (2.68) becomes

$$
\begin{aligned}
\mathcal{L} = &-\frac{1}{4}F^{\mu\nu}F_{\mu\nu} + \frac{g^2 v^2}{2}A^\mu A_\mu + \frac{1}{2}\partial^\mu\chi\,\partial_\mu\chi - \lambda v^2\chi^2 \\
&+ g^2\left(v\chi + \frac{1}{2}\chi^2\right)A^\mu A_\mu - \lambda v\chi^3 - \frac{\lambda}{4}\chi^4,
\end{aligned}
\tag{2.69}
$$

where we have neglected constants, used equation (2.65), and kept all non-linear interaction terms We see that the spurious degree of freedom η has been completely removed[6], leaving four manifest degrees of freedom: the three polarisation states of the vector field, and a remaining massive scalar. This choice of gauge is known as the **unitary gauge** and, as we have just discussed, makes the physical particle spectrum of the theory manifest. Also, we see that the Goldstone boson we saw above has been 'eaten' by the massive gauge boson: what would be a spurious degree of freedom is in fact to be identified with the extra polarisation state needed for a massive gauge boson, rather than a massless one.

2.6 The Standard Model of particle physics

In the previous sections, we have seen two key ideas: (i) forces can be associated with local (non)-Abelian gauge symmetries, all of which involve introducing vector fields; (ii) massive gauge fields can be generated by spontaneous symmetry breaking. For the strong force, the situation is relatively simple, and has essentially already been described in Section 2.4, if quark masses are taken to be inserted by hand. The weak and electromagnetic forces are more complicated, as can be inferred from a number of experimental facts. Firstly, the weak interactions are carried by massive force carriers (i.e. the $W^\pm$ and Z^0 bosons), and thus some form of spontaneous symmetry breaking is needed in order to generate their masses. Secondly, the weak interactions are known to violate parity invariance, namely symmetry of a theory under inversion of all vectors according to[7]

$$
V \to -V.
$$

This in turn has consequences for fermion couplings and masses. Given a massive fermion spinor Ψ, we may define states of definite helicity according to

$$
\Psi_{L,R} = \mathbb{P}_{L,R}\Psi,
\tag{2.70}
$$

where we have introduced the **projection operators**

$$
\mathbb{P}_{L,R} = \frac{1 \mp \gamma_5}{2}, \quad \gamma_5 = i\gamma^0\gamma^1\gamma^2\gamma^3,
\tag{2.71}
$$

---

[6] That η can be completely removed should not surprise us: equation (2.67) tells us that it is the phase of the ϕ field, and it is precisely this phase that can be completely fixed by a gauge transformation.
[7] Here we must carefully distinguish between vectors and *pseudo-vectors* such as the angular momentum $L \to x \times p$. The latter necessarily remain invariant under a parity transformation.

for left- and right-handed states, respectively. Parity transformations turn left-handed spinors into right-handed ones and vice versa, and thus a violation of parity invariance tells us that the left- and right-handed states of massive fermions in nature must behave differently in the equations describing the theory. In particular, they may behave differently under gauge transformations. One may write the Dirac Lagrangian of equation (2.1) for a single fermion species of mass m in terms of the spinors of equation (2.70), as follows:

$$\mathcal{L} = \bar{\Psi}_L(i\gamma^\mu\partial_\mu)\Psi_L + \bar{\Psi}_R(i\gamma^\mu\partial_\mu)\Psi_R - m(\bar{\Psi}_R \Psi_L + \bar{\Psi}_L \Psi_R). \quad (2.72)$$

Here, we see that the left- and right-handed components decouple from each other as far as the kinetic term is concerned. However, the mass term mixes the two components, which potentially violates gauge invariance if they transform differently. The upshot is that fermion masses, similarly to gauge boson masses, must be generated by spontaneous symmetry breaking.

A third complication we will see is that the electromagnetic and weak forces cannot be separated from each other, but must be combined into a single **electroweak theory**. Based on the above discussion, the recipe to be followed to construct this theory is as follows:

1. Identify the (non-)Abelian gauge group corresponding to the electroweak forces. The matter particles will have charges associated with this gauge group.
2. Decide how the matter particles (leptons and quarks) are acted on by this gauge group. For QCD, we saw that each quark corresponded to a triplet, whose members had the same flavour but different colours. For the electroweak theory, we must find the possible multiplets that the gauge group can act upon, and then figure out how to arrange the various left- and right-handed fermions into these multiplets, which are then acted on by appropriate representations of the gauge group.
3. We must introduce a scalar field with a non-zero vacuum expectation value, in order to generate gauge boson and fermion masses via spontaneous symmetry breaking. Indeed, there may be a multiplet of scalar fields, acted on by a particular representation of the gauge group.

It is clear that there are many possible choices at each stage of the above procedure, and it is ultimately experiment that must decide which is correct. For this reason, the correct electroweak theory took many decades to establish. It was first written down in the 1960s, and rewarded with a Nobel Prize (to Sheldon Glashow, Abdus Salam and Steven Weinberg) in 1979. Rather than reconstruct the long and rather tortuous steps that led to the theory, we will simply summarise the main details here.

The gauge group of the electroweak theory is found to be SU(2) × U(1). Unlike the previous examples we have seen, this is called a **semi-simple Lie group**, containing two distinct subgroups, each with their own associated charges. The charge associated with the SU(2) subgroup is called **weak isospin**, and the various matter particles form multiplets which are acted on by SU(2) transformations. To find what these multiplets are, we can make an analogy with the study of angular momentum in quantum mechanics. In the latter case, a particle with total angular momentum J (in units of $\hbar$) is associated with a multiplet of $(2J + 1)$ independent states. Each component of the multiplet can be labelled by the component of the (vector) angular momentum in some direction, which is conventionally taken to be the z-direction. Writing this component as J_z, one has

$$J_z \in \{-J, -J + 1, \ldots, J - 1, J\}.$$

A special case is that of $J = 1/2$, in which case one finds a doublet of states:

$$\zeta = \begin{pmatrix} | \uparrow \rangle_z \\ | \downarrow \rangle_z \end{pmatrix},$$

where $| \uparrow \rangle_z$ and $| \downarrow \rangle_z$ denote states with $J_z = \pm 1/2$ respectively. These will mix with each other if the coordinate axes (x, y, z) are rotated. More specifically, a rotation acts on the doublet according to

$$\zeta \to R\zeta, \quad R = \exp\left[i \sum_{a=1}^{3} \sigma_a \tau_a \right], \tag{2.73}$$

where θ_i are rotation angles about the individual axes, and $\sigma_i = \tau_i/2$, with $\{\tau_i\}$ the **Pauli matrices**

$$\tau_1 = \begin{pmatrix} 0 & 1 \\ 1 & 0 \end{pmatrix}, \quad \tau_2 = \begin{pmatrix} 0 & -i \\ i & 0 \end{pmatrix}, \quad \tau_3 = \begin{pmatrix} 1 & 0 \\ 0 & -1 \end{pmatrix}. \tag{2.74}$$

Comparing equation (2.73) with equation (2.33), we recognise the matrix R as an element of the rotation group, and the set of matrices $\{\sigma_i\}$ as the generators of rotations. The latter obey the commutation relation

$$[\sigma_i, \sigma_j] = i\epsilon_{ijk}\sigma_k, \tag{2.75}$$

which happens to be the Lie algebra of SU(2). Indeed, using the explicit form of the Pauli matrices in equation (2.74), one can confirm that the abitrary rotation matrix in equation (2.73) is a complex 2×2 matrix with unit determinant (the defining property of SU(2)). The Pauli matrices can therefore be interpreted as generators in the fundamental representation of SU(2).

Let us now return to the discussion of weak isospin. The analogy with angular momentum (also described by an SU(2) group), suggests that we consider an abstract three-dimensional 'isospin space'. Individual fields (and their corresponding particles) can then be classified into multiplets, and labelled by two quantum numbers: the total weak isospin T, and the component of isospin T_3 along the three-direction. It follows that T may take half-integer or integer values, such that for a given T the possible values of T_3 obey

$$T_3 \in \{-T, -T + 1, \ldots, T - 1, T\}.$$

In the electroweak theory, the left-handed fermions form doublets, thus have weak isospin $T = 1/2$. First, there is one doublet for each generation of leptons:

$$\begin{pmatrix} \nu_e \\ e^- \end{pmatrix}_L, \quad \begin{pmatrix} \nu_\mu \\ \mu^- \end{pmatrix}_L, \quad \begin{pmatrix} \nu_\tau \\ \tau^- \end{pmatrix}_L,$$

where the subscript L reminds us that this applies to the left-handed components of the fields only. We see that each left-handed neutrino has $T_3 = 1/2$, and the left-handed electron, muon and tauon all have $T_3 = -1/2$. Next, each generation of left-handed quarks also forms a doublet:

$$\begin{pmatrix} u \\ d \end{pmatrix}_L, \quad \begin{pmatrix} c \\ s \end{pmatrix}_L, \quad \begin{pmatrix} t \\ b \end{pmatrix}_L.$$

Weak isospin gauge transformations act on these doublets analogously to how rotations act on angular momentum states in equation (2.73). In this case, however, the gauge parameters $\{\theta_a\}$ accompanying each generator correspond to angles about the abstract axes of weak isospin space, rather than the physical spacetime that the particles live in. We have yet to worry about the right-handed fermions. These have zero weak isospin, such that the lepton states

$$e_R^-, \quad \nu_{e,R}, \quad \mu_R^-, \quad \nu_{\mu,R}, \quad \tau_R^-, \quad \nu_{\tau,R}$$

as well as the quark states

$$u_R, \quad d_R, \quad c_R, \quad s_R, \quad t_R, \quad b_R$$

are all singlets. Note that this also implies that all right-handed fermions have $T_3 = 0$.

So much for the SU(2) factor in the electroweak gauge group. As stated above, there is a further U(1) subgroup, whose action on fields is similar to the original QED gauge transformations of equation (2.5). In this present case, however, the charge associated with the transformations is called **weak hypercharge** Y, and each field type has a different value. We list the electroweak quantum numbers of all fermion fields (separating out left- and right-handed components as is necessary) in table 2.1. We have also included the electric charge of each particle in this table, and one may verify that the results are consistent with the formula

$$Q = \left(T_3 + \frac{Y}{2} \right). \tag{2.76}$$

This is called the **Gell-Mann–Nishijima relation**, and was of historical importance in formulating the correct theory of electroweak interactions, based on the observed properties of meson decays.

We can now start to build the Lagrangian for the electroweak theory. Leaving aside mass terms for now, we know that the fermion fields should have kinetic terms similar to those of equation (2.50), but where the covariant derivative now includes gauge fields related to the weak isospin and hypercharge. We follow convention in denoting by $\mathbf{D}_\mu$ the covariant derivative acting on a general field. However, the specific form that this derivative takes will depend on the field in question, given the

Table 2.1. Electroweak quantum numbers for all SM matter particles, where L (R) denotes a left-handed (right-handed) state, T the total weak isospin, T_3 its component along the three-axis, Y the weak hypercharge and Q the electromagnetic charge.

Field	T	T_3	Y	Q
$\nu_{e,\,L}, \nu_{\mu,\,L}, \nu_{\tau,\,L}$	1/2	1/2	-1	0
e_L, μ_L, τ_L	1/2	$-1/2$	-1	-1
$\nu_{e,\,R}, \nu_{\mu,\,R}, \nu_{\tau,\,R}$	0	0	0	0
e_R, μ_R, τ_R	0	0	-2	-1
u_L, c_L, t_L	1/2	1/2	1/3	2/3
d_L, s_L, b_L	1/2	$-1/2$	1/3	$-1/3$
u_R, c_R, t_R	0	0	4/3	2/3
d_R, s_R, b_R	0	0	$-2/3$	$-2/3$

differing quantum numbers in table 2.1. For the weak isospin doublet fields, we have[8]

$$\mathbf{D}_\mu = \partial_\mu + ig\sigma_a W_\mu^a + ig'\frac{Y}{2}B_\mu, \tag{2.77}$$

Here the $\{W_\mu^a\}$ are components of a gauge field associated with the SU(2) (weak isospin) symmetry, where a is an index in weak isospin space. This is the analogue of the gluon field in the non-Abelian gauge theory of strong interactions and, as in that case, the gauge field must be contracted with generators in the appropriate representation (here the fundamental representation $\{\sigma_a\}$ that acts on fields with weak isospin $T = 1/2$). We have also introduced an Abelian-like gauge field B_μ associated with the U(1) symmetry, where Y is the weak hypercharge. Finally, g and g' are coupling constants.

We also need the form of the covariant derivative for the singlet fields (i.e. with zero weak isospin). This simply corresponds to equation (2.77) without the second term on the right-hand side, given that there is no need to generate weak isospin transformations on fields that have no weak isospin:

$$\mathbf{D}_\mu = \partial_\mu + ig'\frac{Y}{2}B_\mu. \tag{2.78}$$

To write the kinetic terms for the fermions, we simply use the Dirac Lagrangian of equation (2.1), but with all partial derivatives replaced by their covariant counterparts. To write this compactly, let us introduce the vectors of doublets:

---

[8] We have here included a conventional factor of 1/2 in the definition of the weak hypercharge. Other conventions exist in the literature in which this factor is absent. The values of Y in table 2.1 should then be halved.

$$\mathrm{L} = \left[\begin{pmatrix} \nu_e \\ e^- \end{pmatrix}_L, \begin{pmatrix} \nu_\mu \\ \mu^- \end{pmatrix}_L, \begin{pmatrix} \nu_\tau \\ \tau^- \end{pmatrix}_L \right] \tag{2.79}$$

for the lepton fields, and

$$\mathrm{Q} = \left[\begin{pmatrix} u \\ d \end{pmatrix}_L, \begin{pmatrix} c \\ s \end{pmatrix}_L, \begin{pmatrix} t \\ b \end{pmatrix}_L \right] \tag{2.80}$$

for the quarks. In index notation, L_i and Q_i then refer to the left-handed doublets for each generation of leptons and quarks, respectively. Let us further introduce the vectors of right-handed fields

$$E_R = \begin{pmatrix} e_R \\ \mu_R \\ \tau_R \end{pmatrix}, \quad \nu_R = \begin{pmatrix} \nu_{e,R} \\ \nu_{\mu,R} \\ \nu_{\tau,R} \end{pmatrix}, \quad \mathrm{U}_R = \begin{pmatrix} u_R \\ c_R \\ t_R \end{pmatrix}, \quad \mathrm{D}_R = \begin{pmatrix} d_R \\ s_R \\ b_R \end{pmatrix}. \tag{2.81}$$

The fermion kinetic terms can then be written as

$$\begin{aligned}
\mathcal{L}_{\mathrm{f,kin.}} = {} & i\bar{L}_i\, \gamma^\mu(\partial_\mu + igW_\mu^a\, \sigma^a + ig'Y_L B_\mu)L_i \\
& + i\bar{Q}_i\, \gamma^\mu(\partial_\mu + igW_\mu^a\, \sigma^a + ig'Y_Q B_\mu)Q_i \\
& + i\bar{E}_R^i\, \gamma^\mu(\partial_\mu + ig'Y_E B_\mu)E_R^i + i\bar{\nu}_R^i\, \gamma^\mu(\partial_\mu + ig'Y_\nu B_\mu)\nu_R^i \\
& + i\bar{U}_R^i\, \gamma^\mu(\partial_\mu + ig'Y_U B_\mu)U_R^i + i\bar{D}_R^i\, \gamma^\mu(\partial_\mu + ig'Y_D B_\mu)D_R^i,
\end{aligned} \tag{2.82}$$

where we have utilised the fact—which can be verified from table 2.1—that the groups of fields in each term share a common value of weak hypercharge.

As well as the fermion kinetic terms, we also need the kinetic terms for the newly-introduced gauge fields $\mathbf{W}_\mu$ and B_μ, where

$$\mathbf{W}_\mu \equiv W_\mu^a\, \sigma^a$$

is matrix-valued in weak isospin space. By analogy with the QED and QCD cases of equations (2.13) and (2.58), we would write these as

$$\mathcal{L}_{\mathrm{kin.}} = -\frac{1}{4}B^{\mu\nu}B_{\mu\nu} - \frac{1}{4}W^{a,\mu\nu}W_{\mu\nu}^a, \tag{2.83}$$

where $B_{\mu\nu}$ and $W_{\mu\nu}^a$ are the field strength tensors associated with B_μ and $\mathbf{W}_\mu$ respectively. However, we know that such kinetic terms give rise to *massless* gauge fields, whereas in fact we want to capture the fact that some of the gauge bosons in the electroweak theory are massive. As discussed above, this in turn means that we must introduce a scalar field (or set of them) with non-zero vacuum expectation values, so that we can generate the non-zero gauge boson masses through spontaneous symmetry breaking. We mentioned that Goldstone's theorem implies that there is a massless scalar field (or Goldstone boson) for each degree of freedom of a continuous symmetry that is broken. Furthermore, these Goldstone bosons are then identified with the extra polarisation states of the gauge bosons that become massive.

In the electroweak theory, we require three massive gauge bosons (the $W^\pm$ and Z^0 bosons), and one massless (the photon). As we saw in Section 2.1, the gauge group of electromagnetism is U(1), and so the symmetry breaking that must take place is

$$SU(2) \times U(1)_Y \to U(1)_{EM}. \qquad (2.84)$$

We have here appended a suffix to the U(1) groups to emphasise that they are not the same: the U(1) group on the left-hand side corresponds to weak hypercharge Y, whereas that on the right-hand side is associated with electric charge Q. The dimension of the complete gauge group on the left (i.e. the number of independent degrees of freedom needed to specify an arbitrary gauge transformation) is four, and on the right-hand side is one, so this is consistent with three gauge bosons becoming massive. To break the symmetry, we must introduce a set of scalar degrees of freedom in a well-defined multiplet associated with the gauge group. It turns out that the minimal choice is to introduce an SU(2) doublet of complex scalar fields

$$\Phi = \begin{pmatrix} \phi^+ \\ \phi^0 \end{pmatrix}, \quad \phi^+, \phi^0 \in \mathbb{C}, \qquad (2.85)$$

known as the **Higgs multiplet**. This has weak isospin $T = 1/2$, such that ϕ^+ (ϕ^0) has isospin $T_3 = 1/2$ ($T_3 = -1/2$). Furthermore, the superscripts on the fields $\phi^{+,0}$ indicate what turns out to be their electric charge. Requiring that the Higgs field satisfies the Gell-Mann–Nishijima relation of equation (2.76) then implies that the Higgs doublet has weak hypercharge $Y_\phi = 1$, and we can then write down the gauge-invariant Lagrangian

$$\mathcal{L}_{kin.} = -\frac{1}{4} B^{\mu\nu} B_{\mu\nu} - \frac{1}{4} W^{a,\mu\nu} W^a_{\mu\nu} + (\mathbf{D}^\mu \Phi)^\dagger (\mathbf{D}_\mu \Phi) - V(\Phi), \qquad (2.86)$$

where the covariant derivative acting on the Higgs doublet is

$$\mathbf{D}_\mu \Phi = \left(\partial_\mu + igW^a_\mu \sigma^a + \frac{ig'}{2} B_\mu \right) \Phi, \qquad (2.87)$$

and we have introduced the **Higgs potential**[9]

$$V(\Phi) = -\mu^2 \Phi^\dagger \Phi + \lambda (\Phi^\dagger \Phi)^2. \qquad (2.88)$$

This is a generalisation of the potential in the Abelian Higgs model of equation (2.64). However, the fact that Φ is now a doublet of complex fields means that there is a four-dimensional space of real scalar degrees of freedom, making the potential hard to visualise. Nevertheless, it remains true that for $\mu^2 > 0$ and $\lambda > 0$, the minimum of $V(\Phi)$ does not occur at zero field values, but at a non-zero VEV given by

---

[9] Equation (2.88) turns out to be the only potential function consistent with SU(2) × U(1)$_Y$ gauge invariance, and renormalisability, where the latter concept is discussed in Section 2.11.

$$|\phi| = \sqrt{\frac{\mu^2}{2\lambda}} \equiv \frac{v}{\sqrt{2}}, \qquad (2.89)$$

where we have again introduced a factor of 2 when defining the constant v, for convenience. By analogy with equation (2.67), we may write a general Higgs field, perturbed about the vacuum, as follows:

$$\Phi(x) = \frac{1}{\sqrt{2}}(v + h(x)) \exp\left[\frac{i}{v}\Theta^a(x)\sigma^a\right]\binom{0}{1}, \qquad (2.90)$$

We have chosen an arbitrary direction in the space of complex scalar fields, which is then subjected to an arbitrary SU(2) gauge transformation, plus an arbitrary perturbation of the overall normalisation. This gives four degrees of freedom in total, as is appropriate for the Φ field. Were we to substitute equation (2.90) into equation (2.86) and expand, we would find the fields $\Theta^a(x)$ (associated with the broken SU(2) symmetry) appearing as Goldstone bosons, as occured with the field $\eta(x)$ in equation (2.67). As in the former case, these are spurious degrees of freedom, and we may choose to remove them from the Lagrangian by working in the unitary gauge, in which one sets $\Theta^a(x) = 0$ for all a. The Higgs doublet in the unitary gauge is thus

$$\Phi(x) = \frac{1}{\sqrt{2}}(v + h(x))\binom{0}{1}. \qquad (2.91)$$

Upon substituting this into the Lagrangian of equation (2.86) and carrying out a lot of messy algebra, one finds the following terms for the gauge fields (in addition to their kinetic terms):

$$\mathcal{L}_M = \frac{v^2}{8}\left[(gW_\mu^3 - g'B_\mu)(gW^{3\mu} - g'B^\mu) + 2g^2 W_\mu^- W^{+\mu}\right], \qquad (2.92)$$

where we have defined the combinations

$$W_\mu^\pm = \frac{1}{\sqrt{2}}(W_1^\mu \pm iW_2^\pm) \qquad (2.93)$$

that we will interpret shortly. We see that the fields (W_μ^3, B_μ) (associated with the weak isopin and hypercharge gauge symmetries, respectively) mix with each other upon performing the spontaneous symmetry breaking. We can remove the mixing terms by replacing these fields with linear combinations of them. We do not want to change the overall normalisation of the fields, given that the kinetic terms in equation (2.86) already have the conventional normalisation. Thus, we will instead perform a transformation of the form

$$\binom{W_\mu^3}{B_\mu} = \begin{pmatrix} \cos\theta_W & \sin\theta_W \\ -\sin\theta_W & \cos\theta_W \end{pmatrix}\binom{Z_\mu}{A_\mu}, \qquad (2.94)$$

corresponding to a rotation in the space of fields, and where θ_W is referred to as the **weak mixing angle**. Substituting equation (2.94) into equation (2.92), we find that the mixing terms vanish provided we set

$$\tan \theta_W = \frac{g'}{g}. \tag{2.95}$$

Expanding and combining all mass and kinetic terms for our newly chosen fields, we find the quadratic terms

$$\begin{aligned}
\mathcal{L}_2 = &-\frac{1}{4}F^{\mu\nu}F_{\mu\nu} - \frac{1}{4}Z^{\mu\nu}Z_{\mu\nu} + \frac{1}{2}m_Z^2 Z^\mu Z_\mu \\
&-\frac{1}{2}(\partial^\mu W^{+\nu} - \partial^\nu W^{+\mu})(\partial_\mu W_\nu^- - \partial_\nu W_\mu^-) \\
&+ m_W^2 W^{+\mu}W_{\mu-},
\end{aligned} \tag{2.96}$$

the cubic terms

$$\begin{aligned}
\mathcal{L}_3 = &- \mathrm{i}e \cot \theta_W \Big[\partial^\mu Z^\nu \big(W_\mu^+ W_\nu^- - W_\nu^+ W_\mu^-\big) \\
&+ Z^\nu \big(-W^{+\mu}\partial_\nu W^{\mu-} + W_\mu^-\partial_\nu W_\mu^+ \\
&+ W^{+\mu}\partial_\mu W_\nu^- - W^{\mu-}\partial_\mu W_\nu^+\big]\Big] \\
&- \mathrm{i}e \Big[\partial^\mu A^\nu \big(W_\mu^+ W_\nu^- - W_\nu^+ W_\mu^-\big) \\
&+ A^\nu \big(-W^{+\mu}\partial_\nu W_\mu^- + W^{\mu-}\partial_\nu W_\mu^+ \\
&+ W^{+\mu}\partial_\mu W_\nu^- - W^{\mu-}\partial_\mu W_\nu^+\big]\Big),
\end{aligned} \tag{2.97}$$

and the quartic terms

$$\begin{aligned}
\mathcal{L}_4 = &-\frac{1}{2}\frac{e^2}{\sin^2 \theta_W} W^{+\mu}W_\mu^- W^{\nu+}W_\nu^- \\
&+ \frac{1}{2}\frac{e^2}{\sin^2 \theta_W} W^{+\mu}W^{-\nu}W_\mu^+ W_\nu^- \\
&- e^2 \cot^2 \theta_W \big(Z^\mu W_\mu^+ Z^\nu W_\nu^- - Z^\mu Z_\mu W^{\nu+}W_\nu^-\big) \\
&+ e^2 \big(A^\mu W_\mu^+ A^\nu W_\nu^- - A^\mu A_\mu W^{+\nu}W_\nu^-\big) \\
&+ e^2 \cot \theta_W \big(A^\mu W_\mu^+ W^{-\nu}Z_\nu + A^\mu W_\mu^- Z^\nu W_\nu^+ - W^{+\mu}W_\mu^- A^\nu Z_\nu\big).
\end{aligned} \tag{2.98}$$

Here $F_{\mu\nu}$ amd $Z_{\mu\nu}$ are field strength tensors for the A_μ and Z_μ fields, and we have defined

$$m_W = \frac{gv}{2}, \quad m_Z = \frac{m_W}{\cos \theta_W}, \tag{2.99}$$

as well as the combination

$$e = g \sin \theta_W. \tag{2.100}$$

A lot has happened here, and it is very easy for the overall physics to get lost amongst the tangle of complicated formulae that has arisen. Let us therefore walk through the various types of term step by step. The quadratic terms of equation (2.96) contain kinetic terms for the four vector fields (A_μ, Z_μ, $W_\mu^\pm$). Furthermore, there are mass terms for the $W_\mu^\pm$ and Z_μ fields (n.b. the definition of equation (2.93) implies that $(W_\mu^+)^\dagger = W_\mu^-$, and thus these belong in the same mass term). The fact that there is no mixing between any of these fields in the quadratic terms implies that they will propagate independently, and thus constitute the distinct particles in the theory. The massive $W_\mu^\pm$ fields have two longitudinal degrees of freedom between them, and the Z_μ field will have one. Thus, a total of three longitudinal polarisation states has been created from the spontaneous symmetry breaking, which is what we expected from the fact that there were three Goldstone bosons in equation (2.90).

There is no mass term for A_μ in equation (2.94), and it has thus remained massless. We can then identify A_μ with the photon, an intepretation which is confirmed by studying the cubic terms of equation (2.97). Each group of terms represents an interaction between three vector fields, where the second group shows that A_μ couples to a pair of $W_\mu^\pm$ fields. This is entirely consistent with the $W_\mu^\pm$ fields having electric charge as indicated, if A_μ is the photon and e the electromagnetic coupling. We see that the Z field also couples to a $W_\mu^\pm$ pair, indicating that the Z^μ is also electrically neutral. Charge conservation forbids an interaction involving three $W_\mu^\pm$ fields. Finally, equation (2.98) contains interactions between quartets of vector fields, where charge conservation is respected in each term. We see that there are no self-interactions of the A_μ with itself (i.e. only those involving at least two charged fields). Thus, A_μ indeed represents the (chargeless) photon. The massive vector fields Z_μ and $W_\mu^\pm$ represent the neutral and charged carriers of the weak interaction, and so the symmetry breaking scheme of equation (2.84) has indeed captured the desired properties of electroweak interactions that we sought to reproduce.

Although the Goldstone bosons (i.e. the Θ^a fields in equation (2.90)) associated with the symmetry breaking are spurious degrees of freedom that are manifestly absent in the unitary gauge, the field $h(x)$ does indeed contribute to the Lagrangian, and represents a physical particle state. It is called the **Higgs boson**, and the remaining terms from equation (2.86) involving this field are as follows. First, there are terms involving the Higgs field by itself:

$$\mathcal{L}_{\text{Higgs}} = \frac{1}{2}\partial_\mu h\, \partial^\mu h - \mu^2 h^2 - \lambda v h^3 - \frac{1}{4}\lambda h^4. \tag{2.101}$$

These comparise a kinetic term (correctly normalised for a single real scalar degree of freedom), a mass term, plus cubic and quartic self-interactions. Next, there are interaction terms between the Higgs boson and massive vector bosons:

$$\mathcal{L}_{Vh} = \frac{2h}{v}\left(m_W^2 W^{+\mu}W_\mu^- + \frac{1}{2}m_Z^2 Z^\mu Z_\mu\right)$$
$$+ \left(\frac{h}{v}\right)^2\left(m_W^2 W^{+\mu}W_\mu^- + \frac{1}{2}m_Z^2 Z^\mu Z_\mu\right). \tag{2.102}$$

Together with equations (2.96)–(2.98) and equation (2.101), this completes all terms that arise from the Lagrangian of equation (2.86), after spontaneous symmetry breaking. We have also provided kinetic terms for the matter particles, in equation (2.82). Our final task is to provide mass terms for the latter.

Let us start with the lepton sector. In the original SM, the neutrinos are massless, and so we only have to provide mass terms for the (e^-, μ^-, τ^-), all of whose left-handed parts sit in the lower components of the doublets in equation (2.79). Above, we saw that we cannot simply introduce fermion mass terms by hand, as this will not be gauge-invariant by itself. However, we can form the gauge-invariant combination

$$\mathcal{L}_{l,\mathrm{Yuk.}} = -\sum_i y_i \bar{L}_i \Phi e_R^i + \mathrm{h.c.}, \tag{2.103}$$

where 'h.c.' refers to the Hermitian conjugate, and each y_i is an abitrary real parameter. This is called a **Yukawa interaction**, and it will indeed generate lepton masses after electroweak symmetry breaking. To see this, we can substitute in the form of the Higgs doublet in the unitary gauge, equation (2.91), to obtain

$$\mathcal{L}_{l,\mathrm{Yuk.}} = -\sum_l \left[m_l \bar{l}_L\, l_R + \frac{1}{v}h\bar{l}_L\, l_R\right] + \mathrm{h.c.}, \quad m_l = \frac{y_l v}{\sqrt{2}}, \tag{2.104}$$

where the sum is over the leptons $l \in (e, \mu, \tau)$. The first term in the brackets is a mass term, where the mass of each lepton is proportional both to the Higgs VEV v, and the Yukawa coupling y_l. Note that this does not explain why we see such a range of lepton masses in the SM: instead this question is simply rephrased in terms of why the Yukawa couplings are so different.

The same mechanism as above can be used to give masses to the down-type quarks, given that these are also associated with the lower components of SU(2) doublets. However, the up-type quarks also need masses, and they are projected out by the above construction, so that it must be generalised. To this end, we may use the **conjugate Higgs doublet**

$$\tilde{\Phi} = i\tau_2\Phi^* = \frac{(v+h)}{\sqrt{2}}\begin{pmatrix}1\\0\end{pmatrix}, \tag{2.105}$$

where τ_2 is a Pauli matrix, and the second equality holds only in the unitary gauge. It turns out that equation (2.105) corresponds to a **charge conjugation** operation acting on the Higgs doublet, and thus reverses its quantum numbers. In particular, $\tilde{\Phi}$ has weak hypercharge $Y = -1$, given that Φ has $Y = 1$. Armed with equation (2.105), the most general set of mass terms we can write down for the quark sector is

Table 2.2. Electromagnetic and weak parameters entering the fermion Lagrangian of equation (2.107).

Fermion	Q_f	V_f	A_f
(u, c, t)	$\frac{2}{3}$	$\frac{1}{2} - \frac{4}{3}\sin^2\theta_W$	$\frac{1}{2}$
(d, s, b)	$-\frac{1}{3}$	$-\frac{1}{2} + \frac{2}{3}\sin^2\theta_W$	$-\frac{1}{2}$
$(\nu_e, \nu_\mu, \nu_\tau)$	0	$\frac{1}{2}$	$\frac{1}{2}$
(e, μ, τ)	-1	$-\frac{1}{2} + 2\sin^2\theta_W$	$-\frac{1}{2}$

$$\mathcal{L}_{q,\text{Yuk.}} = - Y_{ij}^d \bar{Q}^i \, \Phi \, D_R^j - Y_{ij}^u \bar{Q}^i \, \tilde{\Phi} \, U_R^j + \text{h.c.}, \tag{2.106}$$

where summation over repeated indices is involved, and $\{Y_{ij}^d, Y_{ij}^u\}$ are arbitrary (complex) parameters for the down- and up-type mass terms, respectively. These coefficients are not fixed from first principles, and must ultimately be measured from experiment. At first sight, this looks like a large number of parameters: given that there are three generations, two separate 3×3 complex matrices constitute 18 parameters. However, it may be shown that many of these parameters can be eliminated by performing linear transformations of the fields, the net effect of which is that the complete Lagrangian for the fermion sector can be written as

$$\mathcal{L}_\text{f} = \sum_f \left[\bar{\Psi}_f \left(i\gamma^\mu \partial_\mu - m_f - \frac{m_f h}{v} \right)\Psi_f - eQ_f \bar{\Psi}_f \, \gamma^\mu A_\mu \, \Psi_f \right.$$
$$- \frac{g}{2\cos\theta_W} \bar{\Psi}_f \gamma^\mu Z_\mu (V_f - A_f \gamma_5)\Psi_f \Bigg]$$
$$- \frac{g}{2\sqrt{2}} \sum_l \left[\bar{l} \, \gamma^\mu (1 - \gamma_5) W_\mu^+ \, \nu_l + \text{h.c.} \right]$$
$$+ \frac{g}{2\sqrt{2}} \left[V_{ij} \, W_\mu^+ \bar{U}^i \gamma^\mu (1 - \gamma_5) D^j + \text{h.c.} \right]. \tag{2.107}$$

Here the first line contains the kinetic and mass terms for all fermion fields Ψ_f, as well as their coupling to the Higgs boson and photon, where Q_f is the relevant electric charge. The second line contains the coupling of all fermions to the Z boson, where V_f and A_f are the so-called **vector** and **axial-vector** couplings, respectively. Their values are listed in table 2.2. There are also the couplings of the leptons (and neutrinos) to the $W^\pm$ bosons, where we have made the role of γ_5 explicit.

The fourth line of equation (2.107) contains the couplings of the up- and down-type quarks to the W bosons[10], where $\{V_{ij}\}$ is a set of parameters called the **Cabibbo–Kobayashi–Maskawa (CKM) matrix**. If this matrix were diagonal (i.e. such that only the terms V_{ii} are non-zero), a given up-type quark would couple only to the down-

---

[10] Our notation for the up- and down-type quarks follows that of equation (2.81), where the absence of the R subscript implies that both right- and left-handed components are included.

type quark within the same generation. However, this is not observed experimentally, so that an up-type quark can change into an arbitrary down-type quark by emitting a W boson (and vice versa), where the probability for this to happen depends upon the appropriate CKM matrix element.

We have finally finished our description of the electroweak theory, whose complete Lagrangian is

$$\mathcal{L}_{\text{EW}} = \mathcal{L}_2 + \mathcal{L}_3 + \mathcal{L}_4 + \mathcal{L}_{\text{Higgs}} + \mathcal{L}_{Vh} + \mathcal{L}_f. \tag{2.108}$$

The full SM is then obtained by adding in the couplings of the quarks to the gluon, and the kinetic terms for the latter, as described in Section 2.4. There is no need to add the quark mass terms discussed previously, as these are already contained within equation (2.108).

The SM is a curious theory. On the one hand, it is strongly constrained by mysterious and powerful theoretical underpinnings (e.g. gauge invariance). On the other, it contains parts that—to be frank—are merely cobbled together to fit experiments. Indeed, the latter must be the case, given that it is impossible for theory alone to tell us what the gauge group of Nature has to be, or which multiplets the fermions have to sit in. Nevertheless, it is difficult to shake off the impression that the SM, with its cumbersome structure, cannot be the final answer for what is really happening in our universe at the most fundamental level. We will return to this subject later in the book.

2.7 Cross-sections and scattering amplitudes

Having now arrived at the SM Lagrangian, we can return to the question of how to calculate cross-sections in modern day colliders. Given the ubiquity of quark and gluon radiation in these experiments, we will focus mostly on QCD unless otherwise stated. Typically, a scattering process has some initial state $|i\rangle$ (e.g. containing two beam particles), and a final state $|f\rangle$, as shown schematically in Figure 2.5. If we assume the interaction between the particles is localised in space and time, then we can think of the initial state (at time $t \to -\infty$) as consisting of incoming plane waves,

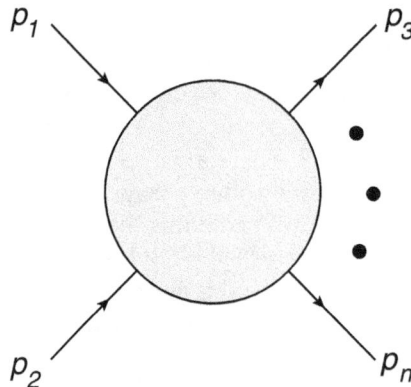

Figure 2.5. A scattering process, in which two beam particles interact, and then produce n final state particles.

associated with a free QFT. Likewise, the final state consists of a bunch of outgoing plane waves at time $t \rightarrow +\infty$, which is also a state of a free QFT. At intermediate times, the particles interact (shown as the grey blob in Figure 2.5), and there we must take the full Lagrangian of the QFT into account. We never have to worry, though, about what the states look like at intermediate times, as the only states we measure are associated with the incoming and outgoing particles.

For our given initial and final states, the probability of a transition from state $|i\rangle$ to state $|f\rangle$ is given by the usual rules of quantum mechanics, as

$$P = \frac{|\langle f|i\rangle|^2}{\langle f|f\rangle \langle i|i\rangle}. \tag{2.109}$$

That is, there is some inner product between the states that measures the overlap between them. One then takes the squared modulus of this to give a real number. Finally, the denominator corrects for the fact that the states themselves may not have unit normalisation. Indeed, it is common to normalise single particle states of given four-momentum p and spin λ according to the formula

$$\langle p', \lambda'|p, \lambda\rangle = (2\pi)^3 2E\delta_{\lambda\lambda'}\delta^{(3)}(\boldsymbol{p} - \boldsymbol{p}'), \quad E = p^0. \tag{2.110}$$

Here the delta functions tell us that states of different momenta and/or spins are mutually orthogonal. Furthermore, the normalisation includes some conventional numerical constants, plus a factor of the energy whose role will become clearer in the following. Equation (2.110) implies that the inner product of a state with itself is given by

$$\langle p, \lambda|p, \lambda\rangle = (2\pi)^3 2E\delta^{(3)}(0)$$
$$= 2E(2\pi)^3 \lim_{p' \to p} \int \frac{\mathrm{d}^3\boldsymbol{x}}{(2\pi)^3} e^{i(p-p')\cdot x}, \tag{2.111}$$

where we have used a well-known representation of the Dirac delta function:

$$\delta^{(n)}(\boldsymbol{p} - \boldsymbol{p}') = \int \frac{\mathrm{d}^n\boldsymbol{x}}{(2\pi)^n} e^{ix\cdot(p-p')}. \tag{2.112}$$

To make sense of the apparently infinite delta-function integral, let us assume that our particle is contained in a cubic box whose sides have length L, such that the volume of the box $V = L^3$. At the end of our calculation we will be able to take $L \rightarrow \infty$, recovering the physical situation we started with. With this trick, equation (2.111) becomes

$$\langle p, \lambda|p, \lambda\rangle = 2EV. \tag{2.113}$$

This is for a single particle state. For an m-particle state, carrying through similar arguments leads to

$$\langle p_1, \lambda_1; p_2, \lambda_2; \ldots; p_m, \lambda_m|p_1, \lambda_1; p_2, \lambda_2; \ldots; p_m, \lambda_m\rangle = \prod_{i=1}^{m} 2E_i V. \tag{2.114}$$

In particular, for the scattering process shown in Figure 2.5, we have

$$\langle i|i \rangle = 4E_1E_2V^2, \quad \langle f|f \rangle = \prod_{k=3}^{n} 2E_kV. \tag{2.115}$$

Next, we concentrate on the numerator of equation (2.109). The quantity $\langle f|i \rangle$ that describes the overlap between the initial and final states is called the **S-matrix**, where the 'S' stands for 'scattering'. On very general grounds this can be decomposed as

$$S_{fi} = \langle f|i \rangle = \delta_{if} + i(2\pi)^4 \delta^{(4)}(P_f - P_i)\mathcal{A}. \tag{2.116}$$

Here the first term on the right-hand side is non-zero only if the initial and final states are equal, and thus corresponds to nothing particuarly interesting happening. In the second term, the factors of i and $(2\pi)^4$ are conventional, and the delta function imposes overall momentum conservation, where

$$P_i = p_1 + p_2, \quad P_f = \sum_{k=3}^{n} p_k$$

are the total four-momenta in the initial and final states, respectively. The remaining quantity $\mathcal{A}$ is called the **scattering amplitude**, where we have suppressed indices i and f (showing that it depends on the nature of the initial and final states) for brevity. We will see later how to calculate $\mathcal{A}$, but for now simply note that the probability of something interesting occuring (i.e. non-trivial scattering between the initial and final states) is given, from equations (2.109), (2.115), (2.116), by

$$P = \frac{[(2\pi)^4 \delta^{(4)}(P_f - P_i)]^2 |\mathcal{A}|^2}{4E_1E_2V^2 \prod_{k=3}^{n} 2E_kV}. \tag{2.117}$$

We can write the square of the delta function as

$$\begin{aligned}
[\delta^{(4)}(P_f - P_i)]^2 &= \delta^{(4)}(P_f - P_i)\delta^{(4)}(0) \\
&= \delta^{(4)}(P_f - P_i) \int \frac{\mathrm{d}^4 x}{(2\pi)^4} e^{i0 \cdot x} \\
&= \frac{TV}{(2\pi)^4} \delta^{(4)}(P_f - P_i),
\end{aligned} \tag{2.118}$$

where we have again used equation (2.112), and introduced a finite time T for the interaction, as well as the finite volume V. Hence, we can rewrite equation (2.117) as

$$P = \frac{[(2\pi)^4 \delta^{(4)}(P_f - P_i)]T|\mathcal{A}|^2}{4E_1E_2V \prod_{k=3}^{n} 2E_kV}. \tag{2.119}$$

from which we see that the probability per unit time is

$$\frac{P}{T} = \frac{[(2\pi)^4 \delta^{(4)}(P_f - P_i)]|\mathcal{A}|^2}{4E_1E_2V \displaystyle\prod_{k=3}^{n} 2E_k V}. \tag{2.120}$$

Note that this is the probability for fixed momenta $\{p_3, p_4, ...p_n\}$ in the final state (for a given initial state). In a real experiment, we have no control over what happens, and so we should sum over all possible values of the final state momenta that sum to the same P_f, to get a total probability per unit time. To see how this works, recall that we have normalised our plane wave states to be in a box with sides of length L. From the de Broglie relationship between momentum and wavenumber

$$\boldsymbol{p} = \hbar \boldsymbol{k} \xrightarrow{\hbar \to 1} \boldsymbol{k}, \tag{2.121}$$

it follows that the momenta in each direction are **quantised**, with

$$\boldsymbol{p} = \frac{2\pi}{L}(n_x, n_y, n_z), \quad n_i \in \mathbb{Z} \tag{2.122}$$

(i.e. we can only fit complete multiples of the particle wavelength in the box). Summing over all possible momentum states then amounts to summing all possible values of the (integer) vector $\boldsymbol{n} = (n_x, n_y, n_z)$, where

$$\sum_{\boldsymbol{n}} = \left(\frac{L}{2\pi}\right)^3 \sum_{\boldsymbol{p}} \xrightarrow{L \to \infty} V \int \frac{d^3 \boldsymbol{p}}{(2\pi)^3}.$$

We then have

$$\sum_{\text{momenta}} \frac{P}{T} = \frac{(2\pi)^4}{4E_1E_2V} \left(\prod_{k=3}^{n} \int \frac{d^3 \boldsymbol{p}_k}{(2\pi)^3 2E_k}\right) \delta^{(4)}(P_f - P_i)|\mathcal{A}|^2. \tag{2.123}$$

We said before that we should only sum over final state momenta that give the right *total* final state momentum. This is fine though, as the delta function will enforce the correct result. Interestingly, we see that the integral over the three-momentum of each final state particle is accompanied by a factor $(2E_k)^{-1}$, whose origin can be traced back to the normalisation of the particle states. It turns out that this is particularly convenient: as you can explore in the exercises, the combination

$$\int \frac{d^3 \boldsymbol{p}_k}{E_k}$$

turns out to be Lorentz-invariant *by itself*. Given that final results for cross-sections have to be Lorentz invariant, it is useful to break the overall calculation into Lorentz invariant pieces.

From known properties of Poisson statistics, the above total probability per unit time will be the same as the event rate for transitions between our initial and momentum-summed final states. To clarify, it is the event rate for a *single* incident/target particle pair. In equation (1.21), we saw that the cross-section is the event rate per target particle, per unit flux of incident particle. Thus, we can convert equation (2.123) into the cross-section by dividing by the flux factor for a single particle. From equation (1.20) (converted to the present notation), thus is given by

$$\mathcal{F}_1 = n_1(v_1 + v_2),$$

where[11]

$$v_i = \frac{|\boldsymbol{p}_i|}{E_i}$$

is the speed of particle i, and n_1 is the number of particles in beam 1 per unit volume. Here we have considered a single particle in volume V, and thus

$$n_1 = \frac{1}{V}.$$

Putting things together, we have

$$\mathcal{F}_1 = \frac{1}{V}\left(\frac{|\boldsymbol{p}_1|}{E_1} + \frac{|\boldsymbol{p}_2|}{E_2}\right).$$

This will combine with the factor $4E_1E_2V$ in equation (2.123) to make the combination

$$F = 4E_1E_2V\mathcal{F}_1 = 4(E_1|\boldsymbol{p}_2| + E_2|\boldsymbol{p}_1|).$$

For collinear beams that collide head-on, one may show (see the exercises) that this can be written as

$$F = 4[(p_1 \cdot p_2)^2 - m_1^2m_2^2]^{1/2}. \tag{2.124}$$

This is called the **Lorentz invariant flux**. It is not the actual flux measured in a given frame, but a combination of factors occuring in the cross-section calculation that happens to be Lorentz invariant, and thus useful! Finally, we have arrived at an important result, namely that the cross-section is given by

$$\sigma = \frac{1}{F}\int d\Phi|\mathcal{A}|^2, \tag{2.125}$$

where

$$\int d\Phi = (2\pi)^4\left(\prod_i \int \frac{d^3\boldsymbol{p}_i}{(2\pi)^3 2E_i}\right)\delta^{(4)}(P_f - P_i) \tag{2.126}$$

---

[11] If you are not familiar with the expression for the speed, recall that the relativistic energy and momenta are given in natural units by $E = m\gamma$ and $\boldsymbol{p} = \gamma m\boldsymbol{v}$ respectively.

is called the **Lorentz-invariant phase-space**, in line with the comments above. It includes a product over all final state particles. Note that all dependence on the intermediate quantities T (a finite interaction time) and volume V (an interaction volume) has completely cancelled out. This is why the above argument, although it contains a number of non-rigorous steps, ultimately ends up giving us the correct answer for the cross-section, that would more properly be obtained from a much more rigorous field theory treatment. We see that calculating cross-sections involves three main steps:

1. Calculate the scattering amplitude $\mathcal{A}$ for a given process, and construct $|\mathcal{A}|^2$.
2. Integrate over the Lorentz-invariant phase-space.
3. Divide by the Lorentz invariant flux factor F.

All of this is easier said than done of course, but we have at least broken down the problem into what appear to be manageable ingredients. In order to go further, we need to know how to calculate the scattering amplitude $\mathcal{A}$.

2.8 Feynman diagrams and rules

For a given scattering process, we can calculate the amplitude using **Feynman diagrams**. Roughly speaking, these are spacetime diagrams showing the history of a given scattering process, and an example is shown in Figure 2.6. This depicts two incoming particles that scatter by exchanging a photon (the wavy line), so that the right of the diagram has a final state consisting of the same particles that came in, but with different momenta. In this way, Feynman diagrams provide a useful way of visualising particle interactions. However, they are not to be taken too literally. In particular, they have Lorentz invariance built in, so really represent a sum over all possible time orderings of a given scattering event, e.g. the diagram in Figure 2.6 represents the case where the first particle emits the photon and the second subsequently absorbs it, and vice versa.

To read Feynman diagrams, we first need to know that different types of line represent different species of particle, and some common conventions are listed in Figure 2.7. On the fermion and scalar lines, the arrow denotes the direction of the

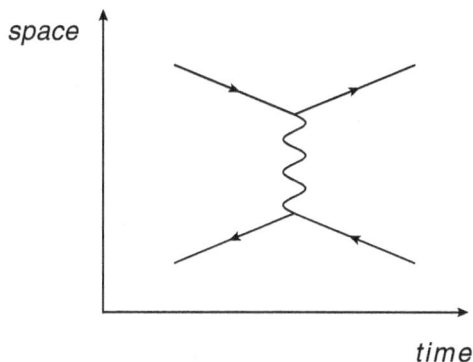

Figure 2.6. An example of a Feynman diagram.

Figure 2.7. Feynman diagram symbols for: (a) a fermion (e.g. lepton or quark); (b) a gluon; (c) a photon, W or Z boson; (d) a scalar (e.g. Higgs boson).

four-momentum. We can then represent *anti-particles* by reversing the arrow, so that this looks like a particle travelling backwards in time. To see why, recall that anti-particles are associated with potential negative energy solutions of relativistic theories. In non-relativistic theories, energy and momentum are related by

$$E = \frac{|\boldsymbol{p}|^2}{2m},$$

which is manifestly positive semi-definite (i.e. $E \geqslant 0$). In a relativistic theory, on the other hand, we have

$$E = \pm\sqrt{|\boldsymbol{p}|^2 + m^2},$$

which can be positive or negative. What QFT does for us is to successfully reinterpret negative energy particles as positive energy anti-particles, where both particles and anti-particles arise as quanta of wavelike solutions of the field equations. A wave has a phase

$$\sim e^{iEt},$$

so a negative energy solution travelling forwards in time is equivalent to a positive energy solution going backwards in time! Hence the reversal of the arrows for an anti-particle on Feynman diagrams.

Note that there are two kinds of line in a Feynman diagram:

(i) *External* lines are associated with (anti-)particles that go off to infinity (or come in from infinity). These are **real** particles, that can be observed in principle.

(ii) *Internal* lines represent particles that are never seen, but are absorbed by (or decay into) real particles. They are called *virtual* particles.

The importance of Feynman diagrams is that they are much more than handy pictures to help us visualise scattering processes. Each one can be converted into a precise mathematical contribution to the amplitude $\mathcal{A}$, using so-called **Feynman rules**. We will not derive the Feynman rules here. Instead, we will simply quote and use the rules we need. We will work in *momentum* rather than *position* space, given that the incoming/outgoing particles at a collider are typically set-up and measured in terms of their definite energy and momentum. Then, the Feynman rules consist of factors associated with each external line, internal line and vertex.

Beginning with the external lines, the rules state that each incoming or outgoing fermion of four-momentum p contributes a basis spinor (or adjoint spinor), whose spinor index will combine with other spinor indices in other parts of the diagram.

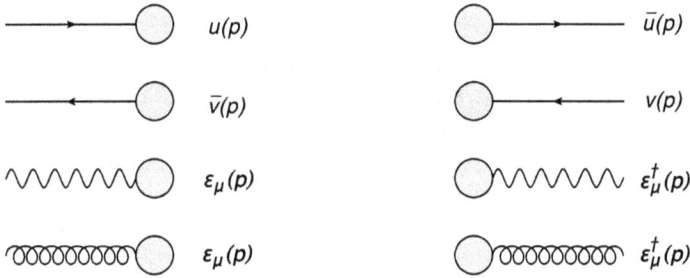

Figure 2.8. Feynman rules for incoming and outgoing (external) particles.

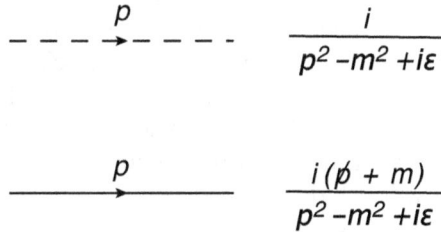

Figure 2.9. Propagators for a scalar particle and a fermion.

This is summarised in Figure 2.8, where we use conventional notation for the basis spinors. Each incoming photon or gluon of four-momentum p contributes a **polarisation vector** $\epsilon_\mu(p)$, which includes information on its polarisation state. Outgoing photons/gluons contribute the Hermitian conjugate of the appropriate polarisation vector. Typically, in all calculations we will end up summing over the possible polarisation and spin states of the external fermions and vector bosons. If we did not do this, we would be calculating an amplitude with fixed spins/helicities. Note that there are no external line factors for scalars.

For each internal line, we need to include a function called the 'propagator', and the relevant functions are shown in Figure 2.9, where in the fermion case we have used the **Feynman slash notation**

$$\slashed{p} \equiv p_\mu \gamma^\mu, \tag{2.127}$$

with γ^μ a Dirac matrix. For completeness, we have also included a small imaginary part

$$i\epsilon, \quad \epsilon \ll 1$$

in the denominator of each propagator. This ensures that the position-space propagator has the correct causal properties. Throughout this book, we will safely be able to ignore this issue, and so will omit the $+i\epsilon$ from now on.

The rules for the propagators involve an inverse power of the squared four-momentum of the particle being exchanged, minus the mass squared. Thus, they have $p^2 \neq m^2$ in general. In other words, virtual particles *do not* necessarily obey the physical relation

$$p^2 = E^2 - |\boldsymbol{p}|^2 = m^2.$$

A common term used for such particles is that they are 'off the mass shell', or just **off-shell** for short. Conversely, real particles are **on-shell**. You will see these terms being used in many research papers and textbooks.

We have so far seen the propagators for scalars and fermions. This also covers anti-fermions, as we simply reverse the direction of the four-momentum. However, what about vector bosons such as photons and gluons? This is a bit more complicated, and in fact the propagators for such particles turn out not to be unique: they change if we make a gauge transformation of the photon or gluon. We can get round this by breaking the gauge invariance, or 'fixing a gauge'. This amounts to imposing extra constraints on the field A^μ, where we consider the (simpler) QED case for now. Then the photon propagator—and thus the results of individual Feynman diagrams—will depend upon the choice of gauge. However, provided we fix the gauge in the same way for all possible Feynman diagrams, gauge-dependent factors will cancel out when we add all the Feynman diagrams together, leaving a gauge-invariant result for the amplitude $\mathcal{A}$. Common choices of constraint that are imposed include

$$n^\mu A_\mu = 0, \tag{2.128}$$

where n^μ is an arbitrary four-vector. This is called an **axial gauge**, as the vector n^μ is associated with a definite axis in spacetime. An alternative choice is

$$\partial^\mu A_\mu = 0, \tag{2.129}$$

which is known as the **Lorenz gauge**. In practice, these constraints can be implemented by adding a so-called **gauge-fixing** term to the QED Lagrangian. Examples include

$$\mathcal{L}_{\text{axial}} = -\frac{\lambda^2}{2}(n^\mu A_\mu)^2 \tag{2.130}$$

for the axial gauge, and

$$\mathcal{L}_{\text{covariant}} = -\frac{\lambda^2}{2}(\partial_\mu A^\mu)^2 \tag{2.131}$$

for the family of so-called **covariant gauges**, which includes the Lorenz gauge as a special case. In each example, the arbitrary parameter λ acts as Lagrange multiplier, enforcing the constraint. Another way to see this is that gauge invariance requires that we be insensitive to the value of λ, which implies

$$\frac{\partial \mathcal{L}}{\partial \lambda} = 0,$$

where $\mathcal{L}$ is the full Lagrangian. The only dependence on λ is in the gauge-fixing term, and thus this equation enforces the gauge-fixing condition. A given virtual photon in an arbitrary Feynman diagram will have Lorentz indices μ and ν associated with its

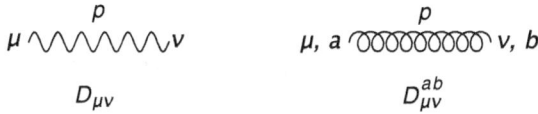

$$D_{\mu\nu} \qquad\qquad\qquad D_{\mu\nu}^{ab}$$

Figure 2.10. (a) The photon propagator will depend on the Lorentz indices at either end of the photon line; (b) the gluon propagator will also depend on the adjoint indices a and b at either ends of the line.

endpoints, as shown in Figure 2.10. The propagator function then depends upon the gauge, as discussed above. In the axial gauge, it turns out to be

$$D_{\mu\nu} = -\frac{i}{p^2}\left[\eta_{\mu\nu} + \frac{p^2 + \lambda^2 n^2}{\lambda^2(n \cdot p)^2}p_\mu p_\nu - \frac{n_\mu p_\nu + p_\mu n_\nu}{n \cdot p}\right]. \qquad (2.132)$$

Note that we can choose to absorb λ into n^μ if we want, a fact that is already evident from the gauge-fixing term of equation (2.130). In the covariant gauge, we have

$$D_{\mu\nu} = -\frac{i}{p^2}\left[\eta_{\mu\nu} - (1 - \lambda^{-2})\frac{p_\mu p_\nu}{p^2}\right]. \qquad (2.133)$$

A special case of this is the so-called **Feynman gauge** ($\lambda = 1$), which has the particularly simple propagator

$$D_{\mu\nu} = -\frac{i\eta_{\mu\nu}}{p^2}. \qquad (2.134)$$

This is by far the simplest choice for practical calculations, and so we will adopt the Feynman gauge from now on. In QCD, the gluon propagator in the Feynman gauge is

$$D_{\mu\nu}^{ab} = -\frac{i\eta_{\mu\nu}\delta^{ab}}{p^2}. \qquad (2.135)$$

This depends on the spacetime indices as before, but also depends in principle on the adjoint (colour) indices associated with the endpoints of a given gluon line (see Figure 2.10). However, we see from equation (2.135) that this colour dependence is trivial, in that the colour charge of the gluon does not change as it propagates.

Such is the simplicity of the Feynman gauge propagator compared with e.g. the axial gauge propagator of equation (2.132), that you may be wondering why people ever bother discussing alternative gauge choices at all. One use of them is that it can be useful to have the arbitrary parameters λ (in covariant gauges) or n^μ (in axial gauges) in the results of individual Feynman diagrams. If any dependence on these quantities remains after adding all the diagrams together, we know we have made a mistake somewhere, as the resulting calculation for the amplitude is not gauge invariant. Thus, alternative gauge choices allow us to check calculations, at least to some extent.

There is also another complication associated with non-axial gauges. The propagator describes how a gluon or photon of off-shell four-momentum p travels

$$\frac{i\delta^{ab}}{p^2 - m^2 + i\varepsilon}$$

Figure 2.11. The propagator for the ghost field in the Feynman gauge.

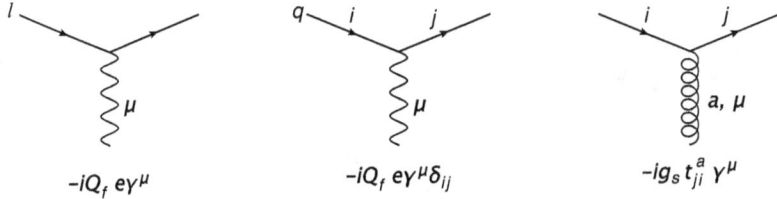

$$-iQ_f e\gamma^\mu \qquad\qquad -iQ_f e\gamma^\mu \delta_{ij} \qquad\qquad -ig_s t^a_{ji} \gamma^\mu$$

Figure 2.12. Feynman rules for the interaction of a lepton l and quark q with a photon/gluon, where i and j denote the colours of the incoming/outgoing quark. Here Q_f is the electromagnetic charge of the fermion in units of the magnitude of electron charge e, g_s the strong coupling constant, γ^μ a Dirac matrix (spinor indices not shown), and t^a_{ji} an element of an SU(3) generator in the fundamental representation.

throughout spacetime. As $p^2 \to 0$ the particle goes on-shell, and we then expect that only two physical (transverse) polarisation states propagate. In covariant gauges (and in particular the Feynman gauge), however, this does not happen: there is an unphysical (longitudinal) polarisation state, that leads to spurious contributions in Feynman diagrams with loops. In non-Abelian theories, these contributions survive even when all diagrams have been added together, such that they have to be removed. It is known that one can do this by introducing a so-called **ghost field**, which is a scalar field, but with fermionic anti-commutation properties. The latter leads to additional minus signs relative to normal scalar fields, that end up subtracting the unwanted contributions, to leave a correct gauge invariant result. Furthermore, the ghost field has an adjoint index a (like the gluon), so that it can cancel spurious contributions for all possible colour charges.

From the above discussion, it is clear that ghosts are not needed in QED, as the spurious contributions only survive in non-Abelian theories. In QCD, ghosts turn out to be unnecessary in the axial gauge, which makes sense: the gauge condition $n^\mu A_\mu = 0$ means that only states transverse to n_μ can propagate, i.e. only physical transverse states are kept in the on-shell limit. In the Feynman gauge, ghosts are indeed required, and the relevant propagator in the Feynman gauge is shown in Figure 2.11.

As well as the Feynman rules for external particles and internal lines, we also need the rules for vertices. A full list of Feynman rules can be found in any decent QFT textbook. Here, we provide a partial list of the more commonly used rules. Those involving fermions and photons/gluons can be found in Figure 2.12. Vertices for the gluon self-interactions, and ghost-gluon interactions, are shown in Figure 2.13. With adjoint indices, four-momenta, and Lorentz indices labelled as in the figure, the three-gluon vertex rule is

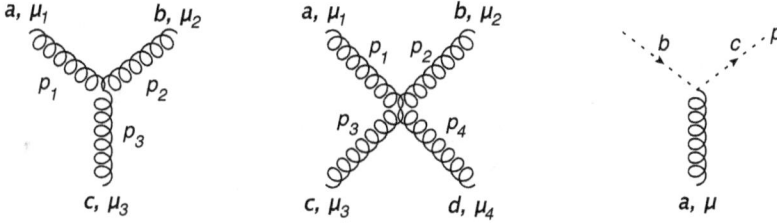

Figure 2.13. Vertices for gluon self-interactions, and the ghost-gluon interaction, in the Feynman gauge. Here $\{a, b, c\}$ label adjoint indices, $\{p_i, p\}$ four-momenta, and $\{\mu, \mu_i\}$ Lorentz indices.

$$
V_3^{abc;\,\mu_1,\mu_2,\mu_3}(p_1, p_2, p_3) = -g_s f^{abc}\big[(p_1 - p_2)^{\mu_3} \eta^{\mu_1\mu_2} + (p_2 - p_3)^{\mu_1} \eta^{\mu_2\mu_3} \\
+ (p_3 - p_1)^{\mu_2} \eta^{\mu_1\mu_3}\big].
\tag{2.136}
$$

The four-gluon vertex is

$$
V_4^{abcd;\,\mu_1,\mu_2,\mu_3,\mu_4}(p_1, p_2, p_3, p_4) = -i g_s^2 \big\{ f^{eac} f^{ebd} \,[\eta^{\mu_1\mu_2} \eta^{\mu_3\mu_4} - \eta^{\mu_1\mu_4} \eta^{\mu_2\mu_3}] \\
+ f^{ead} f^{ebc} \,[\eta^{\mu_1\mu_2} \eta^{\mu_3\mu_4} - \eta^{\mu_1\mu_3} \eta^{\mu_2\mu_4}] \\
+ f^{eab} f^{ecd} \,[\eta^{\mu_1\mu_3} \eta^{\mu_2\mu_4} - \eta^{\mu_1\mu_4} \eta^{\mu_2\mu_3}] \big\}.
$$

Finally, the ghost-gluon vertex is given by

$$
V_{\text{ghost}}^{abc;\mu} = g_s f^{abc} p^{\mu}.
\tag{2.137}
$$

Earlier we referred to g_s as the **strong coupling constant**. We can see why from the fact that all vertices involving gluons have at least one power of g_s. Thus, it represents the strength of interaction between quarks and gluons (or between gluons and other gluons). We can also directly verify the statement made earlier that ghosts decouple in QED: the ghost-gluon vertex of equation (2.137) contains the structure constants f^{abc}, which are zero for an Abelian theory.

Having stated more than the Feynman rules we will need for later, the complete procedure for calculating the scattering amplitude $\mathcal{A}$ is as follows:

1. For a given set of incoming/outgoing particles, draw all topologically distinct Feynman diagrams containing allowed vertices (i.e. corresponding to terms in the Lagrangian for the QFT of interest).
2. Apply the Feynman rules for each external line, internal line, and vertex.
3. Conserve four-momentum at all vertices. If any given four-momentum k remains undetermined, sum (i.e. integrate) over all possible values:

$$
\int \frac{\mathrm{d}^4 k}{(2\pi)^4}.
$$

4. Divide each diagram by a **symmetry factor** corresponding to the dimension of its symmetry group (see your QFT books for details).
5. Include a minus sign for all fermion and ghost loops.

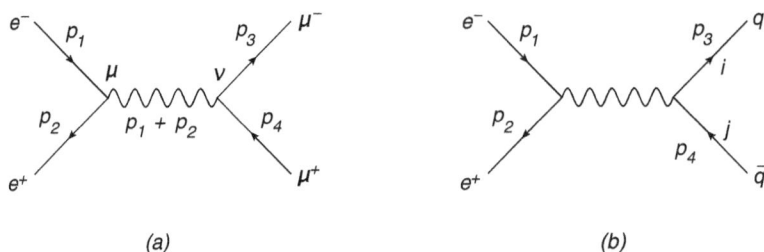

Figure 2.14. (a) Feynman diagram for muon production via electron-positron annihilation; (b) a related Feynman diagram for quark/antiquark production, where i and j label colours.

6. Add all diagrams together to get the scattering amplitude $\mathcal{A}$, where some relative minus signs are needed for fermions (again, see your QFT books).

To really understand how to apply Feynman rules, there is unfortunately no substitute for trying lots of examples, getting confused, looking at a QFT book, trying some more etc. Here, we can at least give a relatively simple (but still complicated!) example. Let us consider the process

$$e^-(p_1) + e^+(p_2) \rightarrow \mu^-(p_3) + \mu^+(p_4) \tag{2.138}$$

in QED, where the quantities in brackets label four-momenta. There is a single Feynman diagram for this process, shown in Figure 2.14(a). Applying the various Feynman rules we have encountered above gives

$$\underbrace{\bar{v}(p_2)}_{\text{Incoming } e^+} \times \overbrace{(-ie\gamma^\mu)}^{e^\pm \text{ vertex}} \times \underbrace{u(p_1)}_{\text{Incoming } e^-} \times \underbrace{\left(\frac{-i\eta_{\mu\nu}}{(p_1 + p_2)^2} \right)}_{\text{Propagator}}$$

$$\times \underbrace{\bar{u}(p_3)}_{\text{Outgoing } \mu^-} \times \overbrace{(-ie\gamma^\nu)}^{\mu^\pm \text{ vertex}} \times \underbrace{v(p_4)}_{\text{Outgoing } \mu^+},$$

so that contracting the Lorentz indices gives

$$\mathcal{A} = \frac{ie^2}{s}[\bar{v}(p_2)\gamma^\mu u(p_1)][\bar{u}(p_3)\gamma_\mu v(p_4)], \quad s = (p_1 + p_2)^2. \tag{2.139}$$

As a further example, we can also consider the similar process

$$e^-(p_1) + e^+(p_2) \rightarrow q(p_3) + \bar{q}(p_4),$$

where a quark–antiquark pair is produced in the final state (with some given flavour q), rather than muons. Applying the relevant Feynman rule from Figure 2.12, one finds

$$\mathcal{A}_{ij} = \frac{ie^2 Q_q \delta_{ij}}{s} [\bar{v}(p_2)\gamma^\mu u(p_1)][\bar{u}(p_3)\gamma_\mu v(p_4)] \tag{2.140}$$

$$= \delta_{ij} Q_q \mathcal{A}_{\text{QED}},$$

where Q_q is the electromagnetic charge of the quark q, and we have labelled by $\mathcal{A}_{\text{QED}}$ the muon amplitude from above.

2.9 Squared amplitudes

As we have seen in equation (2.125), for cross-sections we need the squared amplitude

$$|\mathcal{A}|^2 \equiv \mathcal{A}^\dagger \mathcal{A}. \tag{2.141}$$

If the incoming beams are unpolarised, we need to average over the spin states of the incoming particles. For fermions, this means that we have to insert a sum

$$\frac{1}{2} \sum_{\text{spins}},$$

where we have used the fact that each fermion or antifermion has two independent spin states. For photons or gluons, we may use the fact that these have $(d - 2)$ transverse spin states in d spacetime dimensions, so that averaging gives a sum

$$\frac{1}{(d-2)} \sum_{\text{pols}} \xrightarrow{d \to 4} \frac{1}{2} \sum_{\text{pols}},$$

where the subscript on the summation stands for 'polarisations'. We will see later why we might want to consider values of d that are not equal to 4.

In any given experiment, we have no control over the spin and/or colour of the produced particles. Thus, analogous to the sum over final state momenta that we performed earlier, we have to sum over final state spins and colours. It is then conventional to label by $\overline{|\mathcal{A}|^2}$ the squared amplitude averaged/summed over initial/final state spins and colours, as distinct from equation (2.141), which is for **fixed** spins and colours. Following equation (2.125), we may write the unpolarised cross-section as

$$\sigma = \frac{1}{F} \int d\Phi \, \overline{|\mathcal{A}|^2}. \tag{2.142}$$

In calculating $\overline{|\mathcal{A}|^2}$, a number of tricks are useful. For fermion-related algebra, a useful identity is

$$[\bar{w}_1 \gamma^{\mu_1} \dots \gamma^{\mu_n} w_2]^* = \bar{w}_2 \gamma^{\mu_n} \dots \gamma^{\mu_1} w_1, \tag{2.143}$$

where $w_i \in \{u, v\}$ is a basis spinor or antispinor. For fermion spin sums, we may use the so-called **completeness relations**

$$\sum_{\text{spins}} u(p)\bar{u}(p) = \not{p} + m, \quad \not{p} \equiv p_\mu \gamma^\mu. \tag{2.144}$$

Here we have not shown explicit spinor indices, but note that both sides are a matrix in spinor space. If we do insert spinor indices, it has the form

$$\sum_{\text{spins }\lambda} u_\alpha(p, \lambda)\bar{u}_\beta(p, \lambda) = p_\mu \gamma^\mu_{\alpha\beta} + m\delta_{\alpha\beta}. \tag{2.145}$$

Here μ is a Lorentz (spacetime) index, and $\{\alpha, \beta\}$ are indices in spinor space. The confusing nature of multiple indices is the primary reason why we do not show spinor indices more often. Another completeness relation concerns the sum over the spin states of an antifermion[12]:

$$\sum_{\text{spins}} v(p)\bar{v}(p) = \not{p} - m \tag{2.146}$$

(note a sign difference in the mass term relative to equation (2.144)).

There are a number of useful identities for the traces of strings of Dirac matrices:

$$\begin{aligned} &\text{tr}[\mathbf{I}] = 4; \\ &\text{tr}[\gamma^\mu \gamma^\nu] = 4\eta^{\mu\nu}; \\ &\text{tr}[\gamma^\mu \gamma^\nu \gamma^\rho \gamma^\sigma] = 4[\eta^{\mu\nu}\eta^{\rho\sigma} - \eta^{\mu\rho}\eta^{\nu\sigma} + \eta^{\mu\sigma}\eta^{\nu\rho}], \end{aligned} \tag{2.147}$$

where $\mathbf{I}$ is the identity matrix in spinor space. In the following, we will use the further useful property (see the exercises)

$$\gamma^\mu \not{a} \gamma_\mu = (2 - d)\not{a}, \tag{2.148}$$

where as usual the summation convention is implied.

We also need to sum over gluon/photon polarisation states, which gives us sums of the form

$$\sum_{\text{pols}} \epsilon^\dagger_\mu(p)\epsilon_\nu(p).$$

For a massive vector boson, the sum would be over three physical polarisation states: two transverse, and one longitudinal. This gives the known completeness relation

$$\sum_{\text{pols}} \epsilon^\dagger_\mu(p)\epsilon_\nu(p) = -\eta_{\mu\nu} + \frac{p_\mu p_\nu}{M^2}, \tag{2.149}$$

where M is the mass of the vector boson. To get the corresponding result for (massless) photons and gluons, note that we cannot simply take $M \to 0$ in equation (2.149)! Something strange happens if we do, which shows up as a weird divergence. What this ultimately tells us is that there is no unambiguous answer for the spin sum for massless vector bosons, and indeed this makes sense based on

---

[12] Again we follow convention in omitting an identity matrix in spinor space in the second term.

what we saw earlier, namely that for virtual particles, the states that propagate depend on which gauge we are in. Unphysical states can be cancelled by adding ghosts, and one way to proceed for external photons/gluons is indeed to sum over *all* polarisations (the physical transverse ones, and the unphysical longitudinal one), which turns out to give

$$\sum_{\text{pols}} \epsilon_\mu^\dagger(p)\epsilon_\nu(p) = -\eta_{\mu\nu}. \tag{2.150}$$

We can then correct for the inclusion of the unphysical longitudinal polarisation state by adding external ghosts, using the appropriate vertices. An alternative to this procedure is to introduce an arbitrary lightlike vector n^μ such that

$$n^\mu A_\mu(p) = p^\mu A_\mu(p) = 0, \tag{2.151}$$

where the second equality follows from the covariant gauge condition of equation (2.129). These two conditions ensure that only polarisation states transverse to both n^μ and p^μ survive, which indeed gives the correct number of degrees of freedom ($(d-2)$ in d spacetime dimensions). Then it can be shown that, if we sum over physical polarisations, we get

$$\sum_{\text{pols}} \epsilon_\mu^\dagger(p)\epsilon_\nu(p) = -\eta_{\mu\nu} + \frac{p_\mu n_\nu + p_\nu n_\mu}{p \cdot n}. \tag{2.152}$$

Individual contributions to the squared amplitude (i.e. from a given pair of Feynman diagrams) will depend on the arbitrary vector n^μ. However, given that it is associated with gauge dependence, any dependence on n will cancel out in the final result for $\overline{|\mathcal{A}|^2}$, which can be a useful check on such calculations in practice.

Let us attempt to clarify the above discussion with an example. Earlier we calculated the amplitude for the process of equation (2.138), where the result can be found in equation (2.139). We can now calculate the squared amplitude, averaged and summed over initial/final state spins. First of all, from equation (2.139) we obtain

$$\mathcal{A}^\dagger = -\frac{ie^2}{s}[\bar{v}(p_2)\gamma^\nu u(p_1)]^*[\bar{u}(p_3)\gamma_\nu v(p_4)]^*$$

$$= -\frac{ie^2}{s}[\bar{u}(p_1)\gamma^\nu v(p_2)][\bar{v}(p_4)\gamma_\nu u(p_3)],$$

where we have used equation (2.143). We thus have

$$\overline{|\mathcal{A}|^2} = \frac{e^4}{s^2}\frac{1}{4} \sum_{\text{spins}} [\bar{v}(p_2)\gamma^\mu u(p_1)]\,[\bar{u}(p_1)\gamma^\nu v(p_2)]\,[\bar{u}(p_3)\gamma_\mu v(p_4)]$$

$$\times [\bar{v}(p_4)\gamma_\nu u(p_3)] \tag{2.153}$$

(n.b. each quantity in square brackets is a number, such that the ordering of the sqaure bracketed terms is unimportant). Next, we can use the fermion completeness relations of equations (2.144) and (2.146). Consider for example the combination

$$\sum_{\text{spins}} [\bar{v}(p_2)\gamma^\mu u(p_1)] [\bar{u}(p_1)\gamma^\nu v(p_2)] = \sum_{\text{spins}} \bar{v}_\alpha(p_2)\gamma^\mu_{\alpha\beta} u_\beta(p_1)\bar{u}_\gamma(p_1)\gamma^\nu_{\gamma\delta} v_\delta(p_2),$$

where we have inserted explicit spinor indices. Upon doing so, we can move factors around in this equation, given that each component (labelled by specific indices) is just a number. In particular, we can rearrange the expression as

$$\sum_{\text{spins}} [v_\delta(p_2)\bar{v}_\alpha(p_2)]\gamma^\mu_{\alpha\beta}[u_\beta(p_1)\bar{u}_\gamma(p_1)]\gamma^\nu_{\gamma\delta} = (\not{p}_2)_{\delta\alpha}\gamma^\mu_{\alpha\beta}(\not{p}_1)_{\beta\gamma}\gamma^\nu_{\gamma\delta},$$

where we have used the completeness relations in the second line, and neglected the electron and muon masses for convenience ($m_e \simeq m_\mu \simeq 0$ in any experiment at current energies). We can now note that the remaining spinor indices have arranged themselves into a string representing a matrix multiplication, where the initial and final index δ is the same (and summed over). Thus, we can recognise the final result as a trace in spinor space:

$$\sum_{\text{spins}} [\bar{v}(p_2)\gamma^\mu u(p_1)] [\bar{u}(p_1)\gamma^\nu v(p_2)] = \text{tr}[\not{p}_2\gamma^\mu\not{p}_1\gamma^\nu]. \tag{2.154}$$

Likewise, one finds

$$\sum_{\text{spins}} [\bar{u}(p_3)\gamma_\mu v(p_4)] [\bar{v}(p_4)\gamma_\nu u(p_3)] = \text{tr}[\not{p}_3\gamma_\mu\not{p}_4\gamma_\nu]. \tag{2.155}$$

(again neglecting the fermion masses). As you practice with more squared amplitudes, you will see that spinor factors and Dirac matrices *always* combine to make traces in spinor space, and in fact this must be the case: the squared amplitude is a number, and thus cannot carry any free spinor indices.

Our averaged and summed squared amplitude is now

$$\overline{|\mathcal{A}|^2} \simeq \frac{e^4}{s^2}\frac{1}{4}\text{tr}[\not{p}_2\gamma^\mu\not{p}_1\gamma^\nu]\text{tr}[\not{p}_3\gamma_\mu\not{p}_4\gamma_\nu]$$

$$= \frac{e^4}{s^2}\frac{1}{4}\Big[4\big(p_2^\mu p_1^\nu - p_1 \cdot p_2\eta^{\mu\nu} + p_2^\nu p_1^\mu\big)\Big]\Big[4\big(p_{3\mu}p_{4\nu} - p_3 \cdot p_4\eta_{\mu\nu} + p_{4\mu}p_{3\nu}\big)\Big]$$

$$= \frac{e^4}{s^2}[2(u^2 + t^2) + (d - 4)^2 s^2],$$

where we have used equation (2.147) in the second line, and in the third line introduced the **Mandelstam invariants**

$$s = (p_1 + p_2)^2 \simeq 2p_1 \cdot p_2,$$
$$t = (p_1 - p_3)^2 \simeq -2p_1 \cdot p_3, \tag{2.156}$$
$$u = (p_1 - p_4)^2 \simeq -2p_1 \cdot p_4.$$

Some straightforward but tedious algebra is also needed in order to simplify everything down. We finally obtain

$$\overline{|\mathcal{A}|^2} = \frac{2e^4}{s^2}\left[u^2 + t^2 + \frac{(d-4)s^2}{2}\right] \xrightarrow{d \to 4} \frac{2e^4}{s^2}(u^2 + t^2). \tag{2.157}$$

Note that we can also use this to get the squared amplitude for quark–antiquark production. The only difference in the amplitude for the latter with respect to muon production is the presence of the electromagnetic charge of the quark, and a Kronecker symbol linking the colours of the outgoing quark and antiquark, as we saw in equation (2.140). Then the squared amplitude, summed over colours, is

$$\begin{aligned}
|\mathcal{A}_{ij}|^2 &= \sum_{i,j} \delta_{ij}\delta_{ij}Q_q^2 \, |\mathcal{A}_{\text{QED}}|^2 \\
&= \sum_i \delta_{ii}Q_q^2 \, |\mathcal{A}_{\text{QED}}|^2 \\
&= N_c \, Q_q^2 \, |\mathcal{A}_{\text{QED}}|^2,
\end{aligned} \tag{2.158}$$

where N_c is the number of colours. For QCD, we have $N_c = 3$, but many people leave factors of N_c explicit so that the origin of such factors is manifest. This makes it easier to check calculations.

2.10 Phase-space integrals and the cross-section

As equations (2.125) and (2.142) make clear, to turn the squared amplitude into a cross-section, we must integrate over the final-state phase-space and divide by the flux factor. For either our muon or quark production examples, the two-body final state phase-space is given as a special case of equation (2.126), by

$$\int d\Phi^{(2)} = (2\pi)^4 \int \frac{d^3 p_3}{(2\pi)^3 2E_3} \int \frac{d^3 p_4}{(2\pi)^3 2E_4} \delta^{(4)}(p_3 + p_4 - p_1 - p_2). \tag{2.159}$$

This is Lorentz invariant as we saw before, but to actually carry out the integral we must choose a frame. Let us choose the **lab frame**, in which the incoming electron and positron have equal and opposite momenta. This corresponds to the frame in which the collider's detector is at rest, and examples of such experiments include the old LEP experiment at CERN, and all proposed linear colliders. In this frame, the incoming beams of energy E have four-momenta

$$p_1^\mu = (E, 0, 0, E), \quad p_2^\mu = (E, 0, 0, -E), \tag{2.160}$$

where we have taken the z direction to correspond to the beam axis as usual, and also used the fact that we are neglecting masses. For the outgoing momenta, we can parametrise p_3 in a spherical polar coordinate system:

$$p_3^\mu = (E', E' \sin\theta \cos\phi, E' \sin\theta \sin\phi, E' \cos\theta), \quad 0 \leqslant \theta \leqslant \pi, \quad 0 \leqslant \phi \leqslant 2\pi, \tag{2.161}$$

where E' is the energy. We can then find p_4 from the momentum conservation condition

$$p_1 + p_2 = p_3 + p_4,$$ (2.162)

which gives

$$p_4^\mu = (E', -E' \sin\theta \cos\theta, -E' \sin\theta \sin\phi, -E' \cos\theta).$$ (2.163)

We then find

$$\int d\Phi^{(2)} = \frac{(2\pi)^4}{4(2\pi)^6} \int \frac{d^3 p_3}{E'} \int \frac{d^3 p_4}{E'} \delta^{(4)}(p_3 + p_4 - p_1 - p_2)$$

$$= \frac{1}{16\pi^2} \int \frac{d^3 p_3}{(E')^2} \delta\left(p_3^0 + p_4^0 - p_1^0 - p_2^0\right)$$

$$= \frac{1}{16\pi^2} \int_0^\pi d\theta \int_0^{2\pi} d\phi \int_0^\infty d|p_3| \frac{|p_3|^2 \sin\theta}{(E')^2} \delta(2E - 2E'),$$

where we have used the momentum-conserving delta function to carry out one of the momentum integrals, and inserted the usual volume element in spherical polars. We can use the fact that $|p_3| = E'$, as well as the useful delta-function relation

$$\delta(ax) = \frac{1}{a}\delta(x),$$ (2.164)

to get

$$\int d\Phi = \frac{1}{32\pi^2} \int_0^\pi d\theta \int_0^{2\pi} d\phi \int_0^\infty dE' \sin\theta \delta(E - E').$$ (2.165)

Using the above momentum parametrisation, the Mandelstam invariants of equation (2.156) evaluate to

$$u = -2p_1 \cdot p_4 = -2EE'(1 + \cos\theta), \quad t = -2p_1 \cdot p_3 = -2EE'(1 - \cos\theta).$$

From equation (2.157) we thus have

$$\int d\Phi \,\overline{|\mathcal{A}|^2} = \frac{e^4}{16\pi^2 s^2} \int_0^\pi d\theta \int_0^{2\pi} d\phi \int_0^\infty dE' \sin\theta (2EE')^2$$

$$\times [(1 + \cos\theta)^2 + (1 - \cos\theta)^2] \delta(E - E')$$

$$= \frac{e^4 E^4}{4\pi^2 s^2} \int_0^\pi d\theta \int_0^{2\pi} d\phi \sin\theta [2 + 2\cos^2\theta]$$

$$= \frac{e^4 E^4}{2\pi^2 s^2} 2\pi \int_{-1}^1 dx(1 + x^2), \quad x = \cos\theta,$$

where we have used the fact that the integrand is independent of ϕ to carry out the azimuthal angle integral. Carrying on, we find

$$\int d\Phi \, \overline{|\mathcal{A}|^2} = \frac{e^4 E^4}{\pi s^2} \int_{-1}^{1} dx(1 + x^2)$$

$$= \frac{e^4 E^4}{\pi s^2} \left[x + \frac{x^3}{3} \right]_{-1}^{1}$$

$$= \frac{e^4 E^4}{\pi s^2} \frac{8}{3}.$$

Next, note that

$$s = (p_1 + p_2)^2 = (2E)^2 = 4E^2,$$

so that we finally obtain

$$\int d\Phi^{(2)} \, \overline{|\mathcal{A}|^2} = \frac{e^4}{6\pi}.$$

To get the cross-section, it remains to divide by the Lorentz invariant flux factor, which in the present case (from equation (2.124)) is

$$F = 4\left[(p_1 \cdot p_2)^2 - m_1^2 m_2^2\right]^{1/2} \xrightarrow{m_{1,2} \to 0} 2s.$$

Thus, from equation (2.142) we get

$$\sigma = \frac{e^4}{12\pi s}.$$

It is conventional to write this in terms of the so-called **fine-structure constant**

$$\alpha = \frac{e^2}{4\pi}, \tag{2.166}$$

which finally gives us the total cross-section for muon production in electron/positron scattering:

$$\sigma_{\mu^+\mu^-} = \frac{4\pi\alpha^2}{3s}. \tag{2.167}$$

Note that the fine-structure constant is dimensionless, so that our final result has units of area, as required (n.b. s is a squared energy, and energy has the same dimensions as inverse length in natural units). What we also see very clearly is that the final result is simple, but its calculation definitely is not. This suggests that if we add more particles, we expect things to rapidly get very complicated indeed, and possibly even intractable!

We can use the above result to straightforwardly obtain the cross-section for the quark-antiquark process we considered earlier, based on our observation of equation (2.158). The factors N_c and Q_q^2 carry through the cross-section calculation, leading to the final result

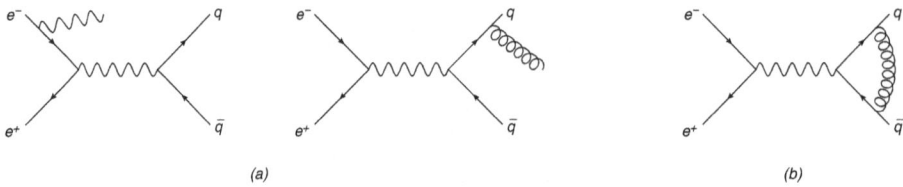

Figure 2.15. (a) Radiation of a real photon or gluon in quark/antiquark production; (b) a virtual correction.

$$\sigma_{q\bar{q}} = \frac{4\pi N_c Q_q^2 \alpha^2}{3s}. \tag{2.168}$$

Summing over quark flavours, we can define the convenient ratio

$$R = \frac{\sum_q \sigma_{q\bar{q}}}{\sigma_{\mu^+\mu^-}} = N_c \sum_q Q_q^2. \tag{2.169}$$

Historically, this ratio was measured using the cross-sections for e^+e^- to hadrons and muons respectively, and was an important verification of QCD, given that it is sensitive both to the number of quark colours and the charges of the quarks.

Note that we have so far only considered Feynman diagrams with the smallest number of vertices for a given initial/final state (i.e. the diagrams of Figure 2.14). However, in general we also need to consider the emission of additional radiation, such as the Feynman diagrams in Figure 2.15(a). This looks like a different process than before due to the presence of the extra gluon or photon. But we may not see the extra particle in the detector if its energy is too low, or if it is very close to one of the other particles, or goes down the beampipe etc. Then the process looks similar to the one we started with. We may also wish to consider an observable in which we sum over any number of extra radiated particles. Then we would have to include all the relevant radiative Feynman diagrams.

As well as real radiation, there are also loop diagrams corresponding to extra virtual particles. An example is shown in Figure 2.15(b), which shows a gluon being emitted and absorbed by the outgoing quark/antiquark pair. This process has the same initial and final state as the one we started with, and so is certainly some sort of correction to it.

Both real and virtual diagrams with extra radiation involve more vertices, and thus more powers of the coupling constants e and g_s. Hence, drawing Feynman diagrams with more loops/legs amounts to an *expansion* in the coupling, and this is called 'perturbation theory'. It is valid only if e and g_s are numerically small enough, otherwise the series is not convergent. The simplest diagrams for any process give the **leading order** (LO) in the coupling. Corrections are then called **next-to-leading order** (NLO), **next-to-next-to-leading order** (NNLO) and so on. For most processes, LO in the QED coupling is sufficient, given that e is numerically small. However, the QCD coupling is larger (it is the strong force after all), so that we definitely have to include higher orders in practical cross-section calculations, in order to get accurate results. Unfortunately, it turns out that as soon as we go to NLO, quantum field theory suddenly becomes infinitely weirder and more complicated! We will study this first for the simpler case of QED, before discussing how to generalise the results to QCD, which is of more practical importance. The results will then allow us to calculate higher-order Feynman diagrams, and thus to begin to describe measurable quantities at particle colliders. We will also have to understand how to calculate processes with incoming (anti)-protons, rather than electrons or positrons.

2.11 Renormalisation in QED

Consider again the scattering process of Figure 2.14(a). At NLO, there will be diagrams involving virtual photons, one of which is shown in Figure 2.16, where we have diligently labelled all four-momenta and Lorentz indices. Applying the Feynman rules for this diagram turns out to lead to a relatively straightforward answer, namely ∞! This is surprising at first sight, and is clearly nonsense. The amplitude is related to a probability, and we cannot have an infinite probability. Let us first see mathematically how the problem arises. If we apply the Feynman rules to Figure 2.16, we find the following expression:

$$
\int \frac{d^4k}{(2\pi)^4} (-ie)^4 \left(-\frac{i\eta_{\alpha\beta}}{k^2} \right) \left(-\frac{i\eta_{\mu\nu}}{(p_1 + p_2)^2} \right) \bar{v}(p_2) \gamma^\beta \frac{i(\not{p}_2 + \not{k} + m)}{(p_2 + k)^2 - m^2} \gamma^\mu
$$

$$
\times \frac{i(\not{p}_1 - \not{k} + m)}{(p_1 - k)^2 - m^2} \gamma^\alpha u(p_1) \bar{u}(p_3) \gamma^\nu v(p_4)
$$

$$
= \frac{e^4}{s} [\bar{u}(p_3) \gamma^\mu v(p_4)] \bar{v}(p_2) \gamma^\alpha I \gamma_\alpha u(p),
$$

where the integral in the first line is over the undetermined loop momentum k, and we have defined the integral (matrix-valued in spinor space)

$$
I = \int \frac{d^d k}{(2\pi)^d} \frac{(\not{p}_2 + \not{k} + m)\gamma_\mu (\not{p}_1 - \not{k} + m)}{k^2 [(p_2 + k)^2 - m^2][(p_1 - k)^2 - m^2]}. \tag{2.170}
$$

The full integral is very complicated, but we can see roughly what happens if all components of the loop momentum k^μ are large. To do this, we can let K denote a generic component of k^μ, and pretend $\not{k} \sim K$ and $k^\mu \sim K$ are just numbers. We can

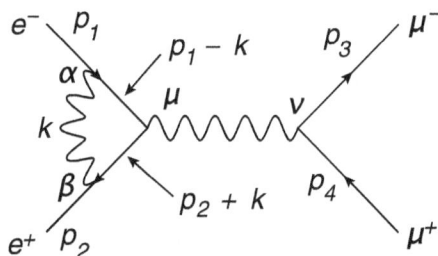

Figure 2.16. Virtual correction to muon production in electron/positron scattering.

then understand how the integrand scales with K. This procedure is called 'power counting', and taking K large yields

$$I \sim \int \mathrm{d}^4 K \, \frac{K^2}{K^6} \sim \int \mathrm{d}^4 K \, K^{-4}, \qquad (2.171)$$

as $K \to \infty$. We see that the integrand behaves like k^{-4} as $k^\mu \to \infty$, but there are four integrals over the components of k^μ. By comparing with the integral

$$\int \mathrm{d}x \, x^{-1} \sim \ln x,$$

we thus expect that the integral I is **logarithmically divergent** for large momenta. A more rigorous treatment confirms this, which is why the Feynman diagram of Figure 2.16 evaluates to infinity. Such divergences are called **ultraviolet (UV)**, as they are associated with high energies/momenta.

Having seen the origin of the infinity, we can now ask what it actually means, and thus whether there is a genuine problem that prevents us using QFT at higher orders. To this end, note that the infinity arises because we integrate up to arbitrarily high values of $|k^\mu|$, which itself involves a particular assumption, namely that we understand QED up to infinite energies or momenta. It is surely quite ridiculous to assert this, given that any real particles that we can measure have some finite momentum, and thus no previous experiments allow us to claim that we understand QED at very high energies. Indeed, at some high enough energy/momentum (equivalent, from the uncertainty principle, to very short spacetime distances), QFT may get replaced with a completely different theory.

The above discussion suggests that we put a cutoff Λ on the momenta/energies that we consider, below which scale we trust QED, but above which we make no assumptions. This will work provided that physics at collider energies is *independent* of physics at much higher energies. This sounds reasonable, and indeed there is a wide family of theories—including the SM—where this 'separation of scales' works. However, the precise cutoff is arbitrary: given a cutoff Λ in momentum/energy, we can clearly choose a different value $\Lambda' < \Lambda$, provided this is still much higher than the collider energy. Then the only way that low energy physics can be independent of the cutoff choice is if the behaviour in Λ can be 'factored out' of the parameters of the theory.

To make this idea more precise, consider writing our original QED Lagrangian as

$$\mathcal{L}_{\mathrm{QED}} = -\frac{1}{4} F_{0\mu\nu} F_0^{\mu\nu} + \bar{\Psi}_0 (i \slashed{\partial} - m_0) \Psi_0 - e_0 \bar{\Psi}_0 A_0^\mu \gamma_\mu \Psi_0. \qquad (2.172)$$

What we are then saying is that each field, mass or coupling can be written

$$A_0^\mu = \sqrt{Z_3} A_R^\mu, \quad \Psi_0 = \sqrt{Z_2} \Psi_R,$$
$$m_0 = Z_m m_R, \quad e_0 = Z_e e_R. \qquad (2.173)$$

Here the square roots are conventional, and the quantities $\{Z_i\}$ contain the divergent behaviour in Λ. This is called **renormalisation**, and the new fields A_R^μ, Ψ_R are called

renormalised fields. Likewise, m_R and e_R are the **renormalised mass and coupling**, respectively. By contrast, the original fields (A_0^μ, Ψ_0) and parameters (m_0, e_0) occuring in the classical Lagrangian are called **bare quantities**. They are *infinite*, but this is not actually a problem: provided we can set up perturbation theory purely in terms of (finite) renormalised quantities, everything we calculate will be finite. Furthermore, there is no cheating or black magic involved in this procedure: if we believe that physics at collider energies should be independent of much higher energies, then it *must* be possible that we can redefine the bare quantities in this way!

The redefinitions of equation (2.173) are all very fine, but they are not unique. The quantities $\{Z_i\}$ depend on the regulator we used to make the loop integral finite (in this case the cutoff scale Λ). Furthermore, we are free to absorb any finite contributions into the $\{Z_i\}$, and not just the contributions that become infinite as the regulator is removed ($\Lambda \to \infty$). Put another way, the renormalised quantities $(A_R^\mu, \Psi_R, m_R, e_R)$ are not unique. Different choices are called different **renormalisation schemes**. Once we fix a scheme, we can measure the (finite) renormalised quantities from experiment. We thus lose the ability to predict e.g. the value of the electron charge and mass from first principles. But this is after all expected: their values would depend on the underlying high energy theory, which we do not claim to know about.

The complete renormalisation procedure has several steps:

1. *Write the Lagrangian in terms of renormalised fields.* From equations (2.172) and (2.173), one finds

$$
\begin{aligned}
\mathcal{L}_{QED} &= -\frac{1}{4}F_{0\mu\nu}F^{0\mu\nu} + \bar{\Psi}_0(i\slashed{\partial} - m_0)\Psi_0 - e_0\bar{\Psi}_0 A_0^\mu \gamma_\mu \Psi_0 \\
&= -\frac{Z_3}{4}F_{R\mu\nu}F_R^{\mu\nu} + Z_2\bar{\Psi}_R[i\slashed{\partial} - Z_m m_R]\Psi_R - Z_e\sqrt{Z_3}\,Z_2 e_R \bar{\Psi}_R A_R^\mu \gamma_\mu \Psi_R.
\end{aligned}
\tag{2.174}
$$

2. *Expand the renormalisation constants.* At tree-level (i.e. diagrams with no loops), we do not see any high energy divergences, and so we do not need to redefine the bare quantities entering the Lagrangian. This implies that each renormalisation constant satisfies $Z_i = 1$ at leading order. Furthermore, each infinity we encounter happens at some subleading order in perturbation theory, so that we expect each Z_i itself to be a perturbation series in the coupling. We may thus generically write

$$
Z_i = 1 + \delta_i,
\tag{2.175}
$$

where δ_i starts at some nonzero power of e. Plugging equation (2.175) into equation (2.174), one obtains

$$
\begin{aligned}
\mathcal{L}_{QED} = &-\frac{(1 + \delta_3)}{4}F_{R\mu\nu}F_R^{\mu\nu} + (1 + \delta_2)\bar{\Psi}_R[i\slashed{\partial} - (1 + \delta_m)m_R]\Psi_R \\
&- (1 + \delta_e)(1 + \delta_3)^{1/2}(1 + \delta_2)e_R \bar{\Psi}_R A_R^\mu \gamma_\mu \Psi_R.
\end{aligned}
\tag{2.176}
$$

All we have done so far is to redefine quantities in the original Lagrangian, and make it look a lot more complicated. However, we will soon see why this is useful!

3. *Split the Lagrangian into a renormalised Lagrangian, plus corrections.* From equation (2.176), we can write

$$\mathcal{L}_{\text{QED}} = \mathcal{L}_{\text{R}} + \mathcal{L}_{\text{c.t.}}, \tag{2.177}$$

where

$$\mathcal{L}_{\text{R}} = -\frac{1}{4}F_{R\mu\nu}F_R^{\mu\nu} + \bar{\Psi}_R[i\slashed{\partial} - m_R]\Psi_R - e_R\bar{\Psi}_R A_R^\mu \gamma_\mu \Psi_R \tag{2.178}$$

looks like the original Lagrangian, but with all quantities replaced by their *renormalised* versions. The correction term in equation (2.177) is given by

$$\mathcal{L}_{\text{c.t.}} = -\frac{\delta_3}{4}F_{R\mu\nu}F_R^{\mu\nu} + i\delta_2\bar{\Psi}_R\slashed{\partial}\Psi_R - (\delta_2 + \delta_m + \delta_2\delta_m)\bar{\Psi}_R m_R \Psi_R$$
$$- [(1 + \delta_e)(1 + \delta_3)^{1/2}(1 + \delta_2) - 1]e_R\bar{\Psi}_R A_R^\mu \gamma_\mu \Psi_R, \tag{2.179}$$

and is $\mathcal{O}(\delta_i)$, thus starts at subleading order in perturbation theory. Equation (2.179) is called the **counterterm Lagrangian**, and each δ_i is called a **counterterm**[13].

4. *Calculate Feynman diagrams with renormalised fields, using a regulator.* In calculating with the Lagrangian of equation (2.177), there are two types of Feynman diagram. Firstly, we have diagrams involving only the Lagrangian $\mathcal{L}_{\text{R}}$. These look exactly the same as the diagrams we obtained from the original Lagrangian, and we can make them finite using e.g. a momentum cutoff Λ, as we saw for the example of Figure 2.16. The diagrams would diverge as $\Lambda \to \infty$.

Secondly, there are Feynman diagrams containing *new* Feynman rules, from the counterterm Lagrangian $\mathcal{L}_{\text{c.t.}}$. Given that the counterterm Lagrangian involves terms of the same form as the original Lagrangian, these Feynman rules will be similar to the original Feynman rules, but with combinations of counterterms in front. Two examples are shown in Figure 2.17, and correspond to Feynman rules

$$-ie\gamma_\mu[(1 + \delta_e)(1 + \delta_3)^{1/2}(1 + \delta_2) - 1]$$

and

$$i[\slashed{p}\delta_2 - (\delta_m + \delta_2 + \delta_m\delta_2)m_R]$$

respectively.

---

[13] You may also see a different terminology in some textbooks, in which the coefficients of each of the terms in equation (2.179) is called a counterterm, rather than each δ_i. However, the two sets of counterterms are simply related by a redefinition.

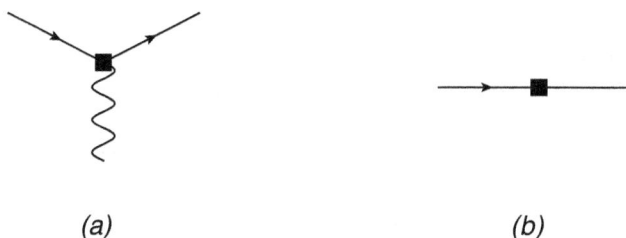

Figure 2.17. Example counterterm Feynman rules.

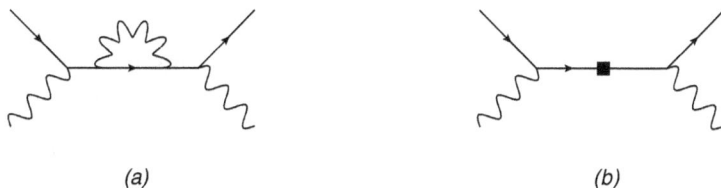

Figure 2.18. (a) A UV divergent diagram at NLO; (b) A counterterm diagram at the same order.

5. *Choose the counterterms to remove divergences.* At each order in perturbation theory, we have a set of regulated Feynman diagrams, where the amplitude depends on the undetermined counterterms δ_i. It turns out that we can choose each δ_i such that they subtract off the contributions from diagrams that diverge as the regulator is removed. As an example, consider the graph of Figure 2.18(a). This is UV divergent, where we may regulate the divergence with our usual cutoff Λ. However, we must also include the diagram of Figure 2.18(b), which includes a counterterm on the internal line. Then we can choose δ_2 and δ_m such that the divergence from Figure 2.18(a) is completely removed, so that the sum of the two diagrams is finite as $\Lambda \to \infty$. Remarkably, once we calculate enough Feynman diagrams to fix all of the counterterms, *all* diagrams at that order in perturbation theory are finite! Furthermore, it can be proven that this procedure can be generalised to arbitrary order in perturbation theory. That is, including higher-order terms in the finite set of counterterms δ_i is sufficient to remove all divergences, to all orders.

6. *Remove the regulator.* Once we have calculated all (regulated) diagrams for a given process, including the counterterm graphs (where we have fixed the values of the counterterms), there are no potential divergences left, and we can safely remove the regulator. We are left with finite perturbative corrections to QFT, that can be compared with experiment!

Let us try to briefly summarise what we have done:

- Loop integrals in QFT have UV divergences, associated with virtual particles with high energy/momentum (or short distances).
- These divergences can be removed by *renormalising* fields, couplings and masses, as in equation (2.173). The factors Z_i can be written as one plus a

counterterm δ_i in each case, where the latter starts at subleading order in perturbation theory.

- We can choose the counterterms to remove all UV divergences, so that Feynman diagrams for *renormalised* fields are finite.
- The renormalised masses/couplings can be measured from experiment, once a choice is fixed for the counterterms.

From now on, we will always be working with renormalised fields, masses and couplings, and so we will drop the subscript R, for brevity.

2.12 Dimensional regularisation

In the previous section, we used the example of a momentum cutoff Λ to regulate Feynman loop integrals. Although this is relatively simple to understand from a conceptual point of view—especially given the way that UV divergences arise—such a regulator is almost never used in practice. This is mainly due to the fact that it is not convenient, in that it violates symmetries of the amplitude. Firstly, it may violate Lorentz invariance, depending on how the cutoff is applied to the different components of the loop momentum. Secondly, it violates gauge invariance. To see this, note that a cutoff essentially separates momentum modes of the field into two groups, according to whether the momentum is below or above the cutoff. A gauge transformation (e.g. equation (2.11)) will induce a change in the field that, in momentum space, will have arbitrary modes in general. Thus, any particular separation into 'low' and 'high' energy modes is clearly gauge-dependent.

By far the most common regulator that people actually use is called **dimensional regularisation**. It was developed by Gerard 't-Hooft and Martinus Veltman, who later won the Nobel prize for showing that the SM could be renormalised to all orders in perturbation theory. This is perhaps all the evidence we need that dimensional regularisation can be very useful! To see the basic idea, recall that we saw earlier that UV divergences occur due to integrals of the schematic form

$$\int^\Lambda \mathrm{d}^4k\, k^{-4} \sim \ln\Lambda, \quad \Lambda \to \infty.$$

It follows that if we instead calculate

$$\int^\infty \mathrm{d}^d k\, k^{-4},$$

where $d < 4$, there will be no (UV) divergence. Normally, it only makes sense to calculate such integrals if d is an integer, i.e. it represents the number of components of a position in spacetime. However, it turns out that the results of *all* Feynman integrals in QFT end up being a continuous function of d. Thus, even if we start with d only being defined at integer values, we can choose to use any non-integer value of d we like! A common choice is to set

$$d = 4 - 2\epsilon, \quad \epsilon > 0, \quad \epsilon \ll 1. \tag{2.180}$$

Here the factor of 2 is a convention, and the physical number of spacetime dimensions is approached as $\epsilon \to 0$. You may be (understandably) puzzled as to how one even calculates an integral in $4 - 2\epsilon$ dimensions! However, all of the details needed for calculating such integrals have been worked out in full. The main one is that UV divergences show up as **poles** (inverse powers) in ϵ.

There is another important feature of dimensional regularisation, related to the fact that the QED coupling e has mass dimension

$$[e] = \frac{4 - d}{2} \qquad (2.181)$$

in d spacetime dimensions (see the exercises). Thus, it is dimensionless in $d = 4$, but has a mass dimension otherwise. To keep all mass dimensions explicit, it is helpful to modify the coupling via

$$e \to \mu^{(4-d)/2}e \equiv \mu^\epsilon e, \quad [\mu] = 1. \qquad (2.182)$$

In words: one can keep the coupling dimensionless, at the price of introducing an arbitrary mass/energy scale μ. Thus, there are really two parameters in dimensional regularisation: the dimension shift ϵ, and the scale μ. It is actually the first of these that is the analogue of the (dimensionful) parameter Λ in the cutoff regularisation that we used previously, as ϵ controls the divergent behaviour. The role of μ remains mysterious for now, but will hopefully become clear in what follows.

2.13 Running couplings and masses

In dimensional regularisation, the one-loop counterterms for the coupling e and mass m are as follows:

$$\delta_e = \frac{e^2}{24\pi^2}\left(\frac{1}{\epsilon} + c_e\right), \quad \delta_m = -\frac{3}{8\pi^2}\left(\frac{1}{\epsilon} + c_m\right). \qquad (2.183)$$

Here c_e and c_m are arbitrary constants, corresponding to the fact that one may absorb finite contributions into the counterterms, alongside the formally divergent part (i.e. the terms $\sim \epsilon^{-1}$). Different choices for (c_e, c_m) constitute different **renormalisation schemes**, and common choices include

$$c_e = c_m = 0,$$

which is known as **minimal subtraction**, or the MS scheme. More commonly used is **modified minimal subtraction** (also known as the $\overline{\text{MS}}$) scheme, which has

$$c_e = c_m = \ln(4\pi) - \gamma_{\text{E}},$$

where $\gamma_{\text{E}} = 0.5772156\ldots$ is the **Euler–Mascheroni constant** (a known mathematical quantity). These numbers look a bit bizarre when plucked out of the air like this, but it turns out that all integrals in dimensional regularisation involve factors of

$$(4\pi e^{-\gamma_{\text{E}}}\mu^2)^\epsilon,$$

which generates the above terms at $\mathcal{O}(\epsilon^0)$. Choosing to absorb these contributions into the counterterms then simplifies the form of higher-order perturbative corrections.

Leaving the factors of (c_e, c_m) arbitrary for generality, the bare coupling is related to its renormalised counterpart by

$$e_0 = \mu^\epsilon e(1 + \delta_e) = \mu^\epsilon e\left[1 + \frac{e^2}{24\pi^2}\left(\frac{1}{\epsilon} + c_e\right)\right] + \mathcal{O}(e^4). \tag{2.184}$$

However, the bare coupling corresponds to the original coupling entering the classical Lagrangian, before any quantum corrections have been computed. It thus knows nothing about the regulators we use to make loop integrals finite, which amounts to the condition

$$\mu\frac{de_0}{d\mu} = 0, \tag{2.185}$$

where we have included an additional factor of μ to match the way this equation is usually presented. As you can explore in the exercises, equations (2.184) and (2.185) then imply

$$\mu\frac{de(\mu)}{d\mu} \equiv \beta(e), \tag{2.186}$$

where (taking $\epsilon \to 0$)

$$\beta(e) = \frac{e^3}{12\pi^2} + \mathcal{O}(e^5). \tag{2.187}$$

This is called the **beta function**, and is defined by equation (2.186). Formally speaking, equation (2.186) tells us how the renormalised coupling changes if we make a different choice of the parameter μ entering regulated Feynman integrals. However, we can interpret this a lot more physically as follows. Imagine we make some choice $\mu = Q_0$, where Q_0 is some energy scale. Then we can measure the value of $e(Q_0)$ from experiment. We could instead have chosen μ to be some different energy scale Q_1, in which case we could have measured $e(Q_1)$. From equation (2.186), the two results are related via

$$\int_{e(Q_0)}^{e(Q_1)} \frac{de}{\beta(e)} = \int_{Q_0}^{Q_1} d\ln\mu,$$

from which one may show (see the exercises)

$$e^2(Q_1) = \frac{e^2(Q_0)}{1 - \frac{e^2(Q_0)}{6\pi}\ln\left(\frac{Q_1}{Q_0}\right)}. \tag{2.188}$$

We see that the renormalised 'coupling constant' e is not constant after all, but *varies with energy*. The absolute value of e at some scale cannot be predicted from first principles—we gave up the ability to do this when we claimed that high energy physics should not influence low energy physics, in setting up renormalisation. However, how the coupling changes with energy is *completely predicted* by QFT perturbation theory! To see what the dependence on energy looks like, the shape implied by equation (2.188) is shown in Figure 2.19. We see that the QED coupling *increases* with energy scale or, from hand-wavy uncertainty principle arguments, *decreases* with distance scale. There is even a nice (albeit rather silly) physical interpretation of this. Consider a point charge (e.g. a single real electron) located somewhere in space. In a quantum theory, virtual $e^{\pm}$ pairs can be created in the vacuum, provided they disappear after a suitable timescale dictated by the uncertainty principle. These $e^{\pm}$ pairs would form dipoles, that would line up, and effectively 'screen' the charge as we move away from it, as shown in Figure 2.20. In other words, quantum effects lead to the coupling (i.e. the strength of the charge of the electron) decreasing with distance, which is precisely the behaviour we found above.

Note from equation (2.188) that the coupling apparently diverges when

$$\frac{e^2(Q_0)}{6\pi} \ln\left(\frac{Q_1}{Q_0}\right) = 1 \quad \Rightarrow \quad Q_1 = Q_0 \exp\left[\frac{6\pi}{e^2(Q_0)}\right].$$

Experiments tell us that

$$\frac{e^2(Q_0)}{4\pi} \simeq \frac{1}{137} \quad \text{at} \quad Q_0 \simeq 91 \text{ GeV},$$

which in turn implies that e diverges at $\simeq 1 \times 10^{91}$ GeV. This is known as the 'Landau pole', and does not tell us in practice that the coupling would diverge: as the coupling gets larger, we would have to take higher orders into account in the beta

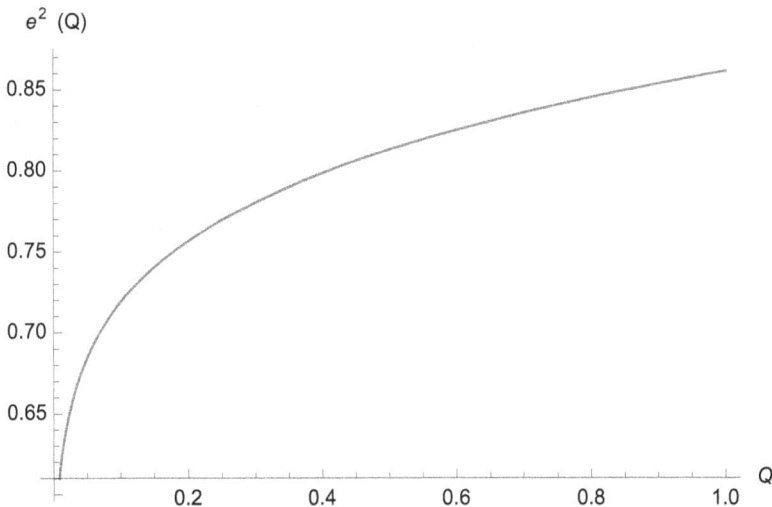

Figure 2.19. Behaviour of the QED coupling with energy scale.

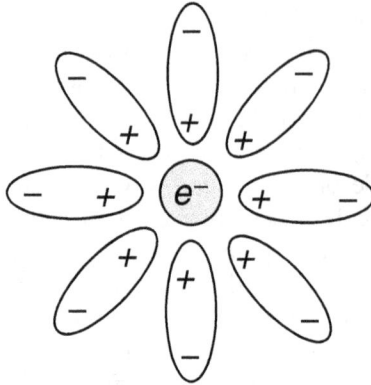

Figure 2.20. Virtual electron–positron pairs can screen a real charge, so that the coupling decreases at large distances.

function, which may modify the behaviour. At some point, perturbation theory may cease to be valid at all. This argument does, however, act as a rough guide as to the energy scale at which QED may break down, and has to be replaced with something else (that may or may not be a QFT). In the SM, QED does not exist in isolation, but mixes with the weak force at energy scales $\mathcal{O}(10^2)$ GeV. However, there remains a coupling constant which diverges at high energy. This energy is much higher than the Planck scale at which we expect quantum gravitational effects must be taken into account, and thus the Landau pole is not a problem in practice.

We have seen that the renormalised coupling depends on an energy scale, and the same argument can be used for renormalised masses. That is, the bare mass must be independent of the dimensional regularisation parameter μ:

$$\mu \frac{\mathrm{d}m_0}{\mathrm{d}\mu} = 0, \tag{2.189}$$

which turns out to imply an equation of the form

$$\mu \frac{\mathrm{d}m(\mu)}{\mathrm{d}\mu} = \gamma_m m(\mu), \tag{2.190}$$

where m is the renormalised mass, and γ_m is called its **anomalous dimension**. From similar arguments to those applied to the coupling, we see that masses also vary with energy scale, where this variation is entirely predicted by perturbation theory.

The variation of couplings and masses with energy is also known as **running**, so that $e(\mu)$ and $m(\mu)$ are sometimes referred to as the **running coupling** and **running mass**, respectively.

2.14 Renormalisation in QCD

In the previous section, we discussed renormalisation in QED in detail. QCD is also a renormalisable theory, where this renormalisation is, unsurprisingly, much more complicated (e.g. there are many more Feynman diagrams, and more terms in the

Lagrangian). However, the process is conceptually similar to QED, leading to running of the strong coupling, quark masses etc.

A notable difference with respect to QED is how the coupling behaves with energy. Conventionally, the beta function of QCD is defined in terms of the parameter[14]

$$\alpha_S = \frac{g_s^2}{4\pi},$$
(2.191)

via

$$\mu^2 \frac{d\alpha_S}{d\mu^2} = \beta(\alpha_S)$$
(2.192)

(n.b. note the square of the scale μ is used, rather than the scale itself). One may write the beta function in perturbation theory as

$$\beta(\alpha_S) = -\alpha_S \sum_{n=0}^{\infty} \beta_n \left(\frac{\alpha_S}{4\pi}\right)^{n+1},$$
(2.193)

which defines the perturbative coefficients β_n. The first of these is found to be

$$\beta_0 = 11 - \frac{2n_f}{3},$$
(2.194)

where n_f is the number of quark flavours. That is, one has

$$\beta(\alpha_S) = -\frac{\alpha_S^2}{4\pi}\left(11 - \frac{2n_f}{3}\right) + \mathcal{O}(\alpha_S^3).$$

Unlike QED, this corresponds to a coupling that *decreases* with energy, as shown in Figure 2.21. Measured data from a variety of sources agree with the prediction very well, which in particular verifies the idea that quarks become weakly interacting at high energies. This property is known as **asymptotic freedom**, and won the Nobel prize for David Gross, Frank Wilczek and David Politzer in 2004.

At low energies, the coupling becomes strong. Combined with measured values, the LO expression turns out to diverge at energy scales $\sim\mathcal{O}(10^2)$ MeV. This makes perfect sense: in nature, free quarks are never observed, but are confined in **hadrons**, which have no net colour. Hadron masses start at a few hundreds of MeV, exactly where α_S becomes strong! If we try to pull quarks apart, the strength of the interaction increases, trapping them inside the hadron. However, if we accelerate hadrons to very high energies, the interactions between quarks and gluons become weak, and we can use perturbation theory to describe them. Thus, asymptotic freedom is particularly important, in that it will allow us to describe scattering processes with incoming (anti-)protons.

---

[14] Confusingly, both α_S and g_s are commonly referred to as the **strong coupling constant**.

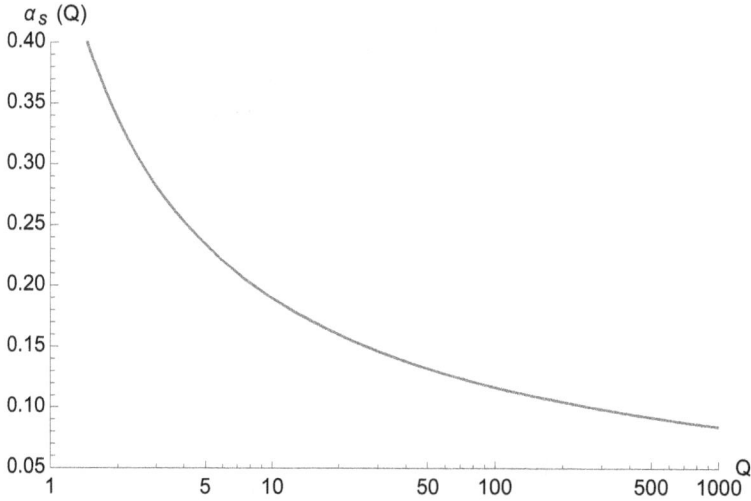

Figure 2.21. Behaviour of the QCD running coupling $\alpha_S(Q)$ with energy scale Q.

As in QED, masses run in QCD due to quantum effects. For practical calculations, the running of the top and bottom (anti-)quark masses are particularly important, as they can be numerically significant. Both the running couplings and masses depend on the scale μ, which we will label as μ_R from now on and refer to as the **renormalisation scale**. If we were to calculate a given observable X to all orders in perturbation theory, the dependence on μ_R would completely cancel (i.e. physical observables must be independent of μ_R, which is arbitrary). However, if we truncate to a given order in α_S (including the beta function), we have neglected subleading terms that depend on the renormalisation scale. Thus, we find a dependence on μ_R, which is formally related to higher orders in α_S. To make this more precise, dimensional analysis tells us that the perturbation series for X must have the form

$$X = \sum_{n=0}^{\infty} \alpha_S^n(\mu_R^2)\, c_n\left(\left\{\frac{p_i \cdot p_j}{\mu_R^2}\right\}, \left\{p_i \cdot p_j\right\}\right),$$

(2.195)

where $\{p_i\}$ are the four-momenta entering the scattering process. At $N^m LO$ in perturbation theory, we should use the running coupling calculated at $N^{m-1} LO$ (e.g. at LO the coupling does not run at all; the running starts at NLO). Then the μ_R variation is a $N^{m+1} LO$ effect.

This is all very fine, but the question remains of what we should choose for the arbitrary scale μ_R, and at first sight it looks like our predictions are completely arbitrary. However, the key point to realise is that the scale dependence is *directly related* to the fact that we have neglected higher orders in perturbation theory, and so the size of the scale variation should be comparable with the size of the terms we neglect. A common practice is then to choose $\mu_R^2 = Q^2$, where Q^2 is some representative squared energy scale in the process (e.g. the centre-of-mass energy in a $e^{\pm}$ collider, or the top quark mass in top quark pair production). One can then vary μ_R within some range of the default choice e.g.

$$\frac{Q}{2} \leqslant \mu_R \leqslant 2Q, \tag{2.196}$$

and use the resulting envelope for observable X to estimate the size of higher-order corrections, and thus a **theoretical uncertainty**.

2.15 Parton distributions

So far we have only calculated cross-sections with incoming electrons and/or positrons. In the SM model, these are fundamental particles, with no substructure. It was not immediately clear how to generalise our calculations to deal with incoming **protons**, which are composite particles made of fundamental quarks and gluons. This is necessary if we want to calculate anything to do with the LHC, both of whose beams are comprised of protons. However, in the previous section we saw that (anti-)quarks and gluons are **asymptotically free** i.e. they become weakly interacting at high energies. This suggests that, for fast-moving protons, individual quarks/gluons can leave the proton and collide with each other. This is essentially the crude picture that we drew in Figure 1.8, and the aim of this section is to make this more precise.

For a very fast-moving proton of momentum P_i, the (anti-)quark/gluon that emerges will have some momentum p_i that is collinear with the proton, i.e.

$$p_i = x_i P_i, \quad 0 \leqslant x_i \leqslant 1, \tag{2.197}$$

where x_i is then the **momentum fraction** of the proton carried by the emerging particle. We can use perturbative QCD to calculate the cross-section for the (anti-)quarks and gluons, as these are fundamental particles. We then expect that the proton cross-section will have the general form

$$\sigma = \sum_{i,j \in \{q,\bar{q},g\}} \int_0^1 dx_1 \int_0^1 dx_2 \, f_i(x_1) f_j(x_2) \, \hat{\sigma}_{ij}. \tag{2.198}$$

Here $\hat{\sigma}_{ij}$ is the cross-section for (anti-)quarks and/or gluons, where i and j label the species of colliding particle. The function $f_i(x_k)$ labels the probability to find a particle of type i with momentum fraction x_k inside incoming proton k. Then to get the total cross-section, we have to sum over all possible momentum fractions, which takes the form of integrals over the variables $\{x_i\}$. Equation (2.198) predates QCD, and indeed was first proposed by Feynman in the 1970s. He called it the 'parton model', where his 'partons' were some set of postulated constituents of the proton. It was later realised that these partons correspond to the (anti-)quarks and gluons of QCD, but to this day the word 'parton' is used to mean a general (anti-)quark or gluon, as it is useful to have such a word! Likewise, $\hat{\sigma}_{ij}$ in equation (2.198) is referred to as the **partonic cross-section**. The functions $f_i(x_k)$ are called **parton distribution functions** (PDFs). They involve strongly interacting physics (e.g. they are sensitive to

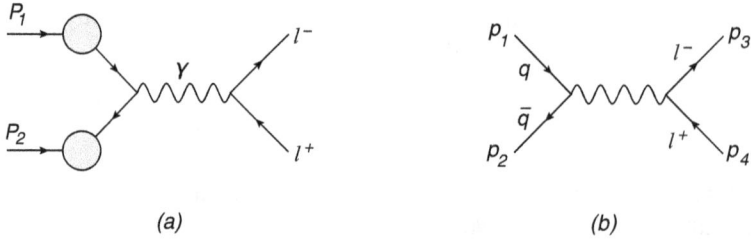

Figure 2.22. Drell–Yan production of a vector boson, which decays to a lepton pair: (a) full process, with incoming protons; (b) LO partonic cross-section.

how partons are confined within hadrons), and thus cannot be calculated in perturbation theory. However, they can be measured in experiments, and then used to predict the results of future experiments. This is a whole subfield in itself, and there are a number of collaborations worldwide that extract parton distributions from data.

When it was first proposed, the parton model was indeed just a model used to fit experimental data. However, it can be fully justified (with a few caveats) from first principles in QCD, and in particular something called the **operator product expansion**. Here, however, we shall merely try to show you why trying to calculate the partonic cross-section *unavoidably* leads to requiring the existence of parton distribution functions.

As an example, let us consider so-called **Drell–Yan production** of a virtual photon, that decays to a lepton pair. The Feyman diagram for this process (complete with incoming protons!) is shown in Figure 2.22(a), and the corresponding LO partonic cross-section in Figure 2.22(b), where we have focused on a given quark flavour q. Note that this is almost identical to the process $e^- e^+ \rightarrow \mu^- \mu^+$ of Figure 2.14, whose squared amplitude (summed and averaged over spins) is given by equation (2.157). To get the LO amplitude for DY production, we may first note that

$$\mathcal{A}_{\text{DY}} = \delta_{ij} Q_q \mathcal{A}_{\text{QED}},$$

where $\mathcal{A}_{\text{QED}}$ is the QED process, i and j the colour indices of the incoming quark and anti-quark, respectively, and Q_q the electromagnetic charge of the quark. When forming the summed and averaged squared amplitude, we must include an average over the incoming colours, given that we have no control over e.g. what colour of quark comes out of the proton. This gives us

$$\overline{|\mathcal{A}_{\text{DY}}|^2} = \frac{1}{N_c^2} \sum_{i,j} \delta_{ij} \delta_{ij} Q_q^2 \overline{|\mathcal{A}_{\text{QED}}|^2}$$

$$= \frac{1}{N_c^2} \sum_{i} \delta_{ii} Q_q^2 \overline{|\mathcal{A}_{\text{QED}}|^2} \qquad (2.199)$$

$$= \frac{2 Q_q^2 e^4}{N_c} \frac{u^2 + t^2}{s^2},$$

where, as usual, N_c is the number of colours. This quantity is perfectly well-defined, and you may thus come to the conclusion that parton distributions are something we can choose to add, but whose existence remains 'hidden' in perturbation theory. However, a problem arises if we consider NLO radiation, an example diagram for which is shown in Figure 2.23. The extra gluon creates a new internal line, whose propagator will contain a factor (taking the quark to be approximately massless)

$$\sim \frac{1}{(p_1 - k)^2}.$$

One may parametrise momenta as

$$p_1^\mu = (|\boldsymbol{p}_1|, 0, 0, |\boldsymbol{p}_1|), \quad k^\mu = (|\boldsymbol{k}|, 0, |\boldsymbol{k}|\sin\theta, |\boldsymbol{k}|\cos\theta),$$

where θ is the angle between the gluon and incoming quark (shown schematically in Figure 2.23). Then the above propagator factor becomes

$$\frac{1}{-2p_1 \cdot k} = \frac{1}{-2|\boldsymbol{p}_1||\boldsymbol{k}|(1 - \cos\theta)}.$$

To get the cross-section, we have to integrate over the phase-space of all final state particles, but the amplitude clearly *diverges* if

$$|\boldsymbol{k}| \to 0 \quad \text{and/or} \quad \theta \to 0.$$

The first of these is called a **soft divergence**, where 'soft' refers to a gluon whose four-momentum components are all vanishingly small relative to some other momentum scale in the problem (here the energy/momentum of the incoming quark). The second is called a **collinear divergence**, as it arises when the angle between the radiated gluon and the incoming quark vanishes. Collectively, these singularities are referred to as **infrared (IR) divergences**, as they are associated with a region of low energy/momentum. For the collinear case, this is the **transverse momentum** of the radiated gluon relative to the quark direction. By their nature, IR divergences are distinct from the UV divergences that we already encountered, that are removed via renormalisation. The latter only occured in loops, and were ultimately not a problem because they involved high energies, which we could not claim to know anything about. IR divergences, on the other hand, involve **real** particles, and also energies that we certainly do probe at colliders. At first glance, then, we seem to have a

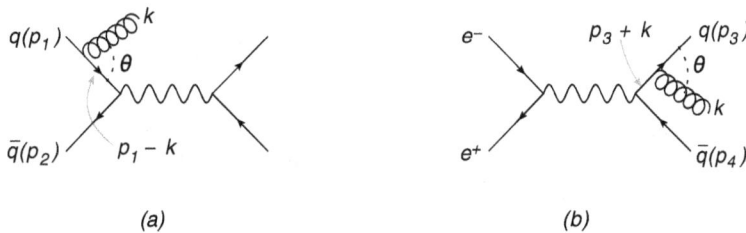

(a)   (b)

Figure 2.23. (a) Radiation of a gluon from an incoming quark leg; (b) final state radiation from an outgoing quark.

genuine problem, namely that all cross-sections involving real radiation will be infinite!

Above we saw that infrared divergences occur when we have radiation from incoming particles. However, the problem is more general than this, and also occurs for radiation from final state particles. An example is shown in Figure 2.23(b), in which a gluon is radiated off the final state quark in the process of Figure 2.14. This produces an extra propagator, which involves a factor

$$\frac{1}{(p_3 + k)^2} = \frac{1}{|\boldsymbol{p}_3||\boldsymbol{k}|(1 - \cos\theta)},$$

where again θ is the angle between the quark and gluon. Studying this final state radiation in more detail will provide clues as to how to proceed. First, note that in the cross-section, we must integrate over the final-state phase-space, which will behave as follows:

$$\int d\Phi \, \overline{|\mathcal{A}|^2} \sim \int \frac{d^3 k}{2|\boldsymbol{k}|(2\pi)^3} \frac{1}{|\boldsymbol{k}|^2},$$

where the factor on the far right-hand side comes from squaring the extra propagator we have found above. We thus find an integral

$$\sim \frac{d^3 k}{|\boldsymbol{k}|^3}$$

as $|\boldsymbol{k}| \to 0$, whose lower limit (zero) produces a logarithmic divergence.[15] As for UV divergences, a practical way to regularise such divergences is to use **dimensional regularisation**. In this case, we must raise the number of dimensions slightly, to reduce the divergence as $|\boldsymbol{k}| \to 0$. Thus, we can set $d = 4 - 2\epsilon$ as before, but where now $\epsilon < 0$ rather than $\epsilon > 0$. This is in fact one of the main reasons why dimensional regularisation is so widely used: we can use it to simultaneously regularise UV and IR divergences, where the necessary sign of ϵ tells us which type of divergence we are dealing with!

Figure 2.23(b) is a contribution to the process

$$e^- + e^+ \to q + \bar{q} + g. \tag{2.200}$$

Including all contributions (e.g. radiation from the anti-quark as well as from the quark), the total cross-section for real radiation turns out to be

$$\sigma_{\text{real}} = \sigma_0 \frac{\alpha_S C_F}{2\pi} H(\epsilon) \left(\frac{2}{\epsilon^2} + \frac{3}{\epsilon} + \frac{19}{2} - \pi^2 \right) + \mathcal{O}(\alpha_S^2), \tag{2.201}$$

where σ_0 is the LO cross-section, and $H(\epsilon)$ is a known function satisfying $H(0) = 1$. We cannot simply return to four spacetime dimensions by taking $\epsilon \to 0$ in equation (2.201), as the result is infinite! This is precisely because of the infrared divergences

---

[15] There is no logarithmic divergence at high values of k, as these will be cut off by the physical centre-of-mass energy.

that we described above. Furthermore, we see that there is a **double pole** in ϵ, corresponding to the fact that the emitted gluon can be both soft *and* collinear. The resolution of this problem turns out to be that, contrary to what we might think, we have not calculated a physically well-defined observable. To see why, note that as the radiation becomes exactly soft ($k^\mu \to 0$) or collinear ($k^\mu \propto p_3^\mu$ or $k^\mu \propto p_4^\mu$), the radiative process of equation (2.200) becomes **physically indistinguishable** from the non-radiative process

$$e^- + e^+ \to q + \bar{q}. \tag{2.202}$$

That is, a real particle detector has some finite energy resolution, and so cannot detect an infinitesimally soft gluon. Nor can it separate two particles that are infinitesimally close together. This suggests that we should add all diagrams at this order in perturbation theory that correspond to the non-radiative process, namely the virtual corrections in Figure 2.24. Their contribution to the total cross-section turns out to be

$$\sigma_{\text{virtual}} = \sigma_0 \frac{\alpha_S C_F}{2\pi} H(\epsilon)\left(-\frac{2}{\epsilon^2} - \frac{3}{\epsilon} - 8 + \pi^2\right) + \mathcal{O}(\alpha_S^2). \tag{2.203}$$

If instead of either equation (2.200) or equation (2.202) we consider the process

$$e^- + e^+ \to q + \bar{q} + \text{radiation}, \tag{2.204}$$

where we do not specify precisely how many particles the radiation must contain, we can obtain the total cross-section at this order in α_S by simply adding together equations (2.201) and (2.203). Then the total result, including the LO term, is

$$\sigma_{\text{tot}} = \sigma_0\left[1 + \frac{3\alpha_S C_F}{4\pi}\right] + \mathcal{O}(\alpha_S^2), \tag{2.205}$$

which is perfectly well-behaved, so that we can take $\epsilon \to 0$. What we have learned is that not all theoretical observables are physically meaningful: only those that are **infrared safe**. Such observables include a sum over all physically indistinguishable states, and it makes sense that we have to do this, as we cannot measure only some indistinguishable states, and not others! Another way of defining an infrared safe observable is that it should be well-behaved upon adding an additional soft particle, or splitting a parton into two collinear ones. Fixing the number of partons in the final state, as we did above in equation (2.200) fails this definition, and thus $\sigma_{q\bar{q}g}$ is not IR safe by itself. This is why equation (2.201) contained infrared divergences.

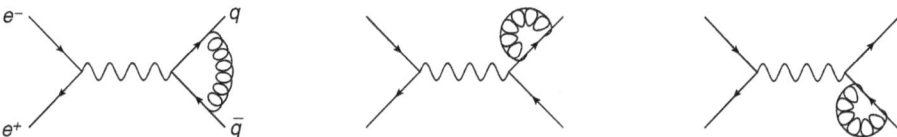

Figure 2.24. Virtual corrections to quark–antiquark pair production.

At higher orders, it is tempting to conjecture that if we add all virtual diagrams to real emission diagrams (including diagrams in which some gluons are real, and some virtual), the total cross-section is finite. For QED, this indeed turns out to be correct, a result known as the **Bloch–Nordsieck theorem**. For a non-Abelian theory, however, the results are—as is becoming depressingly familiar—more complicated. It turns out that IR singularities indeed cancel, provided we include **initial states** with any number of particles, such as that shown in Figure 2.25. This result, known as the **Kinoshita–Lee–Nauenberg (KLN) theorem** is interesting, but not terribly useful for actual calculations at hadron colliders which, after all, only have two incoming beams! It follows that, if we try to construct IR safe observables at hadron colliders, we will have uncancelled IR singularities associated with **initial state radiation**. Thus, partonic cross-sections by themselves become unphysical. This by itself is not a problem though: according to equation (2.198), partonic cross-sections must be combined with PDFs. The latter come to the rescue!

To see how, let us return to the example of radiation in Drell–Yan production (Figure 2.23(a)). Including also radiation from the incoming anti-quark, the total squared matrix element, averaged and summed over colours and spins, is found to be

$$\overline{|A_{q\bar{q}g}|^2} = \frac{C_F}{N_c} \frac{2e^4 Q_q^2 g_s^2}{p_1 \cdot k \, p_2 \cdot k}$$
$$\frac{[(p_1 \cdot p_3)^2 + (p_2 \cdot p_4)^2 + (p_2 \cdot p_3)^2 + (p_1 \cdot p_4)^2]}{p_3 \cdot p_4}. \tag{2.206}$$

Given that we expect a problem if the additional gluon is collinear with one of the incoming momenta, let us take $k \| p_1$ and see what happens. However, we cannot just set $k \propto p_1$, as then the squared amplitude will diverge. Instead, we must find a way to parametrise k^μ, which will allow us to smoothly approach the limit in which it becomes collinear to p_1. To this end, a useful trick is a so-called **Sudakov decomposition**, which starts with the observation that it is generally true that we can write

$$k^\mu = c_1 p_1^\mu + c_2 p_2^\mu + k_T^\mu,$$

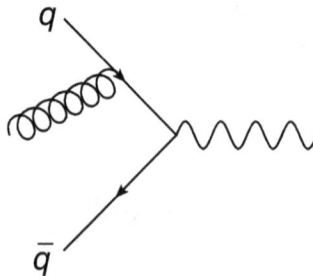

Figure 2.25. An initial state involving three partons.

where $k_T = (0, \mathbf{k}_T, 0)$ is a four-vector that is transverse to both p_1 and p_2, i.e.

$$k_T \cdot p_1 = k_T \cdot p_2 = 0. \tag{2.207}$$

It is conventional to choose

$$c_1 = (1 - z), \quad 0 \leqslant z \leqslant 1, \tag{2.208}$$

where we will interpret the variable z shortly. The on-shell condition $k^2 = 0$ then implies

$$2p_1 \cdot p_2 (1 - z)c_2 - k_T^2 = 0,$$

and thus

$$k^\mu = (1 - z)p_1^\mu + \frac{k_T^2}{s(1 - z)}p_2^\mu + k_T^\mu, \quad s = (p_1 + p_2)^2. \tag{2.209}$$

The collinear limit corresponds to the emitted gluon having no momentum transverse to p_1, i.e. $\mathbf{k}_T \to 0$. Then we see that $k^\mu = (1 - z)p_1^\mu$, so that $(1 - z)$ represents the momentum fraction of the incoming quark that is carried by the gluon in the collinear limit. If we substitute equation (2.209) into the matrix element and keep the leading terms as $\mathbf{k}_T \to 0$, we find (see e.g. the exercises at the end of the chapter)

$$\overline{|\mathcal{A}_{q\bar{q}g}|^2} \xrightarrow{k_T \to 0} \frac{2g_s^2 C_F}{k_T^2} \frac{(1 + z^2)}{z} \overline{|\mathcal{A}_{q\bar{q}}(zp_1, p_2)|^2}. \tag{2.210}$$

Note that the factor on the right-hand side is the LO cross-section, but evaluated with a shift $p_1 \to zp_1$ in the incoming quark momentum. Moreover, the phase-space of the three-particle final state factorises as

$$\int d\Phi^{(3)} = \int \frac{d^3 k}{(2\pi)^3 2|k|} \int d\tilde{\Phi}^{(2)}, \tag{2.211}$$

where the first integral on the right-hand side corresponds to the additional gluon, and the remaining integral is that of the LO process, but with $p_1 \to p_1 - k$. This in turn implies that the partonic cross-section for the radiative process has the form

$$\hat{\sigma}_{q\bar{q}g} \xrightarrow{k \| p_1} \int_0^1 dz \int \frac{d|k_T|^2}{|k_T|^2} \hat{P}_{qq}(z)\hat{\sigma}_{q\bar{q}}(zp_1, p_2), \tag{2.212}$$

where

$$\hat{P}_{qq}(z) = \frac{\alpha_s C_F}{2\pi} \frac{(1 + z^2)}{1 - z}. \tag{2.213}$$

In words: the cross-section including an extra gluon **factorises** in the collinear limit $k \| p_1$, where the additional gluon is described by the function $\hat{P}_{qq}(z)$, where z represents the momentum fraction of the incoming quark that is carried by the quark after the gluon emission. This multiplies the LO cross-section with shifted kinematics $p_1 \rightarrow zp_1$. There is then an integral over all possible momentum fractions z, and gluon transverse momenta $\boldsymbol{k}_{\mathrm{T}}$. The collinear singularity we were expecting shows up in the $|\boldsymbol{k}_{\mathrm{T}}|^2$ integral, as this variable approaches zero.

According to equation (2.198), the hadronic cross-section in this limit is given by

$$
\sigma \xrightarrow{k \| p_1} \int_0^1 \mathrm{d}x_1 \int_0^1 \mathrm{d}x_2 f_q(x_1) f_{\bar{q}}(x_2)
$$
$$
\left[\hat{\sigma}_{q\bar{q}}(p_1, p_2) + \int_0^1 \mathrm{d}z \int \frac{\mathrm{d}|\boldsymbol{k}_{\mathrm{T}}|^2}{|\boldsymbol{k}_{\mathrm{T}}|^2} \hat{P}_{qq}(z)\, \hat{\sigma}_{q\bar{q}}(zp_1, p_2) \right],
$$

(2.214)

where the first term in the square brackets is the LO contribution. Given that $p_1 = x_1 P_1$ (where P_1 is the proton momentum), we can shift $x_1 \rightarrow x_1/z$ in the second term, which after some work gives

$$
\sigma \xrightarrow{k \| p_1} \int_0^1 \mathrm{d}x_1 \int_0^1 \mathrm{d}x_2\, \hat{\sigma}_{q\bar{q}}(p_1, p_2) f_{\bar{q}}(x_2)
$$
$$
\left[f_q(x_1) + \int_{x_1}^1 \frac{\mathrm{d}z}{z} \int \frac{\mathrm{d}|\boldsymbol{k}_{\mathrm{T}}|^2}{|\boldsymbol{k}_{\mathrm{T}}|^2} \hat{P}_{qq}(z) f_q\left(\frac{x_1}{z}\right) \right].
$$

(2.215)

This way of writing the formula is particularly interesting, as the collinear singularity is explicitly associated with a combination involving only the quark PDF. We can then note that the PDFs are not physical by themselves: the only directly measurable quantity is the hadronic cross-section σ, which combines the PDFs with the partonic cross-section. Thus, we are free to move contributions from the partonic cross-section into the PDFs if we want to. By doing so, we can remove the collinear singularity!

Let us call the quark PDF before any modification a 'bare PDF', f_q^0, so that the above formula becomes

$$
\sigma \xrightarrow{k \| p_1} \int_0^1 \mathrm{d}x_1 \int_0^1 \mathrm{d}x_2\, \hat{\sigma}_{q\bar{q}}(p_1, p_2) f_{\bar{q}}^0(x_2)
$$
$$
\left[f_q^0(x_1) + \int_{x_1}^1 \frac{\mathrm{d}z}{z} \int \frac{\mathrm{d}|\boldsymbol{k}_{\mathrm{T}}|^2}{|\boldsymbol{k}_{\mathrm{T}}|^2} \hat{P}_{qq}(z) f_q^0\left(\frac{x_1}{z}\right) \right],
$$

which is just a relabelling. Then we can define a modified PDF f_q via

$$
f_q^0 = f_q - \int_0^{\mu_{\mathrm{F}}^2} \frac{\mathrm{d}|\boldsymbol{k}_{\mathrm{T}}|^2}{|\boldsymbol{k}_{\mathrm{T}}|^2} \int_{x_1}^1 \frac{\mathrm{d}z}{z} \hat{P}_{qq}(z) f_q\left(\frac{x_1}{z}\right) + \mathcal{O}(\alpha_{\mathrm{S}}^2),
$$

(2.216)

where μ_{F} is called the **factorisation scale**. Equation (2.216) amounts to absorbing the collinear divergence as $|\boldsymbol{k}_{\mathrm{T}}| \rightarrow 0$ into the bare PDF. If we then write the hadronic cross-section in terms of modified PDFs at this order in perturbation theory, it becomes

$$\sigma \xrightarrow{k\|p_1} \int_0^1 dx_1 \int_0^1 dx_2 \hat{\sigma}_{q\bar{q}}(p_1, p_2) f_{\bar{q}}(x_2)$$
$$\left[f_q(x_1) + \int_{x_1}^1 \frac{dz}{z} \int_{\mu_F^2} \frac{d|k_T|^2}{|k_T|^2} \hat{P}_{qq}(z) f_q\left(\frac{x_1}{z}\right) \right]. \tag{2.217}$$

In writing this formula, we have used the fact that we do not need to modify the anti-quark PDF in this limit, so have simply set $f_{\bar{q}}^0 \equiv f_{\bar{q}}$. We see that equation (2.217) no longer has a collinear singularity as $k\|p_1$. It has been removed by the factorisation scale μ_F, which acts as a cutoff on the transverse momentum integral. However, there is still a potential problem in equation (2.217): we must integrate over all values of z, but the function $\hat{P}_{qq}(z)$ of equation (2.213) has a divergence as $z \to 1$. We said that $(1 - z)$ represents the momentum fraction of the incoming quark momentum that is carried by the gluon in the collinear limit, and thus $z \to 1$ corresponds to the gluon being soft, with $k^\mu \to 0$. We indeed expect a singularity in this limit. However, it turns out to cancel when we add virtual diagrams. This is a long calculation by itself, but the result is quite simple: virtual corrections have the effect of modifying

$$\hat{P}_{qq}(z) \to P_{qq}(z) = \frac{\alpha_S C_F}{2\pi} \left[\frac{1 + z^2}{(1 - z)_+} + \frac{3}{2}\delta(1 - z) \right],$$

which contains a type of mathematical distribution called a **plus distribution**. This is defined by its action on an arbitrary test function $f(x)$:

$$\int_0^1 dx \frac{f(x)}{(1 - x)_+} \equiv \int_0^1 dx \frac{f(x) - f(1)}{1 - x}. \tag{2.218}$$

After this modification, the hadronic cross-section is completely free of infrared divergences associated with the incoming quark. A similar modification to the bare anti-quark PDF will get rid of all soft and collinear singularities associated with the incoming anti-quark.

Note that the modified PDFs depend on x_i (the momentum fraction of the proton carried by the parton), and μ_F (the factorisation scale). To interpret the latter, we can think of the proton as being some weird quantum cloud of (anti)-quarks and gluons. As a single parton emerges, it emits radiation, where the radiated particle will have different transverse momenta k_T. At some point, this transverse momentum will be high enough that $\alpha_S(|k_T|^2)$ is small (i.e. the coupling decreases at higher scales). Then we can use perturbation theory to describe the radiation. For low transverse momenta, α_S is strong, and perturbation theory no longer works. We can then think of μ_F as some 'dividing scale', that separates perturbative from non-perturbative physics, i.e. if $|k_T| > \mu_F$, then a radiated parton is perturbative, but if $|k_T| < \mu_F$ it is non-perturbative, and thus sensitive to the strongly coupled dynamics that holds the proton together. Such emissions should then be part of the PDF rather than the partonic cross-section, and indeed this is exactly what the modification of equation (2.216) does.

Above, we considered a quark or anti-quark emitting a collinear gluon, namely the **splitting process**

$$q(p) \rightarrow q(zp) + g((1 - z)p)$$

and similarly for the corresponding antiquark splitting. We could also have other splittings, where the full list is shown in Figure 2.26. It turns out that squared matrix elements involving such splittings *always* factorise in the collinear limit, where the relevant so-called **splitting functions** at $\mathcal{O}(\alpha_S)$ are

$$P_{qq}(z) = \frac{\alpha_S C_F}{2\pi}\left[\frac{1 + z^2}{1 - z}_+ + \frac{3}{2}\delta(1 - z)\right]; \tag{2.219}$$

$$P_{gq}(z) = \frac{\alpha_S C_F}{2\pi}\left[\frac{1 + (1 - z)^2}{z}\right]; \tag{2.220}$$

$$P_{qg}(z) = \frac{\alpha_S T_R}{2\pi}[z^2 + (1 - z)^2]; \tag{2.221}$$

$$P_{gg}(z) = \frac{\alpha_S}{2\pi}\left\{C_A\left[\frac{z}{(1 - z)_+} + \frac{1 - z}{z} + z(1 - z)\right]\right. $$
$$\left. + \delta(1 - z)\frac{(11 C_A - 4 n_f T_R)}{6}\right\}. \tag{2.222}$$

We derived $P_{qq}(z)$ above, and will not bother deriving the other splitting functions explicitly. However, they follow a similar method. One may simply replace $q \rightarrow \bar{q}$ in these expressions to get the corresponding splitting functions for antiquarks. Then the fully general expression for modifying *any* bare PDF is

$$f_i^0 = f_i\left(x_i, \mu_F^2\right) - \sum_j \int_0^{\mu_F^2} \frac{d|k_T|^2}{|k_T|^2} \int_{x_i}^1 \frac{dz}{z} P_{ij}(z) f_j\left(\frac{x_i}{z}\right), \tag{2.223}$$

where the sum includes all splittings consistent with the chosen PDF. The factorisation scale μ_F is arbitrary. However, once we have fixed a choice, we can measure the PDFs from data. In fact, we can say a lot more than this, as we discuss in the following section.

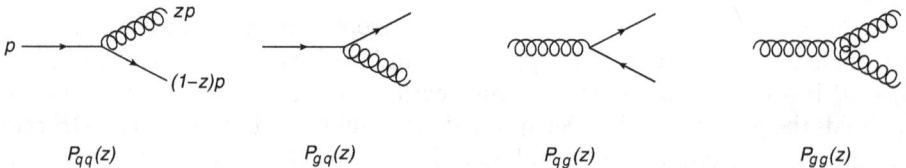

Figure 2.26. Possible parton splittings, and their associated splitting functions.

2.16 The DGLAP equations

The bare PDFs that we started with in equation (2.198) know nothing about the factorisation scale μ_F, that only entered when we defined the modified PDFs. This in turn implies the condition

$$\frac{d f_i^0}{d\mu_F^2} = \frac{\partial f_i\left(x_i, \mu_F^2\right)}{\partial \mu_F^2} - \sum_j \frac{1}{\mu_F^2} \int_{x_i}^1 \frac{dz}{z} P_{ij}(z) f_j\left(\frac{x_i}{z}, \mu_F^2\right) = 0,$$

where we have used equation (2.223). Rearranging, we find

$$\mu_F^2 \frac{\partial f_i\left(x_i, \mu_F^2\right)}{\partial \mu_F^2} = \sum_j \int_{x_i}^1 \frac{dz}{z} P_{ij}(z) f_j\left(\frac{x_i}{z}, \mu_F^2\right), \qquad (2.224)$$

which are known as the **Dokshitzer–Gribov–Lipatov–Altarelli–Parisi equations**, or simply the **DGLAP equations** for short. They tell us that, although we cannot predict the PDFs from first principles, *how they change* with factorisation scale μ_F is entirely calculable in perturbation theory! We can interpret the role of μ_F as follows. Equation (2.217) (generalised to include the full modification of equation (2.223)) tells us that in the collinear limit, hadronic cross-sections depend on combinations such as

$$f_i(x_i, Q^2) + \sum_j \int_{\mu_F^2}^{Q^2} \frac{d|k_T|^2}{|k_T|^2} \int_{x_i}^1 \frac{dz}{z} P_{ij}(z) f_j\left(\frac{x_i}{z}\right),$$

where Q^2 is some process-dependent upper limit of $|k_T|$ (i.e. in practice any emitted gluon has a finite upper limit on its transverse momentum). If we want to keep perturbative corrections fairly small, it makes sense to choose $\mu_F \sim Q$, i.e. to associate the factorisation scale μ_F with some hard energy or momentum scale in the process. We can then interpret the DGLAP equations as telling us how the PDFs vary with energy scale. Here, we used a momentum cutoff in $|k_T|$ to regularise collinear singularities, which also allowed us to interpret the factorisation scale μ_F. However, we could have used a different regulator, and we would still have found the existence of μ_F, and the DGLAP equations. As discussed before, the most commonly used regulator for partonic cross-sections is dimensional regularisation. In that case, the scale μ_F arises as the scale that is used to keep the coupling g_s dimensionless in d dimensions. Furthermore, one can choose to absorb an arbitrary finite contribution to the partonic cross-section into the PDFs, as well as the pure collinear singularity. Different choices constitute different so-called **factorisation schemes**, and a common choice is to use dimensional regularisation and the $\overline{\text{MS}}$ scheme, in which a particular choice of numerical constants is removed. Once such a choice is fixed, the partons can be measured from data.

The above discussion will hopefully remind you of something—the idea of factorisation is very similar to renormalisation. In both cases, divergences are absorbed by redefining bare quantities. Then, the modified/renormalised quantities cannot be predicted from first principles, but their change in energy scale is calculable in perturbation theory. This analogy is nice to think about, but it should be remembered that the nature of the divergences in factorisation (IR) and renormalisation (UV) are fundamentally different. Consequently, the factorisation scale μ_F and μ_R are different in principle, although can be chosen to be the same in practice. Dependence on μ_F cancels between the partonic cross-section and the (modified) partons, up to the order in perturbation theory that we are working at. Thus, at a given order $\mathcal{O}(\alpha_S^n)$, dependence on μ_F and μ_R is $\mathcal{O}(\alpha_S^{n+1})$. Varying μ_F and μ_R independently in some range

$$\frac{\mu_0}{2} \leqslant \mu_F, \mu_R \leqslant 2\mu_0$$

for a suitable default choice μ_0 then gives a measure of theoretical uncertainty (i.e. an estimate of higher-order corrections).

Our original interpretation of the PDFs is that they represent the probability to find a given parton with a defined momentum fraction inside the parent proton. Once higher-order QCD corrections are taken into account and we choose a particular factorisation scheme, we lose this interpretation in general. For example, the DGLAP equations imply that at a sufficiently low scale, the PDFs may become negative. This is not a problem in principle—hadronic cross-sections are obtained by combining the partons with partonic cross-sections, and will be positive provided we are at energy scales where perturbation theory is valid. Nevertheless, physical properties of PDFs survive in the form of **sum rules**. For example, if we look at the up quark PDF, it will naïvely have contributions from two sources: (i) the two up quarks that are meant to be in the proton; (ii) additional up quarks due to the froth of $u\bar{u}$ pairs that can be made due to quantum effects. Given that matter and antimatter are always created in equal amounts, it follows that the second source of up quarks should be equal to the anti-up distribution $\bar{u}(z, \mu_F^2)$. Thus, subtracting the latter from the up distribution, we can form the so-called **valence up-quark distribution**

$$u_V\left(x, \mu_F^2\right) = u\left(x, \mu_F^2\right) - \bar{u}\left(x, \mu_F^2\right), \tag{2.225}$$

where integrating over all momentum fractions for a fixed scale μ_F^2 should tell us that there are indeed two 'net' up quarks:

$$\int_0^1 \mathrm{d}x\, u_V\left(x, \mu_F^2\right) = 2. \tag{2.226}$$

Similarly, for the down quark we have

$$\int_0^1 \mathrm{d}x\, d_V\left(x, \mu_F^2\right) = 1. \tag{2.227}$$

If the partons indeed represent probability densities (i.e. per unit longitudinal momentum fraction), the total momentum fraction carried by all partons must be equal to one:

$$\int_0^1 \mathrm{d}x\, x \sum_i f_i\left(x, \mu_F^2\right) = 1, \tag{2.228}$$

where the sum is over all flavours of quark and anti-quark, and also includes the gluon. This sum rule survives even at higher orders in perturbation theory in arbitrary factorisation schemes. The reason for this is that higher terms in perturbation theory merely constitute splitting of the partons into other partons, which always conserves momentum. Put another way, the DGLAP equations at arbitrary order can always be shown to preserve equation (2.228).

Despite the complicated nature of the above calculations, we have still only given a simplified treatment of PDFs! A fully rigorous treatment shows that hadronic cross-sections actually have the form

$$\sigma = \sum_{i,j} \int_0^1 dx_1 \int_0^1 dx_2 \, f_i\left(x_1, \mu_F^2\right) f_j\left(x_2, \mu_F^2\right) \hat{\sigma}_{ij}\left(x_i, \mu_F^2, \{p_i \cdot p_j\}\right) + \mathcal{O}\left(\frac{\Lambda_{QCD}^2}{Q^2}\right), \quad (2.229)$$

which differs from the naïve parton model of equation (2.198) in that it includes terms depending on the ratio of the energy scale Λ_{QCD} at which the QCD coupling becomes strong, and the typical hard energy scale Q of the scattering process. These additional terms are called **power corrections**, and we will not worry too much about them.

2.17 Global fits of parton distributions

In the previous two sections, we have seen that cross-sections at hadron colliders require the partonic cross-section to be convolved with PDFs, where the latter are not themselves calculable from first principles. In practice, this means that we must measure the PDFs from experiments, such that they can be used to construct cross-sections for subsequent experiments. As each new particle collider has come online since the late 1970s onwards, an increasingly large collection of datasets has been assembled, which have been used to constrain the PDFs ever more precisely.

As may already be clear, the above phrase *measuring the PDFs* is a significant understatement, that hardly begins to do justice to the complicated effort needed to establish the (anti-)quark and gluon distributions of the (anti-)proton! For a start, the various PDFs $\{q_i, \bar{q}_i, g\}$ are each functions of both the longitudinal momentum fraction x, and the factorisation scale μ_F, where the latter is typically associated with a characteristic energy scale Q in the scattering process in which the PDFs are being utilised. This scale can be different depending on the application, and thus one needs to provide the values of the PDFs in the two-dimensional parameter space (x, Q^2) (n.b. the squared energy scale Q^2 is conventionally used). There are a number of collaborations worldwide that do this, and they typically present computer code that can provide a value of a PDF for any desired value of the arguments. This might be done, for example, by having the code read in a **PDF grid** consisting of the values of the PDFs at a set of discrete points in the (x, Q^2) plane, and then to interpolate between these values according to some algorithm.

The main global collaborations producing PDFs are the CTEQ collaboration in the US, the MSHT collaboration in the UK (formally known variously as MRS, MRST, MSTW and MMHT), NNPDF (where the first two letters stand for 'Neural Network'), and ABJM. How the partons are extracted from data differs slightly depending on the collaboration. With the exception of NNPDF (to be discussed below), the various groups use something like the following procedure. First, we may note that the DGLAP equations of Section 2.16 tell us that if we know the PDFs at some low (squared) energy scale Q_0^2, we can predict their values at a higher scale Q^2. This suggests the following algorithm:

1. Choose a **starting scale** Q_0^2 which is high enough for perturbation theory to be valid, but lower than the characteristic energy scale of all scattering processes for which data is available. A typical value is $Q_0 = 1$ GeV.

2. At the starting scale, parametrise the PDFs as a function of x. For example, early parton fits used functions such as

$$xf_i(x, Q_0^2) = (1 - x)^{\eta_i}(1 + \epsilon_i x^{0.5} + \gamma_i x)x^{\delta_i}. \qquad (2.230)$$

Here $f(x, Q_0^2)$ may be an individual PDF, or a combination that is particularly physically meaningful or well-constrained. The power-like forms as $x \to 0$ and $x \to 1$ reflect known theoretical prejudice about how PDFs should behave in these limits, but the quantities $(\eta_i, \epsilon_i, \gamma_i, \delta_i)$ are free parameters, where there will be one set of these for each PDF (or given combination of PDFs) f_i. More modern fits may use superpositions of such functions (thus allowing partons to become negative, as is theoretically allowed), or more complicated functions such as superpositions of known special polynomials.

3. For each dataset with characteristic energy scale Q, evolve the PDFs from the starting scale Q_0 up to Q using the DGLAP equations. This can be done numerically, for given values of the parameters $\{\eta_i, \epsilon_i, \gamma_i, \delta_i\}$. Then, the PDFs at scale Q can be used to obtain the cross-section (or other observable) for the dataset of interest, and a goodness of fit measure constructed which measures the difference between theory and experiment. One example is the well-known χ^2 function, and the total goodness of fit measure is obtained by summing over the results for each dataset.

4. One then optimises the goodness-of-fit measure across the parameter space $\{\eta_i, \epsilon_i, \gamma_i, \delta_i\}$, which involves repeating step 3 whilst varying the potentially large number of parameters that enter the initial PDF parametrisation.

This is called a global (parton) fit, where the word 'global' here means that one tries to fit all free parameters simultaneously, using many different types of data. It is clearly a highly intensive computational task, not least given that contemporary PDF fits contain a corncuopia of datasets, from the LHC and previous colliders. Furthermore, the above algorithm can be implemented at any given order in perturbation theory, where the same order must be chosen for the DGLAP splitting functions as for the other theory input to the fit (e.g. partonic cross-sections, running couplings and masses etc). The state of the art for such fits is NNLO, but collaborations continue to produce partons at (N)LO. This is due to the fact that theory predictions that are used for many scattering processes remain limited to lower orders in perturbation theory, and one should arguably use a consistent set of partons[16].

Parton fitting is a highly dynamic subfield of high energy physics, involving constant theoretical developments in order to improve the precision of the PDFs one obtains. Often, groups will produce a number of different PDF sets, where differing assumptions have been made in the fitting procedure. These PDFs will then be suited to a particular purpose, and it is important for experimentalists to know which PDFs they should be using, and when. It is worth dwelling on a couple of issues that can be important when choosing and using PDFs.

PDF uncertainties

There are clearly a number of sources of error in extracting the PDF distributions from data:

- The datasets one uses have error bars associated with each data point, comprising both statistical and systematic uncertainties. There may also be correlations in uncertainty between different points within the same dataset, or between different datasets.
- Traditional goodness of fit measures (e.g. the χ^2 mentioned above) assume Gaussian uncertainties for the data, which may not actually be the case. Furthermore, datasets might be incompatible due to historically poorly understood systematic effects, with no way of telling which dataset(s) one should ultimately trust.
- There may be a systematic bias introduced due to the fact that the parametrisation used for the PDFs at the starting scale is insufficiently flexible to include all features of the genuine (unknown) PDFs.

For such reasons, all collaborations that produce PDFs include a mechanism for calculating their uncertainties. They might provide a single set representing the best estimate of the central value of each PDF, together with additional sets that represent the uncertainty envelopes arising from e.g. a fixed increase in the χ^2 arising in the fit. Detailed discussion of PDF uncertainties can be found in the publications produced by each collaboration, although it is worth noting that the various systematic uncertainties outlined above prompted the NNPDF collaboration to use a different fitting procedure to that described in the previous section. Their approach involves simulating a number N_{rep} of replicas of the combined datasets entering the fit, obtained by randomly varying each data point within its uncertainty band. Then, a neural network (a type of fancy interpolating function with a large number of free parameters, see Section 5.8.2) is trained to each data replica, so as to parametrise the PDFs at the starting scale. This in turn results in N_{rep} replica sets of PDFs, which can be used to obtain central values and uncertainty bands, thereby alleviating the problems of parametrisation bias, incompatible datasets, and non-Gaussian errors. However, broadly similar results for PDF central values and uncertainties are obtained by other fitting groups.

---

[16] Strictly speaking, it is not incorrect to use e.g. NLO partons, when calculating a hadronic cross-section at LO. However, the opposite way around would indeed be incorrect.

Treatment of heavy quarks

The up, down and strange (anti-)quarks have masses which are lower than the energy scale Λ_{QCD} at which QCD becomes non-perturbative. However, this is not true for the charm and bottom quarks, whose masses are roughly 1.3 GeV and 4.6 GeV respectively. As for the light quarks, these quarks can be produced in scattering processes. An example of this is shown in Figure 2.27, which shows the production of charm quark pairs at an e^-p collider, as happened at e.g. the HERA collider throughout the 2000s. The fact that the charm quark mass m_c is above Λ_{QCD} means that one is perfectly entitled to include it in perturbative calculations. Upon doing so, the cross-section for the process of Figure 2.27 turns out to involve a term

$$\sim \ln\left(\frac{Q^2}{m_c^2}\right),$$

where, as usual, Q^2 is a characteristic (squared) energy scale in the process. This term is clearly troublesome if Q^2 becomes big: in that case, the logarithm becomes large, and overcomes the smallness of the coupling constant. Furthermore, the problem can be shown to get worse at higher orders in perturbation theory, in that higher powers of such logarithms occur, so that as Q^2 becomes large relative to m_c, one can no longer safely truncate the perturbation expansion.

To understand where such large logarithms come from, we can recall the results of Section 2.15, which said that splitting of gluons into massless quarks leads to collinear singularities, that have to be absorbed into the parton distribution functions of the incoming proton(s). In the present case, Q^2 becoming much larger than m_c means that we should effectively be able to neglect the charm quark mass, at the expense of incurring additional collinear singularities. These are precisely the logarithmically singular terms mentioned above: our inability to take the charm quark mass to zero indicates the presence of a collinear singularity, which is thus being regulated by m_c. Understanding the origin of the large logarithms then allows us to solve the problem: if collinear singularities are removed by defining parton densities, we simply have to define new parton distributions for the charm and

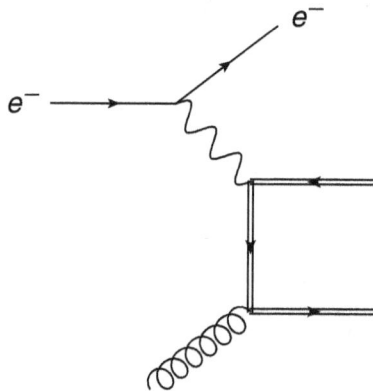

Figure 2.27. Perturbative production of a charm quark pair (double line) at an e^-p collider.

bottom (anti)-quarks[17]. Then, we can absorb the collinear singularities at large Q^2 into the new PDFs, and evolve them in scale according to the appropriately generalised DGLAP equations.

Given the above discussion, we can consider two cases of dealing with heavy quarks. Firstly, we can refuse to define partons for them at all, and generate them purely perturbatively, with full mass dependence. This is called a **fixed-flavour number scheme** (FFNS), and can be an excellent approximation to the truth provided we are not dealing with energy scales that are too large relative to the heavy quark masses involved. Furthermore, heavy quark mass effects can be important for certain observables. A second possibility is that we can treat the heavy quarks perturbatively up to a **threshold scale**, at which we turn on the heavy quark PDFs. The number of (anti)-quark flavours is then different at different energies, and thus this approach is referred to as a **variable flavour number scheme** (VFNS). There will be a different threshold scale for each heavy quark, and at a given scale the number of flavours will increase from e.g. n_f to $n_f + 1$. Given that physics should ultimately be independent of which approach we choose (up to a given order in perturbation theory), we must then carefully match the descriptions at the threshold scale, for which a number of choices and technicalities must be considered (e.g. what to use for the differential cross-sections and DGLAP splitting functions below and above the threshold scale; how to include the heavy quarks in the running coupling etc). By now, all parton collaborations are including some sort of VFNS in their PDF fits, and a detailed discussion can be found in their respective publications.

Further reading

There are a large number of textbooks on quantum field theory, spanning many decades. Examples include:

- 'Quantum Field Theory', L H Ryder, Cambridge University Press. This includes accessible chapters on topics that are not usually covered in introductory QFT books (e.g. SUSY, non-perturbative solutions).
- 'Quantum Field Theory', G Sterman, Cambridge University Press. This is a good book for QCD calculations, containing in particular a very complete discussion of IR singularities, and dimensional regularisation.
- 'An Introduction to Quantum Field Theory', M E Peskin and D V Schroeder. This is a classic text, that is often used by lecturers. It is particularly convenient for looking things up in.
- 'Quantum Field Theory and the Standard Model', M D Schwartz, Cambridge University Press. This is an excellent modern text, with a good coverage of advanced topics, especially those relating to QCD.
- 'The Quantum Theory of Fields', S Weinberg, Cambridge University Press. This is a monumental three-volume work by one of the great intellectual

---

[17] Partons for the top quark are not necessary. It is much more massive than either the charm or bottom quarks, and is also incredibly short-lived.

heavyweights of fundamental physics. It is encyclopaedic in content, and is excellent in providing a historical context.

Some dedicated books on gauge field theory include the following:

- 'Gauge Theories in Particle Physics: A Practical Introduction', I J R Aitchison and A J G Hey, CRC Press. This two-volume set eloquently introduces gauge theory without the full complications of QFT.
- 'Introduction to Gauge Field Theory', D Bailin and A Love, Taylor and Francis. A classic book, that covers many essentials of gauge theory in a concise and elegant manner.

For parton distributions in particular, the following books are recommended:

- 'The Structure of the Proton: Deep Inelastic Scattering', R G Roberts. A small and beautiful book!
- 'Deep Inelastic Scattering', R Devenish and A Cooper-Sarkar, Oxford University Press. Contains a mine of information about modern parton fits, from two renowned practitioners.

Exercises

2.1 (a) Show that the local gauge transformation of an electromagnetic field $A^\mu(x)$ acts on the electrostatic and magnetic vector potentials as

$$\phi \to \phi - \frac{1}{e}\frac{\partial \alpha}{\partial t}, \quad A \to A + \frac{1}{e}\nabla\alpha,$$

where α is an arbitrary function of space and time.

(b) Hence show that the electric and magnetic fields

$$E = -\nabla\phi - \frac{\partial A}{\partial t}, \quad B = \nabla \times A$$

are gauge invariant.

2.2 (a) Define the electromagnetic field strength tensor $F^{\mu\nu}$ in terms of the gauge field A^μ.

(b) Hence prove the Bianchi identity

$$\partial^\alpha F^{\mu\nu} + \partial^\nu F^{\alpha\mu} + \partial^\mu F^{\nu\alpha} = 0.$$

(c) Show that the Bianchi identity implies the Maxwell equations

$$\nabla \times E = -\frac{\partial B}{\partial t}, \quad \nabla \cdot B = 0.$$

2.3 (a) Explain why the set of rotations in two spatial dimensions forms a group.

(b) Is this a discrete group or a Lie group? If the latter, how many generators do you expect?

(c) Explain why the matrices

$$U(\theta) = \begin{pmatrix} \cos\theta & -\sin\theta \\ \sin\theta & \cos\theta \end{pmatrix}$$

form a representation of the group.

(d) Show that an infinitesimal rotation can be written as

$$I + i\theta T + \mathcal{O}(\theta^2),$$

where I is the identity matrix, and

$$T = \begin{pmatrix} 0 & i \\ -i & 0 \end{pmatrix}.$$

(e) Hence show that the matrix $U(\theta)$ can be written as

$$U = \exp[i\theta T],$$

and interpret this result.

2.4 A possible choice of generators for SU(3) is the set of so-called Gell-Mann matrices

$$T^a = \frac{\lambda^a}{2},$$

where the matrices $\{\lambda^a\}$ are given explicitly by

$$\lambda^1 = \begin{pmatrix} 0 & 1 & 0 \\ 1 & 0 & 0 \\ 0 & 0 & 0 \end{pmatrix}, \quad \lambda^2 = \begin{pmatrix} 0 & -i & 0 \\ i & 0 & 0 \\ 0 & 0 & 0 \end{pmatrix}, \quad \lambda^3 = \begin{pmatrix} 1 & 0 & 0 \\ 0 & -1 & 0 \\ 0 & 0 & 0 \end{pmatrix},$$

$$\lambda^4 = \begin{pmatrix} 0 & 0 & 1 \\ 0 & 0 & 0 \\ 1 & 0 & 0 \end{pmatrix}, \quad \lambda^5 = \begin{pmatrix} 0 & 0 & -i \\ 0 & 0 & 0 \\ i & 0 & 0 \end{pmatrix}, \quad \lambda^6 = \begin{pmatrix} 0 & 0 & 0 \\ 0 & 0 & 1 \\ 0 & 1 & 0 \end{pmatrix},$$

$$\lambda^7 = \begin{pmatrix} 0 & 0 & 0 \\ 0 & 0 & -i \\ 0 & i & 0 \end{pmatrix}, \quad \lambda^8 = \frac{1}{\sqrt{3}}\begin{pmatrix} 1 & 0 & 0 \\ 0 & 1 & 0 \\ 0 & 0 & -2 \end{pmatrix}.$$

(a) For an arbitrary Lie algebra, the generators should satisfy

$$\text{tr}[T^a T^b] = T_R \delta^{ab};$$
$$\sum_a T^a T^a = C_R I,$$

where I is the identity matrix. Find the values of T_R and C_R in this case.

(b) What is the value of the structure constant f^{123}?

2.5 (a) Show that any three matrices $\mathbf{X}$, $\mathbf{Y}$ and $\mathbf{Z}$ obey

$$[X, [Y, Z]] + [Z, [X, Y]] + [Y, [Z, X]] = 0.$$

(b) From the definition of the structure constants

$$[\mathbf{t}^a, \mathbf{t}^b] = if^{abc}\mathbf{t}^c,$$

show that they obey the Jacobi identity

$$f^{abc}f^{cde} + f^{aec}f^{cbd} + f^{adc}f^{ceb} = 0.$$

(c) Hence, show that the matrices

$$(T^a)_{bc} = -if^{abc}$$

obey the Lie algebra

$$[\mathbf{T}^a, \mathbf{T}^b] = if^{abc}\mathbf{T}^c.$$

2.6 The covariant derivative acting on a quark field in QCD transforms according to

$$\mathbf{D}_\mu\Psi(x) \rightarrow \mathbf{U}(x)\mathbf{D}_\mu\Psi(x)$$

under a gauge transformation, where $\mathbf{U}(x)$ is the gauge transformation matrix.

(a) Show that this implies

$$\mathbf{D}_\mu \rightarrow \mathbf{U}\mathbf{D}_\mu\mathbf{U}^{-1}.$$

(b) Hence show that if

$$\mathbf{D}_\mu = \partial_\mu + ig_s\mathbf{A}_\mu,$$

then one must have

$$\mathbf{A}_\mu \rightarrow \mathbf{U}\mathbf{A}_\mu\mathbf{U}^{-1} + \frac{i}{g_s}(\partial_\mu\mathbf{U})\mathbf{U}^{-1}.$$

(c) Show that the field-strength tensor defined by

$$ig_s\mathbf{F}_{\mu\nu}\Psi(x) = \left[\mathbf{D}_\mu, \mathbf{D}_\nu\right]\Psi(x)$$

transforms according to

$$\mathbf{F}_{\mu\nu} \rightarrow \mathbf{U}(x)\,\mathbf{F}_{\mu\nu}\,\mathbf{U}^{-1}(x).$$

(d) Hence show that the kinetic Lagrangian term

$$\mathcal{L} = -\frac{1}{2}\text{tr}\left[\mathbf{F}_{\mu\nu}\,\mathbf{F}^{\mu\nu}\right]$$

is gauge invariant.

2.7 Under a boost with speed v in the z direction, the Lorentz transformation for energy and momentum is

$$E' = \gamma(E - \beta p_z)$$
$$p'_x = p_x$$
$$p'_y = p_y$$
$$p'_z = \gamma(-\beta E + p_z),$$

where $\gamma = (1 - v^2)^{-1/2}$ and $\beta = v$ (in natural units!). Show explicitly that the phase-space measure

$$\frac{d^3 p}{(2\pi)^3 2E}$$

is invariant under this transformation.

2.8 For two incoming collinear particles, one may parametrise momenta according to

$$p_1^\mu = (E_1, 0, 0, |p_1|), \quad p_2^\mu = (E_2, 0, 0, -|p_2|).$$

Hence show that the combination

$$4(E_1|p_2| + E_2|p_1|)$$

that occurs when calculating cross-sections is equal to the Lorentz invariant flux factor

$$F = 4\left[(p_1 \cdot p_2)^2 - m_1^2 m_2^2\right]^{1/2}.$$

2.9 (a) The gluon propagator in an axial gauge (with vector n^μ) is given by

$$D_{\mu\nu}^{ab}(q) = -\frac{i\delta^{ab}}{q^2}\left[\eta_{\mu\nu} + \frac{q^2 + n^2}{(n \cdot q)^2}q_\mu q_\nu - \frac{(n_\mu q_\nu + q_\mu n_\nu)}{n \cdot q}\right].$$

Show that, as $q^2 \to 0$, one has

$$q^\mu D_{\mu\nu}^{ab}(q) = n^\mu D_{\mu\nu}^{ab}(q) = 0,$$

and interpret this result.

(b) What happens in the Feynman gauge?

2.10 Draw all the leading order Feynman diagrams for the process

$$gg \to t\bar{t},$$

i.e. top pair production in gluon–gluon collisions.

2.11 The Dirac matrices $\{\gamma^\mu\}$ are defined by the relation

$$\{\gamma^\mu, \gamma^\nu\} = 2\eta^{\mu\nu},$$

where $\eta^{\mu\nu}$ is the metric of Minkowski space. Prove the result

$$\gamma^\mu \slashed{a} \gamma_\mu = (2 - d)\slashed{a},$$

where $\slashed{a} \equiv a^\mu \gamma_\mu$, and d is the number of spacetime dimensions.

2.12 Consider the QED Lagrangian

$$\mathcal{L}_e = -\frac{1}{4} F^{\mu\nu} F_{\mu\nu} + \bar{\psi}(i\slashed{\partial} - m)\psi - e A_\mu \bar{\psi} \gamma^\mu \psi.$$

(a) What is the mass dimension of $\mathcal{L}_e$ in d spacetime dimensions?

(b) By considering the kinetic term for the gauge field A_μ, show that this has mass dimension

$$[A_\mu] = \frac{d - 2}{2}.$$

(c) By considering the mass term for the electron field, show that the field itself has mass dimension

$$[\psi] = \frac{d - 1}{2}.$$

(d) Hence show that the electromagnetic coupling constant has mass dimension

$$[e] = \frac{4 - d}{2}.$$

2.13 In QED, the bare coupling e_0 is related to its renormalised counterpart e_R by

$$e_0 = \mu^\epsilon e_R Z_e, \quad Z_e = \left[1 + \frac{e_R^2}{24\pi^2}\left(\frac{1}{\epsilon} + c_e \right) + \mathcal{O}(e_R^4) \right],$$

using dimensional regularisation in $4 - 2\epsilon$ dimensions, where μ is the dimensional regularisation scale.

(a) Explain why there is an arbitrary constant c_e in the renormalisation factor Z_e.

(b) Explain why the bare coupling obeys the equation

$$\mu \frac{de_0}{d\mu} = 0.$$

(c) Show that this implies the equation

$$\epsilon + \frac{1}{e_R} \frac{de_R}{d \ln \mu} + \frac{1}{Z_e} \frac{dZ_e}{d \ln \mu} = 0.$$

(d) Show that at leading order in the renormalised coupling, the result of part (c) implies

$$\frac{de_R}{d \ln \mu} = -\epsilon e_R.$$

(e) Use the results of parts (c) and (d) to show that at NLO one obtains (in four spacetime dimensions)

$$\frac{dZ_e}{d\ln\mu} = -\frac{e_R^2}{12\pi^2},$$

and hence

$$\frac{de_R}{d\ln\mu} = \beta(e_R), \quad \beta(e_R) = \frac{e_R^3}{12\pi^2}.$$

(f) By integrating this equation between two scales $\mu = Q_0$ and $\mu = Q_1$, show that a solution is given by

$$e_R^2(Q_1) = \frac{e_R^2(Q_0)}{1 - \frac{e_R^2(Q_0)}{6\pi}\ln\left(\frac{Q_1}{Q_0}\right)},$$

and interpret this result.

2.14 For the top pair production process of problem 2.10, what renormalisation scale would you choose?

2.15 Which of the following observables is infrared safe?

   (i) The number of quarks and gluons in the final state of a given process.
   (ii) The total cross-section for a process, summed over any number of final state partons.
   (iii) The cross-section for a fixed number of partons, where these are required to be separated by a minimum distance ΔR in the (y, ϕ) plane.
   (iv) The distribution of the number of final state jets.

2.16 The squared amplitude (averaged over colours and spins) for the process

$$q(p_1) + \bar{q}(p_2) \rightarrow e^-(p_3) + e^+(p_4) + g(k)$$

is given by

$$\overline{|A_{q\bar{q}g}|^2} = \frac{C_F}{N_c}\frac{2e^4Q_q^2g_s^2}{p_1\cdot k\, p_2\cdot k}\frac{(p_1\cdot p_3)^2 + (p_2\cdot p_4)^2 + (p_2\cdot p_3)^2 + (p_1\cdot p_4)^2}{p_3\cdot p_4}.$$

(a) By introducing the Sudakov decomposition

$$k^\mu = (1-z)p_1^\mu + c_2 p_2^\mu + k_T^\mu,$$

where $k_T\cdot p_1 = k_T\cdot p_2 = 0$, show that c_2 is such that

$$k^\mu = (1-z)p_1^\mu + \frac{k_T^2}{s(1-z)}p_2^\mu + k_T^\mu, \quad k_T^\mu = (0, \mathbf{k_T}, 0), \quad s = (p_1 + p_2)^2.$$

(b) What is the interpretation of the limit $k_T \to 0$? Show that in this limit, one has

$$\overline{|\mathcal{A}_{q\bar{q}g}|^2} \xrightarrow{k_T \to 0} \frac{2g_s^2 C_F}{k_T^2} \frac{(1+z^2)}{z} \overline{|\mathcal{A}(zp_1, p_2)|^2},$$

where $\mathcal{A}(p_1, p_2)$ is the amplitude for the LO process

$$q(p_1) + \bar{q}(p_2) \to e^-(p_3) + e^+(p_4),$$

and is such that

$$\overline{|\mathcal{A}(p_1, p_2)|^2} = \frac{2e^4 Q_q^2}{N_c} \frac{(p_1 \cdot p_3)^2 + (p_1 \cdot p_4)^2}{(p_1 \cdot p_2)^2}.$$

Interpret the variable z.

(c) Explain why the three-body phase-space for the process with the additional gluon can be written as

$$\int d\Phi^{(3)} = \int \frac{d^3 k}{(2\pi)^3 2|k|} \int d\tilde{\Phi}^{(2)},$$

where the two-body factor on the RHS is the phase-space for the LO process, but with $p_1 \to p_1 - k$.

(d) Hence show that the partonic cross-section including radiation can be written as

$$\hat{\sigma}_{q\bar{q}g} \xrightarrow{k \| p_1} \int_0^1 dz \int \frac{d|k_T|^2}{|k_T|^2} \hat{P}_{qq}(z) \hat{\sigma}_{q\bar{q}}(zp_1, p_2),$$

where $\hat{\sigma}(p_1, p_2)$ is the cross-section for the LO process, and

$$\hat{P}_{qq}(z) = \frac{\alpha_S C_F}{2\pi} \frac{(1+z^2)}{1-z}.$$

(e) Explain why this result is divergent as one integrates over $|k_T|^2$, and how this divergence can be removed.

(f) Explain why the remaining result is divergent as $z \to 1$, and how this divergence can also be removed!

2.17 For the unregularised DGLAP splitting functions $\{\hat{P}_{ij}\}$, explain why

$$\hat{P}_{qq}(z) = \hat{P}_{gq}(1-z),$$

and also why

$$\hat{P}_{qg}(z) = \hat{P}_{qg}(1-z), \quad \hat{P}_{gg}(z) = \hat{P}_{gg}(1-z).$$

2.18 The bare PDFs are related to the modified PDFs via

$$f_i^0(x_i) = f_i\left(x_i, \mu_F^2\right) - \sum_j \int_0^{\mu_F^2} \frac{d|k_T|^2}{|k_T|^2} \int_{x_i}^1 \frac{dz}{z} P_{ij}(z) f_j\left(\frac{x_i}{z}, |k_T|^2\right).$$

(a) Explain the meaning of the factorisation scale μ_F.

(b) Explain why the bare PDFs must obey the conditions

$$\frac{df_i^0}{d\mu_F^2} = 0.$$

(c) Hence derive the DGLAP equations

$$\mu_F^2 \frac{\partial f_i\left(x_i, \mu_F^2\right)}{\partial \mu_F^2} = \sum_j \int_{x_i}^1 \frac{dz}{z} P_{ij}(z) f_j\left(\frac{x_i}{z}, \mu_F^2\right).$$

2.19 What factorisation scale would you choose for the top pair production process of problem 2.10?

2.20 Varying the renormalisation and factorisation scales around some default choice $\mu_R = \mu_F = \mu_0$ is a common procedure used to estimate the effect of neglected higher-order corrections. Should one always choose a common scale, or vary them completely independently of each other?

IOP Publishing

Practical Collider Physics

Andy Buckley, Christopher White and Martin White

Chapter 3

From theory to experiment

Putting together everything we have learned so far, we have the following recipe for calculating cross-sections for hadron colliders, at arbitrary orders in perturbation theory:

1. Take your desired beyond the Standard Model (BSM) theory or the SM, and calculate Feynman diagrams with incoming (anti-) quarks and gluons, to get the partonic cross-section $\hat{\sigma}_{ij}$;
2. Remove UV divergences via renormalisation, and fix the renormalisation scale μ_R^2;
3. Absorb initial state collinear divergences in the parton distribution functions (PDFs), and set the factorisation scale μ_F^2;
4. Combine $\hat{\sigma}_{ij}$ with the PDFs, to get the hadronic cross-section.

To get *differential* rather than total cross-sections, we can leave some of the phase-space integrals undone. As we have seen, all of this constitutes a huge amount of work. And unfortunately, the results of such calculations look almost nothing like what comes out of a particle accelerator! First of all, the order in perturbation theory to which we are able to calculate may be insufficient for some observables that are measured. This may apply to total cross-sections, or more subtly to certain bins or groups of bins in differential quantities. Related to this, the incoming/outgoing partons radiate a large number of other partons—many more than can be calculated by using exact Feynman diagrams. Secondly, free partons do not exist, but are confined within colour-singlet hadrons. Thus, the (anti-)quark and gluon radiation must somehow clump together to form hadrons, of which there are many different types.

These typically form well-collimated **jets** of particles, which are measurable in the silicon tracker and hadronic calorimeters. Both of these effects are shown schematically in Figure 3.1, which is still a crude simplification, as we have also ignored the remnants of the colliding protons. Some of the remaining particles go down the

doi:10.1088/978-0-7503-2444-1ch3

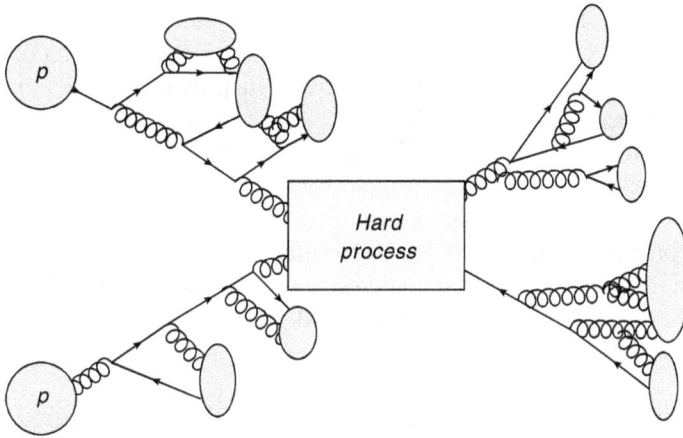

Figure 3.1. (Anti)-quarks and gluons emerge from incoming protons, and radiate before colliding in a **hard process**. Outgoing partons also radiate, and the radiation combines to make colour singlet hadrons (shown as blue blobs).

beam-pipe, but there remains a messy **underlying event** made of hadrons originating from the break-up of the protons. A further problem is that there are multiple protons in each colliding bunch at the LHC, so that scattering events between pairs of protons can overlap. This is called **pile-up**, and the effects increase with the luminosity (i.e. the number of protons per bunch). Finally, the detectors have a finite resolution, and non-trivial coverage (e.g. gaps, cracks etc). Thus, if we are really serious about comparing theory with data, we have to face a huge range of complications!

It is clearly not possible to calculate all of these effects from first principles. Typically, the comparison of theory with data from collider experiments involves:

(i) Inclusion of partial (but exact) analytic results at higher orders in perturbation theory;

(ii) Clever algorithms to estimate higher-order perturbative quantum chromo-dynamics (QCD);

(iii) Phenomenological models to capture non-perturbative phenomena.

Often the results of the second and third of these activities are included in general purpose software packages called **(Monte Carlo) event generators**, where the 'Monte Carlo' label refers to the statistical technique used in the underlying computations. These programs mimic scattering events at colliders, producing **simulated events** consisting of a list of stable particles (e.g. hadrons, photons, leptons...), with associated four-momenta. On the experiment side, the predictions can be corrected for complicated effects, and as a method for comparing data with theory. This often involves running the event generators mentioned above, and thus it is important to understand the systematic uncertainties that are implied by their use. There are various 'levels' at which we can try to compare theory with data:

- Parton level: This consists of having theory calculations with quarks and gluons in the final state, where additional radiation has been corrected for in the data. Furthermore, unstable particles such as top quarks and W bosons may have been reconstructed.
- Particle/hadron level: Such calculations have hadrons or jets in the final state, and all unstable particles will have been decayed. There may also be some modelling of the underlying event.
- Detector/reconstruction level: Here, the theory results differ from particle level in including an additional **detector simulation**, that takes finite resolution effects into account, and possibly also detailed aspects of particular detectors (e.g. where the gaps are).

Clearly, the further down this list we go, the closer a theory calculation gets to what is seen by an experiment. Many physics analysis studies are indeed performed at reconstruction level, particularly searches for new physics as will be discussed in Chapters 9 and 10. But as we will discuss in Chapter 11 there are also ways of taking the 'raw' data, and correcting it back to particle level—or for some purposes parton level, at the cost of some model dependence—so that it can be directly compared with a simpler theory calculation. There are many different uncertainties and ambiguities involved in this process, and you will often see detailed discussions and/or arguments at conferences about what assumptions have gone into a particular experimental analysis or theory calculation, and whether they are valid.

The aim of this chapter is to give a first taste of some of the methods and algorithms involved in state-of-the-art tools that people actually use at colliders like the LHC, and the theory calculations that underpin them. Although we will see some calculations, we will take a mainly schematic approach, that will hopefully be sufficient to help you decide what tool should be used for what purpose.

3.1 Fixed-order perturbation theory

The calculations we have seen so far correspond to parton level (i.e. Feynman diagrams with external quarks/gluons), up to a given order in α_S and possibly other couplings. This is also called **fixed-order perturbation theory**, which distinguishes it from approaches in which we might, for example, sum up certain contributions to all orders in perturbation theory. Although there are no hadrons in the fixed-order final state, this does not actually matter for observables such as total cross-sections, where we sum over all possible final states. Each final state quark or gluon must ultimately end up in some hadron, so whether we sum over final state partons or hadrons is irrelevant. For differential cross-sections, we must be careful only to compare to data that has been corrected back to the parton level.

In terms of what people actually use, leading order (LO) is a very crude approximation, and next-to-leading order (NLO) is the current state-of-the art for many processes. Some processes go beyond this: see table 3.1 for some examples. Note that it is typically SM processes that we want high precision calculations for, rather than BSM processes. For a given new physics signal, we must carefully

Table 3.1. The current highest-calculated perturbative orders for some important processes at the Tevatron and LHC.

Process	State of the art
$pp, p\bar{p} \rightarrow H$	N^3LO
$pp, p\bar{p} \rightarrow t\bar{t}$	NNLO
Drell–Yan	N^3LO
W/Z boson + $\leqslant 4$ jets	NLO

remove SM events that can fake the signal, which are known as **background processes**, or simply **backgrounds**. Typically, we have to understand backgrounds a lot more precisely than signals because they occur in greater quantities: uncertainties in the prediction, if faithfully accounted for, can hence drown out analysis sensitivity to deviations in data from the SM expectations.

The calculation of QCD LO processes has to a large extent been automated. For example, the publicly available codes MadGraph_aMC@NLO (currently in version 5) and Sherpa allow users to specify the scattering process they want to calculate, after which the program can generate events for this process, and calculate total (or differential) cross-sections. A variety of codes also exist for calculating events at NLO for specific processes, such as MCFM, BlackHat, Rocket, and VBFNLO. Another code, Prospino, can calculate NLO total cross-sections for supersymmetric particle processes. NLO automation of arbitrary process types is also well advanced, with both Sherpa and MadGraph_aMC@NLO able to automatically generate events for almost *any* scattering process, including a wide range of BSM models. Not all codes for fixed-order calculations are publicly available, which is particularly the case for calculations at QCD NNLO and beyond. In this case, one can often contact the authors of a particular research paper with requests for numbers, if carrying out a particular analysis.

In practice, all fixed-order calculations must be truncated at some low power of the coupling constant, due to obvious limitations of computing power. It is obviously desirable to include as many orders as possible, but the question naturally arises of whether it is possible—or indeed necessary—to try and include additional information at higher orders, even if this may be incomplete. This is the subject of the following section.

In discussing higher-order corrections, we have mainly focused on those involving the QCD coupling constant α_S. However, there will also be electroweak (EW) corrections, which are becoming increasingly important at the frontier of precision theory predictions. Although the electroweak coupling constants are much smaller than the strong coupling constant, if one calculates to sufficiently high order in α_S, then subleading strong interaction effects can be comparable in size to NLO EW effects, meaning that the latter also have to be included. Automation of these computations is proceeding along similar lines to those for QCD.

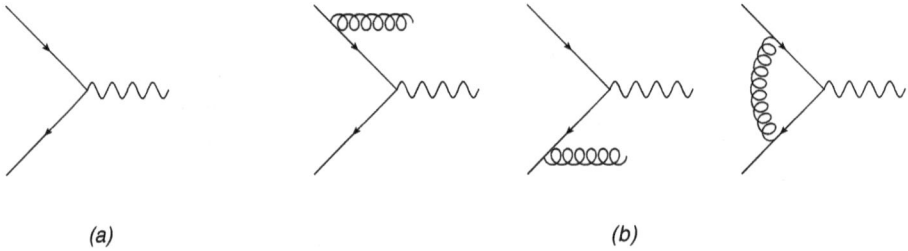

Figure 3.2. (a) LO diagram for DY production of an off-shell photon; (ii) some contributions to the NLO cross-section in the $q\bar{q}$ channel.

3.2 Resummation

There are certain cases in which perturbation theory becomes unreliable. To see how this might happen, let us recall that the perturbation expansion for a general observable X will look something like equation (2.195)[1], where the coefficients of powers of the coupling constant themselves depend on the momenta entering the scattering process. There may then be regions of momentum space where the coefficients of the perturbation series become large, even though the coupling itself may be small. However, reliability of a perturbation expansion relies on both having a small parameter to expand in, and well-behaved coefficients. If the latter criterion fails, we are in trouble!

Let us examine a classic example where this happens, using a process that we have already encountered in Section 2.15, namely Drell–Yan (DY) production of a lepton pair, accompanied by any amount of hadronic radiation. To simplify the discussion, we can further ignore the fact that the virtual photon decays to a lepton pair, and simply consider the process

$$q(p_1) + \bar{q}(p_2) \rightarrow \gamma^*(Q) + H, \tag{3.1}$$

where the notation γ^* implies an off-shell photon, and H denotes any amount of hadronic radiation. The LO Feynman diagram for this process is shown in Figure 3.2(a), and to discuss the cross-section it is conventional to define a dimensionless variable

$$z = \frac{Q^2}{\hat{s}}. \tag{3.2}$$

In words, this is the ratio of the photon invariant mass, and the squared partonic centre-of-mass energy, which can be loosely thought of as the fraction of the final state energy carried by the photon.

Armed with this definition, the differential cross-section corresponding to Figure 3.2(a) with a given quark flavour turns out to be[2]

---

[1] Although strictly we should also add a dependence on the factorisation scale μ_F for a partonic quantity.
[2] We have here quoted the cross-section in four spacetime dimensions. In $4 - 2\epsilon$ dimensions (as would be used in dimensional regularisation at higher orders), one must insert an addition factor of $(1 - \epsilon)$ in equation (3.3).

$$\frac{d\hat{\sigma}_{q\bar{q}}^{(0)}}{dz} = \sigma_0\,\delta(1-z), \quad \sigma_0 = \frac{e_q^2\pi}{3\hat{s}}, \tag{3.3}$$

where e_q is the electromagnetic charge of the quark. Note that this has dimensions of area as required. Furthermore, we should not be surprised about the occurrence of the Dirac delta function, which fixes z to be one: at LO there is only the vector boson in the final state and thus there is no option but for it to be carrying all the energy.

At NLO, there will be a number of different choices for the initial state particles (commonly referred to as different **partonic channels**). Let us focus on the $q\bar{q}$ channel (i.e. the only channel that occurs at LO), for which example Feynman diagrams are shown in Figure 3.2(b), comprising both real and virtual gluon radiation. Combining all contributions and removing all collinear singularities (in the $\overline{\text{MS}}$ scheme), one finds the following result for the differential cross-section:

$$\frac{1}{\sigma_0}\frac{d\sigma_{q\bar{q}}^{(1)}}{dz} \equiv K^{(1)}(z) = \frac{\alpha_S C_F}{2\pi}\left[4(1+z^2)\left(\frac{\ln(1-z)}{1-z}\right)_+ - 2\frac{1+z^2}{1-z}\ln(z)\right. \tag{3.4}$$

$$\left. +\delta(1-z)\left(\frac{2\pi^2}{3}-8\right)\right]. \tag{3.5}$$

We have here expressed the result in terms of the so-called **K-factor**, which divides out the normalisation of the LO cross-section. We see that this contains a single power of the strong coupling α_S as expected at NLO, dressed by a function of z, where the latter may now vary in the range

$$0 \leqslant z \leqslant 1.$$

The physical reason for this is that gluon radiation may carry away energy, so that the virtual photon carries less than the total energy in the final state. For most values of z, the value of the K-factor will be small, constituting a well-behaved correction to the LO process. However, there is an issue for extremal values of z. Examining, for example, the behaviour as $z \to 1$, one finds

$$K^{(1)}(z) \xrightarrow{z\to 1} \frac{4\alpha_S C_F}{\pi}\left(\frac{\ln(1-z)}{1-z}\right)_+ + \cdots, \tag{3.6}$$

which threatens to diverge. Although there is no formal divergence in the integrated cross-section (i.e. the divergence is regularised by a plus distribution, similar to the behaviour of the splitting functions in Section 2.16), it is still the case that this term may lead to a numerically large correction to the total cross-section. If this is true, the validity of perturbation theory appears to be in doubt: a perturbation expansion only makes sense if the coefficients of the parameter we are expanding in are sufficiently small, so that successive terms in perturbation theory converge to a well-defined result. That does not appear to be the case here, and we should at the very least go to one higher order in perturbation theory, and check what happens.

Unfortunately, higher orders in perturbation theory show that the problem gets worse! Indeed, the asymptotic limit of the K-factor at $\mathcal{O}(\alpha_S^n)$ has the form

$$K_{q\bar{q}}^{(n)} \to z \to 12\left(\frac{2\alpha_S C_F}{\pi}\right)^n \frac{1}{(n-1)!}\left(\frac{\log^{2n-1}(1-z)}{1-z}\right)_+ + \ldots \tag{3.7}$$

What we see is that as the power of α_S increases, so does the power of the logarithm of $(1-z)$, which more than compensates for the smallness of the coupling. If we then try to find the total integrated cross-section for the process of equation (3.1), it appears as if all orders in perturbation theory are becoming equally important, so that truncation at a fixed order in the coupling is meaningless.

The solution to this problem is that one can indeed isolate the problem terms of equation (3.7) at arbitrary order in perturbation theory, and sum them up to all orders in the coupling constant. This is known as **resummation**, and is possible because we can often understand the origin of the most sizeable terms, such that their properties can be completely characterised. In the present case, for example, we know that the large logarithmic terms are associated with the kinematic limit $z \to 1$, corresponding to the virtual photon carrying all the energy in the final state. This in turn implies that any additional real (gluon or quark) radiation must be soft i.e. carrying negligibly small four-momentum. From Section 2.15, we know that this is a limit in which infrared singularities occur, which will also be the case when virtual radiation—that is, either soft or collinear to one of the incoming partons—is included. Upon combining all virtual and real radiation, and absorbing initial-state collinear singularities into the parton distribution functions, the formal infrared divergences will cancel. However, one is left with the numerically large terms of equation (3.7), which can be thought of as echoes of the fact that infrared (IR) singularities were once present!

The explanation of the previous paragraph—relating large logarithms to soft and/ or collinear radiation—suggests that this behaviour is not limited to DY production, but will be present in many different scattering processes. Another example is **deep inelastic scattering** at an e^-p collider, in which an off-shell photon combines with a parton from the proton to produce one or more partons in the final state:

$$\gamma^*(q) + f(p) \to \sum_i f_i(k_i). \tag{3.8}$$

Here f and $\{f_i\}$ denote general initial- and final-state partons, respectively, and a LO Feynman diagram is shown in Figure 3.3. In this process, it is conventional to define the so-called **Björken variable**

$$x_B = \frac{Q^2}{2p \cdot q}, \quad Q^2 = -q^2 > 0, \tag{3.9}$$

and one then finds that the differential cross-section in this variable contains terms similar in form to equation (3.7), but with z replaced by x_B, and different constants in front of each logarithm. The kinematic limit giving rise to such large contributions

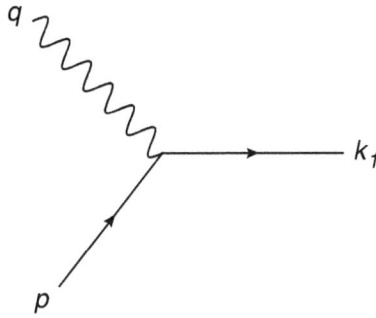

Figure 3.3. LO diagram for the deep inelastic scattering (DIS) process, in which a virtual photon scatters off a quark.

is now $x_B \to 1$, which one may clarify as follows. First, momentum conservation applied to the process of equation (3.8) at a given order in perturbation theory implies

$$p + q = \sum_{i=1}^{n} k_i.$$

Squaring both sides and implementing the definition of equation (3.9), one finds

$$\frac{Q^2(1 - x_B)}{x_B} = \sum_{i,j=1}^{n} |k_i||k_j|(1 - \cos\theta_{ij}),$$

where k_i is the three-momentum of final-state parton i, and θ_{ij} the angle between partons i and j. One now sees that the limit $x_B \to 1$ corresponds to real radiation that is either soft ($k_i \to 0$), or collinear to an existing parton leg ($\theta_{ij} \to 0$).

In general, large logarithms arising from the presence of soft and/or collinear radiation are called **threshold logarithms** (e.g. in the DY case, they corresponded precisely to the off-shell parton carrying all the energy, and thus being produced at threshold). They will arise in any scattering process containing one or more heavy or off-shell particles, and in all such cases we may define a dimensionless **threshold variable**

$$0 \leqslant \xi \leqslant 1,$$

which will be given by ratios of Lorentz invariants in the process of interest, and is such that $\xi \to 0$ at threshold. In the above examples, we had $\xi = 1 - z$ for DY production, and $\xi = 1 - x_B$ for DIS. Note that the definition of such a threshold variable is not unique, and different choices for a given process may exist in the literature[3]. However, once a choice is fixed, the differential cross-section in the threshold variable ξ has the following asymptotic form as $\xi \to 0$:

---

[3] In the above examples of DY and DIS, one may clearly reparametrise the variables z and x_B in such a way that the leading behaviour in the limit z, $x_B \to 1$ remains unchanged.

$$\frac{d\sigma}{d\xi} \sim \sum_{n=0}^{\infty} \alpha_S^n \sum_{m=0}^{2n-1} \left[c_{nm}^{(-1)} \left(\frac{\log^m \xi}{\xi} \right)_+ + c_n^{(\delta)} \delta(\xi) + \cdots \right]. \tag{3.10}$$

We have here neglected an overall normalisation factor, that may itself contain additional factors of (non-QCD) couplings. At each order in the coupling constant, there is a series of large logarithmic terms, each of which is regularised as a plus distribution, and an additional delta function contribution. At any order in perturbation theory, the highest power of the logarithm is $m = 2n - 1$, and such terms are referred to as leading logarithms (LL), as they are usually the ones that are the most numerically significant. Terms which have $m = 2n - 2$ are referred to as **next-to-leading logarithms (NLL)**, and so on. As the order of perturbation theory increases, one becomes sensitive to progressively subleading logarithms (N^mLL).

Given that the NLL terms should be numerically smaller than the LL terms as $\xi \to 0$ (and similarly for the NNLL terms etc), a new form of perturbation expansion suggests itself in the threshold limit: one can sum up the LL terms to all orders in α_S, followed by the NLL terms, and so on. This is orthogonal to conventional perturbation theory, in which one limits oneself to fixed orders in the coupling constant, but includes *all* behaviour in z. Indeed, the difference between these approaches can be understood as a different choice of small parameter: fixed-order perturbation theory (LO, NLO, $\cdots$) involves neglecting terms which are suppressed by powers of the coupling constant, whereas resummed perturbation theory (LL, NLL, $\cdots$) involves neglecting terms which are suppressed by inverse logarithms of the threshold variable, and keeping only particular enhanced terms to all orders in α_S. The latter thus constitutes a 'reordering' of the perturbation expansion, and this is what the prefix 're-' in 'resummation' is intended to signify. Of course, neither of these approaches is ideal. In most situations of interest at colliders, one is faced with the need to calculate a particular total or differential cross-section in perturbation theory, but not necessarily in a kinematic regime where the threshold-enhanced terms are most important. The most common approach is then to use fixed-order perturbation theory to as high an order as possible, and then to supplement the result with additional threshold terms to a given logarithmic order (i.e. where the latter are included to all orders in the coupling)[4]. One hopes that the latter contributions will help improve the estimate for the cross-section, and you will often see such results quoted in experimental analysis papers. For example, if a given theory calculation is described as at NNLO + NNLL order, this means that a fixed-order calculation at next-to-next-to-leading order in $\mathcal{O}(\alpha_S)$ (with full kinematic dependence) has been supplemented with additional terms at arbitrarily high orders, up to and including the third largest tower of logarithms.

There are a number of approaches for resumming threshold logarithms, all of which ultimately make use of the fact that the properties of soft and collinear radiation are, in a sense, independent of the underlying process: the fact that the (transverse) momentum of the radiation is negligible in the threshold limit means, in

---

[4] In combining the fixed-order and resummed results, one must take care not to count any radiation twice.

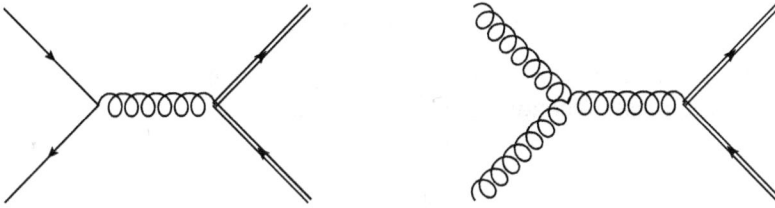

Figure 3.4. LO Feynman diagrams for top quark pair production in QCD.

position space, that the radiation has an infinite Compton wavelength. Thus, it cannot resolve the details of the underlying scattering process, and should somehow factor off. We have seen this already in Section 2.15, where we used the factorising property of collinear radiation to set up the language of parton distribution functions. This **factorisation** allows certain equations to be derived that describe the structure of infrared divergences to all orders in perturbation theory. Finally, the fact that IR divergences are intimately related to the large logarithms appearing in perturbation theory allow the latter to be summed up. In recent years, an approach called **soft collinear effective theory** (SCET) has become popular. It provides a way to rewrite the Lagrangian of QCD to contain separate fields for soft and collinear gluons or (anti-)quarks, and can be used to devise similar resummation formulae to those arising from more traditional approaches. A full discussion of how to derive and apply these resummation approaches is sadly beyond the scope of this book, but a number of excellent review articles and textbooks are available.

Above, we have examined the resummation of large contributions in differential cross-sections with respect to the threshold variable. This is not the only place where you may see threshold resummation being used:

- Upon including higher-order terms in partonic cross-sections, one must also resum corresponding contributions in the Dokshitzer–Gribov–Lipatov–Altarelli–Parisi (DGLAP) splitting functions which describe how the partons evolve. To see this, recall that the threshold logs are partly related to collinear radiation, and the latter will lead to additional singularities at higher orders in perturbation theory, that have to be absorbed into the partons. A first sign of this is that the LO DGLAP splitting functions of equations (2.219)–(2.222) indeed contain large logarithms of the form of equation (3.7), indicating a need for resummation.
- Historically, a number of so-called **event shapes** have been defined, that characterise the topology of different scattering events (e.g. thrust, sphericity etc). These often need to be resummed at extremal values, in order to meaningfully compare theory with data.
- All of the coloured particles in the above examples of threshold resummation were massless, which did not have to be the case. Indeed, a good example of a process with massive coloured particles is that of top quark pair production, whose leading-order Feynman diagrams are shown in Figure 3.4. In this case, two different choices of threshold variable are common in the literature. The first is

$$s_4 = s + t + u - 2m_t^2,$$

where (s, t, u) are the Mandelstam invariants defined in equation (2.156), and m_t the top mass. Secondly, there is the parameter

$$\beta = \sqrt{1 - \frac{4m_t^2}{\hat{s}}}.$$

One may show that both of these tend to zero at threshold, although their behaviour away from the threshold limit differs.

The effect of threshold resummation is usually two-fold. Firstly, it tends to change the central value of a given cross-section. Secondly, it typically decreases the theoretical uncertainty. As described in Sections 2.11 and 2.15, one way to estimate the effect of neglected higher-order contributions is to vary the renormalisation and factorisation scales (independently) around some default scale. The process of resummation includes additional terms that depend on these scales, such that the effect of this scale variation is reduced.

Above we have talked about threshold resummation. However, the idea of resummation is much broader than this, and similar terminology is used to describe the inclusion of any type of enhanced contribution at higher orders. For example, you may have noticed that the splitting functions of equations (2.219)–(2.222), and the DY cross-section of equation (3.5), also become enhanced in the limit of $z, x_B \to 0$ (i.e. the opposite of the threshold limit $z, x_B \to 1$). This enhancement turns out to be related to the so-called **Regge limit** in which the (partonic) centre-of-mass energy becomes much larger than the momentum transfer in a given scattering process, and a large literature exists on how to resum such contributions. Furthermore, for some processes and observables (particularly more differential ones), there may be a much cleaner separation between the kinematic regime where resummation becomes important, and the regime where fixed-order perturbation theory is sufficient. An example of this is shown in Figure 3.5, which shows the transverse momentum $p_T^{\ell\ell}$ of lepton pairs in DY production. In fixed-order perturbation theory, this is fixed to be zero at LO, and at arbitrary fixed order in α_S, there is a divergence as $p_T^{\ell\ell} \to 0$, due to additional radiation becoming soft and/ or collinear with the incoming parton beams. Only upon resumming higher-order logarithms involving $p_T^{\ell\ell}$ does one find a sensible result that matches the data. Whenever you see that a theory calculation has been quoted as being at $N^m LL$ order, it is important to remember that the logarithms being referred to may be different in each case, and so you should carefully check what is being resummed!

In many cases, it is not actually necessary to formally resum (i.e. sum to all orders) enhanced contributions. Instead, one may truncate resummed formulae at a fixed-order in α_S, which amounts to using the enhanced terms as an estimate of a higher fixed-order cross-section. This approach was used e.g. to estimate the top pair cross-section at NNLO, before the full result was known. Finally, we note that the ellipsis in equation (3.10) denotes terms that are suppressed by at least one power of

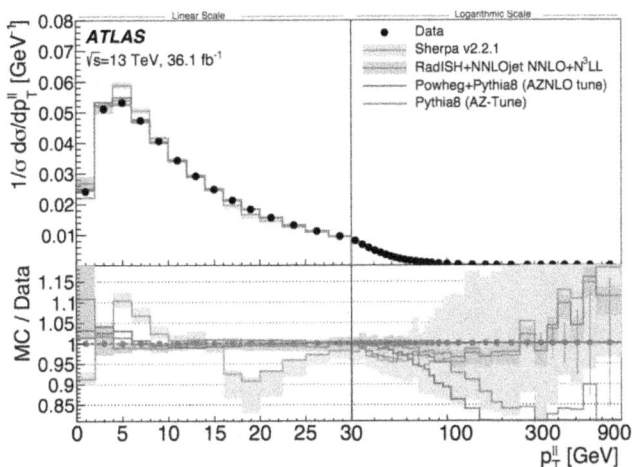

Figure 3.5. The transverse momentum of lepton pairs in DY production, as measured by the ATLAS collaboration. The green curve shows a fixed-order result at NNLO, matched with a resummed calculation at NNNLL order, which agrees well with the data. Reproduced with permisson from *Eur. Phys. J.* C **80** 616 (2020).

the threshold variable. These so-called **next-to-leading power corrections** (NLP corrections) are known to be numerically sizeable in some processes (e.g. Higgs-boson production). Whether or not they can be systematically understood in arbitrary scattering processes is still a matter of investigation, but the results may help to increase our estimates of higher-order cross-sections, where these are not available.

3.3 Parton showers

In the previous section, we saw that one can increase the precision of theory predictions by including higher-order terms in perturbation theory, for certain observables (e.g. total cross-sections, event shapes, differential distributions etc). However, this still does not achieve anything like what is shown in Figure 3.1, namely an accurate model of what scattering events in a collider actually look like. In order to do this, we must include the radiation of a large number of partons, which is much too computationally expensive to do exactly, even using automated tools such as MadGraph or Sherpa. Instead, one may use an ingenious computational algorithm called a **parton shower (PS)**, which was first developed in the 1980s. Over the years, PS algorithms have become highly sophisticated, with entire conferences devoted to the subject. Here, we will describe a simple version of the algorithm, and then briefly summarise currently available codes.

The starting point for PS algorithms is an observation we have already made, namely that radiation is enhanced if it is soft and/or collinear. For simplicity, we will first consider radiation from a final-state parton, and specifically the situation shown in Figure 3.6, in which a final-state parton i splits into partons j and k, carrying momentum fractions z and $(1 - z)$ of parton i, respectively. In the limit in which j

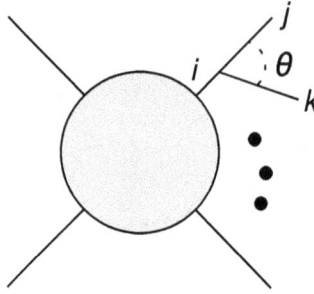

Figure 3.6. An amplitude for n partons, where one parton splits into two. All partons are shown as solid lines, which may be (anti-) quarks or gluons.

and k are collinear ($\theta \to 0$), the partonic differential cross-section for the $(n+1)$-parton process factorises:

$$\mathrm{d}\hat{\sigma}_{n+1} = \mathrm{d}\hat{\sigma}_n \sum_i \int_{k_{\min}}^{k_{\max}} \frac{\mathrm{d}|\boldsymbol{k}_T|^2}{|\boldsymbol{k}_T|^2} \int_{z_{\min}}^{z_{\max}} \frac{\mathrm{d}z}{z} P_{ji}(z). \tag{3.11}$$

Here $\boldsymbol{k}_T$ is the transverse momentum of parton k with respect to j, and P_{ij} the DGLAP splitting function associated with the splitting $i \to jk$, where we have to sum over all possible partons i that can lead to the given final state parton j. We have put a lower limit on the transverse momentum so that the integral converges, and the upper limit will depend on some physical energy scale in the process. Likewise, we have placed limits on the momentum fraction, which will also be dictated by kinematic constraints. Equation (3.11) can be derived similarly to equation (2.212), and clearly has a very similar form. The only difference is that there is no shift in kinematics associated with the n-parton process, due to the fact that the momentum fraction is associated with the final state partons, and not the internal line (in contrast to the case of initial state radiation).

The factorised form of equation (3.11) suggests that the extra parton k is somehow *independent* of the rest of the process, and indeed there is a nice hand-wavy explanation for this, that we already saw in the previous section when discussing threshold resummation. As the transverse momentum of parton k goes to zero, its Compton wavelength becomes infinite. Particles can only resolve things that are smaller than their Compton wavelength, and thus the collinear parton k cannot resolve the details of the underlying scattering process that produced it. It should thus factorise from the n-parton process, as indeed it does.

We have so far argued that the $(n+1)$-parton process factorises in terms of the n-parton one, but we can iterate this argument in the limit in which a whole load of emissions are collinear, as they will all be mutually independent. This gives us a way to approximate cross-sections with a large number of final state partons. Assume that we have some underlying hard scattering process that produces a set of well-separated partons in the final state. If these partons then radiate, we can describe this radiation exactly in the limit in which it is collinear with one of the original hard partons, using the above factorisation. In practice, extra radiation will have some

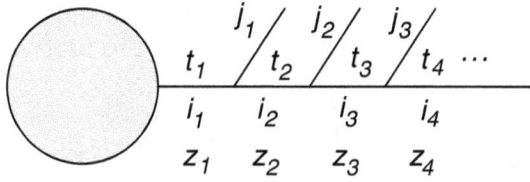

Figure 3.7. A parton emerges from a hard scattering process, and cascades by emitting partons of species $\{j_n\}$, changing its own identity into species $\{i_n\}$ in the process. At each step, the cascading parton has virtuality t_i, and a fraction z_i of its original momentum.

non-zero angle with the hard partons. However, we still know that radiation is *enhanced* in the collinear limit, so is more likely to be close to one of the hard partons. Then the collinear limit gives a reasonable approximation for the extra radiation, even if it is not always strictly collinear.

To see how to generate the extra radiation in practice, we can first note that equation (3.11), and the fact that cross-sections are related to probabilities, allows us to interpret the combination

$$\sum_i \int_{k_{\min}}^{k_{\max}} \frac{\mathrm{d}|k_T|^2}{|k_T|^2} \int_{z_{\min}}^{z_{\max}} \frac{\mathrm{d}z}{z} P_{ij}(z)$$

as the probability to emit a parton with momentum fraction

$$z_{\min} \leqslant z \leqslant z_{\max}$$

of parton i, and transverse momentum

$$k_{\min} \leqslant |k_T| \leqslant k_{\max}.$$

Instead of the transverse momentum, we can also write the above probability in terms of the Lorentz-invariant **virtuality** $t = p_i^2$ of parton i, given that

$$\int \frac{\mathrm{d}|k_T|^2}{|k_T|^2} = \int \frac{\mathrm{d}t}{t}.$$

If parton i emits several partons, it will have some momentum fraction z_i and virtuality t_i before each emission (where the initial momentum fraction will be $z_1 = 1$). Also, the species of the parton may change with each emission, leading to a **parton cascade**, as shown in Figure 3.7. The virtuality will decrease with each emission, as the cascading parton gets closer to being on-shell. Thus, the various virtualities in Figure 3.7 are ordered:

$$t_1 > t_2 > t_3 > \dots.$$

Here we have a situation in which the parton changes virtuality by emitting real, measurable radiation. There is another possibility, namely that the parton emits partons that are not resolvable (due e.g. to finite detector resolution). Between two given virtualities (t_0, t_1), there is then some nonzero probability that the parton emits *no resolvable partons*. This is sometimes called the **no-emission probability** in books and papers, but this is a tad misleading—the parton can still emit, as long as we do not resolve the emissions! In any case, let us denote this probablity by $\Delta(t_0, t_1)$. To find it, note that we can write

$$\Delta_i(t_0, t + \delta t) = \Delta_i(t_0, t)\left[1 - \sum_j \int_t^{t+\delta t} \frac{dt}{t} \int dz P_{ji}(z)\right]. \quad (3.12)$$

This is a large equation, which we can make sense of as follows. The left-hand side is (by definition) the probability that there are no resolvable emissions between virtualities t_0 and $t + \delta t$. The first factor on the right-hand side is the probability that there are no resolvable emissions between t_0 and t. Finally, the quantity in the square brackets is the probability of anything happening (i.e. unity) minus the probability of a single emission between t and $t + \delta t$, where we have used the probabilistic interpretation from above. Provided δt is small enough, we can have only a single emission in the time δt, so that the square brackets are the probability that there is no resolvable emission between t and $t + \delta t$. Putting everything together, equation (3.12) simply amounts to the statement

$$\begin{pmatrix}\text{Probability of no emission} \\ \text{in range } (t_0,\, t + \delta t)\end{pmatrix} = \begin{pmatrix}\text{Probability of no emission} \\ \text{in range } (t_0,\, t)\end{pmatrix}$$
$$\times \begin{pmatrix}\text{Probability of no emission} \\ \text{in range } (t,\, t + \delta t)\end{pmatrix}$$

which is certainly true, provided the emissions are independent. We can use equation (3.12) to solve for the function $\Delta_i(t_0, t)$. First, one may Taylor-expand both sides in δt to give

$$\Delta_i(t_0, t) + \delta t \frac{d\Delta_i(t_0, t)}{dt} + \dots = \Delta_i(t_0, t)\left[1 - \sum_j \frac{\delta t}{t} \int dz P_{ji}(z)\right],$$

which in turn implies

$$\frac{d\Delta_i(t_0, t)}{dt} = -\frac{1}{t}\left(\sum_j \int dz P_{ji}(z)\right)\Delta_i(t_0, t).$$

We have thus obtained a differential equation for $\Delta_i(t_0, t)$, and its solution is

$$\Delta_i(t_0, t) = \exp\left[-\sum_j \int_{t_0}^t \frac{dt'}{t'} \int_{z_{\min}}^{z_{\max}} dz P_{ji}(z)\right]. \quad (3.13)$$

This important result is called the **Sudakov factor**. As stated above, it describes the probability that a parton evolves from some virtuality t_0 to some other virtuality t without emitting any resolvable partons. Note that, although it looks like we have only used properties of real emissions in the above result, it actually includes virtual corrections too. In the derivation, we explicitly used the fact that the probabilty of no resolvable emission, plus the probability of resolved emissions, adds up to one. We need to add virtual *and* real diagrams together in quantum field theory to conserve probability, and thus using the fact that probabilities add up to one amounts to including some appropriate set of virtual contributions in the no-emission probability. We have not specified explicitly the limits (z_{min}, z_{max}) on the momentum fraction integral. These will depend on our criterion for what constitutes a resolvable emission, and can also depend on the virtuality.

A given parton cascade will be characterised by some set of values (z_i, t_i) denoting the momentum fraction and virtuality before each branching. Armed with the Sudakov factor of equation (3.13), we can then associate such a cascade with the following probability:

$$P_{i_1, i_2, \ldots, i_n}^{j_1, j_2, \ldots, j_n}(\{z_i\}, \{t_i\}) = \left[\prod_{m=1}^{n} \Delta_{i_m}(t_m, t_{m+1})\right]\left[\prod_{m=1}^{n} \int \mathrm{d}\tilde{z}_{m+1} P_{i_{m+1} i_m}(\tilde{z}_{m+1})\right]. \qquad (3.14)$$

Here the second square bracket includes the probability of each resolvable emission, where

$$\tilde{z}_{m+1} = \frac{z_{m+1}}{z_m}$$

is the fraction of the momentum of the mth parton that is carried by parton $(m + 1)$. The first square bracket represents the probability that the parton emits no resolvable emissions between each branching. Again, we have used the fact that the branchings can be treated independently, which also enters the derivation of the Sudakov factor. It follows that provided we can generate sets of numbers (z_i, t_i) weighted by the probability of equation (3.14), we can obtain a list of radiated four-momenta as if they had come from a cross-section calculation! This is quite a remarkable achievement, so let us say it again more slowly. We have shown that radiation in the collinear limit can be treated as multiple uncorrelated emissions, where we expect this to provide a reasonable approximation even if the radiation is not quite collinear with the hard partons emerging from the scattering process. Instead of having to calculate complicated Feynman diagrams to work out how the four-momenta of the radiated particles is distributed, we can simply generate pairs of numbers representing the momentum fraction and virtuality of each parton after emission, with some probability distribution (equation (3.14)). This distribution involves fairly simple functions, which can be implemented numerically to generate vast amounts of radiation quickly and easily.

In practice, the final state partons will never reach being completely on-shell, as free partons do not exist. This suggests that we put some lower cutoff t_0 on the allowed virtuality of a parton after branching. Then we can use the following **parton-shower algorithm** to simulate the radiation:

1. Start with a parton of virtuality t_1 and momentum fraction $z_1 = 1$.
2. At each step, generate t_{i+1} by solving

$$\Delta_i(t_{i+1}, t_i) = r_1,$$

where r_1 is a random number uniformly distributed on $[0, 1]$.

3. If $t_{i+1} < t_0$ then stop.

4. Otherwise, after choosing a species for the parton after branching, generate a momentum fraction $\tilde{z}_{m+1} = z_{m+1}/z_m$ by solving

$$\frac{\int^{z_{m+1}} dz' P_{i_{m+1} i_m}(z')}{\int dz' P_{i_{m+1} i_m}(z')} = r_2,$$

where r_2 is a second random number uniformly distributed on $[0, 1]$. Note that the splitting functions contain α_S, and it is conventional to take the virtuality t_m as the appropriate renormalisation scale at each branching. The denominator ensures that the splitting probabilities are normalised to 1 (upon summing over all possible splittings).

5. Generate an azimuthal angle ϕ for parton $(m + 1)$ (e.g. uniformly on $[0, 2\pi]$, or according to some other known distribution).

6. Repeat steps 2–4 for all partons (including emitted partons).

How understandable this is to you depends a bit on how familiar you are with computational methods for generating given probability distributions. It is an example of a **Monte Carlo algorithm**, and the result (after applying momentum conservation at each branching) is a list of four-momenta of emitted particles, and thus a simulated scattering event. This explains the name 'Monte Carlo event generators' for the available software programs that implement such algorithms.

The above algorithm corresponds to evolving downwards in virtuality, for a final state parton. One can use a similar algorithm for *incoming* partons, where one typically evolves *backwards* from the parton entering the hard scattering process, i.e. again from the most off-shell parton, to the most on-shell. For completeness, we note that the Sudakov factor for incoming partons is different to that for final-state partons, in that it includes a ratio of parton distribution functions:

$$\Delta_i(t_0, t) = \exp\left\{-\int_{t_0}^{t_1} dt' \sum_j \int_x^1 \frac{dx'}{x'} P_{ij}\left(\frac{x}{x'}\right) \frac{f_i(x', t')}{f_j(x, t')}\right\}.$$

If you are feeling adventurous you can in principle derive this from everything we have said above!

So far we have used the collinear limit to estimate the effect of radiation, which we justified by saying that radiation is anyway enhanced in this limit, so that we should be capturing the most likely radiated partons by doing so. In fact, this is not quite true. We saw in Section 2.15 that radiation is also enhanced when it is *soft* $k^\mu \to 0$, but not necessarily collinear with any outgoing partons. Such radiation is called **wide-angle soft**, to distinguish it from radiation that is both soft *and* collinear. For simplicity, we can consider a quantum electrodyamics example, in which a hard virtual photon branches into an electron–positron pair. The electron and positron can then radiate, as shown in Figure 3.8. The complete amplitude will be given by the sum of these diagrams, such that the squared amplitude includes the interference between them. If both angles satisfy $\theta_i \ll 1$, we can neglect this interference and treat the emissions as incoherent/independent, as we have discussed in detail above. However, if the angles are not small, then emissions from different legs do indeed interfere with each other, thus are not mutually independent. This then looks like a serious threat to our power shower algorithm, which assumed throughout (for both the resolved emissions and the Sudakov factor), that all emissions are indeed mutually independent! We could simply choose not to include wide angle soft radiation, but the fact that it is enhanced means that we are neglecting potentially large effects, so that our estimate of higher-order radiation is unlikely to be reliable.

It turns out that there is a remarkably simple fix: if we add and square both diagrams in Figure 3.8, the result looks like a sum of independent emissions confined to a cone around each external particle, as exemplified in Figure 3.9. There is a simple (although not exactly rigorous) quantum mechanical explanation: wide-angle photons with $\Delta\theta$ (labelled in the figure) necessarily have a larger Compton wavelength, so cannot resolve the individual electron and positron. Instead, they see the overall net charge of the system, which is zero. Thus, photon emission at large angles is suppressed, a phenomenon known as the **Chudakov effect**.

There is an analogous property in QCD, with the only difference being that the net colour charge of a system of partons can be non-zero. Consider a hard parton k that branches into two partons i and j, as shown in Figure 3.10(a). Each of the partons i and j can radiate, such that summing and squaring the diagrams produces radiation that looks like:

(i) Independent emissons from i and j at angles less than $\Delta\theta$;
(ii) Emission from the parton k at angles greater than $\Delta\theta$, where k has the total colour charge of i and j.

This effect is called **colour coherence**, and implies that soft radiation can be modelled by independent emissions that are strongly ordered in angle, as shown in Figure 3.10(b). Our original PS used virtuality as an ordering variable, but one may show that

$$\int \frac{dt}{t} = \int \frac{d\theta^2}{\theta^2},$$

Figure 3.8. An electron-positron pair emitting wide-angle soft radiation.

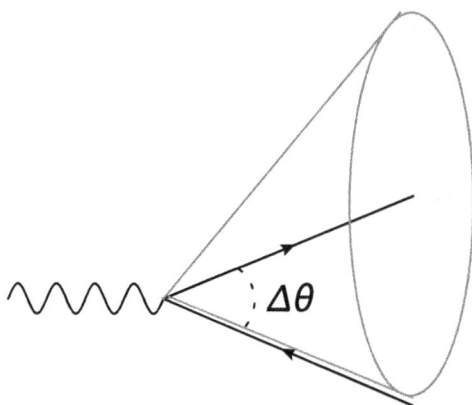

Figure 3.9. Taking interference of wide-angle soft radiation into account, the overall effect is that radiation is confined to a cone around each hard particle.

(a)

(b)

Figure 3.10. (a) A parton splits into two other partons, where the lines may represent (anti-) quarks or gluons; (b) multiple parton radiation can be modelled by independent emissions which are ordered in angle.

where t is the virtuality of the branching parton and θ the emission angle (i.e. the angular separation of the partons after the branching). Thus, by simply replacing t by θ in the above PS algorithm, so-called **angular ordering**, wide-angle soft radation is automatically included!

This is only one way of setting up PSs with correct soft/collinear behaviour. Another is to use an algorithm based on emissions recoiling against a *pair* of partons, rather than a single parton. This is called a **dipole shower**, and has become increasingly popular in recent years for various technical reasons (such as ease of conserving energy and momentum, as well as of incorporating soft gluon interference effects). A number of publicly available Monte Carlo Event Generators implement various PS algorithms, either on specific scattering processes defined in the program, or supplied by the user (e.g. from MadGraph). Such programs are widely used at the LHC, and were also used at previous colliders. The main public codes are Herwig (which offers the choice of an angular ordering angular-ordered PS, or a transverse-momentum-ordered dipole shower), Pythia and Sherpa (which both have transverse momentum-ordered dipole showers).

At the end of the above PS algorithms, the final state as predicted by theory will contain a large number of (anti-)quarks and gluons, in addition to any colour-singlet particles that have emerged from the interaction. However, this is still not what would be seen in an actual collider experiment! For one thing, we know that no free quarks or gluons exist in nature, but are instead confined within hadrons, with no net colour. We must thus attempt to describe the **hadronisation** of final-state partons, which is the subject of the following section.

3.4 Hadronisation

Hadronisation is the process by which a collection of partons (i.e. (anti-)quarks and/or gluons) forms a set of colour-singlet hadrons, comprising (anti-)baryons and mesons. Given that this process involves momentum scales of order of the confinement scale Λ, where the strong coupling α_S becomes strong, it cannot be fully described in perturbation theory. The closer we get to experimental data, the more we must rely on well-motivated phenomenological models. A number of these have been developed, and they usually contain a number of free parameters so that they can be tuned to data, allowing confident prediction of hadronisation effects in subsequent experiments.

Before reviewing the models that are in widespread current use, it is worth pointing out that a more precise approach can be taken in the simple situation in which one considers a single final state hadron h. We can illustrate this using the process of electron–positron annihilation to hadrons, whose leading-order diagram is shown in Figure 2.14(b). Higher-order corrections will involve QCD radiation from the final state (anti-)quark, which may be real or virtual. Ultimately, the final state radiated partons will form hadrons, and in choosing to isolate the particular hadron type h, we can consider the process

$$e^+e^- \to hX, \tag{3.15}$$

where X denotes any other radiated particles (e.g. photons, hadrons) that end up in the final state. Let q^2 be the squared four-momentum—the **virtuality**—of the exchanged vector boson, and x the fraction of the total centre-of-mass energy carried by the hadron h. Then one may define the *fragmentation function*[5]

$$F^h(x, q^2) = \frac{1}{\sigma_0} \frac{d\sigma^h}{dx}, \tag{3.16}$$

where σ_0 is the LO cross-section, and is included for a convenient normalisation. We see that the fragmentation function is directly related to the cross-section for production of a particular hadron, and thus gives us the expected distribution of momentum fractions x of hadrons h in a given scattering process, and at a given virtuality q^2. Clearly we cannot calculate this quantity from first principles, as the hadronisation process is not perturbatively calculable. However, it may be shown that the fragmentation function of equation (3.16) may be decomposed into the following form:

---

[5] Here we have defined a total fragmentation function, irrespective of the polarisation of the exchanged vector boson. One may also define individual functions for each polarisation separately.

$$F^h(x, q^2) = \sum_{i \in q, \bar{q}, g} \int_x^1 \frac{\mathrm{d}z}{z} C_i\left(z, \alpha_S(\mu^2), \frac{q^2}{\mu^2}\right) D_i^h\left(\frac{x}{z}, \mu^2\right), \qquad (3.17)$$

where the sum is over all species of parton (including different flavours), and D_i^h is a *partonic fragmentation function* for each species. The coefficients C_i for each species are calculable in perturbation theory, and at LO equation (3.17) has the interpretation that the total fragmentation function splits into a number of partonic contributions, with D_i^h representing the probability that a given parton i in the final state produces the hadron h. The energy fraction z of the parton i must exceed that of h, hence the requirement that $z > x$. Equation (3.17) closely resembles the factorisation of total cross-sections into perturbatively calculable partonic cross-sections, and parton distribution functions that collect initial state collinear singularities, as discussed in Section 2.15. Indeed, the physics leading to equation (3.17) is very similar. Were one to try and calculate the partonic cross-section for the process of equation (3.15) in perturbation theory, one would find infrared singularities associated with the emission of radiation that is collinear with a given outgoing parton i that could end up in the hadron h. In the total cross-section, these singularities would cancel upon adding virtual graphs, but that is not the case here given that we have chosen to isolate a single particle in the final state, and observables with a fixed number of particles in their definition are not infrared-safe. The only way to get rid of such singularities is to absorb them into some sort of non-perturbative distribution function, and this is precisely the role played by the partonic fragmentation function $D_i^h(x, q^2)$. Analogous to the case of parton distributions, there is an energy scale μ that separates the radiation that is deemed to be perturbative, from that which is non-perturbative and thus is absorbed into the fragmentation function. The arbitrariness of this scale leads to evolution equations for the partonic fragmentation functions, that are similar to the DGLAP equations for parton distributions. One may then measure the fragmentation functions from data.

Whilst the above fragmentation approach is completely systematic, it is clearly insufficient to describe the complete hadronisation of a complex scattering event including large numbers of partons that have to be hadronised. For example, it takes no account of the fact that multiple partons from a given event must end up in a single hadron, in order to produce a colour singlet object. To this end, one may formulate more comprehensive descriptions, albeit at the expense of moving away from more precise field theory statements. All hadronisation models have a number of features in common, e.g.

- The produced hadrons tend to have similar momenta to the partons producing them (a phenomenon known as *parton-hadron duality*). Put another way, hadronisation does not change the directions of the highly energetic partons emerging from the scattering process by very much. This is reassuring, given that hadronisation effects are not calculable from first principles (e.g. from perturbation theory)!

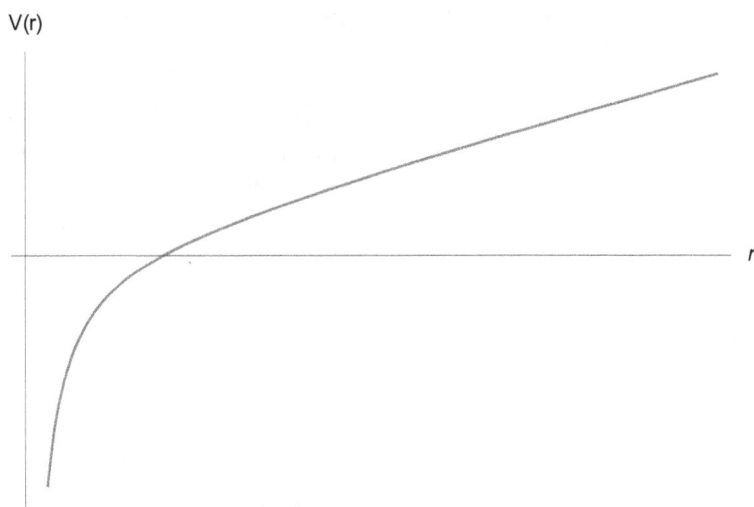

Figure 3.11. Potential between a $q\bar{q}$ pair, as a function of the distance r between them.

- Each model must decide how to divide the colour of the final state partons between (colour-singlet) hadrons. Usually this is done using approximations that reduce the vast number of possible colour flows.
- Each model has a number of free parameters, that can be tuned to data. After tuning a given model to data, it can be used to predict results in subsequent experiments. However, models may have to be retuned if they start to systematically disagree with data, or have additional parameters added.

The two main hadronisation models in current use are described below.

3.4.1 The Lund string model

The **Lund string model** of hadronisation takes as its starting point that the potential between a quark and antiquark has the form shown in Figure 3.11. At short distances, this is inversely proportional to the distance, as in electromagnetism. At larger distances the behaviour is very different, which is possible given the fact that gluons, unlike photons, are self-interacting. The QCD potential has a linear form, which is well-verified by numerical computations (e.g. using lattice QCD). Upon pulling apart a $q\bar{q}$ pair, there is then a constant force between them. Furthermore, the gluon field between the charges becomes increasingly concentrated along the line joining them, and is known as a *flux tube*, or *QCD string*. This is shown in Figure 3.12(a), and the string has a uniform energy per unit length at larger distances, from Figure 3.11. As the distance between a given (anti-)quark pair increases, this energy can become high enough that a new quark and anti-quark can be created. The string then breaks into two, creating two new strings, as shown in Figure 3.12(b).

Consider now the application of this idea to partons emerging from a scattering process after the shower has ended. The set of partons will include $q\bar{q}$ pairs which, as

(a) (b)

Figure 3.12. (a) A QCD flux tube or 'string' connecting a quark and antiquark; (b) at large enough distances, the string can break, creating a new (anti)-quark, and thus two strings.

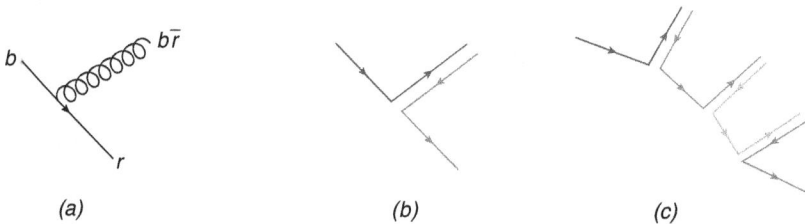

(a)   (b)   (c)

Figure 3.13. (a) A blue quark turns into a red quark by emitting a gluon, where the latter can be thought of as carrying a blue charge, and an anti-red charge; (b) colour flow diagram representing the same situation; (c) colour flow that dominates for multigluon emission in the $N_c \to \infty$ limit, where the individual colours at each gluon emission are randomly chosen.

they separate, will have a QCD string stretching between them. As this string breaks, it creates new (anti-)quarks, which also fly apart, and the string-breaking process iterates until no more breaks are kinematically possible. Hadrons can then be formed by grouping adjacent quarks and anti-quarks. Of course, this will only produce mesons. Baryons can be formed by introducing the concept of a *diquark*, namely a composite state of two quarks (anti-quarks) that has the same colour quantum numbers as a quark (anti-quark). Grouping together a quark with a diquark gives a baryon, and similarly for anti-baryons. Finally, one must consider gluon radiation from quarks. From a colour point of view, a gluon carries both a colour and an anti-colour, as illustrated in Figure 3.13. Considering the colour charges, there will now be a flux tube stretching from the quark to the gluon, and from the gluon to the anti-quark. This looks like a single flux tube with a kink, and thus one may model the effects of gluons by considering the dynamics of kinked strings. A possible complication when multiple gluons are emitted is that there are many possible choices for where the various (anti)-colour charges end up. However, it may be shown in the approximation that the number of colours is large ($N_c \to \infty$) that a single colour flow dominates, namely that of Figure 3.12(c), in which the pattern of Figure 3.13(b) is repeated along the string. The assignment of individual colour charges to hadrons is then unambiguous, and corrections to this approximation turn out to suppressed by a factor $1/N_c^2 \simeq 0.1$, where we have used the physical value $N_c = 3$.

There are a number of free parameters in the string model. For example, the probability for a string to break is described by a certain fragmentation function, which may be modified for different quark flavours (including heavy flavours), for diquarks, or for strings of very low invariant mass. At each string break, the (anti-) quarks are assigned a transverse momentum according to a Gaussian distribution, which may also be modified in certain circumstances. All of the above functions contain parameters whose optimal values may be fit by tuning to data.

3.4.2 Cluster hadronisation

An alternative hadronisation model is based on a property known as *preconfinement*, which states that after a PS, it is always possible to form colour-singlet clusters of the radiated partons. Furthermore, the mass distribution of these clusters turns out to be calculable from first principles. It involves the low-virtuality cut-off scale t_0 of the PS, and the QCD confinement scale Λ. Crucially, however, it does not depend on the hard momentum scale Q of the underlying scattering process, so that the details of cluster formation will be universal i.e. the same in all events.

One way to form clusters is to split any gluons remaining after the shower into $q\bar{q}$ or diquark pairs, according to a suitable procedure (e.g. giving the gluons a mass, and decaying them in a two-body fashion to chosen decay products). In the large N_c approximation, one may show that adjacent (anti-, di-)quarks will have matching (anti)-colours[6], so that they can be straightforwardly associated with a single colour-singlet cluster. Once clusters have been formed, they can be matched to a smoothed-out spectrum of hadron masses. A lighter cluster may be close in invariant mass to a known hadron, and can be identified as such following some minor reshuffling of the kinematics. If there is no known hadron close in mass, one may decay clusters to pairs of hadrons. The probability of a given hadron pair being chosen will depend on the flavour and spin quantum numbers, and the kinematics may be isotropic in the rest frame of the cluster, or modified in some way involving tunable parameters. Heavier clusters (i.e. whose invariant mass exceeds the mass of the heaviest hadron used in a given model) may be decayed into lighter clusters, before proceeding as above.

Of the main general purpose Monte Carlo even generators mentioned above, Pythia uses a string model, and Herwig++ and SHERPA use their own variants of the cluster model. As energies and luminosities of colliders continually increase, hadronisation models must be continually improved, and there is ample scope for new ideas and methods.

3.5 The underlying event

After the algorithms of the previous section have been applied, our theoretical description of a scattering event is starting to look a lot closer to what happens in a collider. However, it is not quite there yet, as there are a number of issues that we have still ignored. These are usually grouped together into the term *the underlying event*, which we are thus taking to mean all physics in a scattering event that is distinct from the hard scattering process, PS, and hadronisation. It is worth bearing in mind, however, that different people use this term in different ways. Furthermore, the term as defined here is, strictly speaking, ambiguous: some of the effects we

---

[6] The domination of a single colour flow in the PS as $N_c \to \infty$ is the same property as that used in the string model, in Figure 3.12(c).

describe below will turn out to be intertwined with the physics already included above, so that there is no clear means of formally separating them!

As might be expected as we get closer to experiment, the issues become very complex, and our means of saying anything concrete about them—using well-established ideas from field theory—dwindles. Our aim here is not to give a highly technical summary of different approaches for coping with the sheer mess of the underlying event. Excellent reviews may be found in the further reading at the end of the chapter, and in any case the state of the art in underlying event modelling is a constantly changing situation. Instead, we will content ourselves with pointing out what some of the main issues are.

3.5.1 Intrinsic transverse momentum

The factorisation theorem for hadronic cross-sections of equation (2.229) assumes that the partons exit the incoming protons with no transverse momentum (k_T) relative to the beam axis. The PS applied to the incoming partons will introduce some transverse momentum, but in general this is found to be insufficient to match what happens in actual experiments. One explanation for this is that there is presumably some **intrinsic** k_T that the partons possess within the proton itself. To illustrate this fact, Figure 3.14 shows the transverse momentum spectrum of lepton pairs produced in DY production at the Tevatron. At LO in perturbation theory,

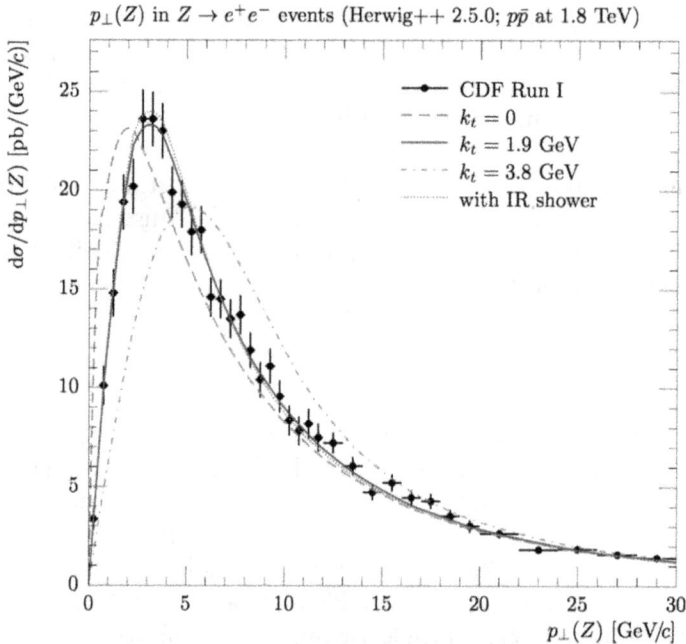

Figure 3.14. The transverse momentum spectrum of lepton pairs produced in DY production at the Tevatron. Data from the CDF experiment is shown alongside various theory predictions, where some amount of intrinsic k_T of the colliding partons is needed in order to match the data. Reproduced with permission from A. Buckley *et al, Phys. Rep.* **504** 145–233 (2011).

this would be zero, and thus this distribution is highly sensitive to the transverse momentum of the incoming partons. One sees that allowing the latter to have intrinsic k_T gives a much better fit to the data, and that this value is not particularly small (i.e. it is appreciably larger than the confinement scale Λ).

One of the simplest models for generating intrinsic k_T for each incoming parton is to assume a Gaussian distribution with a fixed width. One might expect such a model to be correct if the only source of this effect was non-perturbative effects within the proton itself. However, there is clearly a delicate interplay between perturbative and non-perturbative physics, not least in that changing the nature or parameters of the PS will affect the transverse momentum of the incoming partons, thus changing the amount of intrinsic k_T that is needed! More sophisticated models have been developed in recent years to try to take such issues into account.

3.5.2 Multiple partonic interactions

Another deficiency of the standard factorisation formula of equation (2.229) is that it only includes the effects of a single parton from one incoming proton interacting with a single parton from the other. If the transverse distance (i.e. the **impact parameter**) between the incoming protons is small, the overlap between them is large, and **multiple partonic interactions** (MPIs) can occur. There is no well-established factorisation formalism that can fully handle this complication, which is especially unfortunate in that MPIs do appear to be needed to describe experimental data. An example is Figure 3.15, which shows the distribution of the number of charged particles as measured by the ATLAS experiment. The green and blue curves show the theory predictions from a given Monte Carlo event generator, with and without a PS. The latter increases the charged particle multiplicity, as expected given that many (anti-)quarks and gluons will be radiated. However, it is only when the effects of MPIs are included that the prediction lies anywhere near the data. Various models for MPIs have been developed, each of which needs to specify how often MPIs occur as a function of the relative p_T of the protons, which is inversely related to the impact parameter. It must also specify how the scatterings are distributed in the transverse plane. In the simplest models, each parton interaction is treated independently, apart from correlations due to energy and momentum conservation. In more sophisticated approaches, different scatterings may interact with each other, an example of which is shown in Figure 3.16. Care must be taken to make sure that colour flows between such interacting multiple scatterings are accounted for. Furthermore, each interaction should be subjected to parton showering, and hadronisation. There is clearly an interplay between the specific assumptions made in a particular MPI model, and the intrinsic k_T that is needed for the partons inside the proton. Recent years have seen attempts to make the study of MPIs more formal, e.g. by introducing specific factorisation formulae for double parton scattering. This is a rapidly changing area, that is sure to advance further in coming years.

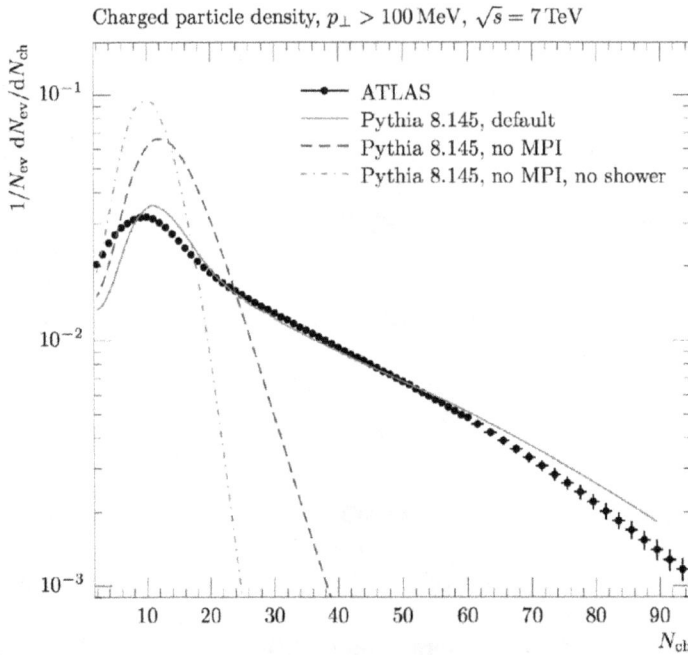

Figure 3.15. Distribution of the number of charged particles with $p_T > 100 MeV$, as measured by the ATLAS experiment at a centre-of-mass energy of $\sqrt{s} = 7 TeV$. Reproduced with permssion from A. Buckley *et al Phys. Rep.* **504** 145–233 (2011).

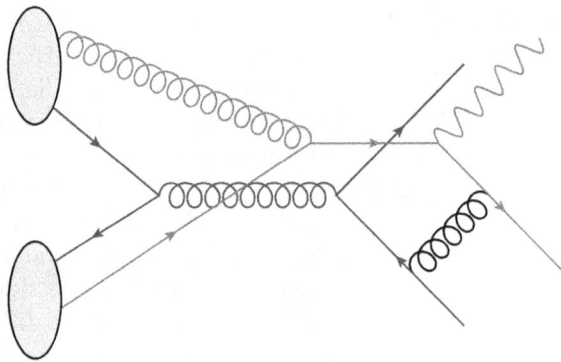

Figure 3.16. Two separate hard scattering processes may occur if two partons emerge from each incoming proton (shown in different colours for convenience). The black gluon links the two processes, so that these are not necessarily independent.

3.5.3 Beam remnants

So far we have talked only about the partons that emerge from the incoming protons. However, what happens to the rest of them, known as the *beam remnants*? In the canonical case in which only single partons emerge and interact, the rest of each proton can be modelled as a diquark, whose flavour quantum numbers are dictated by the parton that has left. Furthermore, the diquark and scattering parton must form a net colour singlet state, and must combine to reconstruct the total energy and momentum of the incoming proton. This includes assigning a transverse momentum to the beam remnant, to offset any intrinsic k_T given to the scattering parton.

The beam remnants must be included when applying hadronisation models to the entire final state, as it is only upon including them that colour is conserved. Furthermore, treatment of the beam remnants clearly overlaps with MPI modelling, given that what looks like a beam remnant from one scattering can turn out to be the seed for a subsequent scattering. It may be necessary, for example, to keep track of colour flows between the beam remnants and different MPI scatterings. There is usually more than one way to do this, and models must decide on a particular colour assignment in any given event.

3.5.4 Colour reconnection

In simple approaches to hadronisation or MPI modelling, it is not possible to exchange colour between distinct hadron clusters or MPI scatterings. Such *colour reconnections* are, for example, subleading in the $N_c \to \infty$ limit that is usually used in assigning colour flows in events. However, these more complicated colour correlations indeed need to be taken into account, as they can have a limiting effect on the precision of some experimental measurements. To give an example, consider the decay of the top quark into a b quark and a $u\bar{d}$ pair, as shown in Figure 3.17. The top quark decays rapidly (i.e. before hadronisation takes place), such that the partons from its decay will potentially be colour-connected with other partons in the same scattering event. The latter may be in the beam remnants or separate MPI scatterings. This in turn leads to a small uncertainty in the invariant mass of the top quark, which can be estimated by varying the parameters in existing colour-reconnection models. An uncertainty of around 0.5 GeV is found, which is very close to the quoted measurement uncertainty on the top mass! The pressing nature of this problem makes it likely that increasingly sophisticated colour-reconnection models will be developed in coming years.

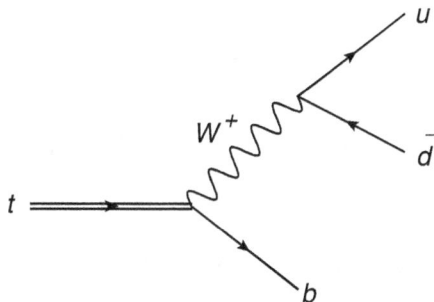

Figure 3.17. Example decay of a top quark.

3.6 Jet algorithms

The output of a Monte Carlo event generator can in principle be compared to data at particle level. However, it is clearly ridiculous to use observables in which all final state hadrons are treated separately, as there can be hundreds/thousands of them. If we look at actual scattering events at colliders, then we find that hadrons typically arrange themselves into collimated **jets** of particles. An example is shown in Figure 3.18, which shows a scattering event recorded by the ATLAS collaboration. The circular diagram shows a cross-section through the detector, with projected tracks of charged particles. The red layer represents the hadronic calorimeter (HCAL), and we see that there are two tight bunches of particles leaving significant energy deposits in the HCAL. Thus, we would classify this as an event containing two jets, or a 'dijet event'.

The formation of jets makes sense from all the QCD that we have so far learned. A given scattering process may produce hard partons in well-defined directions. These then shower, where the radiation is approximately collinear to the original hard particles. At higher orders in perturbation theory, we can get more hard partons, which also shower etc. We then roughly expect a jet for every sufficiently hard parton, where the 'structure' of the jet comes from soft/collinear radiation, plus hadronisation corrections, which will smear out the parton directions. To meaningfully compare data with theory, however, we need to precisely define what we mean by a jet, and procedures for doing this are called **jet algorithms**. Provided we use the *same* algorithm on the theory and experiment sides, we can compare things.

Figure 3.18. A scattering event from the ATLAS detector. Credit https://atlas.cern/updates/news/search-new-physics-processes-using-dijet-events.

To be more precise, a jet algorithm is a systematic procedure for clustering the set of final state hadrons into jets. That is, it is a map

$$\{p_i\} \;\rightarrow\; \{j_l\}$$

from the particle four-momenta $\{p_i\}$ (where i runs from 1 to the number of particles), to some jet four-momenta (where l runs from 1 to the number of jets). There can be more than one particle in each jet, so that the number of jets is less than or equal to the number of particles we started with. As usual in this book, what looks like a simple idea has by now become a hugely complicated and vast subject in itself.

We saw earlier that observables should be **infrared (IR) safe**, meaning that theory results should not change if arbitrary soft/collinear particles are added, which constitutes a physically indistinguishable state. Unfortunately, many early (and widely used) jet algorithms were not IR safe, which caused serious problems when trying to compare theory with data at subleading orders in perturbation theory (where the latter are needed for precise comparisons). Recent years have seen the development of many IR safe algorithms, and a large class includes so-called **sequential recombination algorithms**. They involve associating a distance measure d_{ij} between any two particles i and j. This may be as simple as their separation in the (y, ϕ) plane of Figure 1.10(b), or it may be more complicated. We can also associate a distance d_{iB} between particle i and the beam axis. A typical algorithm is then as follows:

1. Calculate d_{ij} and d_{iB} for all (pairs of) particles.
2. If the minimum

$$\min(\{d_{ij}\},\,\{d_{iB}\})$$

   of the set of distance variables is a d_{ij} variable, combine i and j into a single particle (e.g. add the four-momenta, or use a more complicated **recombination algorithm**).
3. If instead the minimum is a d_{iB} variable for some i, then declare i to be a final state jet, and remove it from the list of particles.
4. Go to step 1, and stop when no particles remain.

Historically, different distance measures have been used in such algorithms. Often they are of the form

$$d_{iB} = \left(p_{Ti}^2\right)^n, \quad d_{ij} = \min\!\left[\left(p_{Ti}^2\right)^n,\,\left(p_{Tj}^2\right)^n\right]\frac{\Delta R_{ij}}{R}, \tag{3.18}$$

for some integer n. Here p_{Ti} is the magnitude of the transverse momentum of particle i, ΔR_{ij} is the distance in the (y, ϕ) plane between particles i and j (see equation (1.36)), and R is a constant parameter of the algorithm called the **jet radius**. Special cases in the literature are $n = 1$ (the k_T **algorithm**), $n = 0$ (the **Cambridge–Aachen algorithm**), and $n = -1$ (the **anti-k_T algorithm**). All such algorithms are IR-safe.

If you apply the algorithm to a set of particle four-momenta, it will indeed generate some set of jet four-momenta. However, it is not at all obvious, having read the above set of steps, what the jets will look like at the end of this! One way to visualise them is to take an example event, and plot the resulting jets in the (y, ϕ) plane. Each jet will have an associated **catchment area**, where particles in this area will end up in that particular jet. Plotting these areas then helps us visualise the shapes such jets carve out in the particle detector. Results for a number of jet algorithms are shown in Figure 3.19, including the algorithms mentioned above. In particular, we see that jets obtained using the anti-k_T algorithm have a nice circular shape, whose radius is roughly equal to the parameter R entering the algorithm—this is because the most energetic particles are naturally clustered first, leading to a very stable centroid around which the soft structure is added up to $\Delta R = R$. It is for this reason that R is often referred to as the **jet radius**. Some of the other jet shapes are more jagged, however, and their constituents can be located *further* than R from the eventual jet centroid, due to its more significant migration (compared to anti-k_T) during the clustering sequence.

Circular jet shapes tend to be preferred by experimentalists, as they can be rapidly approximated by simplified hardware algorithms used in event triggering, and the simple shape makes it easier to correct jet energies for underlying event and pile-up effects. However, other algorithms can be useful for different reasons, particularly since the k_T and CA algorithm correspond, respectively, to the inverses of transverse-momentum ordered and angular-ordered PSs. In the rest of this book, we will simply refer to 'jets of

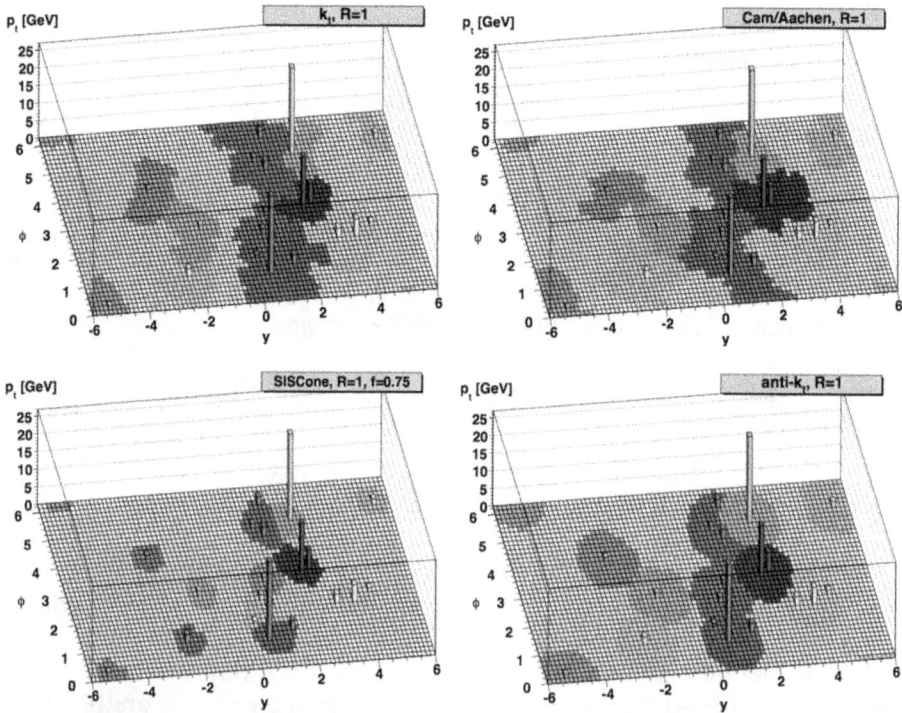

Figure 3.19. The results of clustering a sample event using different sequential recombination jet algorithms, where particles associated with each jet are shown as areas in the (y, ϕ) plane. Reproduced with permssion from *Eur. Phys. J.* C **67** 637–86 (2010).

radius R', where it is understood that these have come from a suitable algorithm with a jet radius parameter R. Sets of jet algorithms are implemented in the publicly available FastJet software package, which can be interfaced with event generators and event reconstruction and analysis codes.

Given a jet, it is useful to be able to define its area in the (y, ϕ) plane, which is more complicated than you might think. Firstly, the natural shape of jets can be non-circular, as discussed above. Secondly, each jet consists of a discrete set of points in the (y, ϕ) plane, representing the four-momenta of individual particles. We are then faced with the problem of how to define a continuous 'area' associated with a set of points, and there are different choices about how to proceed. For example, FastJet implements the following options:

(i) *Active area*. Given that any sensible jet algorithm is IR safe, one may flood any event (simulated or measured) with artificial soft 'ghost' particles, uniformly distributed in the (y, ϕ) plane. The active area of a given jet is then defined according to the number of ghost particles that end up in it.

(ii) *Passive area*. Instead of using a large number of ghost particles, one can add them one at a time, and then check which jet each ghost particle joins. The passive area is defined according to the probability that ghosts end up in a particular jet.

(iii) *Voronoi area*. Given the points in the (y, ϕ) plane defining the constituents of a particular jet, one may form their *Voronoi diagram*. This divides the plane into cells containing a single particle, such that any other point in each cell is closer to that particle than to any other (an example is shown in Figure 3.20). One may the define the *Voronoi area* of a single particle to be the intersection of its Voronoi cell with a circle of radius R (the parameter entering the jet algorithm). Finally, the Voronoi area of the jet is the sum of the Voronoi areas of its constituents.

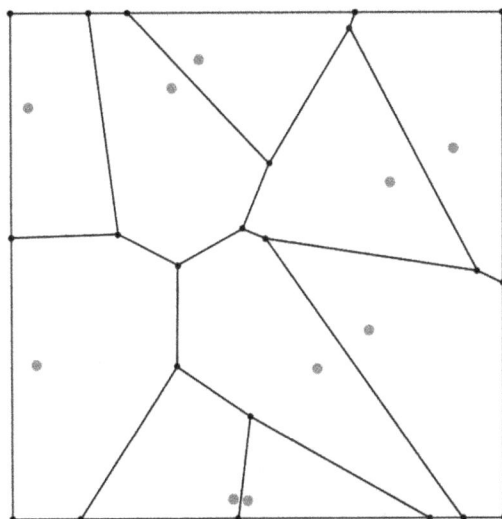

Figure 3.20. Example Voronoi diagram, for a set of points in the (y, ϕ) plane.

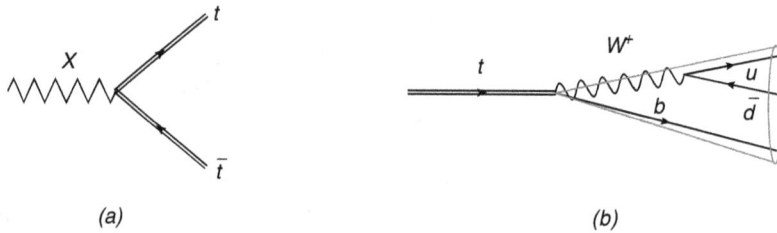

Figure 3.21. (a) Decay of a heavy new physics particle X into a top pair; (b) a highly boosted top, whose decay products are contained in a single fat jet.

These area definitions are different in general, and a given definition may be chosen for particular reasons. For example, passive area measures how sensitive a jet is to noise that is localised in the (y, ϕ) plane, due to its definition in terms of adding single ghost particles. The active area instead measures sensitivity to noise that is more spread out.

In recent years, the study of jets has become increasingly sophisticated. A significant development is the ability to systematically *look inside jets*, to find evidence of intermediate particle decays. Consider, for example, the production of a new physics particle X, that is so heavy that it can decay into a $t\bar{t}$ pair. The (anti-)top quarks may then be highly boosted, such that their decay products are collimated into a single 'fat jet', as shown in Figure 3.21. If one were to recluster the event with a smaller jet radius, one would expect to see three smaller jets inside the original fat jet, whose invariant mass reconstructs the top mass. However, two of these jets (corresponding to the $u\bar{d}$ pair in Figure 3.21(b)) will have an invariant mass that reconstructs the W mass. Looking for such a pattern of *subjets* potentially allows experimentalists to tag top quarks in events, with much greater efficiency than has been done in the past. In general, methods like this are known as *jet substructure* techniques. There are many different algorithms in the literature for examining jet substructure, some of which are implemented in packages such as FastJet. Typically, they combine reclusterings of events with jets of different radii, with additional methods to 'clean up' the constitutents of the fat jet (e.g. by removing softer subjets), in order to make the signal-like patterns of the subjets more visible. We will see explicit examples of substructure observables later on.

3.7 Matching parton showers with matrix elements

As we saw in Section 3.3, PSs can be used to estimate the effect of extra radiation, which makes theory calculations much more like experimental data. An alternative to using PSs is to simply try and calculate higher-order tree-level **matrix elements** (MEs)[7], given that such calculations have been automated in programs like

---

[7] The term 'matrix element' refers to the squared amplitude, and is used extensively throughout the literature on Monte Carlo event generators.

Table 3.2. Relative advantages and disadvantages of two methods for estimating additional radiation: PSs, and higher-order tree-level MEs (squared amplitudes).

Parton shower	Matrix elements
Can include large numbers of partons.	Limited to a few partons only, by computing power.
Bad approximation for widely separated particles.	Exact for well-separated particles.
Exact when particles are close together.	Fails for close particles (collinear singularities).
Includes only partial interference (angular ordering).	Includes all quantum interference.

MadGraph and Sherpa. There are pros and cons of both approaches, as we show in table 3.2.

What we see is that the regions in which the PS and MEs do well are complementary: the former strictly applies when particles are close together, and the latter when they are well-separated. This strongly suggests that we should somehow combine the descriptions, to get the best of both! A number of schemes for doing this exist in the literature. Here, we will describe the one that is implemented in MadGraph.

Naïvely, one might think of simply taking a set of MEs with different numbers of partons, with some appropriate minimum distance between the latter, and showering them. However, a problem arises in that some radiation will be counted twice. Consider the simple case of DY production of a (colour-singlet) vector boson, shown in the top left-hand corner of Figure 3.22. We can generate extra radiation using either the PS, or the ME, and the effect of each is shown by the two directions in the figure. If we simply add together all showered tree-level MEs, we generate multiple configurations with the same number of partons, indicating a double counting of contributions. This double counting is easily removed, in principle at least, by putting a cutoff Q_{cut} on the transverse momentum p_T of each emitted parton (relative to the emitting parton). For $p_T > Q_{cut}$ one should use the ME to generate the radiation, whereas for $p_T < Q_{cut}$ one should use the shower. However, the precise value of Q_{cut} one chooses is arbitrary, and thus it is important to ensure that results are insensitive to the choice. In other words, we must make sure that the ME and PS descriptions smoothly match near the scale Q_{cut}, when removing the double counted contributions. This motivates the following algorithm:

1. Start with a set of MEs generated with 0, 1, 2,... N additional partons, where each one has a minimum transverse momentum with respect to other partons to avoid collinear singularities.
2. Cluster each event into jets, and adjust its probability so that it looks more parton-shower like.
3. Shower the event, and cluster it into jets.

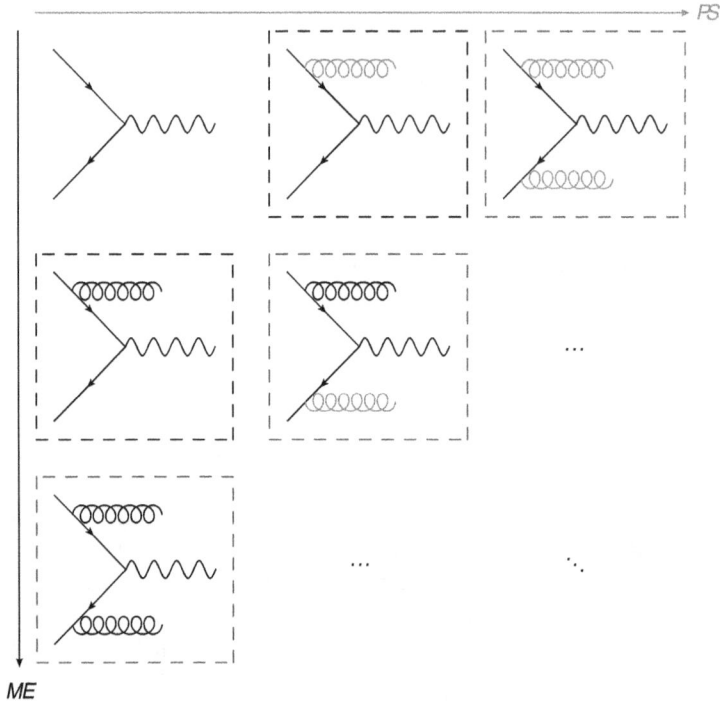

Figure 3.22. Radiation can be generated from MEs and/or PSs. Naïvely adding all such contributions together results in configurations that are counted twice (the dashed boxes).

4. If the original event (before showering) has N jets, keep events with additional jets with $p_T > Q_{cut}$, unless they are harder than the softest jet from the ME.
5. If the original event had $<N$ jets, veto *all* events containing additional jets with $p_T > Q_{cut}$.

Here, step 2 ensures smooth matching between the ME/PS descriptions, and thus insensitivity with respect to the matching scale Q_{cut}. In practice, making the ME probability more PS-like involves reweighting the coupling α_S and/or PDFs, to make scale choices the same as in the PS. Step 4 allows extra jets beyond those generated by the hard ME, but makes sure that the hardest (i.e. most well-separated) jets are all generated by the ME, as they should be. Step 5 again makes sure that the hardest N jets all come from the ME. Note that despite our best efforts, some residual dependence on the matching scale Q_{cut} may remain. It should be chosen so as to give minimal sensitivity when varied.

The above scheme is a variant of so-called **MLM matching**. Another common scheme you will see in the literature is **CKKW matching**, which is implemented e.g. in Sherpa. For a given observable, the difference between matching schemes can be treated as an additional source of theoretical uncertainty, given that it represents an ambiguity in how we describe extra radiation.

3.8 Matching NLO matrix elements

In the previous section, we saw that we can improve the description of well-separated jets by including higher-order tree-level MEs matched to a PS. Although this is certainly an improvement, the results are still only formally accurate to LO in perturbation theory. A full NLO calculation (or higher) would need virtual as well as real corrections to be added, and this is important if we want the total cross-section for our simulated events to be accurate to NLO precision. It is indeed possible to combine NLO MEs with PSs. In doing so, there is a double counting of real radiation that is directly analogous to the double counting encountered in the previous section. However, there is a second source of overlap between the PS and NLO ME, that is a bit more tricky: the **virtual corrections** in the NLO ME overlap with the fact that virtual corrections are included in the Sudakov factor. It is not at all easy to dig out the virtual corrections in the latter, and so the double-counting in this case is very subtle.

Nevertheless, a number of NLO/PS matching schemes have been proposed, of which two are widely used. The first of these is referred to as **MC@NLO**, and was originated by Frixione and Webber. The basic idea is that for a given PS algorithm, one can expand the effect of a single real/virtual emission up to $\mathcal{O}(\alpha_S)$ analytically. Then, one can subtract it from the NLO ME to give a 'modified NLO ME', which by itself is not physical. Finally, one can shower the modified ME, such that the resulting radiation has the double counting explicitly removed. Very schematically, we may write this procedure as

$$\text{MC@NLO} = \text{PS} \times \left[\sigma_{\text{LO}} + \alpha_S \left(\sigma_V - \sigma_V^{\text{PS}} \right) + \alpha_S \int d^4 k \left(\sigma_R - \sigma_R^{\text{PS}} \right) \right]. \quad (3.19)$$

Here the first term in the square brackets is the LO cross-section. The second term is the exact virtual contribution, with the overlap from the PS removed. The final term is the real radiation contribution with the PS overlap removed, integrated over the additional phase space of the extra parton. Finally, the whole contents of the square brackets is interfaced to the PS algorithm. One reason for showing this rather schematic formula is that it makes clear that there are subtractions involved in forming the modified NLO ME. Consequently, some events generated by MC@NLO have **negative probability weights**. This is never a problem in practice, as all generated distributions must be positive within uncertainties. Note that the subtraction terms have to be worked out separately for each PS algorithm, although this only has to be done once in each case. Impressively, an automatic implementation of this algorithm exists (called aMC@NLO), which is part of the publicly available code MadGraph_aMC@NLO. This allows users to specify an arbitrary hard scattering process (up to a couple of caveats), upon which the software automatically creates an MC@NLO event generator! The Sherpa generator can also be used to generate NLO matching to a PS, using a variant of the MC@NLO algorithm.

The second main matching prescription for interfacing NLO MEs with PSs is called POWHEG, by Frixione, Nason and Oleari. In this approach, a special Monte

Carlo event generator is used to generate the hardest radiated parton, that uses the following modified Sudakov factor:

$$\Delta_i \sim \exp\left[-\int_{k_\mathrm{T} > p_\mathrm{T}} \mathrm{d}^4 k \frac{R(\{p_i\}, k)}{B(\{p_i\})}\right].$$ (3.20)

Here the numerator is the exact real-emission ME, which is then normalised to the LO contribution $B(\{p_i\})$. This is then integrated over the momentum of the extra parton, which is enforced to have a higher transverse momentum than the hardest emission. In more simple terms: for the hardest emission, the no-emission probability includes the *full* NLO ME, rather than its collinear limit (which is what a normal PS does). Events from this generator can then be interfaced with a normal transverse-momentum ordered PS. There is no double counting of radiation (or virtual corrections), provided one requires that the PS only radiates partons with transverse momenta *less* than the hardest emission. The POWHEG approach is implemented for a variety of processes in the POWHEG-BOX NLO generator code, to be interfaced with various PS generators.

An advantage of the POWHEG approach is that it works for any transverse momentum-ordered shower. However, for angular ordered showers care is needed in order to get the matching to work. This involves so-called **truncated showers**—see the original literature for details. Furthermore, the POWHEG approach is such that all simulated events have manifestly positive probability weights. However, the exponentiation of the real ME in the Sudakov factor for the hardest emission can create spurious terms at $\mathcal{O}(\alpha_\mathrm{S}^2)$ and beyond, which can be numerically sizeable for some processes.

The difference between the MC@NLO and POWHEG approaches is of formally higher perturbative order, and thus acts as a further source of theoretical uncertainty. Some experimental analyses, for example, will run codes implementing *both* algorithms, to check the systematic uncertainty due to the matching prescription.

In this section, we have seen many different examples of software tools, in increasing order of sophistication. Which tool we should use for a particular analysis very much depends on what we are looking at, and the general answer is that you should use the tool that gives the most accurate answer. General rules include:

(i) If the observable we are calculating is sensitive to lots of jets, we should use higher-order tree-level MEs to get these right.

(ii) If there are few jets (e.g. one extra), but the observable still depends on complicated details of the final state, we should use NLO matched to a PS e.g. MC@NLO or POWHEG.

(iii) If the observable is very inclusive (e.g. a total cross-section), then fixed order perturbation theory may be the best thing to do. Even for differential cross-sections, we can often reweight the integral under them so as to match a higher-order total cross-section calculation.

You may be wondering whether it is possible to combine NNLO MEs with PSs, or to combine multiple NLO MEs with showers etc. Indeed this is possible, and people are starting to develop algorithms for doing so, which are somewhat hampered by current computing power. The basic idea is to try to include as much QCD theory as possible, whilst still having software tools that are fast enough to produce theory predictions that we can compare with data in a suitable timescale (weeks or months).

Further reading

As well as the books already listed in Chapter 2, the following books are useful for advanced topics in collider physics (e.g. resummation, Monte Carlo generators):

- 'QCD and Collider Physics', R K Ellis, W J Stirling and B R Webber, Cambridge University Press. This classic text has lasted many years, and is still worth looking at even if does not include the most recent developments.
- 'The Black Book of Quantum Chromodynamics: A Primer for the LHC Era', J Campbell, J Huston and F Krauss, Oxford University Press. A definitive resource for those wanting to understand how QCD is applied at current collider experiments.
- 'General-purpose event generators for LHC physics', A Buckley *et al*, Phys. Rep. **504** 45–233 (2011) https://doi.org/10.1016/j.physrep.2011.03.005. This is a comprehensive recent review article on contemporary Monte Carlo event generators, with very good discussions of the underlying physics.

Exercises

3.1   (a) What are the advantages of an NLO calculation over an LO one? How about higher orders?

      (b) What are the disadvantages of higher-order calculations?

3.2 Consider a parton i that branches into two partons j and k.

      (a) Assuming the partons j and k to be approximately on-shell and carrying momentum fractions z and $(1 - z)$ respectively, of the energy E_i of parton i, show that the virtuality of parton i is given by

$$t_i \equiv p_i^2 \simeq z(1 - z)E_i^2\theta^2,$$

if θ is small.

      (b) Show further that the transverse momentum of parton k relative to parton j is given by

$$|k_T| \simeq (1 - z)E_i\theta.$$

      (c) Hence show that, at fixed momentum fraction z, one has

$$\frac{\mathrm{d}t_i}{t_i} = \frac{\mathrm{d}|k_T|^2}{|k_T|^2} = \frac{\mathrm{d}\theta^2}{\theta^2}.$$

Figure 3.23. (a) Two seed directions for an iterative cone algorithm, separated by twice the cone radius; (b) the same, but with an additional soft gluon exactly halfway between the two seed directions.

(d) Why might this be useful for a PS algorithm?

3.3 (a) Explain what is meant by the Sudakov form factor $\Delta_i(t_0, t_1)$ for parton species i and virtualities t_0 and t_1.

(b) Explain why this quantity obeys the equation

$$\Delta_i(t_0, t + \delta t) = \Delta_i(t_0, t)\left[1 - \sum_j \int_t^{t+\delta t} \frac{\mathrm{d}t}{t} \int \mathrm{d}z P_{ji}(z)\right]$$

for an outgoing parton, where $P_{ji}(z)$ is a DGLAP splitting function.

(c) Hence show that

$$\Delta_i(t_0, t) = \exp\left[-\sum_j \int_{t_0}^t \frac{\mathrm{d}t'}{t'} \int_{z_{\min}}^{z_{\max}} \mathrm{d}z P_{ji}(z)\right],$$

where suitable limits for z have been imposed. What determines these limits?

3.4 (a) When do you expect the PS approximation of additional radiation to be accurate, and when not?

(b) Why do some PSs order successive emissions by angle, rather than virtuality?

3.5 (a) Why are jet algorithms used in comparing theory calculations with data?

(b) An early example of a jet algorithm is an 'iterated cone algorithm'. This starts by assigning so-called 'seed four-momenta' to the event. Then the four-momenta within a cone of distance ΔR (in the (y, ϕ) plane) of each seed are summed, and the resultant used to define a new seed direction. The process is iterated until all seed directions are stable.

Consider the two seed directions shown in Figure 3.23(a), separated by twice the cone radius ΔR. How many jets will the algorithm return?

(c) Consider adding a single soft gluon to the event, as shown in Figure 3.23(b). If this direction is taken to be a seed, how many jets will the algorithm return now? Hence, is this algorithm IR safe?

3.6 What are the possible sources of double counting of radiation when matching:

(i) PSs with higher-order tree-level MEs?

(ii) PSs with NLO MEs?

3.7 Different software tools for observables at the LHC include: (a) event generators at fixed order in perturbation theory; (b) PSs applied to LO MEs; (c) PSs matched to NLO MEs; (d) PSs matched with higher-order tree-level MEs. Which of these is most appropriate for the following observables, and why?

(i) The total cross-section for Higgs boson production.

(ii) The transverse momentum distribution of the top quark in top pair production, to be compared with parton-level data.

(iii) The rapidity of the fourth hardest jet in top pair production, in association with many jets.

(iv) The distribution of the number of jets accompanying the production of a new physics particle.

(v) The distribution of the transverse momentum of the first hardest additional jet accompanying the production of a new physics particle.

Chapter 4

Beyond the Standard Model

The Standard Model (SM) has done an amazing job of explaining all physics up to (and including) LHC energies, and some calculations in particular match our current experimental measurements with outrageous precision. Nevertheless, no-one in the field of particle physics believes it is the final answer, for a variety of reasons. Firstly, the theory is subject to a number of *theoretical* challenges which, although subjective, give very clear hints that the SM is probably the low-energy limit of a more fundamental theory. There are also a number of *experimental* challenges, including both the parts of the model that have yet to be verified, and a growing number of observed phenomena that cannot be explained by the SM as it stands.

4.1 Theoretical challenges to the Standard Model

The theoretical problems start with the fact that the SM famously does not include gravity, a force for which we lack a quantum description. It is also not clear why the three forces that are included have such different coupling strengths at low energies, and attempts to unify these into a single force at high energies remain popular, taking inspiration from the fact that the electroweak theory already provides a unified description of the electromagnetic and weak forces. When we consider the matter in the SM, we have no particular need for the three generations that are observed, nor why there is such a strange pattern of particle masses. In total, the SM has 19 arbitrary parameters, with no particular reason to select their preferred values *a priori*, a situation that is deeply unsatisfactory to a conscientious theorist.

There are also specific challenges that arise from some technicalities of the theory. The SM Lagrangian is written by including all gauge-invariant terms up to mass-dimension 4, with the latter requirement arising from the fact that such terms will lead to a renormalisable theory. In the quantum chromodynamics (QCD) Lagrangian of equation (2.59), we neglected to include a term of the form

doi:10.1088/978-0-7503-2444-1ch4

$$-\frac{\alpha_S}{8\pi}\theta_{\text{QCD}}F^a_{\mu\nu}\tilde{F}^{\mu\nu,a}, \qquad (4.1)$$

where θ_{QCD} is an unknown parameter, α_S is the strong coupling constant, $F^a_{\mu\nu}$ is the gluon field strength tensor with colour index a, and

$$\tilde{F}^{\mu\nu,a} = \epsilon^{\mu\nu\alpha\beta}F^a_{\alpha\beta}/2$$

is called the **dual field strength tensor**. The extra term of equation (4.1) is allowed by all symmetries of the SM, and receives a contribution from a mechanism known as the **chiral anomaly** that we will not go into in detail.

You should now be wondering why we did not include this term in our earlier discussion of the QCD Lagrangian. It transpires that such a term would give rise to an electric dipole moment for the neutron, in violation of the very stringent observed upper bound that tells us that such a dipole moment does not exist. The only solution in the SM is to tune θ_{QCD} so that the term disappears. This is an example of **fine-tuning**, which occurs whenever we have a parameter that can take a large range *a priori*, but which must be set to a very precise value in order to be compatible with later experimental measurements. Throughout the history of physics, such fine-tuning has usually told us that we are missing a more fundamental explanation. In this specific case, the problem with the missing QCD Lagrangian term is dubbed the **strong CP problem**, and it is the starting point for extending the SM with mysterious new particles called axions.

Another fine-tuning problem of the SM is the so-called **hierarchy problem**, which is particularly relevant to the search for new physics at TeV energy scales. To explain this, let us first recall the results of Section 2.11, in which we found that fermion masses run with energy scale in QFT. Furthermore, this running is logarithmic, which we can find by putting a momentum cutoff Λ on loop integrals, and then examining how the results depend on Λ. It turns out that for *scalar* particles rather than fermions, this running is quadratic. The Higgs boson in the SM is a fundamental scalar. If its mass is governed by some underlying new-physics theory at high energies, we therefore expect the Higgs mass to depend on the energy scale of new physics, Λ, according to

$$m_H^2 \sim \Lambda^2. \qquad (4.2)$$

Small changes in the new-physics scale lead to large changes in the Higgs mass, so it is puzzling that we observe a Higgs mass of $m_H \simeq 125$ GeV. This seems to imply a huge amount of **fine tuning**: if Λ is large, the Higgs mass should also naturally want to be large. If we take Λ being two orders of magnitude larger than m_H as a rule-of-thumb for where the SM begins to be finely tuned, this implies that we expect new physics to be prevalent at energies of $\mathcal{O}(10\,\text{TeV})$, i.e. LHC energies. However, this is a purely theoretical argument, and has a high degree of subjectivity, e.g. why 10 TeV and not 100 TeV?

4.2 Experimental challenges to the Standard Model

At first glance, the experimental status of the SM appears to be rather good. The discovery of the Higgs boson by the ATLAS and CMS experiments in 2012 marked the moment when we had finally observed all of the particles predicted by the theory. However, there are a number of open questions in Higgs physics, which include:

- *Is the Higgs boson really the same as that predicted by the SM?* So far, the measured Higgs boson production and decay rates match the SM predictions. However, the uncertainty on these measurements increases dramatically for some modes, and deviations from SM behaviour may be observed as the uncertainties shrink with the addition of more LHC data. Recall that, in the SM, loop effects must be included when calculating the rate of a given process to a given accuracy. In the SM, we know exactly which particles can act in the loops of our Feynman diagrams, and thus we generally get unambiguous calculations of rates that can be compared to experimental data. If there are new particles in Nature, they would be expected to appear in loops, and thus would change the rates of processes involving particle production and decay. Since any new physics is likely to affect the Higgs sector of the SM, we can thus use precision measurements of Higgs quantities at the LHC to provide a powerful indirect window on the presence of new particles.

- *Is the Higgs field the same as the* SM? The SM Higgs mechanism involves the potential given in equation (2.88). We still have no idea if the Higgs field obeys this equation or not, since the observed Higgs properties could be consistent with a different form generated by new physics. To answer this question, it turns out that we will need to observe the Higgs boson interacting with itself, and then measure the strength of that self-interaction. This is unfortunately exceptionally challenging to do at the LHC. Nevertheless, there is a growing literature that suggests that this is not impossible by the end of the LHC's run. Moreover, such a measurement is a very strong motivation for a future, higher-energy collider.

- *Are there extra Higgs bosons?* The SM contains a doublet of Higgs fields, but there is no reason at all why there should not be more. A model with two Higgs doublets gives rise to two extra neutral Higgs bosons (conventionally referred to as H and A), and two electrically-charged Higgs bosons H^+ and H^-. Although current searches have not uncovered evidence for these particles, much of the predicted mass range remains unexplored at the LHC, and will be covered in more detail over the next decade or so.

In addition to these open questions, there is a growing number of experimental results that the SM cannot explain. For example, we know that the universe is made almost entirely of matter (rather than antimatter), and although we have a mechanism by which this can occur in the SM, the numbers do not work out well enough to explain the matter abundance that we actually observe. The field of cosmology has entered a precision era, telling us that most of the matter in the Universe is in the form of 'dark matter', which cannot be comprised of any of the particles of the SM. Observations instead suggest the existence of a new particle that interacts with SM particles with a strength similar to that of the weak force, with a mass of $\simeq 1$ TeV. Such particles are known as WIMPs, for 'weakly interacting massive particles'[1]. Any new

---

[1] In a burst of acronym whimsy, WIMPs were originally named as the complement to the alternative dark-matter hypothesis of astrophysical 'massive compact halo objects', or MACHOs.

physics associated with this particle can be expected to manifest itself at similar energies, giving rise to a potential collection of new particles at the TeV scale.

Cosmology also tells us other interesting things. For example, most of the energy budget of the Universe is comprised of a mysterious 'dark energy' component that probably requires new particle physics at high energy scales to explain. It is more or less accepted that the Universe also underwent a rapid expansion in its first moments, which is called **inflation**. Such behaviour cannot easily be explained without extending the SM with new fields. The growing field of **particle cosmology** relates cosmological observations to the various high-energy particle physics models that might explain them.

Finally, neutrinos are assumed to be massless in the SM, but it has been known for decades that they have a small mass. Generating this mass technically requires an extension of the SM.

It is worth noting that we could have extended this list of experimental challenges to include tentative observations that are not yet confirmed. These include measurements of various properties of b-meson decays that currently show substantial tension with the SM, and the current best measurement of the magnetic moment of the muon which does not agree with the SM prediction. In both cases, there are theoretical uncertainties associated with non-perturbative QCD calculations that are very hard to pin down, with the result that experts are still split on whether these measurements allow us to definitively reject the SM. Nonetheless, there are some b-meson decay observables that are much less affected by systematic uncertainties, and these also show persistent anomalies.

4.3 Searching for beyond the SM physics

Assuming that we are convinced of the need for a new theory of particle physics that has the SM as a low energy limit, how should we go about trying to discover that new theory? There are two basic approaches, and it is not clear at the outset which will first generate a dramatic discovery. The first is to perform an **implicit search** for new physics by measuring many SM processes with high precision, looking for processes that do not match their corresponding theoretical predictions. This is an *indirect* approach to searching for new physics which, although it can probe high energy scales, gives relatively little information about what the new physics is in the case of discovery. At the same time as testing new-physics theories, it enhances our description of the Standard Model itself, by revealing areas where our heuristic models of e.g. non-perturbative physics must be improved. The second approach is to perform an **explicit search** for the *direct production* of new particles in the proton–proton collisions of the LHC, which has the potential to reveal much more about the type of new particles produced, and their interactions with SM and other particles. We shall here briefly describe these two options, leaving experimental details to Part II.

4.3.1 Implicit searches for new physics

We saw in Chapter 2 that cross-sections and decay rates can be calculated in perturbation theory by including diagrams of successively higher order. If there are new particles in Nature, they may appear in the loops of the higher-order diagrams of processes with incoming and outgoing SM particles, even if they are too massive to be produced directly in LHC collisions. Implicit searches for new physics can therefore access energy scales much higher than those that we can access in explicit searches.

In practice, the implicit search programme at the LHC is performed by measuring a huge range of observables in an attempt to find cracks in the SM. There are many different types of observable that one may choose to measure, including:

- **Inclusive cross-section measurements:** Given the SM, one can calculate the total cross-section for the production of specific particles in the final state (e.g. a single W boson, a pair of W bosons, a top and anti-top quark, etc). For each process, one can then isolate it in the LHC data, and extract a measurement of the production cross-section. Comparison of the theoretical and experimental results then provides a powerful test of the SM. In Figure 4.1, we show various inclusive cross-section measurements performed by the ATLAS experiment, and it can be observed that they match the theoretical predictions closely. Isolation of each process typically relies on selecting specific decay modes of the particles in the final state, for example leptonic decay modes of the Z boson. The measurements are then corrected for the known branching fractions of the particles in the SM. Note also that measurements have been performed with the LHC running at different

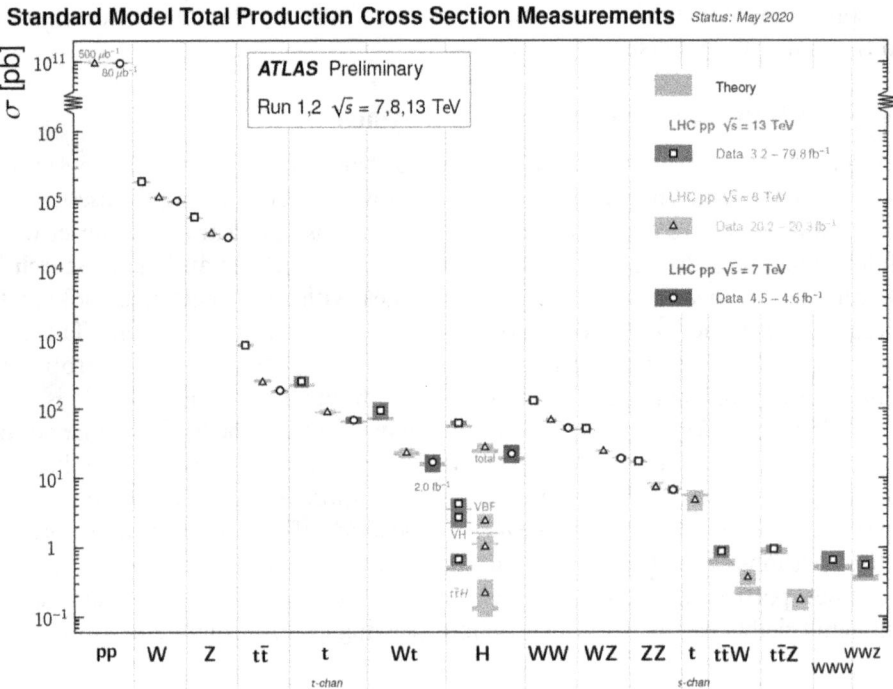

Figure 4.1. Summary of SM total production cross-section measurements performed by the ATLAS collaboration, as of June 2020. The measurements are corrected for branching fractions, and are compared with the corresponding theoretical predictions. CREDIT: https://atlas.web.cern.ch/Atlas/GROUPS/PHYSICS/PUBNOTES/ATL-PHYS-PUB-2020-010/fig_01.png

centre-of-mass energies, which allows an extra test of the SM behaviour. In reality, measurements of inclusive cross-sections test not only our knowledge of perturbative SM calculations, but also our knowledge of non-perturbative effects as encoded in parton distributions and Monte Carlo generators.

- **Differential cross-section measurements:** We saw in Chapter 1 that beyond the SM (BSM) physics would change the kinematics of particle production, which would modify the differential cross-sections of processes that are disturbed by the new physics. Measurements of differential cross-sections are also an essential component of improving theory calculations in regions of energy and momentum that have not been probed before, since the adequate description of these measurements may require higher-order perturbative calculations, or new tunings of the heuristic models that exist within Monte Carlo generators.

- **Particle masses:** The large number of W bosons, Z bosons and top quarks produced at the LHC makes it an ideal machine for measuring the masses of these particles with unprecedented precision, in addition to the fact that the LHC is the first collider that allows us to measure the mass of the Higgs boson.

- **Particle decay rates:** Physics beyond the SM can be expected to modify the decay rates of SM particles, with the Higgs boson decay rates being a particularly unexplored avenue for finding cracks in the SM due to the large remaining experimental uncertainties in several of the decay modes.

- **SM couplings:** The coupling constants of the SM can be extracted from a variety of LHC measurements, at different effective centre-of-mass energies, which allows us to also extract information on the running of the coupling constants. Since BSM physics would change the renormalisation group equations of the SM, and thus predict a different running of the coupling constants, such measurements provide a very generic way of searching for new particles.

4.3.2 Direct production of new particles

Compared to indirect measurements, searching for the direct production of new particles in the proton–proton collisions of the LHC has the potential to reveal much more about the type of new particles produced, and the interactions with SM and other particles. There are several possible options for new particle production at the LHC. For example, a new particle might be produced on its own, or in multiples, with the latter case requiring extra four-momentum. Alternatively, it might be the case that a particle can only be produced alongside one or more SM particles, one or more new particles, or some combination of the two. Given a theory of BSM particle physics, it is possible to calculate the cross-section of each possible production process, revealing which processes have a high enough rate to be visible at the LHC. For a particular production rate σ_{NP}, and integrated luminosity $\mathcal{L}_{\mathrm{int}}$, we can immediately tell if a process is likely to provide a viable discovery by calculating the expected number of events assuming the best possible (but unrealistic) case that all events pass the trigger conditions and are reconstructed with 100% efficiency. This is given by $N = \sigma_{NP} \times \mathcal{L}_{int}$, and if we obtain only a handful of events, it is not possible to discover that process using the assumed integrated luminosity. If, instead, N is greater than $\mathcal{O}(100)$ events, then we might wish to consider the scenario in more detail.

The next step is to consider how our new particles decay. There are three broad categories of particle decay that must be considered separately in LHC searches:

- **Prompt decay to visible particles:** It might be the case that a particle decays at the interaction point purely to SM particles that are visible to the ATLAS and CMS detectors, possibly by first producing intermediate particles that themselves decay visibly. In this case, the invariant mass formed from the sum of the four-momenta of the final decay products has a peak at the mass of the new particle, and we can perform a **resonance search** as described in Chapter 9.

- **Prompt decay to one or more invisible particles:** Many theories of physics beyond the SM feature a particle that has a small interaction strength with SM fields, either due to interactions via the weak force, or via a new force with similar strength to the weak force. Such particles would leave the ATLAS and CMS detectors without being seen, therefore giving rise to missing transverse momentum. Such particles may be produced directly, in which case one must make use of initial state radiation to see them at all, or they may be produced in the decays of SM particles or new particles. In both cases, one must rely on special techniques for semi-invisible particle searches that we describe in Chapter 10.

- **Non-prompt decay:** The new particles predicted by some BSM theories may not decay promptly, meaning that they leave the interaction point before decaying. If the lifetime of a particle is sufficient for it to leave the ATLAS or CMS detector before decaying, and it does not carry colour or electric charge, the phenomenology is identical to that of semi-invisible particle searches. However, if the particle does carry colour or electric charge, or it decays within the detector volume, one can generate a host of novel signatures, such as significantly displaced decay vertices, tracks with unusual properties, or short track segments. We briefly summarise techniques for such searches in Chapter 10.

In all of the above cases, the *signal* of new particle production does not occur in isolation, but instead is hidden in a *background* of SM processes that produce a similar detector signature. To take a trivial example, if we predict that a new particle will, once produced, decay immediately to two photons, we can look for evidence of the particle by looking at all LHC events that contain two photons. But this sample will be contaminated by *any* physics process that produces two photons, and we must find a way to distinguish the events arising from the new particle from the more boring events of the SM. The situation is not improved in the case of semi-invisible decays, since missing four-momentum can be produced by any SM process that results in neutrinos. To complicate things further, the backgrounds for new-physics searches at the LHC typically give us between three and six orders of magnitude more events than the signals we are looking for, based on the considerable hierarchy of cross-sections for SM processes that we saw in Figure 4.1.

4.4 Possible new-physics theories

We have now described various ways of searching for physics beyond the SM, but we have not yet explored any concrete ideas for what the new theory of particle physics might be. There are in fact an infinite number of theories that one might construct to supercede the SM, and theorists have spent the decades since the 1960s exploring a large number of them. Usually, the guiding principle in the design of these new theories has been a particular deficiency of the SM, such as the lack of a compelling dark matter candidate, or the hierarchy problem. Nevertheless, it remains very difficult to propose a theory that has no data validating its existence, and it is equally hard to search for a theory when you do not know exactly what you are looking for. In this section, we will review the most popular theoretical ideas, starting with the most generic.

4.4.1 Effective field theories

Imagine that we wish to proceed purely from the intuition that the SM is the lower energy limit of a more fundamental theory, but we wish to remain as agnostic as possible about what the higher energy theory is. Let us also assume that the energy scale of new physics is *above* our collider energy. We will now demonstrate that there is a natural way to parameterise the effects of the new physics at LHC energies, without ever having to be explicit about the extra particles in the new theory, through the use of an **effective field theory**.

Let us start with a particularly instructive example, namely the production of some new X particle of mass M_X via the process of Figure 4.2, where for the sake of argument we will take the decay products to be leptons.

In order to calculate the details of this process, we will need some Feyman rules, and it is sufficient for our purposes to assume that these are QED-like, so that the new particle couples to quarks and leptons according to the vertices of Figure 4.3. We will also take the propagator for the X particle to be the usual massive propagator:

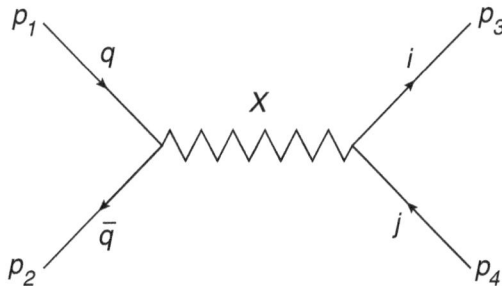

Figure 4.2. The production, and subsequent decay, of a new-physics particle.

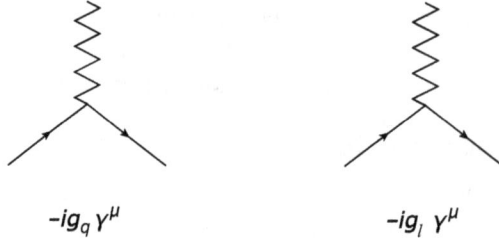

$$-ig_q \gamma^\mu \qquad\qquad -ig_l \gamma^\mu$$

Figure 4.3. Feynman rules for our example new-physics theory.

$$D_{\mu\nu} = -\frac{i}{q^2 - M_X^2}\left[\eta_{\mu\nu} - \frac{q_\mu q_\nu}{M_X^2}\right],$$

where q is its (off-shell) four-momentum. The diagram of Figure 4.2 then leads to the scattering amplitude

$$\mathcal{A} = \frac{g_q g_l}{s - M_X^2}[\bar{v}(p_2)\gamma^\mu u(p_1)][\bar{u}(p_3)\gamma^\nu v(p_4)]\left[\eta_{\mu\nu} - \frac{q_\mu q_\nu}{M_X^2}\right], \tag{4.3}$$

with

$$s = (p_1 + p_2)^2, \quad q^\mu = (p_1 + p_2)^\mu = (p_3 + p_4)^\mu.$$

The second term in the propagator in equation (4.3) (i.e. involving $q_\mu q_\nu$) vanishes if we substitute in for q_μ and q_ν and use the Dirac equations

$$\bar{v}(p_2)\not{p}_2 = \not{p}_1 u(p_1) = 0,$$

where we have taken the quarks and leptons to be approximately massless. Then equation (4.3) simplifies to

$$\mathcal{A} = \frac{g_q g_l}{s - M_X^2}[\bar{v}(p_2)\gamma^\mu u(p_1)][\bar{u}(p_3)\gamma_\mu v(p_4)]. \tag{4.4}$$

Now let us look what happens if we take the mass of the new particle to be much larger than the collision energy, i.e. $M_X^2 \gg s$. We can then Taylor expand the prefactor in equation (4.3):

$$\frac{1}{s - M_X^2} = -\frac{1}{M_X^2}\left[1 + \mathcal{O}\left(\frac{s}{M_X^2}\right)\right],$$

to give

$$\mathcal{A} = \frac{c}{\Lambda^2}[\bar{v}(p_2)\gamma^\mu u(p_1)][\bar{u}(p_3)\gamma_\mu v(p_4)]. \tag{4.5}$$

Here we have defined the (dimensionless) constant

$$c = g_q g_l,$$

and also introduced the energy scale Λ at which new physics occurs, which in the present case is simply

$$\Lambda = M_X.$$

Interestingly, the amplitude of equation (4.5) looks like it comes from a *single* Feynman rule, associated with the interaction Lagrangian

$$\mathcal{L} \sim \frac{c}{\Lambda^2}[\bar{\Psi}_q \gamma^\mu \Psi_q][\bar{\Psi}_l \gamma_\mu \Psi_l]. \tag{4.6}$$

Here Ψ_q and Ψ_l are the quark and lepton fields, respectively, and equation (4.6) is an example of a **four-fermion operator**, in which a pair of fermions and a pair of antifermions interact at the same point in spacetime. Provided we are at energies that are much less than Λ, we can use this interaction instead of the full theory involving the X particle. Indeed, there is actually a historical example of this: Fermi's original theory of weak nuclear decay in the 1930s involved just such four-fermion interactions. However, the theory breaks down at energies around the W boson mass, above which we have to use the full electroweak theory that was developed by Glashow, Weinberg and Salam in the 1970s.

We can make some further useful comments about this example:

- The effective interaction at low energy involves the SM fields only. Thus, we do not need to know the precise nature of the new physics to know that the new physics is there! It will show up as new interactions involving the SM fields.
- Assuming that the coefficient $c \lesssim 1$ (i.e. that the couplings in the new-physics theory are small enough to be perturbative), the interaction will be weak if Λ is large. Thus, the effective interaction of equation (4.6) will constitute a small (but non-zero) correction to the SM.
- Because of the inverse square of the new-physics scale in equation (4.6), the mass dimension of the four-fermion operator is 6 (recall that the Lagrangian has mass dimension 4 in four spacetime dimensions).
- Higher terms in the Taylor expansion of the propagator involve higher inverse powers of the new-physics scale, i.e. Λ^{-4}, Λ^{-6} and so on. Thus, they can be systematically neglected.

We have considered a particular example here, but the above story turns out to be fully general: *any* new physics looks like a set of effective interactions at energies $\ll \Lambda$, where Λ is the energy scale of the new physics. All of these interactions involve operators of mass dimension > 4, and containing SM fields only. Thus, we can account for the presence of new physics by modifying the SM Lagrangian as follows:

$$\mathcal{L}_{\text{BSM}} = \mathcal{L}_{\text{SM}} + \sum_{n=1}^{\infty} \sum_i \frac{c_i^{(n)}}{\Lambda^n} \mathcal{O}_i^{(n)}. \tag{4.7}$$

Here the first sum is over all mass dimensions greater than four, and the second sum is over a set of possible operators $\mathcal{O}_i^{(n)}$ that have that mass dimension. Each operator will be accompanied by some dimensionless **Wilson coefficient** $c_i^{(n)}$, and an inverse power of the new physics scale[2]. For each mass-dimension, there is a finite set of independent operators involving gauge-invariant combinations of SM fields, that can be classified on very general grounds, without assuming a particular new physics theory, e.g. in the four-fermion example above, we could simply postulate that there might be a four-fermion operator, without assuming that this comes from the X particle in the specific new-physics model we considered. At dimension-five, there is a single operator one can write down, that generates neutrino masses and mixings. At dimension-six, there are 59 independent operators. The choice of them is not unique, as one can use the equations of motion of the SM fields to replace operators in a given basis by superpositions of others. Different operator bases exist in the literature.

A given new-physics theory would completely fix the values of the coefficients and new-physics scale, i.e. the combinations

$$\frac{c_i^{(n)}}{\Lambda^n}.$$

We saw this explicitly in our four-fermion example: the coefficient c was fixed by the couplings of the X particle, and the new-physics scale was fixed by the mass M_X. If we do not know what the new-physics theory is, however, we can just assume that all of the operators at a given mass dimension may be present, and then fit their coefficients to the observed data, using the statistical techniques discussed in Chapter 5. Any conclusive measurement of $c_i^{(n)} \neq 0$ for some i and n would then constitute a discovery of new physics! We would then have some work to do in finding out *which* new-physics theories could reproduce the measured coefficient values. Nevertheless, this is a very nice way to search for new physics, as it involves almost no assumptions about what the new physics is. The only thing we must assume is that the energy of the new physics is sufficiently above the collider energies we are working at, so that we can sensibly truncate the expansion in the new-physics scale. The depressing lack of clear new physics at the LHC so far suggests this is a reasonable assumption to make.

The description of equation (4.7) is called an **effective field theory** (EFT) because it involves effective interactions that parametrise the effects of the underlying high-energy theory. This framework is currently being widely investigated by both theorists, and experimentalists working within the ATLAS and CMS collaborations. In particular, measurements of the top quark and Higgs boson are sensitive to a smaller number of operators, making fits to data easier. Most EFT fits currently focus on the Higgs and top-quark sectors (separately), for this reason. Typically, only operators up to dimension-six are included as these have been fully classified. It is then conventional to define the combinations (where we restrict to dimension-six and thus drop the (n) superscript):

---

[2] Here we have assumed a common new-physics scale as a way of organising the expansion. We could have taken a different new-physics scale for each operator. However, we could then simply absorb the difference in the scales into the coefficients $c_i^{(n)}$, leading to the same result as equation (4.7).

$$\bar{c}_i = \frac{c_i v^2}{\Lambda^2}, \tag{4.8}$$

where $v \simeq 246$ GeV is the Higgs vacuum expectation value. Given that new physics is expected to have something to do with the origin of the Higgs boson, this is a particularly convenient way to measure deviations from the SM, i.e. each $\bar{c}_i$ is dimensionless, and $\bar{c} \simeq 1$ would indicate new physics around the energy scale of electroweak symmetry breaking, provided the couplings in the new physics theory are $\mathcal{O}(1)$. This is just a choice, however, and constraints can be obtained using a different normalisation. What is clear, though, is that we cannot independently constrain the new physics scale and the operator coefficients: they always occur together in the EFT expansion. A typical procedure for measuring the $\{\bar{c}_i\}$ is then as follows:

1. Calculate the Feynman rules from each higher dimensional operator.
2. Choose observables which are sensitive to the new Feynman rules (e.g. total and differential cross-sections, decay widths...), and find data for them.
3. Calculate theory predictions for each observable, including the new Feynman rules.
4. For all data points X, construct a goodness of fit measure. A common one is

$$\chi^2(\{c_i\}) = \sum_X \frac{\left(X_{\text{th}}(\{c_i\}) - X_{\text{exp}}\right)^2}{\Delta_X^2}, \tag{4.9}$$

where the numerator contains the squared difference between the theory and experimental values at each data point X, and the denominator contains the theory and experimental uncertainties added in quadrature:

$$\Delta_X^2 = \Delta_{\text{th}}^2 + \Delta_{\text{exp}}^2. \tag{4.10}$$

Here the theory uncertainty includes all of the sources of uncertainty that we have seen throughout the previous chapters, e.g. renormalisation and factorisation scale variations, quoted uncertainties on the parton distribution functions, ambiguities in matching of matrix elements to parton showers, etc. The experimental uncertainty is sometimes quoted with separate statistical and systematic uncertainties, which can be added in quadrature themselves to produce a total. For differential cross-sections, experiments sometimes present correlation matrices describing the interdependence of uncertainties associated with different bins of the distribution. These correlations can be simply included by modifying equation (4.9) to include the quoted correlation matrix.

5. Maximise the likelihood function $\mathcal{L} = \exp[-\chi^2(\{\bar{c}_i\})]$ to find the best-fit point, using the frequentist approach to inference described in Chapter 5.

The result of such a fit will be a set of measured values $\{\bar{c}_i\}$, with some uncertainty bands. An example is shown in Figure 4.4, from an EFT fit in the top-quark sector.

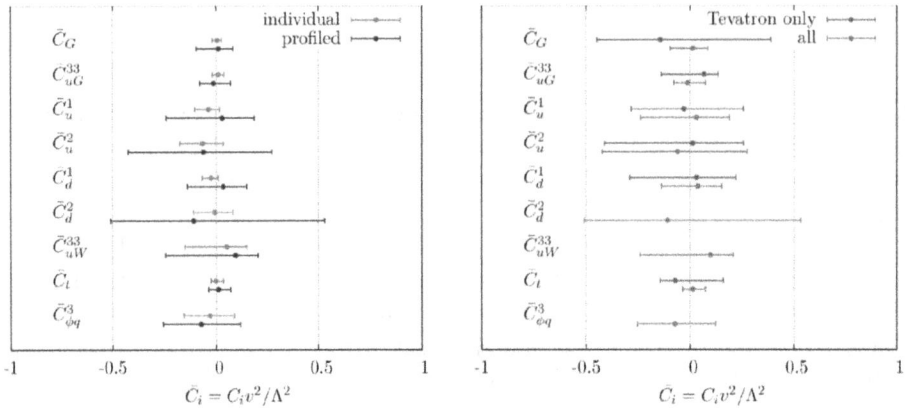

Figure 4.4. Constraints on operator coefficients from a fit of an effective field theory in the top-quark sector.

The various quantities shown are coefficients of combinations of effective operators that affect top-quark behaviour, and they have been obtained in two ways:

   (i) **Individual constraints** have all operator coefficients but one set to zero;

   (ii) **Profiled constraints** correspond to varying all coefficients simultaneously in the fit[3].

The second of these options is the more realistic one, given that all EFT operators are potentially present. However, it is conventional to present results for both choices, which itself can act as an important consistency check. Note that Figure 4.4 contains the results of both LHC and Tevatron data (left panel), and data from the Tevatron only (right panel). This illustrates the dramatic impact that LHC data is already having on constraining new physics.

One can also look at correlations between different operators (i.e., whether a high value for one coefficient might induce a high value for another). To this end, Figure 4.5 displays some two-dimensional slices through the space of operator coefficients, where the different coloured shapes represent the bounds on the value of coefficients arising from data, at different statistical confidence levels. The red star is the best fit point arising from the fit which, interestingly, is not the SM! However, all best fit points are statistically compatible with the origin: they lie within the dark blue shape, whose edge constitutes being one standard deviation away from the origin. It will be very interesting to see how such results evolve as more data from the LHC is implemented in the fits. This is particularly so given that EFT effects are expected to be more visible at higher energies. To see why, note that dimensional analysis implies that the factors Λ^{-n} involving the new physics scale Λ in the EFT expansion must be compensated by extra powers of energy/momentum in the operator itself—note that momenta appear as derivatives in position-space. These momenta end up in the Feynman rules, and often have the effect of boosting

---

[3] A related approach is to use all coefficients but to average rather than fit over the unshown ones: these are called marginalised constraints.

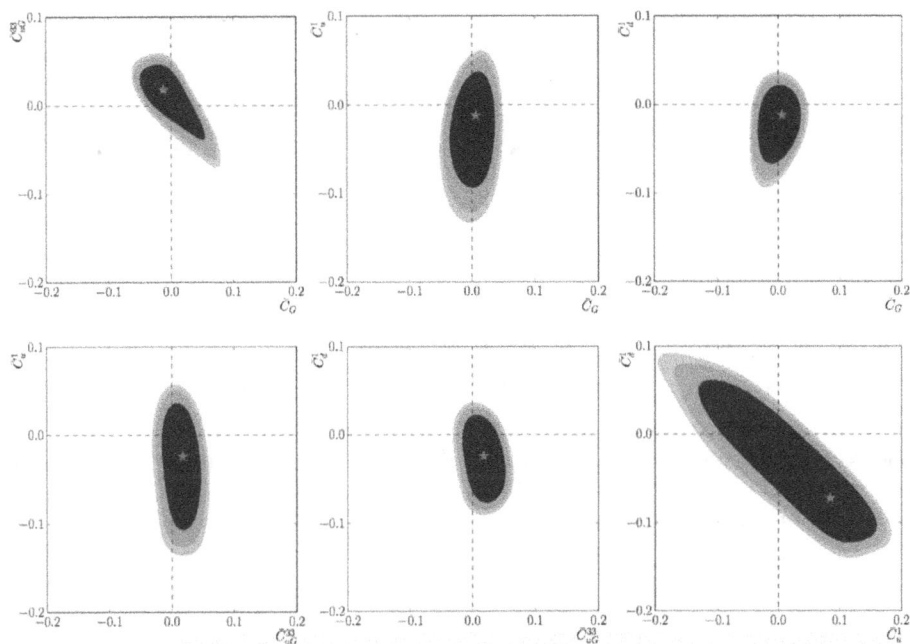

Figure 4.5. Two-dimensional slices through the operator coefficient space, showing constraints on pairs of operators. The red star is the best fit point, and the SM corresponds to the origin in each plot.

particles produced by the new effective vertices, relative to SM production. An example of this effect is shown in Figure 4.6, which shows the invariant mass of the top pair, and the transverse momentum of the top quark, in top pair production. The effect of including a higher dimensional operator (actually, a four-fermion operator in this case!) is to raise the tail of the distribution, producing more boosted top quarks, even though the total cross-section (the integral under the curve) does not change much. For this example, the data seems to favour the SM only, but in other examples the EFT effects are not yet constrained. Note that it follows from this that the jet substructure methods of Chapter 9 would be particularly useful for probing EFT corrections, as they are designed to analyse jets in the boosted kinematic regime.

4.4.2 Grand unification

What if we do not want to remain agnostic about the high scale physics that supercedes the SM? In that case, understanding more deeply how the SM was constructed in the first place often helps to suggest the ways that it might be extended to solve its deficiencies.

For our first example, consider the fact that the SM is a gauge theory, whose full gauge group is $SU(3) \times SU(2) \times U(1)$. The $SU(3)$ part corresponds to QCD, and the remaining $SU(2) \times U(1)$ gives rise to the electroweak forces. In total, there are three separate coupling constants, whose renormalised values run with energy. A schematic plot of this running is shown in Figure 4.7, and we see that the couplings

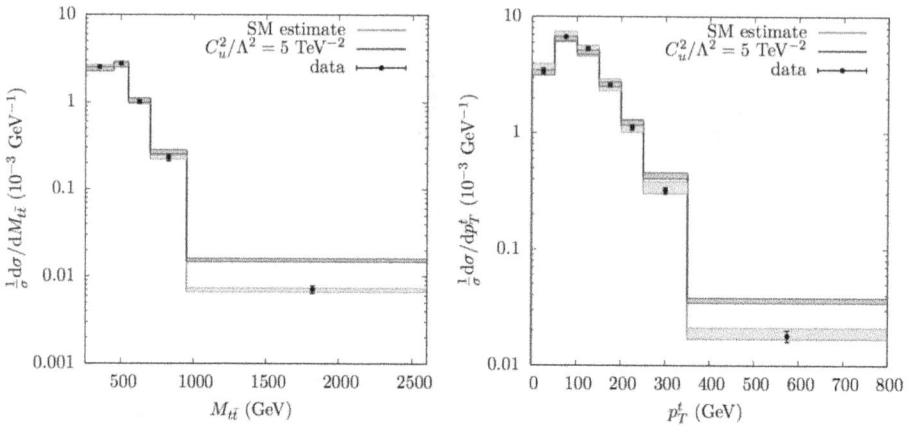

Figure 4.6. The invariant mass of the top-pair, and the transverse momentum of the top quark, in top pair production. The red distributions are obtained using the SM only, and the blue distributions include an additional four-fermion operator.

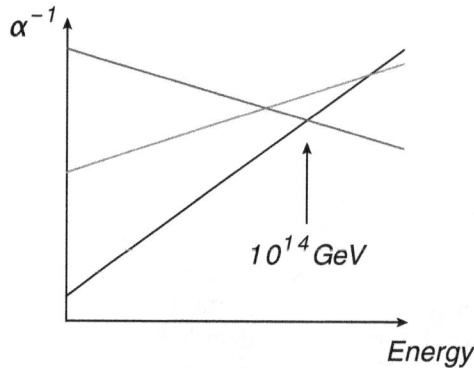

Figure 4.7. Variation of the inverse couplings of the SM with energy, showing that they do not quite meet each other at high energy.

get close to each other, but do not quite meet, at an energy scale of $\simeq 1 \times 10^{14}$ GeV. If they did meet, this could indicate that the three forces that we see at low energy are in fact multiple facets of a *single force*, such that the SM forces are unified into a single theory. This is conceptually very appealing, not least because it avoids having to answer the question of why the SM has three separate forces in it. Furthermore, the whole history of physics is that disparate phenomena are often unified through a few simple fundamental principles. One example of this is the unification of electricity and magnetism, which we now know is a consequence of gauge symmetry!

The idea that the SM forces could emerge from a single theoretical description is called **grand unification**, and such theories are known as **grand unified theories** (GUTs). This scenario was originally explored in the 1970s, and the original GUT theory proposed a single gauge group SU(5), which breaks down to SU(3) × SU(2) × U(1) at low energies. We have seen that non-Abelian gauge transformations act on vector-valued fields, and the vectors in the SU(5) theory contain different SM fields in the same vector, which then mix up under SU(5) gauge transformations. When one tries to put the SM fields into SU(5) vectors such that the correct quantum numbers emerge at low energies, one finds that quarks and leptons have to be in the *same* vector, so that e.g. a quark can change into a lepton by emitting an SU(5) gauge boson. This leads to the decay of the proton, and one can in principle calculate the decay rate from the SU(5) theory, which gives a direct bound on the energy scale at which the unification happens. Unfortunately, there are very strong constraints on proton decay, which rule out the simplest form of SU(5) unification. However, the idea has survived in other, more general, scenarios.

Although there are a lot of prototype GUT theories, it is possible to collect some common features. Firstly, the interactions that allow quarks to turn into leptons are mediated by particles that correspond either to the gauge bosons of the unified gauge group, as in our SU(5) example above, or the Higgs sector of the unified theory. The former are called **vector leptoquarks**, whilst the latter are called **scalar leptoquarks**. The vector leptoquark masses are typically of the order of the scale at which the SM gauge couplings unify, and they can only be directly accessible at the LHC if this unification scale is low enough. For the case of scalar leptoquarks, one can write a low energy effective field theory that describes their interactions, independently of the exact physics at higher energies. For vector leptoquarks, one must add assumptions regarding the higher-scale physics, since the couplings of vector leptoquarks to the SM gauge sector are not completely fixed by their quantum numbers under the SM gauge group. If leptoquarks have a mass of the order of a few TeV, i.e. within the LHC energy reach, they can be produced either singly or in pairs, through strong interactions. Whereas the cross-section for pair production is mostly model-independent, single leptoquark production always depends on details of the GUT model parameters. Experimental searches for both pair and single leptoquark production are based on their decay to a jet and a lepton, which can be reconstructed as a resonance. Many grand unified theories also include heavy sterile neutrinos that can be used to generate masses for the SM neutrinos. Such heavy neutrinos are also popular targets for LHC searches, which target processes such as $pp \to W \to N\ell \to W\ell\ell \to \ell\ell jj$, where N denotes the sterile neutrino, ℓ denotes a charged lepton, and j denotes a jet.

A number of GUT frameworks predict the existence of charged W' or neutral Z' vector bosons and/or a plethora of different scalar states, including such exotica as doubly-charged scalars. These can all be searched for using resonance searches.

4.4.3 Compositeness

Another way to extend the SM comes from revisiting the assumption that particles are fundamental, and thus completely point-like. This could fail at very short distances (equivalent to high energy scales): the SM particles could have a **substructure**, analogous to how the proton and neutron were once thought to be fundamental, until they were found to contain quarks and gluons. No collider has yet seen any evidence for fermion compositeness, but each increase in centre-of-mass energy of a collider allows us to zoom slightly further into the SM fermions. We can refer to the energy scale at which we would first start resolving substructure as Λ, which is referred to as the **compositeness scale**, and which can be thought of as the scale associated with interactions between the hypothesised particle constituents of the SM fermions.

The LHC signatures of fermion compositeness depend on Λ. For example, if Λ is much higher than the centre-of-mass energy of colliding quarks at the LHC, then compositeness will appear at the LHC as an effective four-fermion contact interaction (two of which fermions are the incoming quarks), similar to that of the effective field theory that we described earlier. One could assume various forms for this interaction, but it is common to restrict experimental studies to searching for evidence of specific subsets of the general behaviour, such as assuming that one only has flavour-diagonal, colour-singlet couplings between quarks. In that case, one can write the effective Lagrangian

$$\mathcal{L}_{qq} = \frac{2\pi}{\lambda^2}\Big[\eta_{LL}(\bar{q}_L\gamma^\mu q_L)(\bar{q}_L\gamma_\mu q_L) + \eta_{RR}(\bar{q}_R\gamma^\mu q_R)(\bar{q}_R\gamma_\mu q_R)$$
$$+ 2\eta_{RL}(\bar{q}_R\gamma^\mu q_R)(\bar{q}_L\gamma_\mu q_L)\Big], \tag{4.11}$$

where the subscripts L and R refer to the left- and right-handed quark fields, and η_{LL}, η_{RR} and η_{RL} are parameters that can take values of -1, 0 or $+1$. Different choices of the η parameters correspond to different contact interaction models, that can be constrained separately from LHC observations. This Lagrangian supplies us with the method of searching for quark compositeness in the case of high Λ: we look for events with two jets, and compare the kinematics of the jets in the final state with those expected in the case of the SM. We could do this using a variety of kinematic variables, but a well-motivated choice is to look at the dijet angular distribution

$$\frac{1}{\sigma_{\text{dijet}}}\frac{d\sigma_{\text{dijet}}}{d\chi_{\text{dijet}}},$$

where $\chi_{\text{dijet}} = \exp|y_1 - y_2|$, and y_1 and y_2 are the rapidities of the two jets. This looks rather obscure, but it can shown that, in the limit of massless scattering quarks,

$$\chi_{\text{dijet}} = \frac{1 + |\cos\theta^*|}{1 - |\cos\theta^*|},$$

where θ^* is the polar scattering angle in the centre-of-mass frame of the scattering quarks. For normal QCD dijet processes, $d\sigma_{\text{dijet}}/d\chi_{\text{dijet}}$ is approximately uniform, whereas composite models predict a distribution that is strongly peaked at low values of χ_{dijet}. This makes it a very effective distribution for revealing evidence of compositeness. Note that one could easily extend equation (4.11) with terms that feature leptons and quarks, or leptons and leptons, in which case one can also search for compositeness in dilepton final states.

Compositeness models also predict the existence of excited quarks and leptons, which are analogous to the excited states of atoms or nuclei that you may be familiar with from your undergraduate studies. These states are usually referred to as q^* and

ℓ^* for the quark and lepton case, respectively, where an excited fermion is a heavy fermion that shares quantum numbers with one of the existing SM fermions. The mass of the excited fermions is expected to be similar to the compositeness scale, whilst their interactions with SM particles are model-dependent. Thus, if Λ is within LHC reach, one could search for these excited states directly. For both excited quarks and excited leptons, a search can be performed by looking for a narrow resonant peak in the invariant mass distribution of the excited fermion decay products. For excited quarks, these decay products are typically two jets (instigated by a quark and a gluon), a jet and a photon, or a jet and a weak gauge boson. For excited leptons, one can instead search for decays to a lepton and a photon, or to a lepton and a weak gauge boson.

A separate class of composite theories comes from considerations of the hierarchy problem, which ultimately results from the fact that the Higgs boson is posited to be a fundamental scalar. It transpires that the quantum corrections which shift the Higgs boson mass to very large values (unless we apply fine-tuning) can be evaded if the Higgs boson is a composite particle. Several theories of a composite Higgs boson have been developed over the past few decades, and we will here briefly describe how a particular modern variant evades the hierarchy problem. The best way to explain it is to take a quick detour into an effect that is encountered in QCD, called **chiral symmetry breaking**.

Let us start with a quick revision of some basic QCD facts. Firstly, in the early universe (at high temperatures), quarks and gluons existed as the relevent degrees of freedom in the quark-gluon plasma. At some point, when the temperature of the Universe dropped below $T_C \sim \Lambda_{\mathrm{QCD}}$, the quarks were confined into hadrons, and the Universe became full of composite states of strongly-coupled particles. This is called the **QCD phase transition**. The mesons contain a quark/anti-quark pair, and their masses are typically clustered around the scale Λ_{QCD}. The pions, however, are very much lighter, and it will turn out that the explanation for this could also help to explain why the Higgs boson is very much lighter than might naïvely be expected.

We will proceed by writing the QCD Lagrangian in a simplified form, including only up and down quarks:

$$\mathcal{L} = \frac{1}{4}(F_{\mu\nu}^a)^2 + i\bar{u}\slashed{D}u + i\bar{d}\slashed{D}d - m_u\bar{u}u - m_d\bar{d}d. \qquad (4.12)$$

We will further assume for simplicity that these quarks are massless, which is in fact a good approximation in the present case, since the only relevant fact is that the masses are much smaller than Λ_{QCD}. In that case, and using the fact that the chiral components of the spinors are given by $\psi_{R/L} = \frac{1}{2}(1 \pm \gamma_5)\psi$, we can write the Lagrangian as

$$\mathcal{L} = \frac{1}{4}(F_{\mu\nu}^a)^2 + i\bar{u}_L\slashed{D}u_L + i\bar{u}_R\slashed{D}u_R + i\bar{d}_L\slashed{D}d_L + i\bar{d}_R\slashed{D}d_R. \qquad (4.13)$$

This form has a curious property. It is invariant under separate SU(2) rotations of the left- and right-handed quark components:

$$\begin{pmatrix} u_L \\ d_L \end{pmatrix} \rightarrow g_L \begin{pmatrix} u_L \\ d_L \end{pmatrix}, \quad \begin{pmatrix} u_R \\ d_R \end{pmatrix} \rightarrow g_R \begin{pmatrix} u_R \\ d_R \end{pmatrix}, \tag{4.14}$$

where g_L and g_R are SU(2) transformations from the two groups SU(2)$_L$ (acting only on left-handed components), and SU(2)$_R$ (acting only on right-handed components). We can write the total symmetry as SU(2)$_L$ × SU(2)$_R$, and this is called a **chiral symmetry** since it acts differently on left- and right-handed components. It can be shown[4] that a general transformation of this kind can be written as

$$\begin{pmatrix} u \\ d \end{pmatrix} \rightarrow \exp\{i\theta_a \tau^a + \gamma_5 \beta_a \tau^a\} \begin{pmatrix} u \\ d \end{pmatrix}, \tag{4.15}$$

for the doublet of Dirac spinors u and d. If we set $\beta_a = 0$ and vary θ_a, we get a set of transformations called isospin (not to be confused with the weak isospin that we saw in Chapter 2!), which form a diagonal subgroup of the group SU(2)$_L$ × SU(2)$_R$.

After the QCD phase transition, the ground state of QCD turns out to have[5] a non-zero vacuum expectation value for the quark bilinears $\bar{u}u$ and $\bar{d}d$ given by

$$\langle \bar{u}u \rangle = \langle \bar{d}d \rangle \sim \Lambda_{\text{QCD}}^3. \tag{4.16}$$

No matter how this arose (and we have yet to prove this from QCD itself), we can note that it is not invariant under the chiral symmetry, but it is invariant under the isospin subgroup that operates on the left- and right-handed components in the same way. Thus, the chiral symmetry of QCD is spontaneously broken as SU(2)$_L$ × SU(2)$_R$ → SU(2)$_{\text{isospin}}$. This symmetry breaking comes associated with Goldstone bosons, and it is exactly these Goldstone bosons that we identify with pions. In the limit of exact chiral symmetry, these pions would be massless, but the small explicit breaking of chiral symmetry that results from the up and down quarks having a small mass means that the pions are instead **pseudo-Goldstone bosons** that have a finite (but small) mass. The beauty of effective field theory is that, as in our previous examples, we can write a low-energy theory of the pions without knowing the full details of the higher-energy physics, which means that we can determine most of how the pions interact purely from the application of effective field theory, combined with our knowledge of the symmetry breaking pattern.

How does this relate to the Higgs boson? Imagine that we have a new, strongly-coupled sector of the theory of particle physics, which adds new QCD-like phenomena at higher energies. Thus, there would be new fields like quarks and gluons that kick in at high energies, but below a certain temperature those states would confine into composite objects. Imagine also that this theory has some accidental symmetry which is spontaneously broken to a subgroup, along with a small degree of explicit symmetry breaking. We would get a tower of resonances that correspond to the 'mesons' and 'baryons' of this new sector, and we would also get

---

[4] The form of a general chiral transformation is a special case of Lie's theorem of equation (2.33), where the two terms in the exponent on the right-hand side of equation (4.15) constitute different generators (recall that γ_5 contains a factor of i).

[5] Evidence for this proposal comes from phenomenological models, as well as lattice QCD studies.

pseudo-Goldstone modes from the symmetry breaking that could play the role of the Higgs boson. This is the essence of how modern composite Higgs theories work, although the details become very complicated very quickly. The effect on LHC physics is to modify the couplings of the Higgs boson, and also to introduce new exotic particles such as vector-like quarks which have spin-1/2, and transform as triplets under the SU(3) group, but which have exotic quantum numbers compared to SM quarks. LHC searches for composite Higgs models thus partly rely on trying to find the production and decays of these exotic new states, and partly on shrinking the error bars on the Higgs decay modes.

4.4.4 Supersymmetry

Making the Higgs composite is one way to solve the hierarchy problem, but there is another popular method. The SM is based heavily on symmetry, of which the two main kinds are:

(i) **Poincaré symmetry**, described by the **Poincaré group** of Lorentz transformations plus translations. These symmetry transformations are associated with spacetime degrees of freedom (e.g. positions and momenta).

(ii) **Gauge symmetries**, described by Lie groups acting on an abstract **internal space** associated with each field.

Given how useful symmetry has been in guiding the construction of the SM, it is natural to ponder whether one can make an even more symmetric theory, which combines spacetime and internal symmetries in a non-trivial way. Unfortunately, something called the **Coleman–Mandula theorem** tells us that this is impossible, and that the only way to combine symmetries of types (i) and (ii) above is as a direct product,

Symmetry group of any theory = (Poincaré) × (Internal Lie group),

with no mixing between them. Note that both of the symmetries are described by Lie groups (i.e. the Poincaré group is itself a Lie group). However, the Coleman–Mandula theorem only applies if we require the total symmetry group to be a Lie group. There are in fact other types of interesting mathematical structure, and we can use them to build a theory instead. In particular, one can extend the Poincaré group to something called the **super-Poincaré** group, whose associated algebra is called a **graded Lie algebra** rather than a standard **Lie algebra**. It has the normal Poincaré group as a subgroup, but also includes extra transformations that relate bosonic and fermionic degrees of freedom. To implement this structure, we have to extend spacetime to something called 'superspace', which has additional fermionic (anticommuting) coordinates, in addition to the usual spacetime ones.

If we build the SM on such a space, we get something that looks just like the SM in four conventional spacetime dimensions, but where every bosonic field has a fermionic counterpart, and vice versa. Whether or not we consider 'superspace' to be real or not is up to us—at the very least, we can regard it as a convenient mathematical trick for generating QFTs with more symmetry in them.

The symmetry between bosonic and fermionic fields is called **supersymmetry** (SUSY), and the partner for each SM particle is called a **superpartner**, with various different naming conventions, e.g. you will see **squarks**, **sleptons**, **sneutrinos** (all spin 0). The boson superpartners have spin 1/2, and are typically labelled with the suffix '-ino', e.g. the **wino**, **bino**, **gluino**, **Higgsino**, etc. Note that these partners are defined before electroweak symmetry breaking, so that they correspond to partners of the unbroken SM fields.

There are different ways of making the SM supersymmetric, and the usual variant that people study is called the **minimal supersymmetric SM** (MSSM). As the name suggests, this is the supersymmetric theory that requires a minimum of extra particle content beyond the SM. In the SM, particles are embedded in various representations of the symmetry group of the SM. SUSY theories work in the same way, except that the irreducible representations of the SUSY algebra are now called **supermultiplets** rather than multiplets. Each supermultiplet contains both boson and fermion states, which are the superpartners of each other, and the members of the same supermultiplet must have the same mass, electric charges, weak isospin and colour degrees of freedom. This means that the gauge couplings of the new SUSY particles will be the same as those of their SM partners. One can immediately ask what the simplest possibilities for supermultiplets are. In the MSSM, the various possible combinations of particles in a supermultiplet can always be reduced to combinations of **chiral supermultiplets** (containing a 2-component *Weyl fermion* and two real scalars that are represented by a single complex scalar field), and **gauge supermultiplets** (containing a massless spin-1 gauge boson and a massless spin-1/2 Weyl fermion). These both satisfy a rule stating that the number of bosonic degrees of freedom in a supermultiplet must equal the number of fermionic degrees of freedom and, in addition, it can be shown that the fermions in a gauge supermultiplet must have the same gauge transformation properties for left-handed and right-handed components. This is the reason that SUSY requires new particles: the particles of the SM alone cannot form the entire particle content of the MSSM, as all of the quarks and leptons of the SM have left- and right-handed parts that transform differently under the gauge group, and hence they must all be in chiral supermultiplets. New particles are needed to fill the fermionic degrees of freedom in the gauge supermultiplets, and there is also a need for new particles that fill the bosonic degrees of freedom in the chiral supermultiplets.

The supermultiplets of the MSSM are shown in tables 4.1 and 4.2. Note that the left-handed and right-handed components of the SM quarks and leptons are separate two-component Weyl fermions, and hence each must have its own scalar partner. These **squarks** and **sleptons** are denoted by putting a tilde over the corresponding SM partner, and the subscripts L and R refer to the handedness of the SM partner (the sparticles have spin 0 and hence are neither right- nor left-handed). The gauge interactions of the sparticles are the same as their SM partners.

The Higgs sector of the MSSM requires some explanation, as it is more complicated than that of the SM. It is clear that the Higgs boson of the SM must exist in a chiral supermultiplet given that it has spin 0, though a less obvious fact is that one actually requires two Higgs doublets to complete the MSSM. One of the

Table 4.1. The chiral supermultiplets of the MSSM.

Names		Spin-0	Spin-1/2
squarks, quarks Q		$(\tilde{u}_L \ \tilde{d}_L)$	$(u_L \ d_L)$
$\times 3$ families	$\bar{u}$	$\tilde{u}_R^*$	$u_R^\dagger$
	$\bar{d}$	$\tilde{d}_R^*$	$d_R^\dagger$
sleptons, leptons L		$(\tilde{\nu}_L \ \tilde{e}_L)$	$(\nu_L \ e_L)$
$\times 3$ families	$\bar{e}$	$\tilde{e}_R^*$	$e_R^\dagger$
Higgs, Higgsinos H_u		$(H_u^+ \ H_u^0)$	$(\tilde{H}_u^+ \ \tilde{H}_u^0)$
H_d		$(H_d^0 \ H_d^-)$	$(\tilde{H}_d^0 \ \tilde{H}_d^-)$

Table 4.2. The gauge supermultiplets of the MSSM.

Names	Spin-1/2	Spin-1
gluino, gluon	$\tilde{g}$	g
winos, W bosons	$\tilde{W}^\pm \ \tilde{W}^0$	$W^\pm \ W^0$
bino, B boson	$\tilde{B}^0$	B^0

reasons for this results from the general structure of supersymmetric theories, in which it can be shown that only a $Y = +1/2$ Higgs chiral supermultiplet can have the Yukawa couplings necessary for charge $+2/3$ quarks, and only a $Y = -1/2$ Higgs can give the right couplings for charge $-1/3$ quarks and charged leptons. This gives us the two complex $SU(2)_L$ doublets H_u and H_d shown in table 4.1, containing eight real, scalar degrees of freedom. When electroweak symmetry breaking occurs in the MSSM three of them form Goldstone bosons which become the longitudinal modes of the Z_0 and $W^\pm$ vector bosons, and the remaining five give us Higgs scalar mass eigenstates consisting of one CP-odd 8 neutral scalar A_0, a charge $+1$ scalar H^+ and its conjugate H^-, and two CP-even neutral scalars h^0 and H^0. Whilst the masses of A^0, H^0 and $H^\pm$ can be arbitrarily large, one can set an upper bound on the h^0 mass; the observed Higgs mass of 125 GeV is consistent with this upper bound. Had it been observed to have a larger mass, the MSSM would already have been excluded.

The physical particles that exist in SUSY, and which we aim to discover at the LHC, do not match the fields shown in tables 4.1 and 4.2. Instead, the fields mix in the MSSM to produce the following physical eigenstates:

$$H_u^0, H_d^0, H_u^+, H_d^- \to h^0, H^0, A^0, H^\pm \ \text{(Higgs)}$$

$$\tilde{t}_L, \tilde{t}_R, \tilde{b}_L, \tilde{b}_R \to \tilde{t}_1, \tilde{t}_2, \tilde{b}_1, \tilde{b}_2 \ \text{(stop/sbottom)}$$

$$\tilde{\tau}_L, \tilde{\tau}_R, \tilde{\nu}_\tau \to \tilde{\tau}_1, \tilde{\tau}_2, \tilde{\nu}_\tau \ \text{(stau)}$$

$$\tilde{B}^0,\ \tilde{W}^0,\ \tilde{H}_u^0,\ \tilde{H}_d^0 \to \tilde{\chi}_1^0, \tilde{\chi}_2^0, \tilde{\chi}_3^0, \tilde{\chi}_4^0 \ \text{(neutralinos)}$$

$$\tilde{W}^\pm,\ \tilde{H}_u^+,\ \tilde{H}_d^- \to \tilde{\chi}_1^\pm, \tilde{\chi}_2^\pm \ \text{(charginos)}$$

where the degree of mixing in the squark and slepton sectors is typically proportional to the mass of the associated fermion, and thus assumed to be largest for the third family.

Although the original motivation was rather formal (i.e. the Coleman–Mandula theorem mentioned above), the MSSM turns out to have some remarkable features:

- The couplings in the MSSM unify at high energy, in contrast to the behaviour of Figure 4.7. Furthermore, there is a mechanism (albeit usually put in by hand) to get rid of proton decay.
- There is a natural dark matter candidate, in that the lightest superpartner is stable.
- The hierarchy problem is resolved, as Higgs mass corrections become logarithmic rather than quadratic, thus dramatically reducing the sensitivity of the Higgs mass to the scale of new physics.
- Our only known candidate theory for quantum gravity plus matter (string theory) seems to require SUSY to be consistent.

Just one of these features would be nice, but to have all of them feels extremely compelling to many people. For this reason, SUSY theories have received a massive amount of attention over the past 30 years or so.

If SUSY were an exact symmetry of Nature, the sparticles introduced above would have the same masses as their SM counterparts and would have been seen already in collider experiments. Sadly, this is not the case, and there remains no direct experimental evidence for supersymmetry. Any valid supersymmetric theory must therefore introduce a mechanism for supersymmetry breaking, and one can introduce spontaneous symmetry breaking in a way directly analogous to the treatment of electroweak symmetry breaking in the SM. Many models of spontaneous symmetry breaking have been proposed, and there is no general consensus on which is the correct mechanism. One option is simply to parameterise our ignorance, and write the most general gauge-invariant Lagrangian that explicitly adds SUSY-breaking terms, and for the MSSM it can be shown that this adds 105 free parameters to the SM, in the form of the masses of the new particles, and various CP-violating phases and mixing angles. Some of these are already constrained to be near-zero by experimental measurements (such as measurements of the lifetime of the proton), and yet others have no effect on the behaviour of superpartners at the LHC. There are roughly 24 parameters that are required to describe LHC phenomenology, and these are often compressed to 19 parameters by, for example, assuming that the first and second generation of squarks and sleptons have equal masses (which would not be clearly distinguishable at the LHC in the case of squarks). This 19-parameter model is frequently referred to as the **phenomenological MSSM** (pMSSM), and it is common to interpret the results of LHC sparticle searches in either the pMSSM, or some subset of its parameters.

Other attempts at describing SUSY breaking make speculative assumptions about the physics at high energy scales, before relating this to LHC measurements by using the appropriate set of renormalisation group equations that describe the running of the sparticle masses and couplings. Models include supergravity scenarios, anomaly-mediated SUSY breaking, and gauge-mediated SUSY breaking, and we refer the reader to the literature for the details.

4.4.5 Extra dimensions

Before concluding this chapter, let us consider one final way to solve the hierarchy problem. We saw that the Higgs mass acquires a divergence in the SM that is quadratic in the cut-off scale that we assume for the theory. In principle, the highest scale that we might expect for this cut-off would be the Planck scale $M_{\mathrm{Pl}} \sim 10^{16}$ TeV, which is believed to be the scale at which gravity becomes strong, and its quantum effects become important. It is this large discrepancy between the Planck scale and the electroweak scale (i.e. the scale of the W and Z boson masses) that makes gravity so weak compared to the electroweak and strong interactions.

Attempts to build a quantum theory of gravity, such as string theory, have inspired a variety of theories that contain extra spatial dimensions. Naïvely, these would seem to contradict the obvious fact that we live in only three spatial dimensions. However, there are two main possibilities that allow for the presence of extra dimensions without breaking the reality of our common experience. Firstly, the extra dimensions may be curled up on very small scales, so that we do not see them. Secondly, there are theories in which we live on a 4D spacetime slice (or **brane**) of a larger dimensional world. The relative weakness of gravity then results from its propagation through the 'bulk' containing the extra dimensuions, which makes it appear artificially weak on our brane. Most of the work on LHC searches for extra dimensions is focussed on the following models:

- The **Large Extra Dimensions** model of Arkani-Hamed, Dimopoulos and Davil (ADD), which posits the existence of n compact, extra dimensions with radius R. The important feature of this model is that the Planck mass that we are used to, M_{Pl}, is a feature of our brane only, and it is related to the fundamental Planck mass M_D by the formula

$$M_D = \left(\frac{M_{\mathrm{Pl}}^2}{R^n} \right)^{-1/(2+n)} \tag{4.17}$$

The hierarchy problem is resolved by assuming that $M_D \sim 1$ TeV, i.e. that it actually *is* close to the electroweak scale. The name *Large* results from the fact that R could be of the order of 1 mm to 1 mm. More precisely, the radius of the extra dimensions is given by

$$R = 2 \times 10^{-17+32/n} \text{ cm}, \tag{4.18}$$

with only $n = 1$ excluded by experiment.

- The **Warped Extra Dimensions** model of Randall and Sundrum (RS) which posits a single new compact extra dimension that is warped with a curvature of $k \sim M_{Pl}$. Only gravity can propagate in the extra dimension, and the model consists of a 5D bulk with one compactified dimension, and two 4D branes, called the SM and gravity branes. It can be shown that the relationship between the fundamental Planck mass and our usual one is now given by:

$$M_D = M_{Pl}e^{-k\pi R}. \tag{4.19}$$

If $R = 1 \times 10^{32}$ cm, we get $M_D \sim 1\text{TeV}$, which again eliminates the hierarchy problem.

Theories with extra spatial dimensions turn out to have a variety of consequences at the LHC. The first class of observations results from the propagation of **gravitons** through the compact extra dimensions. It can be shown that the virtual exchange of these states would appear as massive new resonances, called **Kaluza-Klein (KK) excitations**. In the ADD model, the mass of these resonances is given by

$$m_k^2 = m_0^2 + k^2/R^2, \quad k = 0, 1, 2, 3, 4, \dots \tag{4.20}$$

Note that these are regularly spaced and, for large R, the KK resonances are almost continuous. At the LHC, one can search for the direct production of these KK graviton resonances via processes such as $q\bar{q} \to gG$, $qj \to qG$ and $gg \to gG$, which would appear as a monojet signature given that the graviton escapes without interacting with the detector. One can also look for s-channel KK graviton exchange, with a decay to diboson or dilepton resonances.

One can also search for KK graviton resonances in the warped extra-dimension scenario, but the details are very different. Rather than being evenly-spaced, the KK graviton masses are now given by $m_n = x_n k e^{-k\pi R}$, where x_n are the roots of Bessel functions. It is usually the case that only the first resonance (i.e. $n = 1$) is accessible at the LHC, and it has a narrow width given by k/M_{Pl}. On the plus side, it has a coupling to SM particles that is proportional to $1/M_{Pl}e^{-k\pi R}$, which is much stronger than in the ADD model. Further phenomenology in the warped extra dimension case comes from extending the simplest RS model to include the possibility that SM fields can also propagate in the bulk. If that were possible, *all* of the SM fields would create KK towers of resonances, which gives us much more to search for!

For the ADD model, there is a second class of observations that is particularly exciting, namely the search for the formation of microscopic black holes. These are able to form once the collision energy rises above a certain threshold M_{thresh}, which is above M_D, but typically well below M_{Pl}. Black holes that are produced with an energy far above M_{thresh} (called the **semi-classical** case) would decay to a high multiplicity final state via Hawking radiation, and one can search for the production of many high-p_T objects. If the production is instead near the threshold, the theory suggests that a quantum black hole would form, which decays to a two-body final state. Although no actual resonance is produced, the kinematics in the final state mimic a resonance, producing a broad bump at a given invariant mass of the black hole decay products.

In this chapter, we have provided a brief summary of the kinds of new-physics theories that could be measured at current and forthcoming collider experiments. In order to understand how to analyse such experiments, however, we need to know much more about how to interpret what we are doing. This unavoidably leads us to the subjects of probability and statistics, which we examine in detail in the following chapter.

Further reading

- An excellent, LHC-centric review of grand unified theories can be found in 'GUT Physics in the Era of the LHC' by Djuna Croon, Tomás Gonzalo, Lukas Graf, Nejc Košnik and Graham White (*Frontiers in Physics*, Volume 7).
- Many students over the years have learned supersymmetry from "A Supersymmetry Primer" by Stephen Martin (available for free at https:// www.niu.edu/spmartin/primer/). An excellent textbook for beginners that covers the collider phenomenology in detail is "Weak Scale Supersymmetry: From Superfields to Scattering Events" by Howard Baer and Xerxes Tata (Cambridge University Press).
- A superb introduction to composite Higgs theories is given in "Tasi 2009 lectures: The Higgs as a Composite Nambu-Goldstone Boson" by Roberto Contino, available for free at https://arxiv.org/abs/1005.4269.
- A pedagogical review of LHC searches for extra dimensions can be found in the ever-reliable Particle Data Group review (see https://pdg.lbl.gov/2020/ reviews/rpp2020-rev-extra-dimensions.pdf for the most recent article from Y Gershtein and A Pomarol).

Exercises

4.1 Consider a scalar particle of mass m with a quartic self-interaction, as shown in Figure 4.8.

  (a) If the Feynman rule for the self-interaction vertex is proportional to some self-coupling λ, explain why the diagram leads to an expression

$$\sim \lambda \int \frac{d^4 k}{k^2 - m^2}.$$

  (b) Let Λ be the scale at which the theory is expected to break down (e.g. where new physics may enter). Explain why the mass counterterm for the scalar field behaves as

$$\delta m^2 \sim \Lambda^2.$$

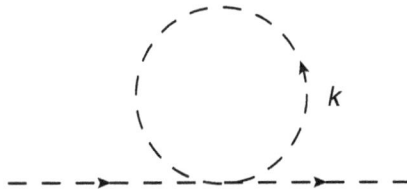

Figure 4.8. Loop diagram in a scalar theory with a quartic self-interaction term.

(c) Explain why this is a potential problem for the SM.

(d) How does this problem show up if dimensional regularisation is used, rather than a momentum cutoff?

4.2

(a) Explain what is meant by the idea of grand unification.

(b) Why might such theories lead to proton decay?

(c) How would you put constraints on the lifetime on the proton?

4.3 Consider that we live in five spacetime dimensions, with spatial coordinates (x, y, z, x_5).

(a) Explain why a free scalar field Φ of mass m living in this space satisfies the equation

$$\left(\frac{\partial^2}{\partial t^2} - \nabla_5^2 + m^2\right)\Phi(x, y, z, x_5, t) = 0,$$

where

$$\nabla_5^2 = \frac{\partial^2}{\partial x^2} + \frac{\partial^2}{\partial y^2} + \frac{\partial^2}{\partial z^2} + \frac{\partial^2}{\partial x_5^2}.$$

(b) Now consider that the extra coordinate x_5 is compactified to form a circle of radius R. Explain why the field can then be written as

$$\Phi(x, y, z, x_5, t) = \sum_{n=-\infty}^{\infty} \Phi_n(x, y, z, t) \exp(inx_5/R).$$

(c) Hence show that the scalar field equation from part (a) implies the infinite set of equations

$$\left(\frac{\partial^2}{\partial t^2} - \nabla_4^2 + m^2 + \frac{n^2}{R^2}\right)\Phi_n(x, y, z, t) = 0.$$

where

$$\nabla_4^2 = \frac{\partial^2}{\partial x^2} + \frac{\partial^2}{\partial y^2} + \frac{\partial^2}{\partial z^2}.$$

Interpret this result.

IOP Publishing

Practical Collider Physics

Andy Buckley, Christopher White and Martin White

Chapter 5

Statistics for collider physics

Statistics is fundamental to drawing conclusions in particle physics. From designing reconstruction algorithms, to optimising the expected performance of an analysis, to projecting data measurements into theory models (or vice versa), everything is built on a foundation of probability and statistics. And unlike many subjects, this probabilistic nature is not just a practical necessity, but a fundamental one: when the partonic final-state of a particle collision is compatible with several classes of contributing amplitudes, there is no way in-principle to measure 'which process happened' with a sufficiently advanced detector: the only way to test models is via probability distributions and their comparisons.

It is hence worth taking some time to construct the main structures and concepts on which so much of our reasoning will rely, from the perspective of a particle physicist rather than a formal mathematician or abstract statistician. More specific applications of these ideas will return in the physics-oriented Chapters of Part II. Unfortunately, the study of probability is made difficult by the fact that we live in a society that fundamentally misunderstands it, a misunderstanding that manifests itself in several ways. Firstly, we grow up with a vocabulary that includes words like 'probable' and 'likely', without ever really defining what they are or what they mean. Secondly, we develop a very poor understanding of risk from an early age, due to being bombarded by media that abuse statistics on a daily basis to convince us that completely expected events are extraordinarily unusual. As a result, many straightforward results in probability theory will initially seem to be counter-intuitive.

5.1 What is probability?

Let us start with a concise definition of what the word 'probability' means. Although the concept has existed in colloquial form for centuries, the modern mathematical definition can be traced back to Kolmogorov's 1933 text *Foundations of the Theory of Probability*. We first assume that some random events E_i are occurring, such that, if one of them happens, the others do not happen. Such events are called **exclusive**.

The probability P of a particular event E_i occurring is then defined by the following **Kolmogorov axioms**:

1. $P(E_i) \geqslant 0$ for all i
2. $P(E_i \text{ or } E_j) = P(E_i) + P(E_j)$
3. $\sum_\Omega P(E_i) = 1$

where Ω is the set of all possible events. These axioms may be familiar to you from your previous study of probability. What may be less familiar is the fact that these really define an abstract, mathematical quantity, and anything that satisfies these axioms can be said to be a probability. The definition is therefore not unique!

Kolmogorov's analysis is based on set theory, and the three axioms above can be used to derive a series of properties that any definition of probability must satisfy. For example, imagine that we have two sets of events A and B, that are non-exclusive subsets of our total event set Ω. Non-exclusive means that some of our fundamental events E_i may be in both A and B. The probability of an event occurring which is either in A alone, B alone, or both A and B is then given by

$$P(A \text{ or } B) = P(A) + P(B) - P(A \text{ and } B), \qquad (5.1)$$

where A and B denotes events in both A and B, and $P(A)$ means 'an event has occurred which is in the set A'. Below, we will shorten this to 'A has occurred'. It is necessary to subtract the last term to avoid double-counting events that are in both A and B when we calculate the probability on the left-hand side.

Conditional probability and Bayes' theorem

Now imagine that an event E_i is known to belong to the set B. We can define the **conditional probability** that the event also belongs to the set A as $P(A|B)$, which is pronounced 'the probability of A given B'. Conditional probability is defined via the relation

$$P(A \text{ and } B) = P(A|B)P(B) = P(B|A)P(A). \qquad (5.2)$$

If the previous occurrence of B is irrelevant to the occurrence of A, then $P(A|B) = P(A)$, and we say that the sets A and B are *independent*. Equation (5.2) then tells us that the independent sets satisfy

$$P(A \text{ and } B) = P(A) \times P(B). \qquad (5.3)$$

Notice that equation (5.2) gives us a way to invert conditional probabilities, since

$$P(B|A) = \frac{P(A|B)P(B)}{P(A)}. \qquad (5.4)$$

This important result is known as **Bayes' theorem**, and it is true of any definition of probability that meets the Kolmogorov axioms. We can write a more general form of Bayes' theorem as follows. Imagine that our events E_i must belong to one and only one of a series of sets $C_1, \ldots, C_N$. If B is any event, then we may write

$$P(B) = \sum_k P(B|C_k)P(C_k),$$

so that Bayes' theorem can be written as

$$P(C_i|B) = \frac{P(B|C_i)P(C_i)}{\sum_k P(B|C_k)P(C_k)}, \tag{5.5}$$

where the left-hand side now represents the probability of an event in the set C_i occurring, given that B has occurred.

5.1.1 Frequentist and Bayesian probabilities

Up to now, we have been dealing only with an abstract definition of probability. To do anything in practice, however, we need a working definition of probability, and we have two popular options to choose from.

The **frequentist probability** $P(x)$ is the asymptotic proportion of trials for which an assertion is true, i.e. if x is true for n out of an asymptotically large number N of trials, then

$$P(x) = \lim_{N \to \infty} \frac{n}{N}. \tag{5.6}$$

By construction, P therefore lies between 0 and 1. Frequentist probabilities can be both calculated from theoretical rate predictions, and assembled statistically by real or computational counting experiments. Note that, by this definition, the frequentist probability can only exist for repeatable experiments. This means that, whilst one can happily discuss the frequentist probability of getting a certain number of heads in one hundred tosses of a coin, we cannot assign a frequentist probability to a physical constant taking a certain value.

The **Bayesian probability** allows a probability to be assigned to non-repeatable experiments, whilst still satisfying the Kolmogorov axioms. The trick is to swap the concept of frequency for something else, which we will term the **degree of belief**. Defining this is both difficult and an active area of research, but for our purposes it is sufficient to equate 'degree of belief' with 'how much money would you be prepared to bet on the outcome?'. Absolute belief that E_i will occur corresponds to a Bayesian probability of one, whilst absolute belief in non-occurrence corresponds to a Bayesian probability of zero. There is necessarily a degree of subjectivity to a Bayesian probability, since it depends on the state of knowledge of the observer of the event, and this will in general change as the observer gains new knowledge. Nevertheless, the concept has real utility, and is in wide use in the physical sciences.

The difference between these two working definitions of probability has led to two independent schools of probability, naturally called the frequentist and Bayesian schools. Just to confuse you, the distinction has nothing to do with the applicability of Bayes' theorem *per se*; equation (5.4) is perfectly valid for sets of repeatable discrete events A and B for both frequentists and Bayesians. What separates a frequentist from a Bayesian is only that the former starts from the frequentist definition of probability, and the latter starts from the Bayesian definition, and this in turn leads to different spheres of applicability for Bayes' theorem.

We were very careful in the preceding paragraph to note that Bayes' theorem is fine for both schools if we have repeatable, discrete events. Imagine instead that we have a theory, and we want to talk about the probability of the theory being true. A Bayesian can simply write down $P(\text{theory})$, knowing that this represents the degree of belief in the theory. A frequentist, however, cannot write anything at all, since there is no repeatable experiment that can be used to define the frequentist probability of the theory being correct. As a consequence, the frequentist cannot apply Bayes' theorem[1].

Let's think a little more about what the Bayesian can do. After the observation of some data, Bayes' theorem can be written as

$$P(\text{theory}|\text{data}) = \frac{P(\text{data}|\text{theory})P(\text{theory})}{P(\text{data})}. \tag{5.7}$$

Some common terminology can be introduced here: the left-hand term is called the **posterior probability** (or just 'the posterior') and $P(\text{theory})$ is the **prior probability** (or just 'the prior'), sometimes denoted with a special π symbol. The other terms in the numerator and denominator are formally known as the **likelihood** and the **evidence**, respectively. The evidence and prior appear to be of equal importance—if they are equal then the posterior and likelihood are identical—but in practice we only have one observed dataset and hence $P(\text{data})$ is fixed, which means that the evidence is just a normalisation factor. The likelihood is the object of primary importance for much statistical inference, and as such is also often given a special symbol, $\mathcal{L}$. Given some data, we can usually calculate $P(\text{data}|\text{theory})$, and indeed we will see examples of this later. What we usually want to know, however, is the probability that our theory is correct given the data, $P(\text{theory}|\text{data})$. We now see that this can only be known if we have specified our prior degree of belief in the theory, $P(\text{theory})$, which may be difficult or impossible to quantify. This, in turn, makes our desired posterior probability subjective. There does exist an *objective* Bayesian school, that claims to have methods for choosing a suitable prior, but the controversy that surrounds such methods renders the point moot for the foreseeable future. As it is, frequentist statisticians reject Bayesian methods due to their apparent subjectivity, whilst Bayesians reject frequentist methods because we cannot use the frequentist probability definition for many cases of real physical interest.

In this book, we will take a rather more pragmatic view. Neither the frequentist nor Bayesian schools of thought are 'correct'. They are simply different definitions of probability that are consistent with the abstract mathematical definition, and a choice of one or the other frames the sorts of question that you are able to ask and answer. Provided you think carefully about what you are doing in any particular analysis, it is easy to avoid criticism from genuinely knowledgeable colleagues. However, you need to be prepared for the fact that you will be working in a world

---

[1] However, frequentist interpretations of experiments can constrain the *estimates* of theory parameters without invoking the concept of a degree of belief. The resulting expressions may hence differ on a purely philosophical level, affecting how they can be used more than the algebraic form of the probability.

where not everyone has grasped the distinction between the two schools, and where terms such as 'likelihood', 'posterior' and 'probability' are used interchangeably in ways that are obviously incorrect, and where elements of frequentist likelihoods may be informally referred to, somewhat erroneously, as 'priors'.

In particle physics, frequentist approaches tend to be canonical because the field's natural scenario is repeatable experiments from high-statistics trials. One of the main motivations for using particle colliders, rather than waiting for nature to supply high-energy collisions, is that a high flux of essentially identical collisions occur in a very controlled environment. By comparison with e.g. astrophysical particle observations, which may reach higher energies but are not under human control, the well-defined initial state reduces many systematic effects to the point where the collisions themselves are most naturally viewed in a frequentist picture. Following this, much of the basic statistics applied to the counting of events falling into bins in physical observables is treated in a frequentist perspective. But for processes where there is significant uncertainty—particularly rare ones where the observed statistics are low—careful treatment is essential. In particular, cosmology is concerned with our single universe, leading to a prevalence of Bayesian methods in that field, and a wealth of interesting discussions in the rich area where collider physics and astrophysics results intersect.

5.1.2 Probability for continuous random variables

We have thus far formulated probability by considering discrete, exclusive events (and sets of those events). The probability of a given event occurring was written as $P(E_i)$, and some set of events A could itself be interpreted as an event, which occurred if any of its constituent events occurred.

In physics, however, we are generally interested in modelling events with more than one possible outcome. In this case, we can again denote by Ω the set of all possible outcomes, and we can label the elements of the set by a variable which takes a specific value for each element of the set. Such a variable is called a **random variable**. The individual elements of Ω might actually be described by several variables, in which case we can either require multiple random variables to specify an event, or use a single random variable which is a multidimensional vector. The difference is only semantic. The random variables can be either discrete (in which case our previous description holds), or continuous. The discrete case clearly corresponds to our previous formulation, except that we are now referring to different outcomes of a single event, rather than treating the different outcomes as distinct events (the set theory looks the same).

5.1.3 Density functions (probability, cumulative, marginal)

For continuous random variables, we need to develop some new concepts. As an example, imagine an experiment whose outcome is characterised by a single, continuous random variable x. Our set Ω corresponds to the set of possible values that x might take in our experiment, but the outcome is unknown in advance because the process is random. We can then ask 'what is the probability of observing a value within the infinitesimal interval $[x, x + dx]^{2}$'. Note that, if the experiment is repeatable, this has a perfectly fine frequentist definition. We can define the **probability density function** (pdf) $f(x)$ via:

$$P(x \text{ is in the interval } [x, x + dx]) = f(x)dx. \tag{5.8}$$

For a frequentist, $f(x)dx$ gives the fraction of times that x is observed within the range $[x, x + dx]$ in the limit that our experiment is repeated an infinite number of times. $f(x)$ itself is not constrained to be less than one everywhere, but the total integral over all x values must be one:

$$\int_{\Omega} f(x)dx = 1. \tag{5.9}$$

The probability for the random variable to take on a value less than or equal to x is given by the **cumulative density function** (cdf), defined as:

$$F(x) = \int_{-\infty}^{x} f(x')dx'. \tag{5.10}$$

Discrete probabilities can also be obtained by integration of the pdf:

$$P(x \text{ is between } x_0 \text{ and } x_1) = \int_{x_0}^{x_1} f(x)dx \tag{5.11}$$

$$= F(x_1) - F(x_0). \tag{5.12}$$

Note that the 'probability of $x = y$' for any y is zero, since $P(y \text{ is in the interval } [y, y + dx]) = f(y)dx$ (the integral collapsing to the point $x = y$), and in an informal sense that makes mathematicians uncomfortable, $dx = 0$. The physical take-away from this is that it is meaningless to talk about the probability of x taking a specific value, and hence finite probabilities always have to be associated with a finite range of x values, however small.

For any given pdf, there are various useful quantities that can be defined. The first is the **mode**, which is the value of x at which the pdf is maximised (for a 1D example). One can also use the cdf to define the *quantile of order α*, x_α, which is the value of x such that $F(x) = \alpha$. This is equal to the inverse cumulative distribution $x_\alpha = F^{-1}(\alpha)$. For example, the **median** is equal to $x_{1/2}$.

Now imagine that the outcomes of the experiment are characterised by two random variables x and y. We can define the **joint pdf** $f(x, y)$ via

$$\begin{aligned} P(x \text{ is in the interval } &[x, x + dx] \text{ and} \\ y \text{ is in the interval } &[y, y + dy]) = f(x, y)dxdy. \end{aligned} \tag{5.13}$$

The normalisation condition is now:

---

[2] Note that we have used x here both for the random variable, and for the range that it might take. We hope that the meaning will remain clear from the context.

$$\int\int_\Omega f(x, y)\mathrm{d}x\mathrm{d}y = 1, \tag{5.14}$$

where Ω is, as usual, the total set of outcomes. If we project a two-dimensional pdf down to, say, the x-axis, we define the **marginal density function** of x as

$$f_x(x) = \int f(x, y)\mathrm{d}y. \tag{5.15}$$

We can also define the marginal density function of y, via

$$f_y(y) = \int f(x, y)\mathrm{d}x. \tag{5.16}$$

Higher-dimensional pdfs can similarly be reduced to lower dimensional ones by marginalising over one or more of the random variables, and we will see this in practice when we later discuss sampling of multidimensional pdfs.

Something else we can do is to take a slice through a multidimensional pdf. Working with our 2D example $f(x, y)$, imagine that we take the slice $x = x_0$. In this case, we would expect to get a 1D function (of y), since x is now fixed to a constant value. The 1D distribution we end up with is called the **conditional density function** of y given that $x = x_0$, given by

$$f(y|x_0) = \frac{f(x_0, y)}{\int f(x_0, y)\mathrm{d}y} = \frac{f(x_0, y)}{f_x(x_0)}, \tag{5.17}$$

where we have used the definition of the marginal density function of x. The denominator exists to normalise the conditional distribution so that it gives unity when integrated over y.

If we have a joint pdf $f(x, y)$, marginal density functions $a(x)$ and $b(y)$, and conditional density functions $p(x|y)$ and $q(x|y)$, we can write Bayes' theorem for continuous variables as

$$q(y|x) = \frac{p(x|y)b(y)}{a(x)}. \tag{5.18}$$

5.1.4 Expectation values and moments

If we have a pdf defined for one random variable x, we can define the **expectation value** of some function $g(x)$ as

$$E(g) = \int_\Omega g(x)f(x)\mathrm{d}x, \tag{5.19}$$

where Ω specifies the entire range of x. The left-hand side is a single number which gives the expectation value of the function $g(x)$, and the right-hand side makes it clear that the pdf $f(x)$ is used a weighting function to determine the contribution of each value $g(x)$ to the integral. When $g(x) = x$, this expectation value has a special name, called the **mean** of x:

$$\mu = \langle x \rangle = \int_\Omega x f(x) \mathrm{d}x. \tag{5.20}$$

We can also define the **variance** $\sigma^2(x)$ of the pdf $f(x)$ as

$$\sigma^2 = E[(x - \mu)^2] = \int_\Omega (x - \mu)^2 f(x) \mathrm{d}x. \tag{5.21}$$

Expanding the bracket on the right-hand side and using equation (5.20), one finds that the variance is equal to $E[x^2] - \mu^2$. As the notation suggests, σ, the square root of the variance, is an important quantity reflecting the degree of spread of values in the distribution: this is called the **standard deviation**.

The mean and the variance are actually special cases of more general quantities called **moments**, which can be used to characterise the shapes of pdfs. $E(x^n)$ defines the nth algebraic moment, whilst $E[(x - \mu)^n]$ gives the nth central moment. We thus see that the mean is the first algebraic moment, and the variance is the second central moment.

5.1.5 Covariance and correlation

If we have a 2D pdf $f(x, y)$, for two random variables x and y, we can generalise our definition of the expectation value of a function $g(x, y)$ via

$$E[f(x, y)] = \int \int g(x, y) f(x, y) \mathrm{d}x \mathrm{d}y. \tag{5.22}$$

Note that we can now define a mean and variance for both x and y, and these are given by

$$\mu_x = E[x] = \int \int x f(x, y) \mathrm{d}x \mathrm{d}y \tag{5.23}$$

$$\mu_y = E[y] = \int \int y f(x, y) \mathrm{d}x \mathrm{d}y \tag{5.24}$$

and

$$\sigma_x^2 = E[(x - \mu_x)^2] \tag{5.25}$$

$$\sigma_y^2 = E[(y - \mu_y)^2]. \tag{5.26}$$

We can also define a two-dimensional **covariance**, given by:

$$\mathrm{cov}(x, y) = E[(x - \mu_x)(y - \mu_y)] = E[xy] - E[x]E[y], \tag{5.27}$$

which generalises the one-dimensional variance to a form that captures the spread of values between the x and y axes. A purer view of the extent of this interaction between the variables can be found via the **correlation coefficient**, given by

$$\mathrm{corr}(x, y) = \rho(x, y) = \frac{\mathrm{cov}(x, y)}{\sigma_x \sigma_y}. \tag{5.28}$$

The denominator of this last expression ensures that the correlation coefficient is between -1 and 1. We say that x and y are *independent* if their pdf factorises, and thus $f(x, y) = f(x)f(y)$. From equation (5.27), the covariance and correlation coefficient then vanish for independent variables.

When we have more than two random variables, we can define analogous covariances and correlation coefficients for each marginal 2D joint distribution. We can in fact write a matrix with the elements $\text{cov}(x_i, x_j)$ (for a set of variables x_i), which is called the **covariance matrix**: when we later discuss covariance and correlation in more depth we will always be referring to this n-dimensional version.

5.1.6 Standard probability density functions

There are, of course, an infinite number of probability density functions, even if we restrict ourselves to 1D, i.e. functions in one parameter only. But several forms are of particular importance because of their ubiquity: here we will focus on the *Gaussian (or Normal) distribution* and its relatives; the *Poisson* and *Binomial* probability-mass distributions; and the more physics-specific *Breit–Wigner (or Lorentzian, or Cauchy)*, and *Crystal-Ball* distributions.

Gaussian/Normal distribution: The Gaussian distribution is the most heavily used object in all of statistics—including physics applications such as statistical mechanics, and path-integral quantum mechanics. There are two reasons for this ubiquity: firstly its mathematical convenience, and secondly its position as the asymptotic distribution obtained under repeated sampling from most physically reasonable distributions. The 1D Gaussian pdf in x is

$$\mathcal{N}(x; \mu, \sigma) = \frac{1}{\sqrt{2\pi\sigma^2}} \exp \frac{(x - \mu)^2}{2\sigma^2} \tag{5.29}$$

where the pre-factor is for normalisation, and the μ and σ parameters are location and width parameters. Indeed, as the symbols suggest, they are the mean and standard deviation, as you can verify by integration:

$$\mu \equiv \langle x \rangle = \int_{-\infty}^{\infty} dx \, x \, \mathcal{N}(x; \mu, \sigma),$$

and

$$\sigma^2 \equiv \langle x^2 \rangle - \langle x \rangle^2 = \int_{-\infty}^{\infty} dx \, x^2 \, \mathcal{N}(x; \mu, \sigma) - \mu^2.$$

We plot this distribution for various parameter values in Figure 5.1, where we indeed see that the distribution peaks at $x = \mu$, and widen as σ increases.

The mathematical convenience of the Gaussian results from its squared exponent: under probability combination rules, independent Gaussians are multiplied together,[3] with the effect that their quadratic exponents are summed—exactly as in Pythagoras' Theorem. This gives a neat geometric interpretation, as well as

---

[3] We shall shortly address the generalisation to non-independent Gaussians.

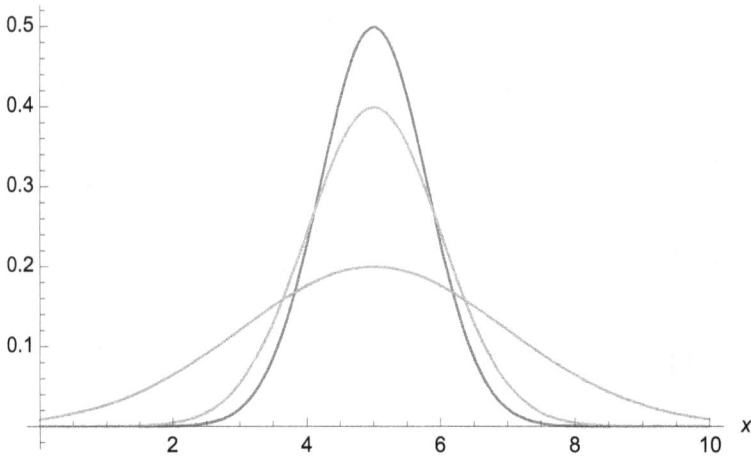

Figure 5.1. The Gaussian distribution of equation (5.29), where $\mu = 5$ and $\sigma = 0.8$ (blue), $\sigma = 1$ (orange) and $\sigma = 2$ (green).

allowing the sum of multiple Gaussians to be replaced with a single Gaussian in an elliptical radial-distance variable $r^2 = \sum_i (x_i - \mu_i)^2/\sigma_i^2$. This convenience re-appears as the ease of combination of Gaussians: the convolution of a Gaussian with width σ_1 by another Gaussian of width σ_2 is itself a Gaussian, with width defined by the Pythagorean quadrature of the two inputs: $\sigma_{tot}^2 = \sigma_1^2 + \sigma_2^2$.

The natural ubiquity of the Gaussian comes from a more rarefied place: the **central limit theorem** (CLT). This states that the sum of variables randomly sampled from different, not necessarily Gaussian, distributions with finite variance tends towards being Gaussian-distributed: the Gaussian is the **attractor distribution** for finite-variance distributions. As this applies to sums of variables, it also naturally applies to mean values: together, these are common sources of measurement variables, and so in the absence of any other indications it is often a good guess that uncertainties will tend toward a Gaussian distribution. We will later see explicit examples of other important distributions which limit to a Gaussian in scenarios relevant to physics measurements.

Double-Gaussian distribution: The double-Gaussian is very commonly used in physics, most often to add 'long tails' to a central Gaussian distribution. It is composed of, as the name suggests, two Gaussians: these share a common mean μ but have distinct widths $\sigma_{1,2}$, and a mixture parameter $a \in (0, 1)$ which governs how much of the probability density lives in each Gaussian:

$$\mathcal{N}_2(x; \mu, \sigma_1, \sigma_2, a) = a\mathcal{N}(x; \mu, \sigma_1) + (1 - a)\mathcal{N}(x; \mu, \sigma_2) \tag{5.30}$$

$$= \frac{1}{\sqrt{2\pi}}\left[\frac{a}{\sigma_1}\exp\frac{(x - \mu)^2}{2\sigma_1^2} + \frac{1 - a}{\sigma_2}\exp\frac{(x - \mu)^2}{2\sigma_2^2}\right]. \tag{5.31}$$

Notably, following the argument above about Gaussian convolution, a double-Gaussian cannot be motivated from the simultaneous application of two random

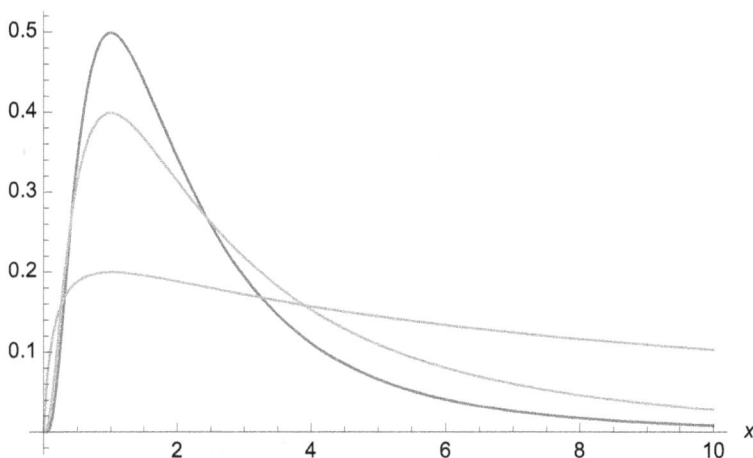

Figure 5.2. The log-normal distribution, shown for various values of the width parameter σ.

processes in the CLT limit (which would result in a broader single-Gaussian). Instead we must assume that there are two distinct statistical populations, or that we are just using the function for convenience.

Log-normal distribution: Our last distribution directly related to the Gaussian is the log-normal: a Gaussian distribution in the logarithm of the parameter x. Like the double-Gaussian it is rarely motivated *a priori*, but often used to force positivity in cases where the infinitely long negative tail of a naïve Gaussian invalidates the domain of the described variable—for example a positive particle momentum, mass, or production rate. While convenient, the price paid is a skewing of the distribution, which for distributions intended to be within a few standard deviations of the physical boundary at zero can be very abrupt indeed. This skewing can be seen in Figure 5.2, in which we plot the form of the log-normal distribution for various parameter values. We will not dwell on this distribution any further, but it is good to be aware of its availability—and side-effects if used without due care.

Poisson distribution: The Poisson distribution gives the probability density for observation of a number of events k, produced by a stochastic process with constant rate λ, i.e. an expected number of events λ in whatever continuous interval of time, distance, integrated luminosity, etc is being considered. It has the form

$$\mathcal{P}(k; \lambda) = \frac{\lambda^k e^{-\lambda}}{k!}. \tag{5.32}$$

Note the discreteness of k required by the factorial function on the denominator[4]: the Poisson is strictly a *probability mass function* (pmf) rather than a distribution, but we can be a bit sloppy with terminology provided we keep this restriction in mind. In

---

[4] The Poisson can also be generalised to continuous number of observations, via the relation $n! = \Gamma(n + 1)$, but this is rarely used in physics applications. Even when events are continuously *weighted*, the 'right thing' is usually to use discrete Poisson statistics and multiply the event weight back in afterwards.

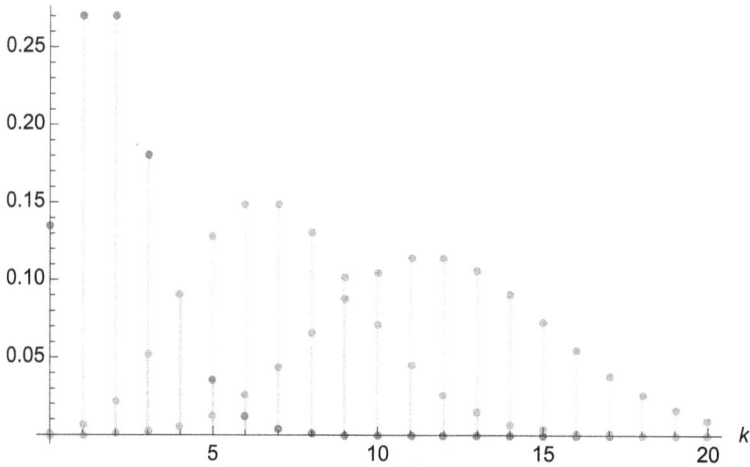

Figure 5.3. The Poisson distribution, shown for $\lambda = 2$ (blue), $\lambda = 7$ (orange) and $\lambda = 12$ (green).

fact, it is something of a hybrid: a continuous function in λ for fixed k, and discrete in k for fixed λ.

It is entertaining to see how this works, since if we fix k, the Poisson has to integrate to 1 over all positive λ, and for fixed λ a sum over all positive k must also result in 1: both directions, continuous and discrete, must be well-normalised probability distributions. We leave this as an exercise for the reader, who will find the $k!$ appearing from iterated integrations by parts, and the exponential from its infinite series representation.

It is worth also taking a moment to appreciate what extraordinary internal stress this function is under: three rapidly diverging functions wrestling each other for supremacy. It is remarkable that the Poisson distribution is finite at all, but it is: in the end, the exponential and factorial suppress the large-k and large-λ aspirations of the diverging power law. This isn't just anthropomorphic whimsy, but a serious computational issue: computing Poisson likelihoods easily runs the risk of numeric overflows or inaccuracies in handling and combining very large and very small numbers. For this reason, and because of applications in statistical limit-setting to be shown shortly, we often find the Poisson log-likelihood (or log-pmf) to be a more convenient form:

$$\ln \mathcal{P}(k; \lambda) = k \ln \lambda - \lambda - \ln(k!). \qquad (5.33)$$

A very important and useful feature of the Poisson distribution is its limiting behaviour for large rates (and also numbers of samples, as the two are correlated), i.e. $\lambda \to \infty$. As λ becomes larger and the peak of the distribution moves further from zero, the asymmetry forced by positivity reduces, and for $\lambda \gtrsim 10$ the shape is approximately symmetric and Gaussian-like. We can see this in Figure 5.3, which plots the Poisson distribution for various parameter values. With further increases in λ the discrete nature of the Poisson distribution becomes unresolvable—especially as, computationally, histograms are invariably used. This approach to a Gaussian is the CLT in action, as mentioned earlier. In this limit, the Poisson behaves as a pseudo-continuous

Gaussian distribution with mean[5] $\mu = \lambda$ and $\sigma = \sqrt{\lambda}$. Note that the *relative width* of this stochastically assembled distribution, $\sigma/\mu = 1/\sqrt{\lambda}$, reduces with larger rates: *large numbers of observed events produce better-defined results*. This is an example of the *Law of Large Numbers*: that the average of a large number of samples from a distribution will become asymptotically closer to the expectation value.

Binomial distribution: The binomial distribution is another discrete distribution of great importance. It describes the distribution of total scores from a set of binary tests, i.e. given a set of yes/no questions, at the end of n questions the number k to have, e.g. the 'yes' result is given by the binomial distribution. The binomial is the simplest case of a more general **multinomial distribution**, where the tests can now return multiple answers, but due to difficulty of manipulation the multinomial is rarely used directly.

The binomial pmf is

$$\mathscr{B}(k; n, P) = \binom{n}{k} P^k (1 - P)^{n-k} \tag{5.34}$$

$$= \frac{n!}{k!(n - k)!} P^k (1 - P)^{n-k} \tag{5.35}$$

where P is the probability of success in each test (e.g. the 'yes' probability in the binary questions) and therefore $(1 - P)$ is the probability of failure/'no'. The combinatorial prefactor is called the combination function or binomial coefficient, and counts the number of ways of distributing k successes and $(n - k)$ failures in n trials. To see where it comes from, note that the number of ways of assigning a given list of yes and no outcomes to a set of n trials is $n!$, given that we have n choices for the first entry in the list, $(n - 1)$ for the next and so on. However, we have then overcounted the total number of distinct outcomes, as we can separately reorder the assignment of success outcomes amongst themselves $k!$ ways, and the number of failures $(n - k)!$ ways. The resulting distribution, in k for various n and fixed P, is illustrated in Figure 5.4.

As for the Poisson pmf, the large-number limit is interesting. Indeed, we can obtain the Poisson distribution as a particular large-n limit of the binomial. To see how this works, imagine that we have a finite time-interval divided into n smaller, equal-sized bins, each with a success probability P_n. If we increase the number of bins while ensuring that the mean success rate for the interval, $\mu = nP_n$ remains fixed, then in the continuum limit, $n \to \infty$, the binomial distribution behaves as a Poisson with mean rate $\lambda = \mu$

$$(k; n, P) \xrightarrow{\; n \to \infty, \, nP \to \mu \;} (k; \mu) \,. \tag{5.36}$$

As the Binomial pmf in this specific limit approaches a Poisson pmf, it should be no surprise that the Central Limit Theorem again contrives to make a large number of samples distribute as a Gaussian, unsurprisingly with mean $\mu = nP$, and with and standard deviation $\sigma = \sqrt{nP(1 - P)}$. We can see this limit behaviour in Figure 5.4,

---

[5] Recall that we used the fact that the mean of a Poisson distribution was the same as the (constant) event rate in Chapter 1, when deriving the form of the cross-section.

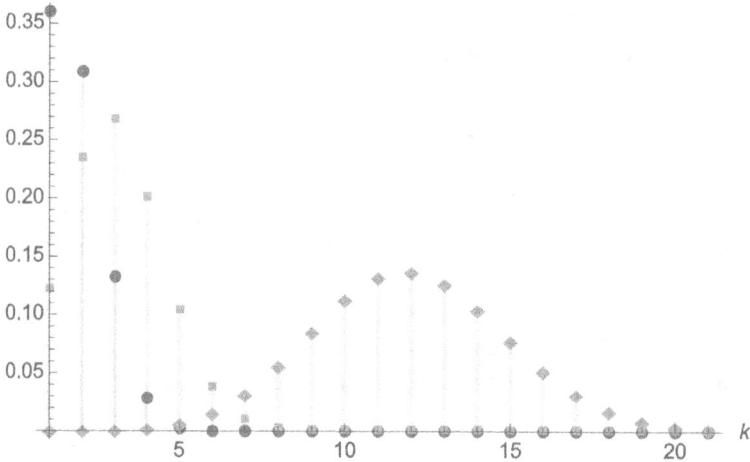

Figure 5.4. The binomial distribution, shown for $p = 0.3$ and $n = 5$ (blue), $n = 10$ (orange) and $n = 40$ (green).

as the number of trials n increases. We may also note the scaling of the relative width of the distribution, $\sigma/\mu = \sqrt{(1 - P)/nP} \propto 1/\sqrt{n}$—the Law of Large Numbers ensuring that the distribution becomes better defined with increasing statistics.

The binomial and Poisson distributions are hence closely related to one another and to the Gaussian—you may find it useful to view the binomial and the Poisson distributions as essentially the same concept, with one reflecting statistics aggregated over a continuous period, and the other over a set of discrete trials.

Cauchy distribution: A particularly interesting pdf is the *Cauchy* distribution, given by:

$$f(x; x_0, \Gamma) = \frac{1}{\pi} \cdot \frac{\Gamma}{\Gamma^2 + (x - x_0)^2} \tag{5.37}$$

This has an undefined expectation value and in fact all other moments are also undefined. The parameters x_0 and Γ represent the mode and the half-width, half-height locations, respectively. Particle enthusiasts will notice that this distribution has the same form as the function that describes the width of a decaying resonance, in which context it is commonly referred to as the *Breit–Wigner* distribution. It is also referred to as a *Lorentzian* distribution, and we plot its form in Figure 5.5.

Chi-squared distribution: If $x_1, \ldots, x_N$ are independent random variables, each of which has a Gaussian pdf given by $\mathcal{N}(x_i; \mu = 0, \sigma = 1)$, then the sum of the squares of the numbers has a *chi-squared* distribution, given by:

$$\chi^2(x; k) = \frac{\left(\frac{x}{2}\right)^{(k/2)-1} e^{-x/2}}{2\Gamma\left(\frac{k}{2}\right)} \tag{5.38}$$

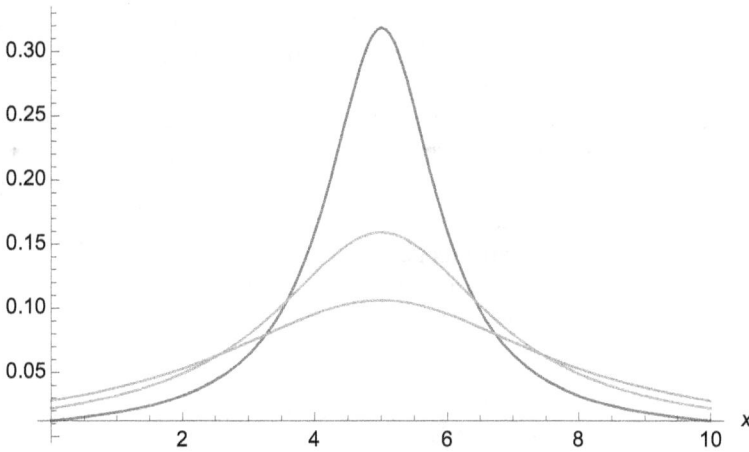

Figure 5.5. The Cauchy distribution, with $x_0 = 5$ and $\Gamma = 1$ (blue), $\Gamma = 2$ (orange) and $\Gamma = 3$ (green).

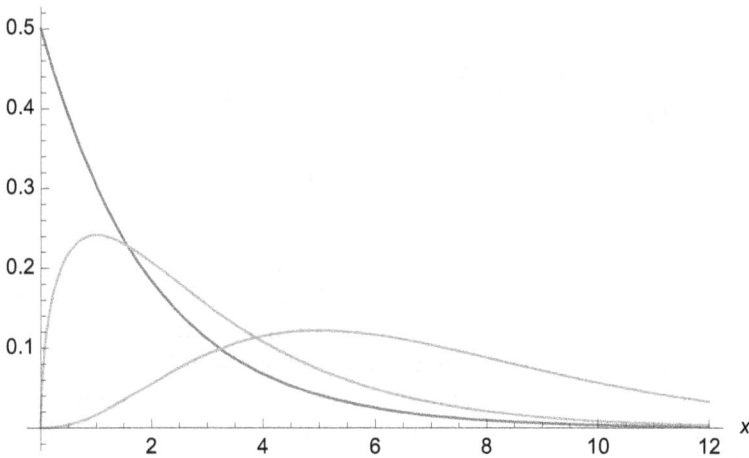

Figure 5.6. The χ^2 distribution, with $k = 2$ (blue), $k = 3$ (orange) and $k = 7$ (green).

This falls monotonically for $k = 1$ and $k = 2$, and peaks at $x = k - 2$ for higher k, which behaviour may be seen in Figure 5.6. The mean is $\langle x \rangle = k$ for all k, whilst the variance is $2k$.

Composite distributions: Experimental physics distributions are rarely 100% pure: they consist of an ideal core corresponding to one or more fundamental processes, which are then distorted by secondary effects, measuring apparatus, imperfect reconstruction techniques, etc.

Examples of this abound—the Double-Gaussian distribution is already a composite of a sort, although we have here grouped it along with the Gaussian and other related distributions for coherence. But as noted in our discussion of that distribution, it is composite in the relatively simple sense of summing two distributions to reflect the contribution of multiple core processes. The influence of processes which distort the whole core distribution, generally by convolutions, is more subtle and it is these which we will focus on here.

The first such distribution, since it is heavily used in particle physics, is the **Crystal-Ball distribution**, named after the classic experiment that popularised it. It is essentially a Gaussian distribution with one 'heavy tail', typically on the negative side of the Gaussian peak as motivated by wanting a distortion capturing the effects of energy loss in detector interactions and read-out. Such an effect could be encoded, for example, by convolution of a truncated exponential or power law with a Gaussian, but convolutions are typically not expressible in analytic closed form and instead the Crystal-Ball function is a piecewise combination of a power-law and Gaussian: its pdf is

$$f_{CB}(x; \alpha, n, \bar{x}, \sigma) = \frac{1}{N} \cdot \begin{cases} \exp-\dfrac{(x - \bar{x})^2}{2\sigma^2} & \text{for } \dfrac{x - \bar{x}}{\sigma} > -\alpha \\[2ex] A \cdot \left(B - \dfrac{x - \bar{x}}{\sigma} \right)^{-n} & \text{for } \dfrac{x - \bar{x}}{\sigma} \leqslant -\alpha \end{cases} \qquad (5.39)$$

where

$$A = \left(\frac{n}{|\alpha|} \right)^n \cdot \exp\left(-\frac{|\alpha|^2}{2} \right), \qquad B = \frac{n}{|\alpha|} - |\alpha|,$$

$$C = \frac{n}{|\alpha|} \cdot \frac{1}{n - 1} \cdot \exp\left(-\frac{|\alpha|^2}{2} \right), \qquad D = \sqrt{\frac{\pi}{2}} \left(1 + \mathrm{erf}\left(\frac{|\alpha|}{\sqrt{2}} \right) \right), \qquad (5.40)$$

and the normalization factor $N = \sigma(C + D)$.

It can be seen that the α parameter controls the point where the Gaussian and power law distributions meet (in units of Gaussian standard deviations σ from the Gaussian mean $\bar{x}$), and n is the absolute value of the power law exponent. The A–D coefficients, controlling the relative admixtures of the two distributions, arise from requiring continuity and differentiability on the piecewise transition. This form is by construction good for modelling intrinsically near-Gaussian distributions, and provides efficient evaluation and sufficient flexibility for many real-world cases. We show an example of this distribution in Figure 5.7.

Myriad convolved distributions can also be concocted, but most are non-analytic. A good, and useful, example is the **Voigt profile** distribution, used extensively in lab and astrophysical spectroscopy, which encodes symmetric **smearing** of a Lorentzian profile: it is hence useful for particle physicists to know about this distribution and the availability of both explicit numerical implementations and good analytic approximations, e.g. for modelling resolution effects on particle mass resonance peaks. Adding further complexity is often needed, however, and here numerical programming techniques become important to perform, e.g. convolution of a Lorentzian with an asymmetric Crystal-Ball: software packages like ROOT, SciPy, and lmfit provide systems for such calculations, should manual evaluation by repeated sampling from the component distributions be unsuitable.

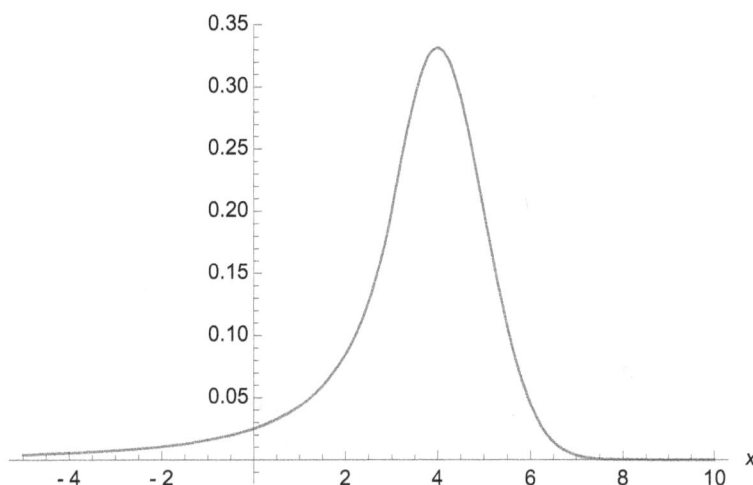

Figure 5.7. Example Crystal-Ball distribution, in which a Gaussian at larger parameter values is joined onto a power-law behaviour at low values.

5.2 Correlations

In the previous section, we saw a number of examples of probability density functions, that are commonly used in statistical applications. However, a real physical dataset—consisting of various combinations of measured variables—will have its own very complex pdf, that depends on various parameters. The dimensionality of this parameter space can be huge: for example, confounding theory parameters like the factorization and renormalization scales, pdf error sets, and parton shower ambiguities; parametrisations of uncertainties in the detector response and reconstruction; and finally the parameters of the fundamental physics model being tested, e.g. particle masses, couplings and spins. Nevertheless, we can at least distinguish two different types of parameter in the pdf. The **parameters of interest** (POIS) are those that correspond to the physics we are interested in (e.g. the mass of the Higgs boson). **Nuisance parameters** are those which are associated with the uncertainties, and in most cases we know something about them already (e.g. we might know that the energy measured by our detector is off by a certain amount, but we know that this discrepancy roughly follows a Gaussian distribution with a narrow width).

The pdf is in general distributed in an arbitrary way through the high-dimensional space of all these parameters, and thus not necessarily structured or aligned with any of them. We cannot work in general with such an arbitrary form, but fortunately can approach it systematically in terms of a moment expansion, starting with the mean (first moment) and the variance (second central moment). In principle one can carry on through third and higher moments until the whole distribution is characterised, but in practice we usually stop at second order. These first two moments can be interpreted conveniently as describing a multivariate Gaussian distribution, which functions as a second-order approximation to the full pdf.

As we have considered it so far, a space of k variables will have k variances σ_i^2, each computed via the second moment $\langle x_i^2 \rangle = \int dx_i x_i^2 f(x)$, but this only characterises how the marginalised pdf is spread in one variable x_i at a time: as discussed, the pdf need not have any particular alignment with the choice of parameter axes and this measure may not be particularly illuminating. A more informative second-moment structure is the **covariance matrix**:

$$\text{cov}(x)_{ij} \equiv \Sigma_{ij} = \langle x_i x_j \rangle - \langle x_i \rangle \langle x_j \rangle. \tag{5.41}$$

This is a very important generalisation of the variance, which includes *mixtures* of parameters x_i and x_j and hence characterises the (second-order) **correlations** induced between them via the pdf. Note that the set of single-variable variances lies on the diagonal of the covariance matrix, $\text{cov}(x)_{ii}$: the covariance hence naturally replaces the variance, although some authors use a separated nomenclature such that 'covariance' refers only to the off-diagonal $i \neq j$ terms of this matrix.

Computing the covariance matrix for an analytic pdf is done as usual via integrals of the first and second moments of the distribution. Note that by construction the covariance matrix is symmetric, but not necessarily positive: *negative elements in the matrix tell us that when x_i tends to be larger than average, x_j tends to be smaller than average*—and *vice-versa*. This is a **negative correlation**, and by contrast positive covariance entries characterise *positively correlated* variables.

5.2.1 Covariance and pdf geometry

The covariance is the key object that allows us to build a Gaussian approximation to the full pdf, by generalising our usual one-dimensional Gaussian to its multivariate form:

$$\mathcal{N}(x; \mu, \Sigma) = \frac{1}{\sqrt{2\pi^k \det\Sigma}} \exp\left\{ -\frac{1}{2}(x - \mu)^T \Sigma^{-1}(x - \mu) \right\} \tag{5.42}$$

$$= \left(\prod_i^k \frac{1}{\sqrt{2\pi}\sigma_i} \right) \cdot \exp\left\{ \sum_{i,j}^k -\frac{1}{2}(x_i - \mu_i)\Sigma_{ij}^{-1}(x_j - \mu_j) \right\}, \tag{5.43}$$

where Σ^{-1} is the **inverse covariance matrix**, sometimes called the **precision matrix**: the inversion here is equivalent to the placing of the single-dimensional variance σ^2 in the denominator of the 1D Gaussian exponent.

The interpretation of this is not yet clear, as the covariance matrix has in general $k \times k$ non-zero entries. However, as a real symmetric matrix, Σ can always be diagonalised by a similarity transform (basis rotation), to give its 'natural' representation $\Sigma_{\text{diag}} = O^T \Sigma O = \text{diag}(\hat{\sigma}_1^2, \hat{\sigma}_2^2, \ldots \hat{\sigma}_k^2)$, where the $\hat{\sigma}^2$ eigenvalues are the variances of independent 1D Gaussians aligned along the orthogonal eigenvectors of the covariance, as illustrated in Figure 5.8.

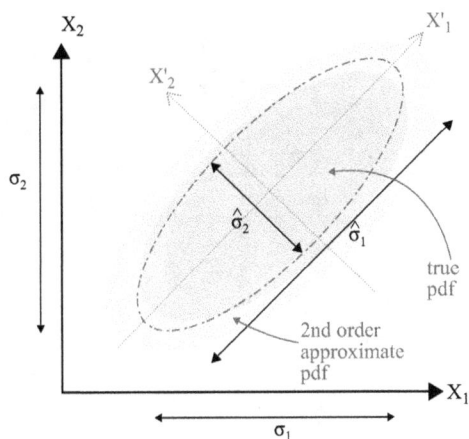

Figure 5.8. Gaussian approximation to a general pdf in two dimensions, with the eigenvectors of the covariance matrix defining the principle directions of independent Gaussians centred on μ, and with variances given by the eigenvalues.

This is the sort of picture to keep in mind when thinking of covariance—the matrix defines an ellipsoid in the space of all parameters, whose axes are rotated with respect to the parameter basis. This interpretation helps us to think intuitively about correlation, and to visualise what the uncertainty along any other other direction in the parameter space is, simply by geometric projection of the ellipsoid boundary on to the new direction vector. But it is important also to remember that this is only the second-order term in a generally infinite moment-expansion series: in particular, the covariance/Gaussian formalism is intrinsically symmetric, with equal deviations from the mean in positive and negative directions along every parameter axis (and their linear combinations) (Figure 5.9).

5.2.2 From covariance matrix to correlation matrix

While the covariance characterises the whole pdf at second-order in the moment expansion, it is not a particularly convenient object to inspect, e.g. via a plotted heatmap because the different variables will typically have very different scales. Consider, for example, a 2×2 covariance matrix between the $x_1 = \eta$ and $x_2 = p_T$ variables, with the p_T in GeV. The rapidities have a maximum range of -5 to 5 in the ATLAS and CMS detectors, while for most processes the transverse momenta will be spread over hundreds or even thousands of GeV. We hence expect many orders of magnitude difference between the covariance terms Σ_{11} (η-squared), Σ_{22} (p_T-squared) and the mixture Σ_{12}: the p_T-squared 2–2 term will dominate, making it hard to see the significant 1–2 correlations between the variables. The arbitrariness of this is made even clearer if we note that we could have listed the momenta in MeV or TeV variables instead, completely changing the matrix numerics.

The solution to this is to standardise the variable scales so that all are comparable, and the obvious way to do so is to divide each term by the standard deviations of the two variables entering it. This gives us the **correlation matrix,**

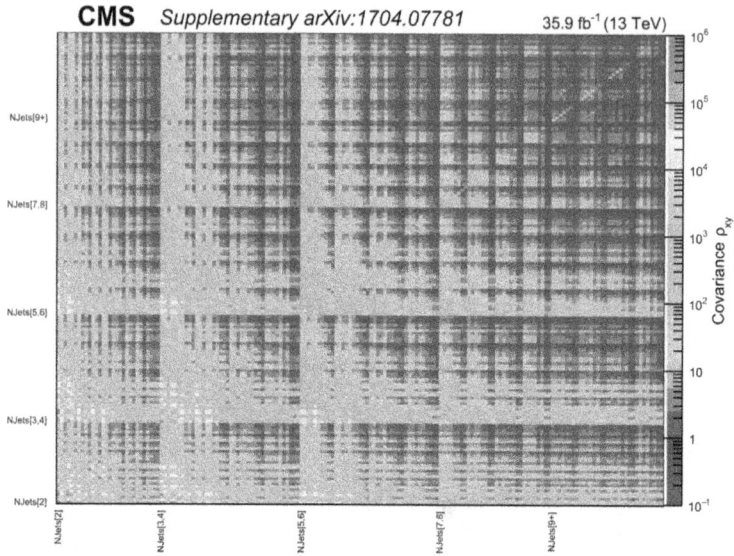

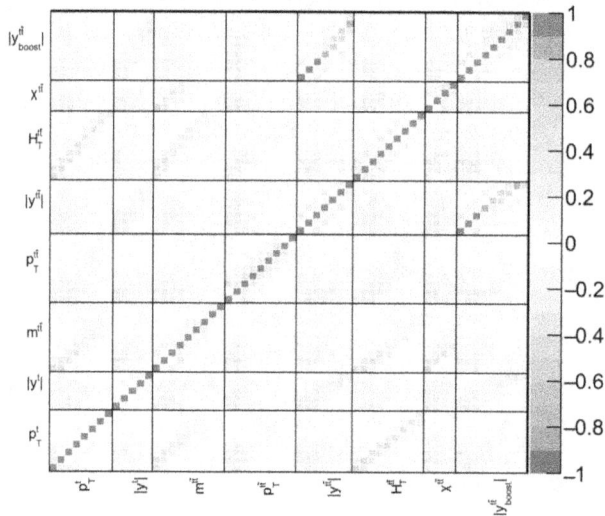

Figure 5.9. A (log-)covariance matrix between the binned signal regions of a CMS search for hadronic signatures of supersymmetry (top), and a correlation matrix between the bins of multiple reconstructed observables in an ATLAS measurement of semileptonic $t\bar{t}$ events (bottom). The influence of the absolute scale of bin populations is evident in the upper covariance figure, where large populations and hence absolute uncertainties for low jet-multiplicity bins in the left and bottom of the plot give way to low populations and absolute uncertainties in the top and right. By contrast, the lower correlation plot has divided out the absolute scales of uncertainties as characterised by the diagonal covariance bins, more clearly exposing the correlation structures of the observables. Reproduced with permission from CMS-SUS-16-033, supplementary figure 9; and ATLAS TOPQ-2018-15, supplementary figure 106.

$$\text{corr}_{ij} \equiv \rho_{ij} = \frac{\Sigma_{ij}}{\sigma_i \sigma_j} = \frac{\Sigma_{ij}}{\sqrt{\Sigma_{ii}\Sigma_{jj}}}, \qquad (5.44)$$

which effectively includes the rotation angles between the variable basis and the diagonal eigenbasis, but has discarded the relative scalings of the component 1D Gaussians.

Correlations, in this second-order approximation and perhaps beyond, are very important objects in particle-physics data-interpretation, as without them we are restricted to using single variables in model testing. The variables that are correlated can be quite abstract: we can have, for example, correlations between:

- different bins of the same histogram;
- bins of *different* distributions;
- observables and parameters (telling us about sensitivities of models to measurements and *vice versa*); and
- model parameters.

allowing linear correlation of observables with either each other variances, but that combinations of them may be well constrained, with small variances.

Correlations between model parameters often show that individual parameters may be unconstrained, with large. We shall see later the use of correlation information in forming likelihood tests, and in methods for estimating statistical 'connectedness' observables.

5.2.3 Uncertainties from covariances

As discussed, the covariance matrix is a useful generalisation of the concept of the variance of a single-variable distribution to the correlated spread of all pairs of variables. In the same way as the variance is a single-number representation of the uncertainty on a variable due to its pdf, and with the same limitations such as implicit error-symmetry, the covariance encodes this information for all variables at the same time. An important question, less obvious than for the 1D variance, is then how to extract and combine uncertainties from covariances. It is our view that, as a physicist dealing either with the measurement or interpretation of precise experimental data, you can never know too much about tricks and techniques for using covariances.

It is extremely useful to realise that the covariance matrix is not inviolable, but may be surgically manipulated to focus attention on subsets of its information. A common example is where you wish to publish correlation data from a likelihood fit without publicly revealing the details of your systematic error model: in this case, simply removing all rows and columns corresponding to nuisance parameters leaves behind a still-symmetric, square covariance matrix illustrating the correlations between the values of the measurement bins arising from the POIs only. In doing so, the nuisance parameters have been effectively eliminated like an integration variable that cannot itself appear in the final integral but whose integrand (cf. the joint pdf) is imprinted upon it. This approach is used, for example, in the Simplified

Likelihood formalism to be discussed in Section 5.8.1. Note that the most extreme case of marginalisation, dropping all rows and columns except for a single variable, gives Σ_{ii}: the total, diagonal variance of variable i as imprinted by the pdf joint structure in the other $k - 1$ parameters.

Non-square covariance subsets can also be useful: for example, we regularly wish to know the contributions of each systematic error component to the value of each measurement bin (or model-parameter extraction, or whatever). Slicing to a rectangular matrix whose first index runs over only the measurement bins, and the second over only the systematics' nuisance parameters, gives a measure of this: this is known as a **cross-covariance matrix**. While useful, this conflates the spread of the POI with the (usually) less interesting spread in the nuisance parameter—they are called nuisances for a reason. Not to worry: we can use 'half' of the correlation matrix construction of equation (5.44) to project out the uncertainty of POI p due to nuisance parameter n:

$$\sigma_p = \frac{|\Sigma_{pn}|}{\sqrt{\Sigma_{nn}}}. \tag{5.45}$$

Combination of correlated uncertainties

We are often faced with the situation that the correlation matrix contains too much information for a particular context, such that we might want to reduce its complexity by combining information associated with a given family of parameters. For example, the detailed parametrisation and decomposition of uncertainties in e.g. jet calibration into nuisance-parameter components is not generally of interest to a phenomenologist reading an experimental paper, nor is the correlation of a theory prediction with, for example, every eigenvector of a pdf error set, of interest to the experimental reader. Instead we want to summarise the error from the whole set of nuisance parameters in that category, without discarding it entirely.

It is tempting to think that this manipulation can be performed simply by taking the rectangular POIs versus nuisances covariance slice discussed previously, and summing the rows of **elementary nuisance** covariances in each category to give a representative covariance between each *group* of nuisances and each of the POIs, i.e.

$$\Sigma_{pN} = \sum_{n \in N} \Sigma_{pn}. \tag{5.46}$$

Indeed, there are instances where such a combination is meaningful e.g. the sum of independent covariance matrices for a set of histogram bins can be combined in this way—each covariance element is effectively a 'σ^2', so their linear sum implements combination in quadrature—but we cannot perform this elegant looking combination more generally for an important and rather educational reason: relative signs.

Covariance-matrix elements can have relative minus signs. This tells us that when the value of one parameter goes up, the other tends to go down, and vice versa. This may have no more significance than the arbitrary sign between the nuisance parameter and its impacts[6]. But when we combine over systematics, we no longer have a single representative nuisance parameter with a meaningful direction, and are instead treating the covariance elements as unsigned indications of the size of an uncertainty. The sum therefore must at least be over the *absolute* values of the covariance elements, $|\Sigma_{pn}|$. Failing to do so would mean that combining two oppositely-signed sources of error would result in a reduced total uncertainty, even though any sign-combination of the two elementary nuisance parameters could have been true.

Another important factor in combining sources of uncertainty within covariance matrices is to ensure that they really are *sources* of error and not consequences. The covariance matrix does not distinguish between correlation and causation, as is trivially seen if one introduces an extra variable which is algebraically related to an existing nuisance parameter: it will acquire the same correlations with the POIs as the original, and it is tempting to sum this apparent contribution into the total uncertainty. However, it represents no new information, but a reflection of an established error source. This example is extreme, but more subtle versions are introduced by performing a fit that correlates elementary nuisance parameters directly as well as through their common influences on the POIs.

The combination of whole covariance matrices is a final useful concept for breaking down groups of uncertainties into orthogonal components, the classic decomposition being into several 'systematic' sources of uncertainty, and one 'statistical' component. In this mode, several smaller $n_{poi} \times n_{poi}$ covariance matrices areestimated, each corresponding to the correlated uncertainties between the POIs induced by a subset of the uncertainties. The limitation of attentionto the correlations between the POIs only is because, by construction, thenuisance parameters are different in each group of uncertainties and hencecannot be combined other than in a trivial block-diagonal form. To compute an effective total POI-covariance matrix, these component covariances are added linearly (cf. standard deviations in quadrature) over the set of error groups C:

$$\Sigma_{ij}^{tot} = \sum_{c \in C} \Sigma_{ij}^{c}. \tag{5.47}$$

5.2.4 Mutual information between variables

In this discussion, our focus has been primarily on how random variables co-vary, for the purposes of estimating, propagating, and minimising uncertainties. But there is another motivation, namely identifying measurable observables which tell us a lot about an unmeasurable one: whether an event-level variable is a good indicator for signs of beyond the Standard Model (BSM) physics, for example. We often speak informally of such observables as being *strongly correlated* with what we want to know, but is that really what we mean?

---

[6] For example, in QCD theory uncertainties, does an 'up' variation mean an increase in coupling, or an increase in scale (which *decreases* the strong coupling)? Either is fine, but the polarity of the labelling choice cannot affect the physical conclusions.

A good counter-example appears in effective-field theories, where a cross-section receives a new-physics contribution in part from the 'EFT-squared' term, $\Delta\sigma \supset |\mathcal{M}_{\text{EFT}}|^2 \propto C^2$, where $\mathcal{M}_{\text{EFT}}$ is the matrix element for the process and C is the Wilson coefficient setting the strength of a new effective interaction, which can be either positive or negative. The measured cross-section is hence directly connected to signs of new physics, which correspond to any $C \neq 0$, but if we compute the correlation coefficient between σ and C, e.g. between -1 and 1, we will be disappointed: despite the cross-section clearly being influenced by C, the fact that that influence is the same for both positive and negative C values means that the overall correlation is zero.

Fortunately there is a better way, which more directly answers the question of how much one variable can tell us about another, or in the jargon *how much information is shared between two variables*. This measure is the **mutual information**, and is a core concept in information theory. While we do not have have the space here for a full introduction to information theory, the topic has come into renewed prominence in recent years through machine-learning methods, and you are unlikely to regret time spent familiarising yourself with its main concepts.

The most important concept in information theory is the **entropy** of a random variable, defined in the Gibbs sense as

$$H = \sum_i P_i \ln \frac{1}{P_i} \tag{5.48}$$

$$\rightarrow \int \mathrm{d}x \, p(x) \ln \frac{1}{p(x)}, \tag{5.49}$$

for discrete and continuous distributions on the first and second lines, respectively. We state here without proof that the entropy of a variable can be viewed as its total information content. This statement may seem more intuitive if you examine the formulae and note that if a single value has probability 1 (so all other $P_i = 0$, or $p(x) \rightarrow \delta(x - x_0)$), then the information content is zero: observing the 100% predictable outcome of a sampling from it tells us nothing. Similarly, a set element or value of x with probability zero contributes no information to the total. The real interest, of course, lies away from these extremes, and has great application to many topics from code-breaking to digital compression.

Let us now consider a joint probability distribution $p(x, y)$ in variables x and y. The mutual information between these two variables is then

$$I(x, y) = \int_X \mathrm{d}x \int_Y \mathrm{d}y \, p(x, y) \ln\left(\frac{p(x, y)}{p(x)p(y)}\right), \tag{5.50}$$

where X and Y denote the domains (ranges of possible values) of the random variables x and y, respectively[7]. The similarity with the entropy is clear, and

---

[7] In a slight abuse of notation in equation (5.50), we use x and y both to denote the random variables themselves, as well as the dummy variables on the right, which take all possible values in X and Y, respectively.

equation (5.50) in fact captures a measure of 'information distance' between the two variables, as expressed through their probability distributions. There is some of the flavour of the covariance projection method here, too: in the argument of the logarithm, the joint probability is normalised by the marginal probabilities of each variable.

Again, some intuition about $I(x, y)$ can be gained through its extreme configurations: if the distributions of events in variables x and y are completely independent of one another, then $p(x, y) = p(x|y)p(y) = p(x)p(y)$ and the logarithm collapses to zero: there is zero mutual information when the variables are completely disconnected. At the other extreme, when one variable is completely dependent on the other, $p(x, y) = p(x) = p(y)$, and the integral collapses to $I(x, y) = H(x) = H(y)$, i.e. the full information content of either variable is shared with the other. The fractional mutual information $I(x, y)/H(x, y)$ is hence a useful normalised measure of how useful one variable is as a measure of the other: a powerful technique, without the pitfalls of correlation, for assessing which variables have the most *a priori* power for parameter inference, signal/background discrimination, or detector-bias corrections.

We should note, however, that mutual information is not a silver bullet—we still have to be careful about how the events which define the probability measures are generated. And continuous variables are more awkward in practice than they appear here. Given a finite event sample, we cannot actually integrate but have to sum over a set of discrete bins, whose sizes bias the calculation somewhat. Bias corrections or fitting of integrable functions are needed, and this extra complexity has, thankfully, been already implemented in various publicly available computational data-analysis toolkits and packages.

5.3 Statistical estimators

In practice we rarely have the luxury of working with analytic distributions: by its fundamental nature, experimental particle physics makes observations in terms of events, and conclusions must be made from the aggregate properties of event sets rather than from perfectly smooth distributions. As we cannot run experiments either with infinite luminosity or for an infinitely long time, we cannot know with perfect precision the means, standard deviations, or any other property of the underlying distributions, as these are not directly accessible but can only be approximated by **estimators** subject to **statistical fluctuations**. Fortunately, the Law of Large Numbers means that more events (often referred to informally as 'large statistics', 'more statistics', or similar) have fewer fluctuations and hence are better estimators. Note that estimation is a useful concept for both Bayesians and frequentists, but the two schools may rely on different choices of estimator.

The topic of estimators permeates all interpretation performed with finite statistics, and in particular we are often interested in **maximum likelihood estimators** (MLEs), i.e. those which return the parameter θ that maximises the likelihood $p(D|\theta)$. It is also important that we find **unbiased estimators**, i.e. those which correctly approach the true values given asymptotically large statistics. An estimator which does not asymptotically approach the true value is called **biased**. It is worth noting that biased estimators may converge more quickly, i.e. the uncertainty on the estimator value down-scales more rapidly with number of events than the best unbiased estimator: in some applications this may be the critical factor, but here we focus on the unbiased variety.

5.3.1 Summary estimators

First we consider unbiased statistical estimators for whole event sets, starting with the **sample mean** of the dataset,

$$\tilde{\mu} = \tilde{E}[x] \equiv \frac{1}{N} \sum_i^N x_i, \tag{5.51}$$

for N events, each having value x_i. That this is a suitable estimator of the mean is perhaps unsurprising, since you have undoubtedly seen the statistical mean many times before. However, it is worth taking a moment to appreciate that it is not *a priori* obvious. Similarly, a **sample variance** is derived directly from the pdf from following equation (5.21):

$$\tilde{\sigma}^2 = \sigma^2 = \tilde{E}[x^2] - \mu^2 \tag{5.52}$$

$$= \frac{1}{N} \sum_i^N (x_i^2) - \mu^2. \tag{5.53}$$

The sample standard deviation trivially follows this as $\tilde{\mu} = \sqrt{\tilde{\sigma}^2}$.

In practice, however, this is not quite what we need: it assumes perfect knowledge of the distribution mean, μ, but far more usually we will instead have the estimator $\tilde{\mu}$, estimated using the *same* data sample. Simply replacing $\mu \to \tilde{\mu}$ in equation (5.52) does not give an *unbiased* estimator, since one degree of freedom in the dataset (effectively one event) is 'used up' in evaluating the sample mean $\tilde{\mu}$ before computing the variance relative to it. The fix to this is simple: replace the N factor in equation (5.52) with $N - 1$, or equivalently multiply by $N/(N - 1)$:

$$\tilde{\sigma}^2_{\text{corr}} = \langle \tilde{x}^2 \rangle - \langle \tilde{x} \rangle^2 \tag{5.54}$$

$$= \frac{1}{N - 1} \sum_i^N x_i^2 - \tilde{\mu}^2. \tag{5.55}$$

This replacement is called **Bessel's correction**, and the result is the unbiased estimator for the population variance. But interestingly it does not fully unbias the standard deviation estimator—in fact there is no general expression for an unbiased sample standard deviation that can be used with all distributions! In practice this is rarely of importance, but is worth being aware of if you are unable to avoid small sample numbers (e.g. in studies of rare processes where despite large total data samples, the number of signal candidates to pass selection criteria remains small).

All these concepts are essentially finite-statistics retreads of what we have already discussed in probabilistic pdf/pmf language. You can convince yourself that in the limit of infinite statistics, these quantities are unbiased, e.g. $|\tilde{\mu}-\mu| \xrightarrow{N=\infty} 0$. But we are led to a new concept: the uncertainty of the estimator resulting from the finiteness of the statistics. The canonical example of this is the uncertainty on the estimator of the mean, $\tilde{\mu}$, known as the **standard error on the mean**. This is given by

$$\hat{\sigma} = \sigma/\sqrt{N}, \tag{5.56}$$

where again we see the classic $1/\sqrt{N}$ asymptotic approach to perfect accuracy. This expression tells us that the estimator of the mean has an uncertainty which will disappear with sufficient statistics, and that intrinsically wider distributions require higher statistics to achieve a given estimator precision. Technically, we should be talking in terms of the estimator of the standard error, $\tilde{\hat{\sigma}} = \tilde{\sigma}/N$, but this notation is all getting rather heavy and we will shortly drop it except when essential for clarity.

5.3.2 Histograms as distribution estimators

This brings us neatly to formalising physics **histograms**, i.e. binned distributions over a population of collider events. The ideal physics histogram of variable x is the differential cross-section $d\sigma/dx$ (or its unit-normalised probability density), where the dx literally means that the bins of the bar chart are infinitely narrow and the curve is hence perfectly smooth. Infinitely thin bins are nice in theory, but in experiment (and in numerical simulations) we need them to be finitely sized so there are finite probabilities of observations falling into each bin (and a finite number of bins). Histograms are hence finite-statistics, finite-resolution estimators of the true, smooth distribution.

The height of a bin is then proportional to the number of fills (or more generally the sum of their weights, if not every fill is considered equally likely) that fell into it, divided by the finite bin width Δx: this division makes the probability of the bin proportional to its area rather than height, which is the formal distinction between a 'bar chart' and a 'histogram'. We can relate the binned density estimate f_{bin} to the ideal differential density f by $f_{\text{bin}}(x) = \int_{x-\Delta x/2}^{x+\Delta x/2} f(x)dx/\Delta x$. Similarly, the normalisation of such a histogram is obtained by summing the areas (i.e. heights ×widths), of the bins—by direct analogy with the smooth integral $\sigma = \int d\sigma/dx \, dx$. Another consequence of finite computer memory and so on is that statistical histograms will have a finite number of bins, i.e. their axis does not use finite-sized bins to cover the entire real number line from $-\infty$ to $+\infty$, but instead has a finite range below which is the **underflow bin**, and above which is the **overflow bin**. Computational implementations of histograms store out-of-range fills in these 'overflows': be sure to remember them as appropriate when calculating integrals or moments over histograms.

The need to divide by bin-width is often ignored with little consequence in particle-physics plots, particularly because the ROOT data analysis software plots bar charts without the width-division, and because uniformly sized bins are the most common choice: in this case the only effect is in normalisation, not shape. However, you should be aware that the width-division is necessary to make a plot physically comparable to theoretical differential cross-sections. Knowing this also frees you up to use non-uniform binnings, which can be extremely useful to get comparable fill populations (and hence statistical uncertainties) in all bins across e.g. a steeply falling differential spectrum as found in 'energy scale' variables like (transverse) momenta and masses.

The same idea generalises to 2D histograms and beyond: the variable measure is now of higher dimensionality, so the relevant probability measure is the fill content of the differential area $dxdy$ in 2D, the volume $dx\,dy\,dz$ in 3D, etc and their finite-bin equivalents $\Delta x \Delta y$, $\Delta x \Delta y \Delta z$, etc. The relevant idealised density is now the corresponding higher derivative, e.g. $d^2\sigma/dx\,dy$ or $d^3\sigma/dx\,dy\,dz$—which may look intimidating on plots but is just particle-physics code for '2D histogram', '3D histogram', and so on, giving a constant reminder that the bin height (or colour-scale) should now be divided by the bin area, volume, or whatever $\prod_i dx_i$ quantity is appropriate for its dimensionality. Be aware that the computer memory needed to handle n-dimensional binnings scales as $N^n_{\text{axis bins}}$, which can easily get out of hand: in practice one- and two-dimensional binnings are almost always all that you need.

In practice we need to be able to represent histograms as graphical data objects, which largely reduces us to at most two- dimensional projections—three-dimensional surface plots are sometimes used, but mostly for aesthetic excitement as they contain no more (and sometimes less) information than a 2D 'heat map' plot, usually with the same colour scheme as was used to paint the surface in the 3D view. Most of the time, '1D' plots of the binned density as a function of one variable are used[8], with multiple lines representing different slices (partial integrals) of the hidden dimensionality. It is much easier to show multiple models or phase-space slices on such a 1-parameter plot than to overlay multiple colour maps or surfaces.

Another very useful data type is the **profile histogram**, which is a compression of one axis of a 2D histogram so that each e.g. x-bin contains the mean value of y in that range of x, $\langle y \rangle = \int_{-\infty}^{+\infty} dy \int_{x_0}^{x_1} dx\, f(x, y)$ (again including the overflows). Profiles are a very useful way for visually characterising how the typical value of one variable depends on another, in a form which allows plotting and comparison of multiple lines on a single graph.

The last feature of histograms to explore in this summary is that the bins of a histogram are typically represented with an indicator of their statistical uncertainty, the **error bar**. It is important to be careful about the meaning of this—despite the name 'error', it does not mean that we might have made a mistake in the measurement (that's what **systematic errors** are for) but that, were we to perform the experiment again with the same finite event statistics, we would get a slightly different result just from randomness alone. Were we to make lots of equal-statistics

---

[8] The plot itself is two-dimensional, of course: when we say a '2D histogram' we mean that there are two independent variable axes, plus the density values as functions of those two variables.

experiment repetitions, we would typically find a Gaussian distribution (thanks to the CLT) with a characteristic standard deviation.

It is this scale that we represent with the error bar on a histogram, and for equal weights it is given by $\sqrt{\sum w} \sim \sqrt{N}$ for bin population N—and divided by the bin-width as usual. Since the total bin content is N, this means that the relative uncertainty decreases as the characteristic $1/\sqrt{N}$ as the bin-population statistics increase. For a profile histogram we are typically more interested in the statistical error on the mean $\langle y(x) \rangle$ than in the intrinsic width of the y-distribution in each x-bin, $\sigma_y(x)$; hence the usual error bar used in profile histograms is instead the standard error $\sigma_y(x)/\sqrt{N}$, again becoming more precise with the square-root of the number of events. We should note, however, that this Gaussian limit for error-bar extraction is not the whole story—we will return later to confidence limits, which among other things give us a more precise (and asymmetric) recipe for the construction of error bars.

One important consequence of this scaling, however, is that it informs the choice of **binning** for a randomly distributed physical variable, both in terms of the raw number of bins, and their generally non-uniform widths. For a population histogram, a good rule of thumb is to achieve roughly equal populations (and hence similar relative errors) in every bin, hence choosing to bin more densely in regions of high population, and to use wide bins on the tails of distributions. The total number of bins is then limited by the need to have a sufficient population in each bin that a Gaussian approximation is usable (if necessary), and to not site bin edges closer together than the detector resolution—a limitation which would show up as troublesome **bin migrations** under propagation of systematic errors. In profile histograms the same ideas apply, but with the equal-population heuristic now modified by the intrinsic width of the y distribution in each x-bin, such that the relative standard error of each bin can be of similar magnitude.

5.3.3 Correlation estimators

We have already considered the definition and interpretation of the covariance from analytic pdfs. In practice it is more likely that particle physicists will need to compute the **sample covariance** estimator, obtained over a dataset of N samples drawn from the (usually *non*-analytic) pdf:

$$\tilde{\Sigma}_{ij} = \tilde{E}[x_i x_j] - \tilde{E}[x_i]\tilde{E}[x_j] \tag{5.57}$$

$$= \frac{\sum\limits_{n}^{N} x_{i,n} x_{j,n}}{N} - \frac{1}{N^2} \sum\limits_{m}^{N} x_{i,m} \sum\limits_{n}^{N} x_{j,n}, \tag{5.58}$$

where $x_{i,n}$ is the value of the ith parameter in event/sample n. In practice we will not distinguish strongly between the true covariance Σ and its estimator $\tilde{\Sigma}$, and will generally use the un-tilde'd symbol.

5.3.4 Efficiency estimators

A very important category of correlated statistic is **efficiencies** : figures which quantify the fraction of some set of (desirable) objects to pass a selection procedure.

The selection of interest could be of individul physics objects, for example hadronically decaying tau leptons selected from hadronic-jet backgrounds, electrons separated from photon and jet backgrounds, b-jet flavour tags separated from light- and charm-jet fakes, or something more nuanced such as picking the *correct* set of b- and light jets for a hadronic top reconstruction from among all the possible combinations. Alternatively, it could be the selection of whole events composed from such objects, e.g. the efficiency of BSM signal events to pass unscathed through the set of selection cuts designed to cut away their main SM-background processes.

For a given sample of N objects of interest, of which n pass the selection, the measured efficiency is obviously

$$\epsilon_{\text{meas}} = n/N. \tag{5.59}$$

But this is usually not really the question we want an answer to: we don't care so much about the specific efficiency of a calibration sample (which in any case might consist of simulated data from a Monte-Carlo (MC) event generator rather than real data), but what it tells us about the true efficiency ϵ we can expect on *other* data to be analysed.

The standard approach is to compute the binomial uncertainty: we treat our N events as being independently selected with efficiency ϵ, and the probability of n being selected is hence

$$P(n; N, \epsilon) = \frac{N!}{n!(N-n)!}\epsilon^n(1-\epsilon)^{N-n}. \tag{5.60}$$

Taking $dP/d\epsilon = 0$ to find the maximum likelihood (or equivalently and more neatly, $d \ln P/d\epsilon = 0$), we find the MLE for the asymptotic efficiency as $N \to \infty$ to be, unsurprisingly, the same form as the measured version: $\tilde{\epsilon} = n/N$.

A more interesting question is what its *uncertainty* looks like[9]. Were we looking purely at an event count, the binomial estimator for the uncertainty on n objects passing the cuts would be $\sqrt{n}$. But this clearly cannot be the case for the efficiency estimator $\tilde{\epsilon} = n/N$: i.e. as $n \to N$, $\tilde{\sigma}_\epsilon \to \sqrt{N}$, placing its 1σ upper limit at $1 + \sqrt{N}$—i.e. greater than 1, which is clearly absurd.

A slightly more careful analysis is needed. We instead consider the binomial variance of ϵ, which we can write as

$$\sigma^2[\epsilon] = \sigma^2[n/N] = \frac{1}{N^2}\sigma^2[n] \tag{5.61}$$

---

[9] Note that the uncertainty on ϵ_{meas} is zero: there is no ambiguity about how many objects passed or failed the selection. Uncertainties only appear when we try to extrapolate our finite-statistics observations to draw conclusions about the true, asymptotic parameters of the system.

$$= \frac{1}{N^2} N\epsilon(1 - \epsilon) \tag{5.62}$$

$$\approx \frac{\epsilon_{\text{meas}}(1 - \epsilon_{\text{meas}})}{N} \tag{5.63}$$

$$= \frac{(n/N)(1 - n/N)}{N}, \tag{5.64}$$

and hence obtain the estimator of the standard deviation,

$$\tilde{\sigma}[\tilde{\epsilon}] = \frac{\sqrt{m(1 - m/N)}}{N}. \tag{5.65}$$

Neatly, this goes to zero at both the $m \to 0$ and $m \to N$ extremes, in such a way that the 1σ error bars never extend beyond the logical efficiency range $[0, 1]$. This is a very useful property to have on our list, being the standard estimator of uncertainty on an estimated efficiency.

But we should be a little critical: it is only by convention that we report 1σ uncertainties, and e.g. the 2σ error bars can go outside the physical range. It is also not true that the uncertainty on the efficiency really goes to zero at the extremes: after all, the number of samples N was finite, and a larger sample might have revealed small deviations from absolute pass/fail behaviour. What we should actually do is construct asymmetric uncertainty estimators, such that the upper uncertainty goes to zero as $n \to N$, and vice versa for the lower uncertainty as $n \to 0$. Such a construction can be performed, for example, using the **Poisson bootstrap** method, but for most purposes the binomial estimators are sufficient, as long as they are not taken too literally.

5.3.5 Likelihood estimators

We often want to test how well some data is described by a model, and other than subjective eyeballing of plots, we need a quantitative measure of the **goodness of fit**. There are many ways in which a goodness of fit can be quantified, but for our typical problem of needing to optimise the parameters of a probabilistic model to a set of statistically noisy data, the obvious choice is the statistical likelihood, $p(\text{data}|\text{model})$, or any function monotonically related to it.

The canonical goodness of fit **test statistic**, is the χ^2 test introduced in 1900 by Karl Pearson. This is a simple and intuitive measure for comparing data to expectations, introduced for comparing data counts $y(x)$ in n_b measurement bins $\{X\}$ to an analytic model $f(x)$ without systematic uncertainties. The original Pearson test statistic is defined as

$$\chi^2 = \sum_{x \in \{X\}} \frac{(y(x) - f(x))^2}{f(x)}. \tag{5.66}$$

The limitation to a counting experiment is responsible for this form, which maybe differs from what you have encountered in physics. To generalise, and to understand it better, we note that in the limit of large statistics the binomial distribution of the expected count in each bin limits to a Gaussian distribution with variance $\sigma^2(x) = f(x)$. The test statistic is then equivalent to

$$\chi^2 = \sum_{x \in \{X\}} \frac{(y(x) - f(x))^2}{\sigma^2(x)}. \tag{5.67}$$

This is more intuitive: it is the squared deviations of observation from expectation, summed over all the bins, with the contribution from each bin scaled relative to its variance. Large deviations in uncertain bins are hence proportionally penalised with respect to the width of the (assumed Gaussian) pdf from which the observations have been sampled. Small values of the χ^2 indicate a model that fits the data well.

But how well? And how do we compare χ^2 statistics from observations with different numbers of bins? A rule of thumb is to quote χ^2/b, but this is overly simplistic. To go further it helps to identify the χ^2 as a simple example of a **log-likelihood** statistic. The likelihood of observing a set of bin values $\{y(x)\}$ with model M is $\mathcal{L} \equiv \prod_{x \in \{X\}} p(y|M)$, and hence if we assume that each bin's random variable $y(x)$ has a Gaussian pdf $\mathcal{N}(y; f(x), \sigma(x))$, the logarithm of the likelihood is

$$\ln p(y|M) = \sum_{x \in \{X\}} \ln \mathcal{N}(y(x); f(x), \sigma(x))$$
$$\sim \sum_{x \in \{X\}} \frac{-(y(x) - f(x))^2}{2\sigma^2(x)} = -\chi^2/2. \tag{5.68}$$

So the χ^2 is (negatively) monotonic to the log-likelihood:[10] any calculation we might have performed with the log-likelihood can instead be performed using $-\chi^2/2$ (provided that the underlying assumption of normally distributed likelihood components is valid). Indeed we shall shortly see an important application of this, in comparison of the test statistic

$$t = -2 \ln \mathcal{L} \tag{5.69}$$

to the chi-squared distribution. This identification also makes clear the close relationship between the χ^2 and Gaussian pdfs.

We can make one final refinement to the χ^2 statistic, using the machinery now at our disposal. In the presentation shown so far, we assumed that the bins x were independent and hence the total likelihood was a simple product of each bin's individual likelihood. As seen in Section 5.2, this is not generally true: the random

---

[10] The attentive reader may wonder why we dropped the constant offset $\sum_x \ln(1/\sqrt{2\pi}\sigma_x)$ arising from the Gaussian normalisation terms. Isn't this important? In fact, the χ^2 test statistic is implicitly comparing the model to the 'saturated model' where the prediction exactly equals the data. As this has the same normalisation constants, the extra terms cancel in the implied comparison. While this is a minor point here, the concept of cancellation between two models is an important one we shall return to.

variables $y(x)$ may be correlated with one another. Fortunately this does not pose any fundamental problem to the χ^2 statistic, which is easily extended using the covariance matrix between the bins, Σ, whose second-order characterisation of correlations captures directly the structure of a multidimensional Gaussian pdf in the space of bin values, with the principle axes now lying along linear combinations of the bin axes. As we have shown that up to a factor of -2, the naïve χ^2 statistic is the summed log-likelihood of the independent Gaussians, it is no surprise that in the presence of bin-correlations, the χ^2 generalises to the similarly scaled exponent of the multivariate normal distribution,

$$\chi^2_{\text{corr}} = (\boldsymbol{y} - \boldsymbol{f})^T \boldsymbol{\Sigma}^{-1}(\boldsymbol{y} - \boldsymbol{f}) \tag{5.70}$$

$$= \sum_{i,j}(y_i - f_i)\Sigma_{ij}^{-1}(y_j - f_j), \tag{5.71}$$

where the summation and $1/\sigma^2$ factors have been handily absorbed into the vector and matrix multiplications, and the inverse covariance $\boldsymbol{\Sigma}^{-1}$ respectively[11].

Equation (5.70) describes a quadratic-sided 'valley' in χ^2 as a function of statistical deviations from the model, assuming that the model is true. We will return to this picture later, with variations in the model parameters away from their fitted maximum-likelihood estimators. A useful result follows that the curvature of the χ^2 well, or more generally that of the negative log-likelihood, can be used to derive an estimator for the covariance. Computing the **Hessian matrix of second derivatives** of the negative log-likelihood gives an estimate of the inverse covariance matrix:

$$-\frac{\partial^2 \mathcal{L}}{\partial f_\alpha \partial f_\beta} = \frac{\partial^2(\chi^2/2)}{\partial f_\alpha \partial f_\beta} = \frac{1}{2}\frac{\partial^2}{\partial f_\alpha \partial f_\beta}\left[\sum_{i,j}(y_i - f_i)\Sigma_{ij}^{-1}(y_j - f_j)\right] = \Sigma_{\alpha\beta}^{-1}. \tag{5.72}$$

This result is exact for a Gaussian likelihood, and is more generally useful as an approximation in the immediate vicinity of the MLE model. Numerical evaluation of these derivatives is used by numerical fitting tools to estimate the covariance in the vicinity of the best-fit point, i.e. the MLE, of a likelihood model.

For the case of a signal discovery, there are some specific likelihood estimators that are useful to know. Imagine that you wish to optimise an analysis designed to search for a particular signal, and you want to know when you will have sensitivity to a discovery. For any given analysis design, you can simulate the number S of expected signal events, and the expected number B of background events. A general statement of the significance s of the hypothetical signal observation is then given by the number of Gaussian standard deviations of the background distribution to

---

[11] An important caveat must be borne in mind when using covariance (and hence correlation) matrices between bins of *normalised* distributions, i.e. whose integral has a fixed value. In that case, knowing the value of $n_{\text{bin}} - 1$ bins fixes the value of the remaining one, resulting in a singular covariance whose inverse cannot be computed. To make statistical progress one must either drop one bin, or re-obtain the unnormalised distribution.

which the signal corresponds, $s = S/\sigma(B)$. If the systematic uncertainty on the background can be neglected, this is given by $s = S/\sqrt{B}$, where we have assumed a Poisson distribution for the background event counts, thus making the standard deviation equal to the square root of the mean. If the systematic uncertainty on the background is non-negligible, a useful rule of thumb is $s = S/\sqrt{B + \delta B^2}$, where δB is the absolute systematic uncertainty. The accuracy of these formula becomes poor in the case of small numbers of background events, where $B < 100$ is sufficient to count as 'small'. In that case, a better simple estimator is $s = \sqrt{2n_o \ln(1 + S/B) - 2S}$, where we now have an assumed observed event count n_o that is allowed to depart from the predicted yields S and B.

More detailed likelihoods, specific to event-counting experiments with systematic uncertainties, will be described in Section 5.8.1.

5.3.6 Weighted-event estimators

To add extra complexity, both due to Monte Carlo sampling techniques and to mismatches between the generated configuration and the real experimental conditions, events need not be equally important. Some may be more representative of the intended distribution than others, and accordingly are given larger **event weights**.

Weights are used as an augmentation to the estimators we have discussed so far, such that the sums and averages in unbiased estimators can be replaced with *weighted* sums and averages. These bias the estimator, such that the moments obtained are not representative of the population of events, but of a different distribution. There are two classes of motivation for doing so: to correct for quantifiable errors in the simulation, and to improve simulation efficiency. Examples of the first class are to have generated events with a pile-up distribution that doesn't agree with the data run, or where the generator or detector description has a known mismodelling. Both of these can be reweighted to 'standard candle' distributions in variables x, such that the weights are

$$w_i = f_{\text{true}}(x_i)/f_{\text{MC}}(x_i), \tag{5.73}$$

where $f_{\text{true}}(x_i)$ is the pdf of the distribution we want to reproduce, and $f_{\text{MC}}(x_i)$ is the Monte-Carlo-generated distribution.

The second class is where the physical distribution gives an inconvenient convergence of errors for the intended application: the classic case here is wanting to study event features (most trivially, the differential cross-section) as a function of transverse momentum, when the p_T spectrum falls rapidly, e.g. like $d\sigma/dp_T \sim p_T^{-4}$. Direct sampling of this cross-section/pdf will produce far more events at low-p_T than are statistically required, even with finer, resolution-limited binning at low-p_T: the statistical errors in the low-p_T bins will be tiny, while on the high-p_T tail they will be huge, and the bins will accordingly need to be made very wide to avoid major statistical fluctuations. Weighting can be used here to balance statistical precision across the p_T range, by sampling the events not from the physical distribution but from an 'enhanced' one such as $p_T^4 d\sigma/dp_T$, which is much flatter across the p_T range being studied; the weights then have to recover the physical spectrum by cancelling the enhancement factor, i.e. in this case $w_i = p_{T,\,i}^{-4}$.

Whatever the motivation for the weights, the biased, weighted sample mean is

$$\tilde{\mu} = \frac{1}{\sum\limits_i w_i} \sum_i w_i x_i, \tag{5.74}$$

equal to the estimated mean of the true distribution rather than the MC one. Less obviously, the weighted variance estimator (including a weighted Bessel correction) becomes

$$\tilde{\sigma}^2 = \frac{\sum\limits_i \left(w_i x_i^2\right)\sum\limits_i w_i - \left(\sum\limits_i w_i x_i\right)^2}{\left(\sum\limits_i w_i\right)^2 - \sum\left(w_i^2\right)}. \tag{5.75}$$

Note that, in both of these cases, what matters is not the total scale of the weights, but how they are distributed across the event set. A global rescaling of the form $w_i = w$ allows the w_i terms to be pulled out of the sums, where they cancel: any physical quantity cannot be sensitive to the overall scale of weights, and indeed different event generators use this freedom to set weight scales variously close to unity, or to around the process cross-section in some generator-specific choice of units.

5.3.7 Effective sample size

A final useful weighted quantity, while not exactly an estimator, is the **effective sample size** (or effective number of entries) in the weighted distribution: as the events are not all equal, it is reasonable to ask how many 'full' events a weighted sample is worth in terms of statistical power. This is defined by considering the standard error on the mean as a function of the variance, $N_{\text{eff}} \equiv \sigma^2/\tilde{\sigma}^2$, to give a quantity known as *Kish's effective sample size*:

$$N_{\text{eff}} \equiv \frac{\left(\sum\limits_i w_i\right)^2}{\sum w_i^2}. \tag{5.76}$$

The behaviour with a constant weight distribution can again be easily checked, in which case it reduces to $N_{\text{eff}} = \sum_i 1 = N$. This is the case even with a global weight multiplication by a negative number. Weighting cannot *increase* the effective number of events, it either preserves them (with a global scaling) or reduces them (with any non-trivial distribution).

At this point it seems apt to clarify which N we have been using in the construction of our summary estimators, as in both experiments and phenomenology we often work with simulated event samples which are then rescaled for equivalence to an experimental dataset of a given integrated luminosity. Which N should we use: the number of generated events or the event yield corresponding to the experimental data period? A little thought should confirm that it is the generated number: in this circumstance what we are trying to do is estimate the Poisson rate of expected events under the given physics model, and it is in-principle possible to reduce the statistical uncertainty on this expected rate to zero by use of an infinitely large sample size: this is exactly the standard error on the mean. Fortunately, the uniform scaling by luminosity and cross-section factors does not change the effective sample size, and so it is a robust choice.

5.4 Parameter inference

The typical end result of a particle-physics statistical analysis is to draw conclusions about model parameters. These parameters can be very low-level, such as the set of cross-section values in a set of kinematic bins, or high-level, e.g. the fundamental steering parameters of the electroweak SM, or of a BSM theory. In this section we discuss the ingredients for inferring maximum-likelihood values for parameters and, more importantly, the ranges of values that we can constrain them to within some statistical precision. This process naturally includes elements of fitting, sampling, and construction of confidence intervals.

Parameter inference is a fairly straightforward and well-defined problem, but the solution is complicated by a number of factors. Firstly, we have already seen that there can be many parameters in any given problem, and finding the best set of parameters therefore becomes a difficult sampling problem. Secondly, the approach to parameter estimation differs for Bayesians and frequentists, and many physicists mix up techniques from both camps with wild abandon, producing results that are sort-of correct but confusing in some cases, and utterly wrong in others. In the following, we will therefore aim to be absolutely clear which camp we are in when we detail a specific method.

Let's imagine that we have made n observations of some quantity x, which together constitute our data $D = \{x_1, x_2, x_3, \ldots, x_n\}$. Our model for explaining the data has the parameters $\theta = \{\text{POIs, NPs}\}$, where we have noted explicitly that θ may contain both parameters of interest and nuisance parameters. We now want to infer the values of θ.

An important quantity for both frequentists and Bayesians is—as already encountered several times—the *likelihood* of the data given a model, $p(\text{data}|\text{model})$. Most theories are not single, inflexible models, but a paramerised family of them—for example, using the θ parameters above. For a fixed set of observed data, the likelihood as a function of the model parameters is a key object: formally this is the **likelihood function**, although we will continue to just use the colloquial simpler form 'likelihood'. As this is such an important quantity for inference, we set it apart from generic probability distributions with a dedicated symbol $\mathcal{L}(\theta; D)$[12]. For the data and model given above, the likelihood is following equation (5.7).

---

[12] We here use this 'semicolon' likelihood-function notation to avoid confusion. In the literature a common convention is to write $\mathcal{L}(\theta|D) \equiv p(D|\theta)$, a strong candidate for the most counterintuitive notation of all time. We prefer not to muddy the water further with an inverted 'bar' notation, while still introducing the common $\mathcal{L}$ symbol.

$$\mathcal{L}(\boldsymbol{\theta}; \boldsymbol{D}) \equiv p(\boldsymbol{D}|\boldsymbol{\theta}) \qquad (5.77)$$

$$= \prod_{i=1}^{n} p(x_i|\boldsymbol{\theta}), \qquad (5.78)$$

It is important to note that a conditional probability density function $p(A|B)$ is a pdf in A but not in B; as the likelihood function is defined with *fixed* $\boldsymbol{D}$ and its free parameter is the condition $\boldsymbol{\theta}$, $\mathcal{L}(\boldsymbol{\theta}; \boldsymbol{D})$ is not a pdf on $\boldsymbol{\theta}$. Instead, it is a collection of values that indicate how likely the already-measured data $\boldsymbol{D}$ is under each possible $\boldsymbol{\theta}$. In particular, $\int \mathcal{L}(\boldsymbol{\theta}; \boldsymbol{D}) \, d\boldsymbol{\theta} \neq 1$: an extreme example of this is that if the data was 100% predictable and *independent* of $\boldsymbol{\theta}$, i.e. $\mathcal{L}(\boldsymbol{\theta}) = 1$, the integral clearly diverges.

On the right-hand side of equation (5.77), we have assumed that we can calculate the likelihood of any particular observation x_i given a set of parameters for the theory. This is true in general, although not necessarily in an analytic form, since the whole point of our model is that it predicts our observed quantity as a function of its parameters. The likelihood of obtaining our complete set of observations $\boldsymbol{D}$ is then given by the product of each of the individual likelihoods.

So far, we have said nothing that wouldn't be equally useful for a frequentist and a Bayesian. However, where we go next depends critically on which one we are.

5.5 Bayesian inference

In **Bayesian inference**, we use Bayes' theorem (equation (5.7)) to convert the likelihood to the posterior probability of the parameters given the data:

$$p(\boldsymbol{\theta}|\boldsymbol{D}) = \frac{\mathcal{L}(\boldsymbol{\theta}; \boldsymbol{D})p(\boldsymbol{\theta})}{\int \mathcal{L}(\boldsymbol{\theta}; \boldsymbol{D})p(\boldsymbol{\theta})\mathrm{d}\boldsymbol{\theta}}. \qquad (5.79)$$

We then have an actual probability density function for the parameters given the observed data. Although this is not known analytically in most cases, it can be sampled, and the samples can be used to perform probability calculations, e.g. evaluating the expectation values of functions of $\boldsymbol{\theta}$. This is conceptually simple, but note that our posterior, and hence our inferences, might strongly depend on the assumed prior on the parameters, $\pi(\boldsymbol{\theta}) \equiv p(\boldsymbol{\theta})$. A Bayesian analysis therefore needs to at least specify the assumed priors clearly, and should be repeated with different choices of prior if there is no motivation for any particular choice. The result of a Bayesian inference calculation will be $p(\boldsymbol{\theta}|\boldsymbol{D})$, which is a multidimensional function of both our parameters of interest and our nuisance parameters. Let's assume that we want to investigate a particular parameter θ_1. A Bayesian does this by calculating the *marginal posterior pdf* of θ_1, which is the pdf of θ_1 regardless of the values of the other parameters[13]:

$$p(\theta_1|\boldsymbol{D}) = \int p(\boldsymbol{\theta}|\boldsymbol{D})\mathrm{d}\theta_2\mathrm{d}\theta_3\dots\mathrm{d}\theta_n. \qquad (5.80)$$

---

[13] Often referred to as 'integrating out' or 'marginalising over' the other parameters.

The key point here is that it is the pdf that is the real result. What if we want to quote the 'value' of θ_1? We could take the mode of $p(\theta_1|D)$, or the mean, or some other summary statistic, but all of these might be unsatisfactory in some circumstances. For example, if we have a spike in the marginal posterior at one value (which is the mode), but a very broad region of probability density elsewhere which has a higher volume, then it is more probable that our parameter exists outside of the spike. The pdf, however, tells us everything, and results will typically be presented in the form of the marginal posterior for each parameter of interest. We can also quote values with uncertainties using the notion of *credible regions*.

5.5.1 Credible intervals and credible regions

For ease of comprehension, we will first examine uncertainty estimation on a single parameter θ_i via the 1D marginal pdf $p(\theta_i|D)$. Our motivation is to identify a range (or even a set of disjoint ranges) in θ_i within which the true value of θ_i falls with a particular probability. This fraction is called the **confidence level** (CL), and the corresponding parameter range is the **credible interval** (CrI). The two are related as follows:

$$\int_{\theta_i \in \mathrm{CrI}} p(\theta_i|D)\mathrm{d}\theta_i = \mathrm{CL}. \tag{5.81}$$

Note that this is ambiguous: there are an infinite number of possible choices of θ_i interval (especially when allowed to be disjoint). Which we prefer depends on the circumstances: for example, if testing for the (non-)existence of a rare process, it may make sense to build the interval on the lower tail of the distribution (integrating rightward from $-\infty$ in the parameter value) to set an upper limit on the rate; without such a context, the highest-priority volume elements might be motivated by their corresponding probability density rather than their parameter values. In general we need a well-motivated **ordering criterion** for sequentially adding parameter-space volume elements to the CrI.

Of the better-motivated orderings, an obvious choice is to add points to the interval starting with those that have the highest posterior probability density $p(\theta_i|D)$ (known as the **maximum *a posteriori* probability** or MAP), and continue 'integrating down the probability density' until CL coverage has been achieved. This scheme can also be naturally generalised to the full (unmarginalised) likelihood, again adding elements to the CrI—but now multidimensional $\mathrm{d}^n\theta$ elements rather than one-dimensional marginal ones—from the mode of the pdf downwards, until probability fraction CL is contained, cf. equation (5.81). This construction is illustrated in Figure 5.10.

Note that disjoint CrIs also naturally emerge from this 'top down' scheme for CrI construction, should the distribution be multimodal, and the secondary modes have high-enough probability densities to be included in the CrI. This is an elegant feature in general, but has an obvious downside should one wish to use the CrIs of POIs to define 'error bars' in graphical or tabulated data: no such contiguous bar can be drawn. This is more a reflection on the relative poverty of that uncertainty

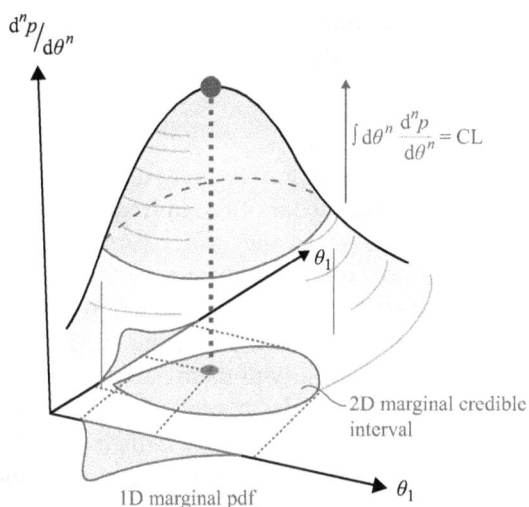

Figure 5.10. Confidence interval construction by projection from the pdf $p(\boldsymbol{\theta}|\boldsymbol{D})$. The vertical axis shows the probability density as a function of two model parameters θ_i and θ_j, which are represented on the two other axes: these could either be POIs or NPs. The global mode of the pdf is shown as a blue dot on top of the pdf bump, and projected into the 2D θ_i–θ_j plane, and on to each axis. The whole pdf is also projected on to the axes, giving the marginal pdfs: the modes of these need not match the global mode. Credible regions can be made either globally (here in θ_i–θ_j) or marginally by integrating from the mode downwards until a fraction CL of the pdf lies in the CrI.

representation[14], but nevertheless it is an important consideration when projecting pdfs into lower dimensional spaces. A similar consideration for error-bar representations is that the mode of the marginal pdf is guaranteed to lie within the marginal CrI, but the global mode of the pdf comes with no such guarantee: while logically consistent, it seems strange for a nominal 'most probable' point to lie outside its own confidence interval. Again, this reflects the fact that the global mode itself is not actually an interesting quantity in general.

This use-case motivates alternative CrI constructions, for example choosing the marginal CrI to be symmetric around the mode of the marginal pdf, with an integral of CL/2 in both the positive and negative deviations from the marginal mode. This suffers the problem of not being consistent between each POI, but depending on priorities this may be an acceptable compromise compared to disjoint CrIs, or a mode outside the CrI region. There is no global best answer to these potential condundrums: perhaps try for the most consistent view, using marginal projections of the globally constructed pdf mode and multidimensional CrI, but be prepared to fall back on more ad hoc constructions for the sake of unproblematic data presentation. In the cases where these complications arise, it is particularly

---

[14] For graphical presentation, a richer view is provided by, for example, violin plots in which the error bar is replaced by a marginal pdf estimate.

important to also publish more complete estimates of the full pdf, since the graphical or tabulated form is obviously not the whole story.

5.6 Frequentist inference

Having covered the Bayesian approach to inference, it is time to investigate frequentist methods. We start with the fact that, in **frequentist inference**, one cannot define a prior on the parameters, since they are assumed to have some unknown 'true' values, and there are no repeatable experiments that allow us to define a frequentist probability of them having particular values. We only have one universe, and in that universe the parameters have a particular value. It therefore makes no sense to discuss the frequentist probability of them *actually* having some other value. Instead, we can only refer to how likely or unlikely it was that we observed our particular data given the parameter values, which is fine since we have repeatable experiments in that case. This, naturally, is quantified by the familiar likelihood function $\mathcal{L}(\boldsymbol{\theta}; \boldsymbol{D})$.

To infer the values of the parameters, we need to find an estimator $\tilde{\boldsymbol{\theta}}$ that is as close as possible to the true vector of parameters $\boldsymbol{\theta}$. We can define various estimators, but by far the most useful is the choice that maximises the likelihood, called the **maximum likelihood estimator** (MLE). It is usually easier to work with the natural logarithm of the likelihood $\ln \mathcal{L}$, which is maximised in the same place, under the condition

$$\frac{\partial \ln \mathcal{L}}{\partial \theta_i} = 0, \qquad i = 1, \ldots, N. \tag{5.82}$$

The MLE can be shown to be *consistent* in the asymptotic limit, which means that it is arbitarily close to the true $\boldsymbol{\theta}$ as the number of observations tends to infinity. In this same limit, and under general conditions, $\tilde{\boldsymbol{\theta}}$ is Gaussian-distributed. The reader should bear in mind that the asymptotic limit may well not apply in practice (particularly for a limited number of observations), which makes $\tilde{\boldsymbol{\theta}}$ a biased estimator in practice.

Having obtained the MLE $\tilde{\boldsymbol{\theta}} = \{\text{POI}s, \text{NP}s\}$, we may again want to quote a value of each parameter of interest. The easiest way to do this is simply to observe the values of the POIsubset in $\tilde{\boldsymbol{\theta}}$. This is called **profiling**, and it corresponds to taking the values of the other parameters that maximise the likelihood in the full parameter space. This is faithful in the sense that the set of MLE POI values identified in this way are fully consistent with one another. However, notice that it is not the same as the set of **marginal maximum likelihoods**, in which all parameters other than θ_i are integrated out, reducing the likelihood to a 1D distribution before estimating θ_i. This set is simpler to visualise but does not necessarily describe a high-likelihood combination, since each marginal estimator can differ from the global estimator to a greater or lesser extent depending on the structure of the likelihood function. The difference between profiling and marginalising is one of the key operational differences between a Bayesian and a frequentist, and it can have a significant impact on the results one obtains, as well as their philosophical meaning.

There is in fact a close relationship between the MLE method and the Bayesian approach. Bayes' theorem tells us that $P(\theta|D) \propto P(D|\theta)P(\theta)$. Ignoring philosophical difficulties for now, this tells us that, if $P(\theta) = 1$, the maximum of the likelihood will be the same as the mode of the posterior. So the MLE can actually be thought of as the peak location for the Bayesian posterior pdf in the case of a flat prior.

How do we obtain the MLE? Although it is possible analytically in limited cases, it is almost always found numerically, and it is common to turn the problem of maximising $\mathcal{L}$ into the equivalent problem of *minimising* $-\ln \mathcal{L}$, where the minus sign ensures that a minimum will be found rather than a maximum. There may easily be $\mathcal{O}(100)$ parameters or more, once nuisances are included, and in addition there are few distributions for which the optimum can be computed analytically; real-world likelihoods tend not to be among them. The exact way in which numerical minimisation is performed is not a topic in which physicists typically have a deep involvement, there being off-the-shelf tools to perform it, either in the HEP *Software:* ROOT-oriented ecosystem, or in e.g. the rapidly growing data-science world outside particle physics. But it is worth noting that function minimisation in high dimensional spaces is generally not computationally straightforward and introduces issues of fit convergence, stability, and the possibility of finding a non-global minimum: manual 'babysitting' of fits is not unusual.

One class of minimisation technique is 'hill-climbing' optimisation, which starts with an arbitrary solution, then makes an incremental change 'step' to the solution based on estimated function gradients. Several specific methods exist to efficiently estimate the gradient, such as the BFGS method which iteratively improves an estimate of the loss function's Hessian curvature matrix. If a better solution is found after a step, another step is made, and so on until the solution stops changing. This means that the search is local, and the algorithm may easily get stuck in local minima of the negative log-likelihood. In a real physics example, it is essential to run multiple minimisations with different initial conditions to check that the results are robust. There is no general way to be sure that the true global minimum has been found via this process.

To overcome the problems of local search strategies, one chooses from a wide range of *global optimisation* algorithms, which should provide more robust behaviour in the case of multimodal functions. One of these is differential evolution which has recently provided lots of exciting results in particle physics and cosmology, proving capable of exploring very complicated likelihood functions in up to 20 parameters. Differential evolution is an example of an **evolutionary algorithm**, where a population of NP individuals or 'target vectors' $\{X_i^g\}$, are evolved through the parameter space, for a number of generations. Here i refers to the ith individual, and g corresponds to the generation of the population. Each X_i represent a vector of parameters, i.e. one point in the parameter space.

The algorithm starts with an initial generation that is selected randomly from within the range on each parameter. A generation is evolved to the next via three steps that are inspired by genetics: mutation, crossover and selection. The simplest variant of the algorithm is know as rand/1/bin; the first two parts of the name refer to the mutation strategy (random population member, single difference vector), and the third to the crossover strategy (binomial).

Mutation proceeds by identifying an individual $\mathbf{X}_i$ to be evolved, and constructing one or more donor vectors $\mathbf{V}_i$ with which the individual will later be crossed over. In the rand/1 mutation step, three unique random members of the current generation $\mathbf{X}_{r1}$, $\mathbf{X}_{r2}$ and $\mathbf{X}_{r3}$ are chosen (with none equal to the target vector), and a single donor vector $\mathbf{V}_i$ is constructed as

$$\mathbf{V}_i = \mathbf{X}_{r1} + F(\mathbf{X}_{r2} - \mathbf{X}_{r3}), \tag{5.83}$$

with the scale factor F a parameter of the algorithm. A more general mutation strategy known as rand-to-best/1 also allows some admixture of the current best-fit individual $\mathbf{X}_{\text{best}}$ in a single donor vector,

$$\mathbf{V}_i = \lambda\mathbf{X}_{\text{best}} + (1 - \lambda)\mathbf{X}_{r1} + F(\mathbf{X}_{r2} - \mathbf{X}_{r3}), \tag{5.84}$$

according to the value of another free parameter of the algorithm, λ.

Crossover then proceeds by constructing a trial vector $\mathbf{U}_i$, by selecting each component (parameter value) from either the target vector, or from one of the donor vectors. In simple binomial crossover (the bin of rand/1/bin), this is controlled by an additional algorithm parameter Cr. For each component of the trial vector $\mathbf{U}_i$, a random number is chosen uniformly between 0 and 1; if the number is greater than Cr, the corresponding component of the trial vector is taken from the target vector; otherwise, it is taken from the donor vector. At the end of this process, a single component of $\mathbf{U}_i$ is chosen at random, and replaced by the corresponding component of $\mathbf{V}_i$ (to make sure that $\mathbf{U}_i \neq \mathbf{X}_i$).

In the selection step, we pick either the trial vector or the target vector, depending on which has the best value for the objective function.

A widely-used variant of simple rand/1/bin DE is so-called jDE, where the parameters F and Cr are optimised on-the-fly by the DE algorithm itself, as if they were regular parameters of the objective function. An even more aggressive variant known as λjDE is the self-adaptive equivalent of rand-to-best/1/bin, where F, Cr and λ are all dynamically optimised.

5.6.1 Exact confidence intervals

Having obtained the MLE for each parameter, how do we quote an uncertainty on it? Since we are now frequentists, let us be clear what this uncertainty means. If we repeated our experiment a number of times (but each time with the same number n of observations of x), we would extract a slightly different $\tilde{\boldsymbol{\theta}}$ each time. The MLE parameters thus have a natural spread, and the uncertainty we want to quote on them is related to that spread. In other words, the uncertainty is related to the distribution of the *estimator*, which we will call $g(\tilde{\boldsymbol{\theta}}|\boldsymbol{\theta}_0)$, where $\boldsymbol{\theta}_0$ represents the unknown true value of the parameters (such that our estimator has some distribution with respect to the true value). $g(\tilde{\boldsymbol{\theta}}|\boldsymbol{\theta}_0)$ is referred to as the **sampling distribution**, and it should be normalised so that it integrates to unity. It is thus equivalent to the pdf that the $\tilde{\boldsymbol{\theta}}$ values are drawn from, assuming some true value of the parameters.

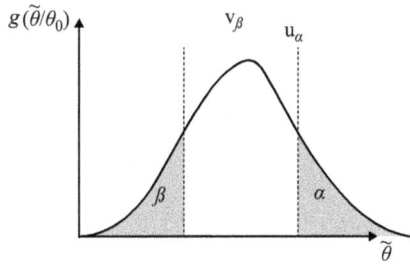

Figure 5.11. Illustration of a hypothetical sampling distribution $g(\tilde{\theta}|\theta_0)$ for a particular choice of the true parameters θ_0, for a 1D example. The distribution can be obtained via Monte Carlo sampling if we lack an analytic expression.

Assume for simplicity that we only have one parameter θ. In the general case, we do not know $g(\tilde{\theta}|\theta_0)$ analytically. However, we can still obtain it by using the **Monte Carlo** method. This consists of simulating a large number of experiments, computing the MLE values each time, then looking at how these results are distributed. We can then define a **confidence interval** (CoI) for each of our parameters, whose defining characteristic is that it contains the true value of the parameter some fixed proportion of the time (e.g. 95%) again given by the **confidence level** (CL). It is worth reading this sentence again several times; many people think (and, even worse, *say in public*) that a confidence interval and a Bayesian credible interval are the same thing. They have quite distinct meanings. A frequentist interval says nothing about the probability of a parameter taking a certain value (which is undefined for a frequentist), but instead tells us the fraction of times that the derived CoI would contain the true value of the parameter a certain fraction of times (given by its CL) in a large number of repeated experiments. Note that the random variable here is not the true value, but the estimated interval itself!.

The classic construction of CoIs is the **Neyman construction**. This is more than a little baroque, so let's work through it piece by piece. First, assume that, for any given true parameter value θ_0, we know $g(\tilde{\theta}|\theta_0)$ by Monte Carlo simulation. A hypothetical distribution for a particular choice of θ_0 is shown for a 1D example in Figure 5.11. We can then choose some value u_α such that there is some probability α to observe $\tilde{\theta} \geqslant u_\alpha$. We can also choose a value v_β such that there is a probability β to observe $\tilde{\theta} \leqslant v_\beta$. We can write

$$\alpha = \int_{u_\alpha(\theta_0)}^{\infty} g(\tilde{\theta}|\theta_0)\,\mathrm{d}\tilde{\theta} = 1 - G(u_\alpha(\theta_0)), \tag{5.85}$$

where G is the cdf corresponding to g, and we have made it clear that various quantities depend on our assumed true value of the parameter θ_0. We can also write a similar expression for β:

$$\beta = \int_{-\infty}^{v_\beta(\theta_0)} g(\tilde{\theta}|\theta_0)\,\mathrm{d}\tilde{\theta} = G(v_\beta(\theta_0)). \tag{5.86}$$

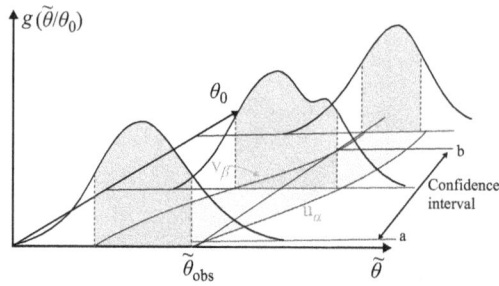

Figure 5.12. Illustration of the Neyman confidence-limit construction.

If we now vary the assumed θ_0, we get different $u_\alpha(\theta_0)$ and $v_\beta(\theta_0)$ values, and we could plot these in the $\theta_0 - \tilde{\theta}$ plane, as shown in Figure 5.12. The region between $u_\alpha(\theta_0)$ and $v_\beta(\theta_0)$ is called the **confidence belt**, and the probability for $\tilde{\theta}$ to be inside the belt is given by

$$P(v_\beta(\theta_0) \leqslant \tilde{\theta} \leqslant u_\alpha(\theta_0)) = 1 - \alpha - \beta. \tag{5.87}$$

We now assume (reasonably) that $u_\alpha(\theta_0)$ and $v_\beta(\theta_0)$ are monotonically-increasing functions of θ_0, which allows us to define the inverse functions $a(\tilde{\theta}) \equiv u_\alpha^{-1}(\tilde{\theta})$ and $b(\tilde{\theta}) \equiv v_\alpha^{-1}(\tilde{\theta})$. We can then show that

$$P(a(\tilde{\theta}) \leqslant \theta_0 \leqslant b(\tilde{\theta})) = 1 - \alpha - \beta. \tag{5.88}$$

Let us now say that we have obtained a particular value of the estimator $\tilde{\theta}_{obs}$ in our experiment. This defines two values of a and b, shown in Figure 5.12, at the points where the horizontal line for $\tilde{\theta}_{obs}$ intersects the confidence belt[15]. The interval $[a, b]$ is then said to be a CoI, at a confidence level of CL $= 1 - \alpha - \beta$. Every point in the interval corresponds to a model with which the observed data is consistent at the given CL. a and b are themselves random variables, since if we repeated our experiment many times, we would get different $\tilde{\theta}_{obs}$ values, which define different values of a and b. The point of this construction is that, under many repeats, θ_0 will be within the CoI in a fraction of the experiments given by $1 - \alpha - \beta$.

You should think of this complex procedure as an elaborate workaround for the fact that a frequentist cannot define a probability that a parameter takes a particular value, but only the likelihood of the observed data under the *hypothesis* of that value. Sometimes, we only want a one-sided interval, in which case we can use a as the lower limit on θ_0 such that $\theta_0 \geqslant a$ with probability $1 - \alpha$, or we can use b as an upper limit on θ_0. In the case of the two-sided intervals, the confidence level $1 - \alpha - \beta$ is not sufficient to uniquely determine how to choose $u_\alpha(\theta_0)$ and $v_\beta(\theta_0)$ for fixed θ_0 (i.e. we could shift $u_\alpha(\theta_0)$ and $v_\beta(\theta_0)$ around

---

[15] More generally, but unusually, there could be more than two intersection points, producing a disjoint interval with the same CL.

and still get the same probability of $\tilde{\theta}$ being between u_α and v_β). There are various prescriptions that can be used (each called an **ordering principle**), including:

- add all parameter values greater than or less than a given value;
- draw a central region with equal probability of the measurement falling above the region as below (e.g. $\alpha = \beta = \gamma/2$) (the **central confidence interval**);
- start from the mode of $g(\tilde{\theta};\theta_0)$ and work downwards, adding points until the region contains the required amount of $\int g(\tilde{\theta}|\theta_0)\mathrm{d}\tilde{\theta}$
- the **Feldman–Cousins prescription**, in which points are added in order of the decreasing likelihood-ratio:

$$R = \frac{g(\tilde{\theta}|\theta_0)}{g(\tilde{\theta}|\theta_{\text{best}})},\tag{5.89}$$

where θ_{best} is the value of θ (of all physically allowed values) that leads to the absolute maximum of $g(\tilde{\theta}|\theta)$.

5.6.2 1D confidence intervals for Gaussian-distributed quantities

We just saw the exact way to define CoIs on a maximum likelihood estimate, and it is both horribly complicated, and computationally expensive (thanks to the requirement of running a large number of Monte Carlo experiments in general). However, we can take a massive shortcut if we know that the quantity we are interested in is drawn from a Gaussian distribution. This is often true for estimators, under general conditions, in the limit of a large number of observations. In this asymptotic limit $g(\tilde{\theta}|\theta_0)$ becomes a Gaussian distribution centered on θ_0, with some variance $\sigma_{\tilde{\theta}}^2$. Assume for now that we know the value of $\sigma_{\tilde{\theta}}$ (we will correct this in a moment). The cdf of $\tilde{\theta}$ is then given by

$$G(\tilde{\theta};\theta, \sigma_{\tilde{\theta}}) = \int_{-\infty}^{\tilde{\theta}} \frac{1}{\sqrt{2\pi\sigma_{\tilde{\theta}}^2}} \exp\left(\frac{-(\tilde{\theta}' - \theta)^2}{2\sigma_{\tilde{\theta}}^2}\right)\mathrm{d}\tilde{\theta}'.\tag{5.90}$$

Assuming that our particular experiment has led to an estimator value of $\tilde{\theta}_{\text{obs}}$, equations (5.85) and (5.86) can be used to implicitly define the CoI $[a, b]$,

$$\alpha = 1 - G(\tilde{\theta}_{\text{obs}}; a, \sigma_{\tilde{\theta}}) = 1 - \Phi\left(\frac{\tilde{\theta}_{\text{obs}} - a}{\sigma_{\tilde{\theta}}}\right)\tag{5.91}$$

$$\beta = G(\tilde{\theta}_{\text{obs}}; b; \sigma_{\tilde{\theta}}) = \Phi\left(\frac{\tilde{\theta}_{\text{obs}} - b}{\sigma_{\tilde{\theta}}}\right),\tag{5.92}$$

where we have used the standard notation for the cdf of the Gaussian function Φ. The actual values of a and b can be obtained by inverting these equations:

$$a = \tilde{\theta}_{\text{obs}} - \sigma_{\tilde{\theta}}\,\Phi^{-1}(1 - \alpha)\tag{5.93}$$

$$b = \tilde{\theta}_{\text{obs}} + \sigma_{\tilde{\theta}}\,\Phi^{-1}(1 - \beta).\tag{5.94}$$

Table 5.1. Values of the confidence level that correspond to choices of $\Phi^{-1}(1 - \gamma/2)$ for a central confidence level, for a 1D, Gaussian-distributed estimator.

$\Phi^{-1}(1 - \gamma/2)$	Confidence level
1	0.6827
1.645	0.90
1.960	0.95
2	0.9544
2.576	0.99
3	0.9973

Here, Φ^{-1} is the inverse function of the Gaussian cdf, which is the **quantile** of the standard Gaussian. Note that $\Phi^{-1}(1 - \alpha)$ and $\Phi^{-1}(1 - \beta)$ tell us how far the edges of our CoI a and b are from $\tilde{\theta}_{\text{obs}}$, in units of $\sigma_{\tilde{\theta}}$. The choice of α and β then allows us to define CoIs at different confidence levels. For a central confidence interval with $\alpha = \beta = \gamma/2$, the confidence level is given by $1 - \gamma$, and a common choice is $\Phi^{-1}(1 - \gamma/2) = 1, 2$ or 3. This means that the edges of the interval are 1, 2 or 3 standard deviations away from $\tilde{\theta}_{\text{obs}}$. Another option is that we choose $1 - \gamma$ to be a round number, so that we get a nice round confidence level (e.g. $95\% \sim 2\sigma$). The values of the confidence level for different values of $\Phi^{-1}(1 - \gamma/2)$ are shown in table 5.1.

The choice $\Phi^{-1}(1 - \gamma/2) = 1$ results in a '1σ' error bar, which allows us to quote the CoI as

$$[a, b] = [\tilde{\theta}_{\text{obs}} - \sigma_{\tilde{\theta}}, \tilde{\theta}_{\text{obs}} + \sigma_{\tilde{\theta}}]. \tag{5.95}$$

If we do not know the value of $\sigma_{\tilde{\theta}}$, the maximum likelihood estimate $\tilde{\sigma}_{\tilde{\theta}}$ can usually be used instead, provided we have a large enough set of observations.

5.6.3 1D confidence intervals from the likelihood function

There is another quick and useful method for obtaining approximate confidence intervals on estimators when the number of observations is large, which is to use the original likelihood function itself. We stated above that the distribution of the estimator becomes Gaussian in the asymptotic limit, which means its pdf is just

$$g(\tilde{\theta}|\theta) = \frac{1}{\sqrt{2\pi\sigma_{\tilde{\theta}}^2}} \exp\left(\frac{-(\tilde{\theta}-\theta)^2}{2\sigma_{\tilde{\theta}}^2}\right). \tag{5.96}$$

However, it can indeed be shown that under the same conditions, the likelihood function itself becomes Gaussian, centered around the maximum likelihood estimate $\tilde{\theta}$:

$$\mathcal{L}(\theta; \boldsymbol{D}) = \mathcal{L}_{\text{max}} \exp\left(\frac{-(\theta - \tilde{\theta})^2}{2\sigma_{\tilde{\theta}}^2}\right). \tag{5.97}$$

Note that we have not proved that the variance of both Gaussians is the same, but this can be shown. The log of the likelihood is then given by

$$\ln \mathcal{L} = \ln \mathcal{L}_{max} - \left(\frac{-(\theta - \tilde{\theta})^2}{2\sigma_{\tilde{\theta}}^2} \right). \tag{5.98}$$

We see that a one $\sigma_{\tilde{\theta}}$ variation of θ around the MLE value leads to a change in the log-likelihood of 1/2 from its maximum value. We can turn this round and *define $\sigma_{\tilde{\theta}}$* as the variation in $\tilde{\theta}$ that results from varying the log-likelihood by 1/2 around its maximum value. The 2-sigma region corresponds to $\Delta \log L = \frac{1}{2}(2)^2 = 2$, and the 3-sigma region corresponds to $\Delta \log L = \frac{1}{2}(3)^2 = 4.5$, so that in general we have

$$\ln \mathcal{L}(\tilde{\theta} \pm N\sigma_{\tilde{\theta}}) = \ln \mathcal{L}_{max} - \frac{N^2}{2}. \tag{5.99}$$

Putting this together with equation (5.95) tells us that the 68.3% central confidence level is given by the values of θ at which the log-likelihood function decreases by 1/2 from the maximum value, which is much easier to think about than the exact confidence interval procedure given by the Neyman construction. In fact, this concept remains useful even if the likelihood function is not Gaussian, since it can be shown that the central confidence interval $[a, b] = [\tilde{\theta}-c, \tilde{\theta}+d]$ is still approximately given by

$$\ln \mathcal{L}(\tilde{\theta}_{-c}^{+d}) = \ln \mathcal{L}_{max} - \frac{N^2}{2}, \tag{5.100}$$

where $N = \Phi^{-1}(1 - \gamma/2)$ is the quantile of the standard Gaussian that corresponds to the desired CL $= 1 - \gamma$.

5.6.4 Multidimensional confidence regions

The above methodology can be generalised to the multidimensional case of m estimators, where our estimators for the parameters are now $\tilde{\boldsymbol{\theta}}$. In the large sample limit, the estimators will be distributed according to a multivariate Gaussian centered on the unknown true values $\boldsymbol{\theta}_0$:

$$g(\tilde{\boldsymbol{\theta}}|\boldsymbol{\theta}_0) = \frac{1}{(2\pi)^{(m/2)}|V|^{1/2}} \exp\left[-\frac{1}{2}(\tilde{\boldsymbol{\theta}}-\boldsymbol{\theta}_0)^T \Sigma^{-1}(\tilde{\boldsymbol{\theta}}-\boldsymbol{\theta}_0) \right], \tag{5.101}$$

where Σ^{-1} is the inverse of the Gaussian covariance matrix.

It is then possible to construct a multidimensional **confidence region** that has the right coverage properties for $\boldsymbol{\theta}_0$ as follows (where we state the recipe without proof).

As in the 1D case, it can be shown that, in the limit of a large number of observations, the likelihood function also takes on a Gaussian form:

$$\mathcal{L}(\boldsymbol{\theta}; \boldsymbol{D}) = \mathcal{L}_{max} \exp\left[-\frac{1}{2}(\boldsymbol{\theta} - \tilde{\boldsymbol{\theta}})^T \Sigma^{-1}(\boldsymbol{\theta} - \tilde{\boldsymbol{\theta}}) \right] \tag{5.102}$$

Table 5.2. For different choices of the confidence level, this table gives the Q_γ value that can be used with equation (5.103) to derive the confidence region.

	Q_γ				
$1 - \gamma$	$m = 1$	$m = 2$	$m = 3$	$m = 4$	$m = 5$
0.683	1.00	2.30	3.53	4.72	5.89
0.90	2.71	4.61	6.25	7.78	9.24
0.95	3.84	5.99	7.82	9.49	11.1
0.99	6.63	9.21	11.3	13.3	15.1

with the same covariance matrix that appears in the pdf for the estimator. Under these conditions, the confidence region with confidence level $1 - \gamma$ can be obtained by finding the values of θ at which the log-likelihood function decreases by $Q_\gamma/2$ from its maximum value,

$$\ln \mathcal{L}(\boldsymbol{\theta}; \boldsymbol{D}) = \ln \mathcal{L}_{\max} - \frac{Q_\gamma}{2}, \tag{5.103}$$

where $Q_\gamma = F^{-1}(1 - \gamma; m)$ is the quantile of order $1 - \gamma$ of the χ^2 distribution with m degrees of freedom. Though this sounds highly obscure, the procedure is actually fairly simple, thanks to the existence of look-up tables for this mysterious Q_γ number. For example, if we want a 90% confidence region, we can use table 5.2 to tell us the Q_γ number that can be used to derive the confidence region using equation (5.103) (i.e. it tells us how far down in the log-likelihood to go when constructing the confidence region).

If we depart from the asymptotic limit, this procedure can once again still be used, but the resulting confidence regions will be only approximate. In fact, for a higher number of estimators m, the approach to the Gaussian limit is slower as we add more observations. It is always possible to use Monte Carlo simulation to derive the actual coverage of the confidence region (assuming that the look-up table cannot be trusted if we are not in the Gaussian regime).

5.7 Sampling

In discussing estimators and parameter inference, we have often assumed that values can be drawn from probability distributions in order to compute resulting statistical moments, or map the shape of the likelihood or posterior-probability function. While nature apparently does this all the time, it is not so obvious that we can efficiently mimic the same process — yet we need to, because sampling is central to particle physics, from event-generation, to parameter inference and hypothesis testing, and to estimating corrections for detector biases.

We begin with the apparently simple (yet still with potential to be ruinously inefficient) task of drawing samples from a one-dimensional analytic or binned distribution, then move to the research frontier of efficiently sampling from high-dimensional distributions with minimal *a priori* knowledge about where the bulk of the probability density lies.

5.7.1 Making a histogram from an arbitrary pdf

It is often useful to be able to obtain a histogram of sampled data from an arbitrary pdf, and this can be achieved using the inverse cdf sampling technique. This is based on the fact that the cdf F is a one-to-one mapping of the domain of the cdf into the interval [0, 1]. Thus, if U is a uniform random variable in the range [0, 1], then $F^{-1}(U)$ has the distribution F. One can thus generate uniform random numbers in the interval [0, 1], and transform them via the inverse cdf, which gives us numbers whose distribution matches the original pdf. A potential issue with this method is that you might not be able to integrate your pdf analytically to get the cdf, and you may have to therefore integrate it numerically. This in turn leads to potential instabilities in the integration and, even when it does work, it is a computationally-expensive way of obtaining samples.

Another method is rejection sampling. Although we cannot sample from the unknown pdf directly, we can sample from a variety of other distributions. We can thus define a proposal distribution $Q(x)$ and sample from that instead. We will insist that $Q(x)$ is easy to sample from, and also that it envelopes the pdf $p(x)$. In other words, for some scaling factor k, $kQ(x) > p(x)$ for all x (i.e. the proposal function is always above the pdf we want to sample from). A sensible choice is $k = \max(p(x)/q(x))$. The rejection sampling procedure then proceeds as follows. We draw lots of random numbers $\{z_i\}$ from the $Q(x)$. For each one, we draw a random number u_i between 0 and $kQ(z_i)$. We then evaluate $p(z_i)$, and we only keep the sample if $u_i < p(z_i)$. Given many samples, this will eventually converge to a set of samples that are distributed according to $p(x)$.

Both these methods are useful standalone, as well as being key building blocks for more sophisticated sampling methods.

5.7.2 Obtaining likelihood or posterior samples

Sampling is crucial for both frequentist and Bayesian analyses, but there are important differences in the types of method that work well in each case. In both schools, one has to decide what to do with non-POI parameters when making distributions projected into subsets of the POIs (for example, 1D or 2D plots of model-parameter compatibility with observations). Frequentist analyses proceed by profiling to find the best-fit point within the parameter space, and the effort is concentrated on finding and understanding the immediate neighbourhood of that best fit point. Bayesian analyses instead marginalise over the non-POIs to lower-dimensional pdfs.

Which is the harder task depends on the nature of the pdf. A relatively well-behaved function with a well-defined global optimum and few local-optima may be easily profiled with a local-optimization strategy, while marginalisation requires a more holistic mapping of all probability density that contributes to the marginalisation integral. But in more problematic spaces with a sparse and multimodal function, locating the global optimum at all becomes a hard problem in need of a very global approach. Here, the marginalisation integral may be easier to perform than locating the global best-fit point. We briefly reviewed sampling strategies for global optimisation in Section 5.6, and hence will here focus our attention on the concrete case of sampling from a Bayesian posterior.

Assume that we have a posterior pdf $p(\theta|D)$ that, in the general case, can be multimodal, with multiple "hot spots" of probability density in the parameter space. It is important that a sampler visits *all* of these parameter regions in order for marginal integrals to correctly converge. A central problem, however, is that as the dimensionality of the parameter space increases, the chance of a uniformly sampled point having a significant probability density becomes vanishingly small, since there is a huge volume of improbable regions. In this section, we will review the worst possible method of sampling (random uniform sampling from the prior), and suggest two superior replacements that receive extensive use in the physics literature.

5.7.2.1 Grid and random sampling

The simplest method for sampling the posterior function is to perform a uniform scan over the parameters θ and, for each value, calculate the posterior by hand by evaluating the likelihood and multiplying it by the prior. At the end, we can ensure the correct normalisation by dividing the posterior value for each point by:

$$p(H) = \sum_{\theta^*} p(H|\theta^*)p(\theta^*) \tag{5.104}$$

where θ^* runs over the samples we have taken. Even in 1D, this will fail to give us an accurate evaluation of the posterior when it is sharply peaked compared to our grid size. In larger numbers of dimensions (meaning more parameters), however, we meet a different problem. If we have D parameters in total, the number of grid scan points at which we need to evaluate the posterior scales as $\mathcal{O}(n^D)$, where n defines the grid resolution. To get anything like a reasonable sampling density of a multidimensional posterior, this becomes a very large number very quickly.

What if, instead of scanning points over a grid of equal resolution, we simply threw samples at random from the prior range on our parameters, and evaluated the posterior at each of those points? We now get a different problem, which is that our samples will mostly be concentrated at the boundary of the prior range when D gets large. To see this, imagine that we randomly choose a variable to be between 0 and 1, and we define P to be probability of not being near the boundary (i.e. not too close to 0 or 1). As we increase D, we can define an n-dimensional boundary, where the probability of a randomly selected point being at the boundary is given by (see Figure 5.13),

$$P(\text{boundary}) = 1 - P(\text{not boundary}) = 1 - P^D . \tag{5.105}$$

This probability tends to 1 as D increases, meaning that our random samples concentrate at the boundary of our sampling range. This in turn leads to biased inferences.

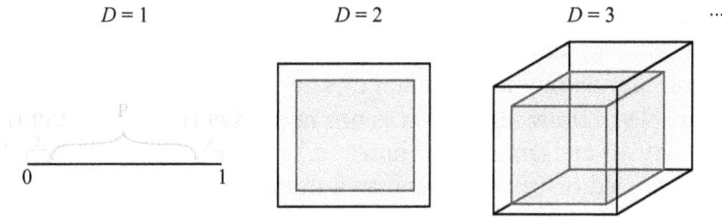

Figure 5.13. Illustration of the "boundary dominance" problem when random sampling in $D \gg 1$ dimensions, where the naïve fraction of the space contained in the outer parts of the parameter ranges, $P(\text{boundary}) = 1 - P^D \to 1$.

A final problem with both random and grid sampling is that it is hugely wasteful to spend lots of time evaluating the posterior in regions of the space where the posterior probabiltiy density is very small. It would be far better to have a method that somehow obtained samples in proportion to the probability density, so that we concentrated our CPU time in interesting regions of the parameter space.

5.7.2.2 Markov Chain Monte Carlo methods

Although this last request sounds too good to be true, it turns out that this is actually possible, through the use of Markov Chain Monte Carlo (MCMC) methods. Imagine that we draw some set of samples $\{X_i\}$ from a posterior distribution, one after the other. In MCMC sampling, we do not worry about the denominator of Bayes' thereom, and we assume instead that we can only sample from the unnormalised posterior distribution (so we never bother calculating the expensive integral in the denominator which is, after all, just a normalisation constant). Each X_i is a D-dimensional vector of numbers that represent each of the parameters in our D-dimensional parameter space. This is an example of a stochastic process, in which we have a set of random variables that are associated with some index, and we can think of the process as describing a system that is changing randomly in time. In our case, a higher number of the index means that the sample was drawn after a variable with a lower number of the index, so we have a concept of "future" and 'past'. Drawing samples corresponds to wandering around the parameter space.

A stochastic process is said to have the Markov property if the conditional probability distribution of future states of the process (conditional on both past and present states) is only dependent on the present state. What this means for our sampling problem is that our sampling process will have the Markov property if the conditional probability of the next sample taking a certain value depends only on the present sample, and not on all of the samples that preceded it. A Markov Chain Monte Carlo (MCMC) algorithm is an algorithm that builds this feature into the selection of the samples, which will turn out to be extraordinarily useful for us.

We will cover only one example, which is the *Metropolis-Hastings Algorithm*. This involves defining a proposal density Q which depends only on the current state of the system, which we will label $X^{(t)}$ (this is just the vector of parameters that defines the current sample). The density $Q(X'; X^{(t)})$ (where X' is the next sample that we will consider adding to our chain of samples) can be any fixed density from which it is possible to draw samples; it is not necessary for $Q(X'; X^{(t)})$ to resemble the posterior that we are sampling in order to be useful. It is common to choose a simple, easily-sampled distribution such as a multidimensional Gaussian.

We then apply the Metropolis-Hastings algorithm as follows. First, we choose a random sample somewhere within the allowed parameter space (i.e. the parameter values fall within the range of the prior. This is our starting value of $X^{(t)}$. We then use the proposal density $Q(X'; X^{(t)})$ to obtain a new sample X'. We choose whether to accept the sample or not, by computing the quantity:

$$a = \frac{p^*(X'|D)Q(X^{(t)}|X')}{p^*(X^{(t)}|D)Q(X'; X^{(t)})}, \qquad (5.106)$$

where $p^*(X_i|D)$ is the unnormalised posterior for the parameter choice X_i. If $a \geqslant 1$ the new state is accepted, otherwise the new state is accepted with probability a (i.e. we can draw a random number on the interval $[0, 1]$, and use that to decide whether we accept the same or not). If we accept the new sample, it becomes the next point in our Markov chain, but if we reject it, we stay at the current point and try another sample. Note that, if Q is symmetric function of $X^{(t)}$ and X', then the ratio of Q factors evaluates to 1, and the Metropolis-Hastings method reduces to the Metropolis method, which involves a simple comparison of the posterior at the two candidate points in the Markov chain.

It can be shown (rather remarkably), that the probability distribution of $X^{(t)}$ tends to the posterior distribution as $t \to \infty$, provided that Q is chosen such that $Q(X'; X) > 0$ for all X, X'. Thus by choosing points via the Metropolis algorithm, we obtain samples from the unnormalised posterior, spending most of our sampling time in regions where the posterior is interesting. This is exactly what we wanted!

Note that the presence of the caveat $t \to \infty$ implies that there is an issue of convergence in the application of the Metropolis algorithm, and this is to be expected from the Markov Chain nature of the method. Each element in the sequence $X^{(t)}$ has a probability distribution that is dependent on the previous value $X^{(t-1)}$ and hence, since successive samples are correlated with each other, the Markov Chain must be run for a certain length of time in order to generate samples that are effectively independent. The exact details of convergence depend on the particular posterior being sampled, and on the details of Q, and hence there is some element of tuning involved in getting the algorithm to run successfully. The initial samples in the chain are part of a "burn-in" phase, where the chain is effectively performing a random walk around the space until it finds an area of interest. It will then start to explore the maximum of the posterior (which might be a local maximum only, and we shall return to this shortly).

The purpose of the definition in equation (5.106) is to ensure that the Markov Chain used in the Metropolis method is reversible. By this it is meant that the chain satisfies the principle of detailed balance, i.e. that the probability $T(X^a; X^b)$ of the chain making a transition from a state X^a to a state X^b is related to the reverse transition via $T(X^a; X^b)P(X^a) = T(X^b; X^a)P(X^b)$. This property is necessary if we require the distribution of samples from the chain to converge to the posterior.

Before we move on, it is worth thinking about the issues with the Metropolis-Hastings algorithm. The choice of proposal function is clearly very important. Imagine that we have a single peak in the posterior, with some characteristic width. Ideally, the step size of the proposal function would be similar to that width. If it is much smaller, you will spend ages random-walking around the space before you find the region of interest. If it is much bigger, you are likely to overshoot the maximum in the posterior each time you make a step, and you will struggle to converge to the posterior within a reasonable number of samples. Unfortunately, you do not know the shape of the posterior in advance in most physics examples! Strategies for getting around this are doing preliminary scans with different step sizes to try and get a rough mapping of the posterior which you can use to guide future scans, or using more fancy MCMC techniques that have adaptive step sizes. The latter have to be created carefully to ensure that detailed balance is preserved, but there are a number of techniques on the market.

Another problem is that, as the number of parameters increases, it can get harder to find regions of interest if they occupy an increasingly small volume of the space. Changing the proposal function helps, since a multidimensional Gaussian is actually quite a spiky function in high numbers of dimensions. Again, however, some preliminary insight (which may come from the physics of the problem) is required if the results are to be optimum.

Finally, there is a classic issue with MCMC techniques, which is that they struggle with multimodal posterior functions, i.e. functions with more than one peak. Imagine that we have two peaks, one that is relatively narrow, and one that is relatively broad. A Metropolis algorithm with a Gaussian proposal of a fixed width is going to struggle here, since the Markov chain might find only one of the peaks, and you will think that it has converged to the full posterior whilst remaining ignorant of the second peak. Running different chains with different proposal densities is one way around this, but it is then difficult to weight the samples correctly in order to obtain the relative height of each peak. If a posterior has many modes, it is better to use another technique.

5.7.2.3 Nested sampling

For multimodal posteriors, there is one particular technique that has been very popular in particle physics and cosmology over the past decade, called nested sampling. If we go back to our Bayes' thereom formula, we have

$$p(\theta|D) = \frac{\mathcal{L}(\theta; D)p(\theta)}{\int \mathcal{L}(\theta; D)p(\theta) \, d\theta} = \frac{\mathcal{L}(\theta; D)p(\theta)}{Z}, \qquad (5.107)$$

where Z is the Bayesian evidence. In MCMC sampling, we threw this away because it is only a normalisation constant, and it cancels in the ratio in equation (5.106).

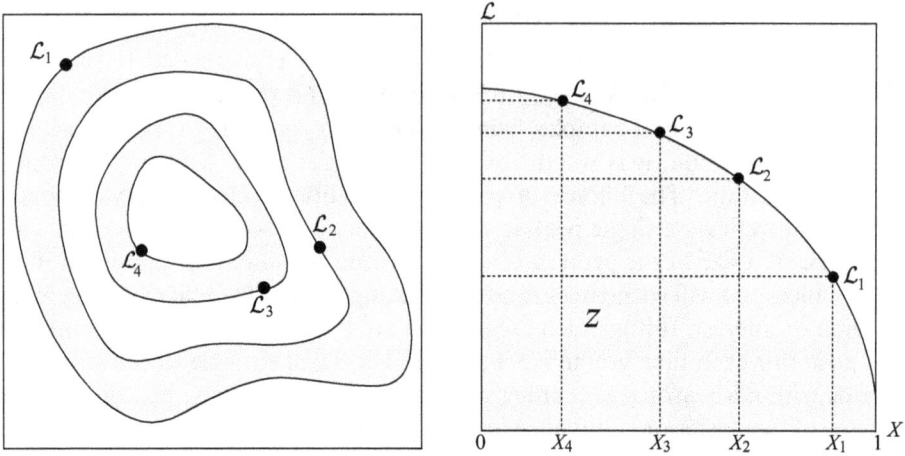

Figure 5.14. Cartoon illustrating (a) the posterior of a two dimensional problem; and (b) the transformed $\mathcal{L}(X)$ function where the prior volumes X_i are associated with each likelihood $\mathcal{L}_i$.

However, it turns out to be a fundamental quantity of interest for model comparison (see Section 5.8.2), and there has therefore been much effort expended in finding quick ways to calculate it for posteriors that are unknown analytically. Rather than targetting posterior samples directly, nested sampling is a method for calculating the Bayesian evidence, subsequently obtaining samples from the posterior as a by-product. It does this by transforming the multi–dimensional evidence integral into a one–dimensional integral that is easy to evaluate numerically. We first define the prior volume X via $dX = p(\boldsymbol{\theta})\mathrm{d}\boldsymbol{\theta}$, so that

$$X(\lambda) = \int_{\mathcal{L}(\theta;D)>\lambda} p(\boldsymbol{\theta}) \, \mathrm{d}\boldsymbol{\theta}, \qquad (5.108)$$

where the integral extends over the region(s) of parameter space contained within the contour $\mathcal{L}(\boldsymbol{\theta}; \boldsymbol{D}) = \lambda$. To get a picture of what this means, imagine that we have two parameters, and the likelihood function defines a hill in those parameters. $\mathcal{L}(\boldsymbol{\theta}; \boldsymbol{D}) = \lambda$ defines a horizontal slice that is at some height up that hill. $X(\lambda)$ is then given by the integral of the prior over the parameters, over the part of the hill that is higher than that slice. Another way to state this is that equation (5.108) gives the cumulant prior volume covering all likelihood values greater than λ. As λ increases, the enclosed prior volume X decreases from 1 to 0.

The definition of the prior volume allows us to write the evidence integral as:

$$\mathcal{Z} = \int_0^1 \mathcal{L}(X)dX, \qquad (5.109)$$

where $\mathcal{L}(X)$ is the inverse of equation (5.108), which is a monotonically decreasing function of X. This is a little obscure, so take a look at Figure 5.14 which shows nested sampling in action. Imagine that we start at the bottom of our likelihood hill. Then X must be equal to 1, since in equation (5.108) we are simply integrating the prior over its full range, and it is normalised as a pdf (i.e. it integrates to unity). Now take a slightly higher likelihood value, $\mathcal{L}_1$. This defines a value of $X = X_1$ that must be slightly smaller, since the prior volume above that likelihood value is now smaller. This is marked on the right-hand side in Figure 5.14. We can keep going in this fashion, choosing higher likelihood values (which define successive shells of the likelihood as we go up the hill), and charting their associated X values on the plot on the right. What we find is that the X values decrease as we decrease the likelihood, until finally the X value goes to zero.

Now, the evidence Z is given by the area under the curve in the right-hand figure, and we can get an approximate value of that using numerical integration. Assume that, for some sequence of values X_i,

$$0 < X_M < \cdots < X_2 < X_1 < X_0 = 1 , \tag{5.110}$$

we can evaluate the likelihoods $\mathcal{L}_i = \mathcal{L}(X_i)$, shown schematically in Figure 5.14. The evidence can then be approximated as a weighted sum:

$$Z = \sum_{i=1}^{M} \mathcal{L}_i w_i, \tag{5.111}$$

with weights w_i given by the trapezium rule as $w_i = \frac{1}{2}(X_{i-1} - X_{i+1})$. An example of a two-dimensional posterior and its associated function $\mathcal{L}(X)$ is shown in Fig. 5.14.

How can one actually perform the summation in equation (5.111)? A neat way to do it is as follows:

- Set an iteration counter to $i = 0$, and draw N 'active' (or 'live') samples randomly from the full prior $p(\boldsymbol{\theta})$. Note that the initial prior volume is thus $X_0 = 1$.
- Sort the samples by descending likelihood, and remove the sample with the smallest likelihood ($\mathcal{L}_0$) from the active set. Replace the inactive point with a new point drawn from the prior, provided it has likelihood $\mathcal{L} > \mathcal{L}_0$. Increment the iteration counter.
- Continue subsequent iterations until the prior volume is traversed.

Once the evidence is found by Monte Carlo integration, the final live points and the full sequence of discarded points can be used to generate posterior inferences, by assigning each point the probability weight

$$p_i = \frac{\mathcal{L}_i w_i}{Z}. \tag{5.112}$$

One can then construct marginalised posterior distributions or calculate inferences of posterior parameters such as means, covariances, etc. You may be wondering how it is that one can draw samples from the prior subject to the condition $\mathcal{L} > \mathcal{L}_i$ at each iteration i. Indeed, this would seem to require advance knowledge of the likelihood function itself, but the whole point is that we do not know it analytically, and can only sample from it!

The MultiNest algorithm that has been used in many physics studies tackles this problem through an ellipsoidal rejection sampling scheme. This operates by first enclosing the live point set within a set of ellipsoids (which may overlap), chosen to minimize the sum of the ellipsoid volumes. New points are then drawn uniformly from the region enclosed by these ellipsoids, which removes the need to sample from the full prior volume at each iteration. The MultiNest algorithm has proven very useful for tackling inference problems in cosmology and particle physics, partly because the ellipsoidal decomposition is useful for dealing with posteriors with curving features (typical of e.g. global fits of supersymmetric models), and for dealing with multimodal posteriors (where ellipsoids that do not overlap can be evolved independently as separate modes).

5.8 Hypothesis testing

In the previous section, we dealt with the case that a specific theory can be assumed to give rise to our dataset, and the only thing that remains is to obtain the values of the free parameters of the theory. However, we may instead ask *how well* our chosen theory describes the data, which is not the same question as 'which set of parameters from my theory best fits the data?'. To see this, consider that a Bayesian can always define and map a posterior probability density function, whilst a frequentist will always find a set of parameters for which the likelihood is maximised. This will occur even if the theory has no relation to the data at all! To evaluate how well a theory fits data, we must instead use the framework of **hypothesis testing**, which includes both tests of a single hypothesis, and comparisons between different hypotheses.

As in the previous chapter, how we proceed depends on whether we are a frequentist or a Bayesian. A significant fraction of hypothesis testing results in high energy physics use the frequentist framework, and thus we will concentrate on that in the bulk of this chapter. We will, however, briefly describe the Bayesian approach in Section 5.8.2.

5.8.1 Frequentist hypothesis testing

In the frequentist approach to hypothesis testing, we shall see that the fundamental question addressed is the following. Given a finite set of observations obtained from a stochastic process, what fraction of all of the possible sets of observations under hypothesis $\mathcal{H}$ are at least as compatible with the hypothesis as the observed set?

Hypotheses and p-*values*

First, assume that we have a set of observed data, $\boldsymbol{D}$, and we want to test a single hypothesis $\mathcal{H}$ that we think may give rise to that data. A test of a single hypothesis is called a **goodness of fit** test. We can write the conditional probability of having made the observation that we did under the hypothesis $\mathcal{H}$ as $p(\boldsymbol{D}|\mathcal{H})$ and, in the case that $\mathcal{H}$ has no free parameters, we refer to $\mathcal{H}$ as a **simple hypothesis**. If instead the hypothesis comes associated with a set of parameters $\boldsymbol{\theta}$, the hypothesis is called a **compound** or

composite hypothesis, and we can write a function $p(D|\theta)$[16]. Under repeats of our experiment, we would observe different datasets, since our measurement is a stochastic process. If the hypothesis $\mathcal{H}$ were true, then we would expect our observed dataset to be somewhere within the statistical fluctuations of the possible datasets generated by the hypothesis. It is then natural to ask whether our particular observed dataset falls within the expected statistical fluctuations or not, and to quantify the level of consistency with the expected behaviour given the hypothesis.

The standard way to do this is to use a test statistic which is a function of the measured data $t(D)$ that maps D either to a single number or a vector. One could in principle use the original data vector itself as the test statistic (i.e. $t = D$), but a statistic of lower dimension is preferred since it reduces the computational overhead without affecting the ability to discriminate hypotheses. For a goodness of fit test, the test statistic is chosen to be a function that quantifies the agreement between the observed data values D, and the values expected from the hypothesis, such that small values of t indicate better agreement (e.g. the χ^2 statistic introduced in Section 5.3.5). Assuming that we have defined a 1D test statistic t, we can define the pdf $f(t(D)|\mathcal{H})$ of the test statistic, under variations of D. We can then state the probability P of observing, given the assumption that $\mathcal{H}$ is true, a result that is as compatible or less with $\mathcal{H}$ than the observed data set. Since larger t values correspond to lower compatibility of the data with the hypothesis (by construction), we can write this as the following integral over the pdf of the test statistic,

$$P = \int_{t_{\text{obs}}}^{\infty} dt\, f(t|\mathcal{H}), \tag{5.113}$$

where t_{obs} is the value of the test statistic observed in the data, and the left-hand side is called the *p*-value of the observation given the hypothesis[17]. A small *p*-value indicates that the measurement made is highly unlikely under the hypothesis, meaning that some alternative hypothesis—which the statistics do not directly help us to construct—may be preferred. The *p*-value tells us the significance of any discrepancy between the observed data and the predictions of the hypothesis, but it is not yet a hypothesis test on $\mathcal{H}$ itself. However, we can define the latter by comparing $(1 - P)$ to a standard threshold, stating that $\mathcal{H}$ is rejected at a confidence level CL if

$$\text{CL}_{\text{obs}} = (1 - P) > \text{CL}. \tag{5.114}$$

Using this equation alongside the definition of the *p*-value, we can write

$$\text{CL}_{\text{obs}} = \int_{-\infty}^{t_{\text{obs}}} dt\, f(t|\mathcal{H}). \tag{5.115}$$

---

[16] Note that this is not philosophically equivalent to the likelihood of the data that we described in Section 5.4. In the case of the likelihood of the data, we take the data as fixed, and we vary the parameters. In the present case, we are taking the hypothesis as fixed, and we interested in variations of the data.

[17] The lower-case *p* is a break with our convention of capital *P* for finite probabilities, but this is the standard notation for *p*-values.

This tells us what fraction of potential datasets, or alternatively re-runnings of the experiment in statistically parallel universes—*assuming* $\mathcal{H}$ to be true—would have resulted in an observation at least as compatible with the hypothesis as the one we actually have.

The chi-squared test

The canonical goodness of fit test for binned data is the *chi-squared* test, which uses the χ^2 test statistic that we first introduced in Section 5.3.5. For now, let us assume that we have a histogram of some quantity x with n_b bins, in which the observed counts $y(x)$ in each bin can be compared to the predictions of a model $f(x)$ which has no free parameters. If the observed counts are Poisson-distributed with mean values $f(x)$, and the number of entries in each bin is more than about 5, then it can be shown that the χ^2 statistic defined in equation (5.66) follows a **chi-squared distribution**,

$$f(z = \chi^2; k) = \frac{1}{2^{k/2}\Gamma(k/2)}z^{k/2-1}e^{-z/2}, \tag{5.116}$$

where k is the number of **degrees of freedom**, which must be positive-definite (and we have $k = n_b$ in our example). This distribution gives the probability density for obtaining a measured value of χ_k^2 if all uncertainties are due to Gaussian statistical fluctuations. The distribution peaks at $x = 0$ for $k \leqslant 2$ then at finite $x = k - 2$ thereafter: the well-known ansatz that a good fit will have $\chi^2/k \sim 1$ results from the fact that the expected value of $\chi_k^2 = k$ if the **null hypothesis** (that the test samples are drawn from the reference distribution) is true.

The width of the χ^2 distribution grows with the degrees of freedom, as $\sigma_k = \sqrt{2k}$, and so deviations from $\chi_k^2/k = 1$ are suppressed like $1/\sqrt{k}$—the usual situation where the absolute error grows but the relative error shrinks, both as the square-root of the sample number. A sound goodness-of-fit test must take the changing width of the pdf into account. We can complete our discussion of the χ^2 test by use of our hypothesis testing methods: a model can be considered compatible with data at confidence level CL (under the assumption of purely statistical, purely Gaussian uncertainty) if its χ_k^2 value is less than the value of X for which

$$\int_0^X f(z = \chi^2; z) \, dz = \text{CL}. \tag{5.117}$$

If we set the confidence limit to the Gaussian 1-sigma level, $\text{CL} = \int_{-\sigma}^{\sigma} \mathcal{N}(x; 0, \sigma) \, dx \sim 0.683$, the corresponding 1σ χ^2 limit is 1 for $k = 1$, and 2.3 for $k = 2$. The first of these leads to a commonly quoted rule of thumb for computing the 1σ uncertainty on a χ^2 fit: find the parameter deviations that increase the χ^2 by $\Delta\chi^2 = 1$ above the fit minimum. Note that this only applies if there is only one degree of freedom, and that strictly the *shift* is only equal to 1 if the best-fit has $\chi_{\text{min}}^2 = 0$: the fact that χ_{min}^2 will by construction be larger than the 68% CL value for

32% of the time is often brushed under the carpet! At the 95% CL, the χ^2 limits are 3.84 for $k = 1$, and 5.99 for $k = 2$.

What exactly is k? It is not just the number of bins, but the number of unconstrained elements in the statistical system. If our model really came from nowhere and has no free parameters, then every bin is free to fluctuate relative to the model. But what if we introduce n_θ parameters $\boldsymbol{\theta}$ to the model and find the combination $\boldsymbol{\theta}_{\text{fit}}$ that fits best? In this case the model and data are no longer decoupled and free to statistically fluctuate relative to each other: if a bin fluctuates upward in data, the fitted model will try to follow it to the extent permitted by the global fit measure. In principle every model parameter subtracts one freely fluctuating bin, reducing the degrees of freedom to $k = n_b - n_\theta$. If one had as many free parameters as bins, i.e. $n_\theta = n_b$ and hence $k = 0$, the fitted model *could* simply consist of setting $f(x) = y(x)$ and then the χ^2 would always be a delta-function at zero. Such models where $f(x) = y(x)$ are known as **saturated models**, and typically have as many parameters as there are data points. In practice real statistical models are not so mercenary, but contain some assumptions, e.g. of physics or at least of parametric smoothness. But the statistical test assumes the worst and the chi-squared distribution is hence not defined for $k < 1$.

Comparing simple hypotheses
Acceptance or rejection of a single hypothesis is an important result, but it is frequently the case that we want to compare two hypotheses (or sets of hypotheses). For example, recall the Higgs boson discovery that we described earlier. Having obtained (at great cost!) a histogram of the diphoton invariant mass for LHC events with two photons in, we can ask whether the histogram is consistent with (a) the SM with no Higgs boson, or (b) the SM with a Higgs boson. In this case, we would typically refer to (a) as the 'background' or 'null' hypothesis $\mathcal{H}_0$, whilst we would refer to (b) as the 'signal' hypothesis $\mathcal{H}_1$. It should be clear that this approach is also relevant for a whole host of signal discoveries, and in each case the null hypothesis is chosen to be something 'boring', like 'consistent with SM expectations', while $\mathcal{H}_1$ represents some more radical possibility. In a hypothesis test of $\mathcal{H}_0$ with respect to $\mathcal{H}_1$, it might be the case that we erroneously do not reject $\mathcal{H}_0$ even though $\mathcal{H}_1$ is true. This is called a **type-II error**, or a **false negative** and we define the statistical power of our test as the probability of not committing a type-II error. We might also reject the hypothesis $\mathcal{H}_0$ even though it is true. This is called a **type-I error** or **false positive**.

We will start with a comparison between two simple hypotheses $\mathcal{H}_0$ and $\mathcal{H}_1$ which are completely specified, with no free model parameters. The **Neyman–Pearson lemma**, derived from a more general case that we will shortly consider, then states that the most powerful test statistic for discrimination between the two hypotheses is the **likelihood ratio**,

$$\lambda(\boldsymbol{D}) = \frac{p(\boldsymbol{D}|\mathcal{H}_1)}{p(\boldsymbol{D}|\mathcal{H}_0)}, \tag{5.118}$$

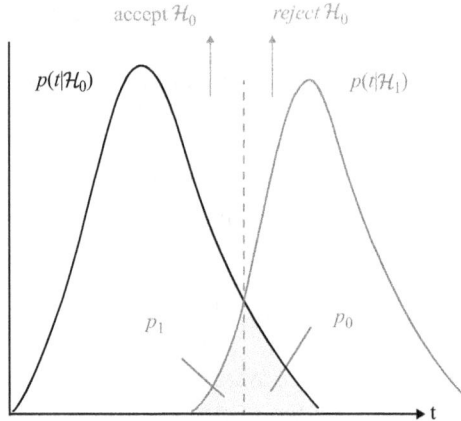

Figure 5.15. Illustration of a frequentist hypothesis test of two hypotheses $\mathcal{H}_0$ and $\mathcal{H}_1$. t stands for a general test statistic, of which log likelihood ratio λ is a particular instance.

or any monotonic relative, most importantly the **log likelihood ratio** (LLR),[18]

$$\text{LLR}(\boldsymbol{D}) = -2 \ln \lambda(\boldsymbol{D}) \tag{5.119}$$

$$= -2[\ln p(\boldsymbol{D}|\mathcal{H}_1) - \ln p(\boldsymbol{D}|\mathcal{H}_0)], \tag{5.120}$$

where the conventional factor of -2 is to be explained on page . The logarithm turning the likelihood ratio into a difference of log-likelihood terms gives rise to an alternative name for this statistic: the **delta log likelihood** (DLL or ΔLL).

The LLR is a particularly important object in hypothesis testing where $\mathcal{H}_1$ represents a new signal of some kind (e.g. testing a theory of particle physics beyond the SM); since a likelihood ratio $\lambda(\boldsymbol{D}) > 1$ reflects a preference for the signal hypothesis over the null, and the logarithm is a monotonic function[19], the -2 prefactor means that more negative LLR values mean the data is more 'signal like', and vice versa.

The situation is illustrated in Figure 5.15, which shows hypothetical pdfs (under changes of the observed data) for a general test statistic t given two hypotheses $\mathcal{H}_0$ and $\mathcal{H}_1$. In our case, $t = \lambda$ (larger values of $t = \lambda$ correspond to more signal-like behaviour). We can define some cut on the test statistic, c for which we will accept H_0 to the left of the cut, and reject it to the right. The probability of a type-I error is then given by

$$\alpha = \int_c^\infty p(t|\mathcal{H}_0)\mathrm{d}t \tag{5.121}$$

---

[18] Note, technically it is a 'log likelihood-ratio', and certainly not a 'log-likelihood ratio'. It's safest to avoid the hyphens.

[19] In the ratio it doesn't matter that $\ln p$ is negative for probabilities $p \in [0, 1]$—and this range anyway does not apply for probability *densities*.

i.e. assuming that $\mathcal{H}_0$ is true, we then take the pdf of the test statistic $p(t|\mathcal{H}_0)$ and we integrate the tail above the cut where we reject $\mathcal{H}_0$. This is called the significance of the test of H_0, in line with our previous section on testing a single hypothesis.

The probability of a type-II error is instead given by taking the pdf that assumes $\mathcal{H}_1$ is true, and integrating over the lower tail to the left of the cut (i.e. where we accept $\mathcal{H}_0$ instead of $\mathcal{H}_1$). The probability of a type-II error is thus given by

$$\beta = \int_0^c p(t|\mathcal{H}_1)\mathrm{d}t, \qquad (5.122)$$

where the lower limit of the integral would instead be $-\infty$ for a test statistic that could take arbitrarily negative values. The statistical power of our test is then $1 - \beta$. We can say that H_0 is rejected at a confidence level CL $= 1 - \alpha$ if the likelihood ratio $\lambda(\boldsymbol{D}) > c$; i.e. if $\mathcal{H}_0$ were true, a large fraction CL of the possible observed datasets would have given a λ below the threshold c, yet one above the threshold was measured. Note that this tells us how to choose c - we need to set it to the value that generates our desired CL for the hypothesis test.

It is instructive to think carefully about how one would actually choose a value of c for a given CL, for general hypotheses: if one does not know the form of the pdf for the test statistic analytically, one can obtain it via Monte Carlo simulation. For the λ statistic above, for example, one can assume that $\mathcal{H}_0$ is true, run a simulation of that hypothesis to get a toy dataset $\boldsymbol{D}'$, and calculate $\lambda(\boldsymbol{D}')$. On repeating this exercise for many toy experiments, one gets $p(\lambda|\mathcal{H}_0)$. The CDF of this can be evaluated numerically, and then inverted to find c for a given CL.

It is also useful to think about why the likelihood ratio construction is powerful. It comes from cancellation of degrees of freedom: in our preceding testing of a single hypothesis, the more random variables (e.g. bins of a distribution) there are, the more sources of statistical fluctuation, the broader the test-statistic distribution, and the harder to exclude the model. But if we have two hypotheses, being compared against the same data, then the same data fluctuations occur in both hypothesis tests and can be effectively cancelled between them. The statistics being finite, this cancellation is not perfect, and so the likelihood functions do not collapse to delta functions (which may be trivially discriminated!), but retain some statistical width and overlap which necessitates the $\lambda(\boldsymbol{D}) < c$ hypothesis-rejection threshold.

You may find it interesting to revisit this explanation and understand its connection to the asymptotic result for nested hypothesis comparison in the following section. Also note that in gaining discriminatory power we lost something else: the ability to determine whether *both* hypotheses are inconsistent with the data: this issue is addressed shortly.

Comparing nested hypotheses: likelihood profiling
So much for simple hypotheses. A more complicated task is where the two hypotheses have free parameters. We will consider the case where the two

hypotheses live in the same parameter space, θ, which turns out to be a necessary condition of a likelihood ratio test. The typical situation is where one hypothesis— conventionally the null, $\mathcal{H}_0$—is fixed, and the other is varied through the space, finding the regions where it is relatively preferred or disfavoured with respect to $\mathcal{H}_0$. This scenario is called a **nested model**, as the fixed hypothesis $\mathcal{H}_0$ is a subset, nested inside the more general space of $\mathcal{H}_1$ hypotheses to be evaluated.

A typical example is the case of a model whose parameters θ include parameters of interest ϕ, and some nuisance parameters ν. If we want to find a p-value for ϕ, we can proceed as above and construct a test statistic t_ϕ, for which larger values indicate increasing incompatibility of the data and some hypothetical choice ϕ. If we then observe a value $t_\phi = t_{\phi,\text{obs}}$ for our particular dataset, the p-value of the hypothesis ϕ is given by

$$t_s(\phi, \nu) = \int_{t_{\phi,\text{obs}}}^{\infty} p(t_\phi | \phi, \nu) \mathrm{d}t_\phi. \tag{5.123}$$

This necessarily depends on ϕ, since ϕ specifies the hypothesis that we are testing. The problem is that it also depends on the nuisance parameters, and we can only reject the hypothesis ϕ if the p-value is less than α for all possible values of the nuisance parameters. We can instead define a test statistic that is approximately independent of ν as follows. First, we define the **conditional maximum likelihood**, or **profile likelihood**, $\mathcal{L}(\phi, \hat{\hat{\nu}})$, where the double hat denotes the value of the nuisance parameters that maximises the likelihood given a particular value of ϕ. The procedure of choosing the nuisance parameter values that maximises the likelihood for a given ϕ is itself called **profiling**. We can then define a test statistic in which the choice of null hypothesis is the 'best-fit' MLE point in the space of θ, for which we have a likelihood $\hat{\mathcal{L}} = \mathcal{L}(\hat{\theta}) = \mathcal{L}(\hat{\phi}, \hat{\nu})$:

$$\lambda_p(\phi) = \frac{\mathcal{L}(\phi, \hat{\hat{\nu}})}{\mathcal{L}(\hat{\phi}, \hat{\nu})} \tag{5.124}$$

This is called the **profile likelihood ratio**, or sometimes (confusingly) just the *profile likelihood*, and it now only depends on ϕ. Usually likelihood profiling is taken a little further: the space of parameters ϕ may itself be larger than can be shown in a 1D or 2D plot, which requires us to do something with the extra parameters ϕ if we want to plot a 1D or 2D CL = 95% limit contour[20]. In this case, we can include the ϕ parameters that we do not wish to plot in the list of parameters that get profiled out.

The qualitatively interesting thing about this construction, is that the null and test hypotheses have different degrees of freedom: for example, if ϕ is a two-dimensional parameter plane, $\mathcal{L}(\phi)$ has two fewer degrees of freedom than $\hat{\mathcal{L}}$ when we perform maximisation over the variables that carry hats or double hats, and hence its statistical goodness of fit is expected to be two degrees of freedom worse. An extremely useful result called **Wilks' theorem** comes into play here: subject to

---

[20] Note that a Bayesian would instead *marginalise* over the extra parameters in projecting a large parameter set down to a 1D or 2D plot, and this difference between profiling and marginalisation is one of the key working differences between frequentist and Bayesian methods in practice.

certain reasonable assumptions[21], and independent of the general form of the likelihood function $\mathcal{L}(\theta)$, the **profile log likelihood ratio**,

$$t_\phi = -2 \ln \lambda(\phi) \tag{5.125}$$

$$= -2 \ln \frac{\mathcal{L}(\phi, \hat{\hat{\nu}})}{\mathcal{L}(\hat{\phi}, \hat{\nu})} \tag{5.126}$$

$$= -2[\ln \mathcal{L}(\phi, \hat{\hat{\nu}}) - \ln \mathcal{L}(\hat{\phi}, \hat{\nu})], \tag{5.127}$$

is distributed according to the chi-squared distribution, with degrees of freedom k equal to the *difference* in degrees of freedom between the global and conditional likelihoods, i.e. $\|\phi\|$. We can see a consistency here: the conditional likelihood $\mathcal{L}(\phi, \hat{\hat{\nu}})$ is bounded between zero and $\mathcal{L}(\hat{\phi}, \hat{\nu})$, hence the ratio is between 0 and 1, the log-ratio between 0 and $-\infty$, and the t_ϕ statistic from 0 to ∞: the domain of the chi-squared distribution. Note that our notation t_ϕ here means 'the test statistic corresponding to the hypothesised parameters ϕ' (we have written that instead of $t(\phi)$ because the notation will help us in another context later).

This effective reduction in degrees of freedom in the test statistic makes the profile LLR test particularly powerful, and it is the go-to statistical test in Large Hadron Collider searches. To assess whether a point in a parameter space is significantly (with confidence CL) excluded relative to the global best-fit, one computes the t_ϕ statistic at that point and compares to the critical value of χ^2 assuming the relevant number of degrees of freedom. If the t_ϕ value is less than the critical value, the point is unexcluded: it fits the data less well than the best-fit point, but only to an extent that could be explained by a fraction CL of datasets generated within the statistical fluctuations. If the cdf is greater than the agreed CL threshold, the model point is excluded. Typically a 95% CL threshold is used in collider physics, as a semi-arbitrary threshold to consider a model excluded. This, just short of a Gaussian 2σ cdf, can be compared with the far stricter 5σ CL typically demanded in order to make a positive discovery: while arbitrary, these thresholds reflect a physics bias that discarding speculative models is a lower-stakes game than adding them to the physics canon.

The CL_s measure

Given the absence of clear signals from the LHC so far, in many models this point may be approximated by the SM configuration—often at a very high-mass or low-coupling limit of the model, where the new physics has no observable effect—but MC models are imperfect, and even with attempts to systematically model (and profile) their failings, it is expected that 'the truth' does not actually lie in the parameter space of our models. How can we deal with the situation where the null

---

[21] A particularly important one is that the main probability density is located well away from physical boundaries in the parameter space, for example a positivity requirement on masses or cross-sections. If working with theories close to such a boundary, Wilks' Theorem cannot be assumed to hold, and an explicit Monte Carlo method or similar must be used.

hypothesis might also be excluded by the data when LLR methods, including the ubiquitous profile likelihood, are explicitly constructed as test statistics *relative* to a presumed-reliable null or best-fit reference model?

An approach known as the CL_s method has proven useful in cases where we are searching for a small new signal in a dataset. We have two possible hypotheses in this case:

- The **background-only hypothesis** that corresponds to the SM without a Higgs boson. We can calculate the number of LHC events with two photons in that results from this theory, which we will call b, and the likelihood of observing a specific number of events in the data will be given by a Poisson likelihood whose mean expected rate is set to b.
- The **signal-plus-background hypothesis** that corresponds to the SM with a Higgs boson. The number of expected signal events in this case, s can be calculated from the Higgs theory, and the likelihood of observing a certain number of events in data is given by setting the mean expected rate of the Poisson distribution to $s + b$. Conventionally, we actually scale the signal expectation by a **signal-strength** parameter μ, such that our mean expected rate is $b + \mu s$. This allows us to test different hypotheses for μ, but we will consider only the case $\mu = 1$ in this section.

Given an observed number of events, we then want to determine which of these two hypotheses is more favoured. For each value of μ, there is a pdf for the log likelihood ratio test statistic under assumed variations of the observed data (exactly as we found above for a general test statistic given that a particular hypothesis was true). To spell this out completely, let's assume that we have observed eight events in data, and that our theory calculations give us $b = 3$ and $s = 4$. For the $\mu = 0$ hypothesis, our likelihood ratio is given by $\mathcal{P}(8; \lambda = 7)/\mathcal{P}(8; \lambda = 3)$. We could then sample repeatedly from a Poisson distribution $\mathcal{P}(k; \lambda = 3)$ (which is our background model), to get different hypothetical numbers of observed events. For each of these we could recalculate the test statistic (i.e. -2 times the log of the likelihood ratio), and the set of these values for different hypothetical datasets sketches out the pdf of the test statistic $f(t_\mu | \mu = 0)$. For the signal-plus-background hypothesis, we can follow a similar procedure, only we now sample our hypothetical datasets from $\mathcal{P}(k; \lambda = 7)$, which gives us the pdf $f(t_\mu | \mu = 1)$. By construction, hypothetical data samples drawn from the null hypothesis $\mu = 0$ will be more null-like (or background-like) and will produce a more positive-valued t_μ distribution than signal-like samples drawn from the $\mu = 1$ hypothesis—but being random processes, it is perfectly possible for the two pdf distributions $f(t_\mu | \mu = 0)$ and $f(t_\mu | \mu = 1)$ to significantly overlap. Indeed, if the signal is small, the background and signal-plus-background hypotheses might not look very different.

Figure 5.16 shows two example test statistic distributions, taken from the search for the Higgs boson at the Large Electron Positron collider at CERN. For an assumed Higgs mass of $115.6 \, \text{GeV}/c^2$, the plot shows the two LLR distributions for the background and signal+background hypotheses (where we note that the original paper uses Q for the LLR instead of t_μ). The actual observed dataset defines a particular value of the test statistic t_μ^{obs}, which is shown as the vertical red line.

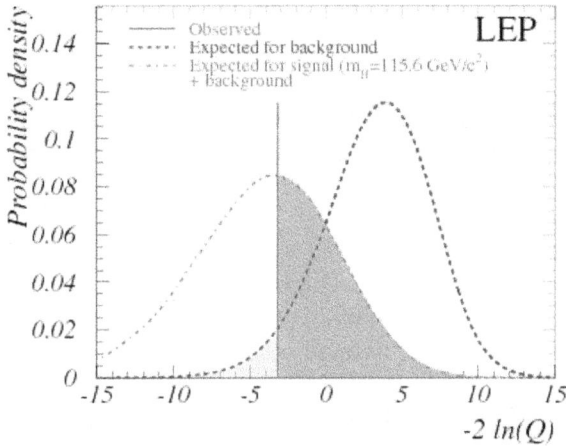

Figure 5.16. Example of the LLR pdfs for the background and signal+background hypotheses, from the Higgs-boson search at the Large Electron Positron collider. Q is the log likelihood ratio $\mathcal{L}(s + b)/\mathcal{L}(b)$ where, in the absence of systematic uncertainties, the likelihood terms here would be Poissonian. Credit: http://cds.cern.ch/record/508857/.

In analogy with equation (5.115), we can define an observed CL value for each hypothesis using

$$\mathrm{CL_{sb}} = \int_{t_\mu^{\mathrm{obs}}}^{\infty} f(t_\mu | \mu = 1) \mathrm{d}t_\mu, \tag{5.128}$$

and

$$\mathrm{CL_b} = \int_{t_\mu^{\mathrm{obs}}}^{\infty} f(t_\mu | \mu = 0) \mathrm{d}t_\mu. \tag{5.129}$$

In both cases, $t_\mu > t_\mu^{\mathrm{obs}}$ means a *less* signal-plus-background-like configuration than that observed in data. Thus, $\mathrm{CL_{sb}}$ represents the frequentist probability of observing a dataset that is at least as signal-plus-background-like as the observed dataset, which is, by definition, the p-value of the signal-plus-background hypothesis. For the background hypothesis, the logic is inverted: the p-value is the fraction of the $f(t_\mu | \mu = 0)$ distribution that is *more* signal-plus-background-like, i.e. has more negative LLR values: $p_b = 1 - \mathrm{CL_b}$. As this 'change of direction' in the p values is confusing, we will instead work with the two right-going $\mathrm{CL_i}$ integrals, $\mathrm{CL_{sb}}$ and $\mathrm{CL_b}$.

If we are in the nice situation that either the background-only or signal-plus-background hypothesis is definitely true, $\mathrm{CL_b}$ should be large most of the time. Even if the signal-plus-background hypothesis is true rather than the background-only hypothesis, $\mathrm{CL_b}$ should be close to 1, because the signal-plus-background LLR distribution produces t_μ values further to the left, *increasing* the value of the right-going $\mathrm{CL_b}$ integral. The problem situation is when the observation unluckily happens to come from a very unusual configuration on the un-signal-plus-background-like right-hand tail of either LLR distribution. In this situation we could easily conclude that the signal-plus-background hypothesis is excluded, because its

p-value CL_{sb} is small—a right-going integral that starts on the right-hand tail of the distribution. But be careful: while t_μ^{obs} is much more compatible with the background than the signal, in absolute terms it has poor compatibility with either. In this case perhaps the CL_{sb} value alone is not the whole story and we should be less willing to exclude signal models when the background model (or, more generally, some null reference hypothesis $\mathcal{H}_0$) itself is on statistically shaky ground.

The classic example of this arises when the expected yield of events in our search process is, say, three events, and none are observed. Whether the expectation arises from the background-only or signal-plus-background model, it only has a $\mathcal{P}(0; 3) \sim 4.98\%$ probability of producing a result with a rate this low. Let's say that our expectation of three events comes from the background model. An observation of 0 events is more background-like than signal-like, but it is the extreme value of 'background-like', sitting on the far right-hand tail of the background-only LLR distribution to such an extent that it is arguably unrepresentative of the background hypothesis at 95% CL! The long-standing consensus in this situation was for a long time that experiments should not publish claims of signal model exclusion in circumstances where the data can also be argued to be incompatible with the null hypothesis. But this seems too harsh: a **low sensitivity** signal process that would produce, say, a signal plus background expectation of four events may be indistinguishable from the background hypothesis, but an 'obvious' signal that leads to 14 signal plus background events can surely still be reasonably excluded. How do we quantify this intuition?

A neat solution, although strictly speaking ad hoc rather than a pure frequentist result, is the CL_s method. In this, the CL reported by the hypothesis test is not CL_{sb} as we would expect, but a modified version,

$$CL_s \equiv \frac{CL_{sb}}{CL_b}, \qquad (5.130)$$

where the labelling is just a mnemonic reminiscent of division rules, rather than meaning that it is actually a CL for signal alone (without background). The effect of this construction is to reduce CL_{sb} by the extent to which CL_b is less than unity: an extreme observation, unrepresentative even of the background model, will have a low CL_b and hence the even lower CL_{sb} will be somewhat inflated (Figure 5.17).

Likelihoods for event observation

So far we have mostly discussed likelihoods, hypothesis testing etc in the abstract. The one exception to this has been our treatment of the χ^2 test statistic, which we motivated as (up to a factor of -2) the logarithm of a composite likelihood built from many Gaussian-distributed random variables—hence contributing a product over bin likelihoods to the total likelihood, or equivalently a sum of log-likelihoods to the total log likelihood. This picture continues to be relevant in our extension to likelihoods more complex than Gaussians.

In this section we extend this treatment, appropriate in the statistical limit where random variables are effectively Gaussian distributed, to a more precise treatment of

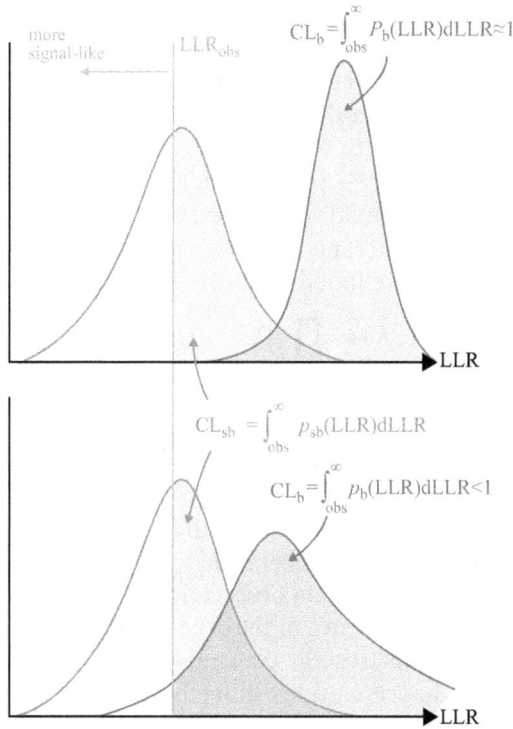

Figure 5.17. Demonstration of the construction of CL_{sb} and CL_b confidence levels from integration of the respective $s + b$ and b LLR distributions, across all entries less signal-like than the observed LLR. The ad hoc $CL_s \equiv CL_{sb}/CL_b$ construction only deviates from CL_{sb} if the $s + b$ and b distributions significantly overlap with each other and the observation, as in the bottom plot.

event-counting likelihoods. For this we use a multi-bin Poisson distribution with expected/observed yields λ_i and k_i for each bin i,

$$\mathcal{L}(k; \lambda) = \prod_i \mathcal{P}(k_i; \lambda_i) = \prod_i \frac{e^{-\lambda_i}\lambda_i^{k_i}}{k_i!}. \tag{5.131}$$

It is worth comparing this to the Poisson examples in the previous two sections. In those cases, we assumed that we had n repeated observations of a variable drawn from a single Poisson distribution. We are now dealing with a histogram, each bin of which has its own Poisson distribution.

Several features of equation (5.131) are immediately worthy of note. First, it is a discrete distribution (a pmf) in k_i, reflecting that real events are observed in integer quantities. If there are non-integer event weights to be applied, they can be absorbed into the expected rates λ rather than the observed yields k. And secondly, in the large-k limit, as its individual Poisson terms tend towards Gaussians, the product in this likelihood tends towards to a multivariate Gaussian.

In searches for new physics, this form is typically made more explicit, expressing the expected bin yields in terms of background-process and signal-process components b_i and s_i respectively, i.e. $\lambda_i = b_i + s_i$. The signal here is generally dependent on the POIs ϕ of the new-physics model, which might be a complex theory in which a detailed calculation is required to obtain the number of signal events (e.g. supersymmetry). For simple hypothesis testing, however, it is common to simply introduce an ad hoc signal strength parameter μ, which scales the signal relative to a nominal model such that $\lambda_i = b_i + \mu s_i$, of which we shall see an example shortly. Both the signal and background rates are likely also subject to systematic uncertainties due to modelling and detector effects, characterised by a set of nuisance parameters ν. Our likelihood expression is then

$$\mathcal{L}(\theta = \{\phi, \nu\}; k) = \prod_i \mathcal{P}(k_i; b_i(\nu) + s_i(\phi, \nu)) \tag{5.132}$$

$$= \prod_i \frac{e^{-(b_i + s_i)}(b_i + s_i)^{k_i}}{k_i!}. \tag{5.133}$$

Our primary use for this likelihood is usually to perform hypothesis testing to label POI regions (in μ or a more general ϕ space) within a model by the (in) compatibility of their $\lambda(\theta)$ with the data observations k, at confidence level CL. Note the mismatch between the dependence of the expected rates λ on the full set of model parameters, θ, against our interest in constraining the POI subset, ϕ: as before, frequentist and Bayesian interpretations differ here, the frequentist view being to work purely with the likelihoods and to eliminate the nuisance parameters ν by optimisation in the profile LLR construction; and the Bayesian being to construct the full posterior probability from the likelihood via Bayes' theorem and priors on the components of θ, and then to marginalise (integrate) over ν.

As it is more common in collider physics, we focus on the frequentist approach via the profile log likelihood ratio,

$$t_\phi = -2 \ln \frac{\mathcal{L}(\phi, \hat{\hat{\nu}})}{\mathcal{L}(\hat{\phi}, \hat{\nu})} \tag{5.134}$$

$$= -2 \sum_i [k_i(\ln \lambda_i(\phi, \hat{\hat{\nu}}) - \ln \lambda_i(\hat{\phi}, \hat{\nu})) - (\lambda_i(\phi, \hat{\hat{\nu}}) - \lambda_i(\hat{\phi}, \hat{\nu}))], \tag{5.135}$$

where we note that the discretising factorials from the denominators of the Poisson pmfs have cancelled between the global and conditional profile (log-)likelihoods since the observed k are the same in both cases. As before, a CL figure may be computed from t_ϕ by a rightward integral from its observed value:

$$\mathrm{CL}_\phi = \int_{t_\phi^{\mathrm{obs}}}^{\infty} f(t_\phi | \phi) \mathrm{d}t_\phi. \tag{5.136}$$

Sometimes one uses a variation of the t_ϕ test statistic, if a signal strength parameter μ is being tested as a single POI (this is then often referred to as t_μ, but we will continue to use the more generic t_ϕ). One motivation for modification is that we often restrict ourselves to models which can only *increase* expected yields; in this case it makes sense to 'freeze' the global maximum likelihood statistic at the conditional value for $\mu = 0$, should the MLE estimator $\hat{\mu} < 0$, giving rise to the test-statistic $\tilde{t}_\phi$. The q_0 test statistic modifies $\tilde{t}_\phi$ to truncate the statistic for testing the $\mu = 0$ hypothesis (whose disproof would indicate discovery of a new process), and a q_μ test statistic is used for placement of upper limits on the strength of searched-for processes. In this presentation we will focus on the canonical t_ϕ test-statistic, whose consequences qualitatively apply also to the variations, but the reader should be aware of their existence and the need in general to use the variant appropriate to their task. A comprehensive presentation can be found in the Asimov dataset paper listed in the Further Reading of this chapter.

Estimating the t_ϕ *probability density*
In our previous discussion of the profile LLR method, we rather glibly asserted that Wilks' theorem rides to our rescue and that the t_ϕ LLR test statistic is chi-squared distributed with the number of degrees of freedom equal to the number of ϕ parameters constrained in the profile likelihood-ratio numerator. But in high-stakes experimental interpretations—for example, the statistical tests of data used to discover the Higgs boson, and which continue to be used for tests of theories of physics beyond the SM—we need to be more careful, and consider the extent to which the asymptotic conditions for strict Wilks' theorem applicability are not satisfied. One way to do this would be to randomly sample from the likelihood functions to build up an estimator of the t_ϕ probability density empirically. Sampling is a wonderfully intuitive way to gain an understanding of statistics, but it is not always the most efficient way, and in fact we can get a long way with a more analytic approach.

If we consider a single POI, $\phi \in \boldsymbol{\phi}$, we can form an LLR statistic t_ϕ solely on this variable. We can then define two different pdfs that are generally useful:

- $f(t_\phi|\phi)$, which is the pdf of the test statistic t_ϕ assuming that the hypothesis ϕ is true. This is what we worked with above to define CL values for testing the hypothesis ϕ, but it is worth parsing again to make sure that you really understand it. For example, consider the case where ϕ is equal to a signal strength parameter μ, we could define a log likelihood ratio for a particular value of μ. We could then look at how this is distributed under the assumption that our data indeed arise from the hypothesis with that value of μ.
- $f(t_\phi|\phi')$, which is the pdf of the test statistic under the assumption that some other hypothesis ϕ' is true. In this case, for example, we could draw sample datasets from the hypothesis ϕ', but still evaluate t_ϕ for each one, which results in a different pdf to $f(t_\phi|\phi)$.

Now assume that the true hypothesis from which our data are drawn has a parameter value equal to ϕ'. We then want to test some other value of the parameter ϕ.[22] Using a result due to Wald it can be shown that, in the limit of asymptotically large

---

[22] A good example is again that of the Higgs search, for which our data are drawn from the hypothesis $\mu = 1$, and we might want to test the hypothesis $\mu = 0$ in order to see if we can reject it.

statistics, $t_\phi \approx (\phi - \hat{\phi})^2/\sigma^2$, where $\hat{\phi}$ is the MLE estimator for ϕ, which is itself Gaussian-distributed with mean ϕ' and variance σ^2. As a result, it can be shown that $f(t_\phi|\phi')$ is distributed following a *non-central chi-squared distribution* for one degree of freedom,[23] with a non-centrality parameter $\Lambda = (\phi - \phi')^2/\sigma^2$. For the special case of $\phi = \phi'$, i.e where the parameter value under test is the correct value, the non-centrality becomes zero, and the central chi-squared distribution result of Wilks' theorem is regained. The strength of the Wald result is that it provides a systematic correction to the Wilks assumption when the assumed parameter value ϕ is erroneous. This result, as for Wilks, generalises to higher numbers of POIs by increasing the degrees of freedom in the (non-central) chi-squared distribution.

We have avoided mathematical depth on this result, since the consequences require several further steps of calculation and details which add little conceptual insight—and more importantly, the procedure is encoded in standard statistics tools such as RooStats. But it is clear that to estimate $f(t_\phi|\phi')$ and hence extract a p-value for the model under test, we need an estimate for Λ, and hence for the variance of $\hat{\phi}$, σ^2.

It turns out that σ needs to be obtained from the covariance matrix of the estimators for all of the parameters, $V_{ij} = \mathrm{cov}[\hat{\theta}_i, \hat{\theta}_j]$, where the θ_i represent both ϕ, and any additional nuisance parameters. If we assume that θ_0 represents the ϕ parameter, this gives us $\sigma^2 = V_{00}$. The inverse of the covariance matrix, V_{ij}^{-1} can, in turn, be calculated from the **Fisher information matrix** $\mathcal{I}(\theta)$, the second derivative matrix (called the 'Hessian') of the log of the original likelihood of the parameters $\mathcal{L}(\theta; D)$:

$$V_{ij}^{-1} = -\mathrm{E}\left[\frac{\partial^2}{\partial\theta_i\partial\theta_j} \ln \mathcal{L}(\theta; D)\right] \equiv \mathcal{I}(\theta) \tag{5.138}$$

$$= \mathrm{E}\left[\left(\frac{\partial \ln \mathcal{L}(\theta; D)}{\partial\theta_i}\right)\left(\frac{\partial \ln \mathcal{L}(\theta; D)}{\partial\theta_j}\right)\right] \tag{5.139}$$

where the expectation value is taken with our single parameter of interest set equal to ϕ', and it is an average over possible realisations of the data.

This is an extremely interesting object. As the second identity above shows, it is the covariance matrix of the set of variables $\partial \ln \mathcal{L}/\partial\theta_i$, which are called the likelihood **scores**. Since

---

[23] The non-central chi-squared distribution has the rather horrible form:

$$f_X(x; k, \lambda) = \sum_{i=0}^{\infty} \frac{e^{-\lambda/2}(\lambda/2)^i}{i!} f_{Y_{k+2i}}(x), \tag{5.137}$$

where Y_{k+2i} follows a chi-squared distribution with $k + 2i$ degrees of freedom. This very non-intuitive formula is stated for completeness only.

$$\frac{\partial \ln \mathcal{L}(\boldsymbol{\theta})}{\partial \theta_i} = \frac{\partial \mathcal{L}(\boldsymbol{\theta})}{\partial \theta_i} \bigg/ \mathcal{L}(\boldsymbol{\theta}), \tag{5.140}$$

they express the relative sensitivity of the likelihood to parameter variations around. $\boldsymbol{\theta}$ An intuitive way to understand the Fisher information is that if the likelihood is sensitive to a parameter, it will sharply peak around the true value $\boldsymbol{\theta}'$: as a result its score—the derivative—will undergo a large shift from large positive to negative slopes as it passes through $\boldsymbol{\theta}'$. The score will hence have a large (co)variance, $\mathcal{I}$, in the vicinity of the true parameter values. (The same picture correctly leads to the conclusion that the mean score is zero at $\boldsymbol{\theta}'$, where the likelihood is maximised, so the second moment is automatically central.) A more geometric picture, cf. the second equality in equation (5.138), is that the Fisher information is the Hessian curvature of the log-likelihood function: indeed, the $\mathcal{I}_{ij}$ matrix can be used as a metric in a differential geometry approach to likelihoods, called **information geometry**.

As an aside, one important application of this object for Bayesian statistical inference is in defining 'uninformative priors' on model parameters, when there is no *a priori* evidence to favour some parameter values over others. It is often assumed that this role is played by a flat (uniform) prior probability density $p(\boldsymbol{\theta})$, but in fact it depends on the statistical model within which the parameter set $\boldsymbol{\theta}$ plays its roles: scale parameters, for example, generally propagate through models differently than location parameters, and e.g. Gaussian models' parameters propagate differently from those of Poisson models. A neat way to view uninformative priors is that they transform trivially under reparametrisation of their model, which can be achieved by use of the Fisher information in the so-called **Jeffreys prior** construction:

$$\pi_J(\boldsymbol{\theta}) \propto \sqrt{\det\{\mathcal{I}(\boldsymbol{\theta})\}} \,. \tag{5.141}$$

Some important examples of Jeffreys priors encountered in physical contexts are: a uniform prior on the mean of a Gaussian; a $1/\sigma$ prior on the standard deviation σ of a Gaussian; and a $1/\sqrt{\lambda}$ prior on the mean rate parameter λ of a Poisson distribution.

The Fisher information is also connected to our immediate problem of estimating the variance of our POI ϕ by the **Cramér–Rao bound**, which states that the inverse of the Fisher information is a lower bound on the variance of any unbiased estimator of the model parameters $\boldsymbol{\theta}$. Maximum-likelihood estimators are unbiased in the large-sample limit, where the Wald result also applies, and so the Fisher information gives us everything we need to construct $f(t_\phi|\phi')$ and extract a p-value from it. This extends to multiple parameters in the natural way.

As an example, consider the multi-bin Poisson likelihood of equation (5.131), and assume that we have some parameters θ_i of the model which will affect the expected rates of the Poisson likelihood. We can evaluate the first and second derivatives as

$$\frac{\partial \ln \mathcal{L}(\boldsymbol{\theta}; \boldsymbol{k})}{\partial \theta_j} = \sum_i \left(\frac{k_i}{\lambda_i} - 1\right)\frac{\partial \lambda_i}{\partial \theta_j} \tag{5.142}$$

$$\frac{\partial^2 \ln \mathcal{L}(\boldsymbol{\theta}; \boldsymbol{k})}{\partial \theta_j \partial \theta_k} = \sum_i \left[\left(\frac{k_i}{\lambda_i} - 1 \right) \frac{\partial^2 \lambda_i}{\partial \theta_j \partial \theta_k} - \frac{\partial \lambda_i}{\partial \theta_j} \frac{\partial \lambda_i}{\partial \theta_k} \frac{k_i}{\lambda_i^2} \right]. \tag{5.143}$$

The maximum likelihood estimators for the parameters are obtained by setting the scores to zero, $\partial \ln \mathcal{L}/\partial \theta_j = 0$ for all j. This occurs, unsurprisingly, when the observations are equal to their model expectations, i.e. when all $k_i = \lambda_i$, the simultaneous modes of all the constitutent Poisson distributions. This motivates the idea of a special, artificial dataset[24] for a given model, in which all observations are equal to the model expectations (in practice to be derived via analysis of a very large Monte Carlo event sample), and hence the maximum-likelihood parameter estimators return the true values. This has become known as the **Asimov dataset**, in homage to the Isaac Asimov short story *Franchise*, where a single voter is chosen as sole representative of a political electorate.

Once we do that, the Asimov dataset allows us to estimate the Fisher information, and hence the $\boldsymbol{\theta}$ covariances, then $f(t_\phi | \phi')$, and finally the p-value beyond the Wilks approximation. Examining the second-derivative term in equation (5.143), we see that it is linear in the observed yields k_i. Hence, its expectation value over lots of datasets is found by just evaluating it with the expectation values of the data, which is the same as choosing the Asimov dataset. We can thus replace the k_i with their Asimov values:

$$\frac{\partial^2 \ln \mathcal{L}_{\mathrm{A}}(\boldsymbol{\theta}; \boldsymbol{\lambda})}{\partial \theta_j \partial \theta_k} = -\frac{\partial \lambda_i}{\partial \theta_j} \frac{\partial \lambda_i}{\partial \theta_k} \frac{1}{\lambda_i}. \tag{5.144}$$

Here, we have defined the Asimov likelihood $\mathcal{L}_{\mathrm{A}}$ as the likelihood that results from using the Asimov dataset. Using this construction, the derivatives of the Asimov likelihood $\mathcal{L}_{\mathrm{A}}(\boldsymbol{\theta}; \boldsymbol{\lambda})$ can be evaluated numerically, and the parameter variances extracted. Alternatively, we can return to Wald's result that $t(\phi) = (\phi - \hat{\phi})^2/\sigma^2$ and note that with the Asimov dataset the ML estimator $\hat{\phi} = \phi'$, hence $t_{\mathrm{A}}(\phi) = (\phi - \phi')^2/\sigma^2 = \Lambda$ directly.

A final important summary result can be obtained from the Asimov dataset. Using the Wald approximation, the distributions of the t_ϕ and other test statistics can be derived. The significance of an observed t_ϕ value can be evaluated from this distribution for a choice of μ, and is generally of the form $Z(\mu) \sim \sqrt{t_\phi}$.[25] As the significance is a monotonic function in the value of the test statistic, the **median expected significance** to be obtained by an experiment—the key metric for design optimisation of an analysis—is that function applied on the median $t_\phi(\mu)$, which we can approximate by its Asimov value $t_{\mathrm{A}}(\mu)$. The Asimov median expected significance of a statistical analysis based on likelihood-ratio test statistics is hence

---

[24] In the sense of a set of binned yields, not a full set of event-sample data.

[25] Actually, t_ϕ has a slightly more awkward form: the q_0 and q_μ variants have exactly this square-root form.

$$\text{med}[Z|\mu] = \sqrt{t_A(\mu)}. \tag{5.145}$$

Simplified likelihoods

The full dependence of an event-counting likelihood on the model parameters θ is typically not known in full detail, as every value of θ would involve generating and processing a high-statistics MC event sample with that model configuration, and processing it through a full detector simulation, reconstruction, and analysis chain. This would take days at best, on large-scale computing systems, when what is required is typically sub-second evaluation times, in order that the likelihood can be optimised after many evaluations. An approximate approach is hence nearly always taken, parametrising expected (both signal and background) event yields as a function of each $\theta_i \in \theta$, based on interpolation between MC templates typically evaluated at values -1, 0, and +1 in the normalised nuisance parameters.

These **elementary nuisance** responses are in general non-linear, making the likelihood a very complex function, and difficult to report as an analysis outcome—although this is a rapidly developing area. In this context, a useful development has been the **simplified likelihood** scheme, which replaces this full complexity with linear responses of expected (background) event yields to an effective nuisance parameter Δ_i for each bin i, with interplay between elementary nuisances absorbed into a covariance matrix between the new effective nuisances. This permits a much simplified explicit likelihood form,

$$\mathcal{L}_S(k; \lambda(\theta_S))$$

$$= \prod_i \mathcal{P}(k_i; \lambda_i = s_i + b_i + \Delta_i) \cdot \frac{1}{(2\pi)^{n/2} \det\{\Sigma\}} \exp\left(-\frac{1}{2}\Delta^T \Sigma^{-1} \Delta\right), \tag{5.146}$$

where the new nuisance parameters directly modify the nominal background event yield, $b_i \to b_i + \Delta_i$, and the new Gaussian term imposes a penalty on such deviations.

This may appear distressingly close to imposition of Bayesian priors in a frequentist likelihood, and it is not unique to simplified likelihoods: similar penalty terms are usually applied to elementary nuisance parameters in full-likelihood fits. It is hence worth taking a little time here to explain why this is acceptable. These constraints should be conceived of as encoding the restrictions on the estimators of the nuisance parameter values imposed by the experiment's calibration procedures—which are based on dedicated studies, effectively in control regions well away from the signal bins[26]. It could be possible to perform (non-simplified) analyses with the calibration distributions and elementary parameters included directly in the final likelihood—but it would be a fantastically complex, computing-intensive, and hard-work approach. Parametrising the effect of such calibration studies on the likelihood

---

[26] This assumption is also worth noting: as more exotic physics signatures are considered, some such as e.g. long-lived particles or 'emerging jets' do indeed overlap with calibration heuristics, mandating a less factorised than usual approach to analysis and calibration.

via nuisance-penalty terms, either simplified or elementary, is hence a pragmatic, and statistically consistent approach. Just make sure not to call them 'priors' or say that they indicate our degree of belief in the true nuisance-parameter values!

As multiple elementary nuisances affect each bin's yield in different ways, potentially combining to reinforce or cancel one-another's effects, the effective nuisances are correlated. This is captured to second order by the covariance matrix $\Sigma_{ij} = E[\Delta_i \Delta_j]$ where the expectation value has been calculated (usually estimated by bootstrap sampling) over the true likelihood, with greater weight on nuisance configurations that combine to give better agreement with the data. The covariance matrix Σ expresses these correlations in the space of Poisson expected rates: the product of Poisson distributions itself cannot be diagonalised since each Poisson is a pmf, and a similarity transformation would not respect their discretisation in k.

5.8.2 Bayesian hypothesis testing

Before we conclude this section, we will provide a brief description of the Bayesian approach to hypothesis testing. A Bayesian can define the posterior probability $P(\mathcal{H}|D)$ of a hypothesis $\mathcal{H}$, and could therefore in principle reject $\mathcal{H}$ if $P(\mathcal{H}|D)$ was small. The problem is that, by Bayes' theorem, $P(\mathcal{H}|D)$ is proportional to the prior belief in the hypothesis $P(\mathcal{H})$, which remains a subjective quantity in general. The way around this is to restrict Bayesian hypothesis testing to *comparisons* of two hypotheses, in which case we can construct a quantity that does not depend on the prior probabilities of the two hypotheses.

Consider two hypotheses $\mathcal{H}_1$ and $\mathcal{H}_2$, which are described by the parameters θ_1 and θ_2 respectively. Some parameters might be common to both hypotheses, and some may not. The full prior probability for each hypothesis can be written as:

$$\pi(\mathcal{H}_i, \theta_i) = P(\mathcal{H}_i)\pi(\theta_i, \mathcal{H}_i), \tag{5.147}$$

where $\pi(\theta_i, \mathcal{H}_i)$ is the normalised pdf of the hypothesis parameters, and $P(\mathcal{H}_i)$ is the overall prior probability for $\mathcal{H}_i$ (i.e. our degree of belief that that particular hypothesis is true). The posterior probability for each hypothesis is then given by:

$$P(\mathcal{H}_i|D) = \frac{\int P(D|\theta_i, \mathcal{H}_i)P(\mathcal{H}_i)\pi(\theta_i|\mathcal{H}_i)d\theta_i}{P(D)}. \tag{5.148}$$

We can then write the *ratio* of the posterior probabilities for the hypotheses as

$$\frac{P(\mathcal{H}_1|D)}{P(\mathcal{H}_2|D)} = \frac{\int P(D|\theta_1, \mathcal{H}_1)\pi(\theta_i|\mathcal{H}_1)d\theta_1}{\int P(D|\theta_2, \mathcal{H}_2)\pi(\theta_i|\mathcal{H}_2)d\theta_2}\frac{P(\mathcal{H}_1)}{P(\mathcal{H}_2)} \tag{5.149}$$

where we can factor out the ratio $P(\mathcal{H}_1)/P(\mathcal{H}_2)$ because the integral is only over the θ_i variables of the hypotheses. Thus, the posterior ratio is equal to the product of the prior ratio, and another factor which is defined to be the **Bayes factor**:

$$B_{12} = \frac{\int P(D|\theta_1, \mathcal{H}_1)\pi(\theta_i|\mathcal{H}_1)d\theta_1}{\int P(D|\theta_2, \mathcal{H}_2)\pi(\theta_i|\mathcal{H}_2)d\theta_2}. \tag{5.150}$$

If the prior probabilities for the hypotheses were equal, then the Bayes factor tells us what the ratio of posterior probabilities would be. Furthermore, if the hypotheses $\mathcal{H}_1$ and $\mathcal{H}_2$ are simple hypotheses (with no parameters), then the Bayes factor is simply the likelihood ratio for the hypotheses. The Bayes factor is thus related to how much the probability ratio of $\mathcal{H}_1$ to $\mathcal{H}_2$ changes because of the data, and it is hence a numerical measure of the evidence supplied by the data in favour of one hypothesis over the other. The numerator and denominator of the Bayes factor are the Bayesian evidences of the respective hypotheses, and thus a Bayesian model comparison amounts to a comparison of the Bayesian evidence for each particular hypothesis (a quantity that can be evaluated numerically using, for example, the nested sampling algorithm described earlier).

Note that, although the Bayes factor is independent of the overall prior on each hypothesis, it still depends on the priors on the model parameters θ_i for each hypothesis. A Bayesian model comparison requires a well-motivated choice for these priors if it is to be useful in practice (or alternatively one can rerun the comparison with different choices of prior to determine how sensitive the final conclusion is to the choice).

5.8.3 Local and global statistical significance

There is one final, very important aspect of hypothesis-testing to touch upon: the risk that comes from performing *many* statistical tests (as we do), then quoting the strongest single significance observed. In exactly the same way that the chi-squared distribution broadens for higher degrees of freedom (capturing that many simultaneously observed data bins give more scope for deviations from a nominal model), if many tests are performed there is an intrinsically higher chance of observing a significant deviation in any one of them.

This 'meta-statistical' effect is generally known as the multiple-comparisons problem (or '*p*-hacking' and 'data snooping', when abused to generate exciting headline results), but in particle physics is more commonly referred to as the **look-elsewhere effect** (LEE). The effect and term rose to particular prominence in the run up to the Higgs boson discovery in 2012, where the SM's agnosticism with regard to the value of the Higgs boson mass m_H meant that many tests had to be performed to cover the spectrum of mass hypotheses. As honest scientists, it is essential that we counter the risk of reporting a high local significance (i.e. from the most discrepant single test), when the more representative number is the lower **global significance** of the whole test suite.

The difference between local and global significance is expressed by a **trials factor** which accounts for the multiplicity of comparisons. In many other fields, this is referred to as 'studentizing' the test-statistic distribution, and corresponds to a scaling by $1/\sqrt{N_{\text{tests}}}$. In a case like the m_H bump-hunt, however, the set of tests is a continuous variable: how many tests does that correspond to?

Given a likelihood-ratio test statistic $q(\theta)$ and a significance threshold c in that test statistic, we define the largest observed test statistic as

$$q(\hat{\theta}) \equiv \max_{\theta}[q(\theta)]. \tag{5.151}$$

The trials factor T is then given by the ratio between the maximum local p-value to the p-value for the excess being seen anywhere in the parameter range,

$$T = \frac{P(q(\hat{\theta}) > c)}{P(q(\theta) > c)}. \tag{5.152}$$

In the asymptotic limit where $q(\theta)$ is χ^2-distributed, the maximum $q(\theta)$ is bounded like

$$P(q(\hat{\theta}) > c) \leqslant P(\chi_k^2 > c) + \langle N(c) \rangle, \tag{5.153}$$

where $\langle N(c) \rangle$ is the expected number of $q(\theta) > c$ 'up-crossings' sampled from the background-only model and $k = |\theta|$, i.e. the expected global p-value will actually be higher by $\langle N(c) \rangle$ than the local maximum would suggest. Computing the trials factor hence requires knowing $\langle N(c) \rangle$, which is not trivial: it is specific to the details of the statistical model, and while it can be estimated by Monte Carlo 'toy' sampling,

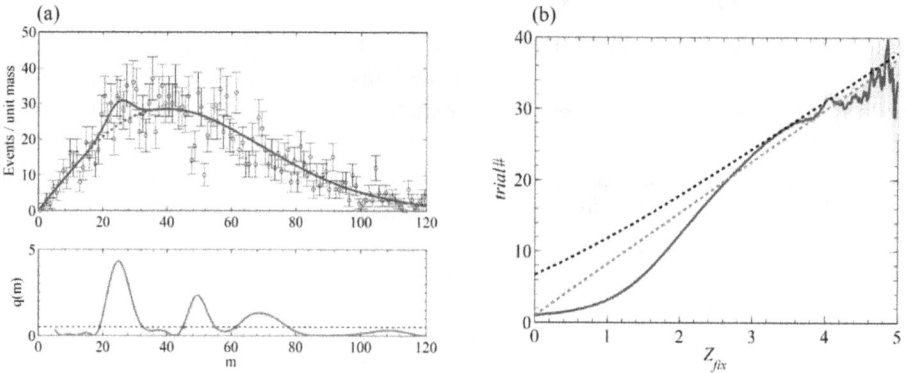

Figure 5.18. Example estimation of the look-elsewhere effect significance correction. Reproduced with permission from Gross and Vitells 2010 *Eur. Phys. J.* C **70** 525–30, arXiv:1 005.189 1. (a) An example background-only toy-MC pseudo-experiment histogram for a hypothetical bump-hunt in mass parameter $m \equiv \theta$. The blue lines show the best-fit total and background components, and the bottom graph shows the distribution of $q(m)$ as compared to the low c_0 reference level indicated by the horizontal dashed line. (b) The resulting trial factor for local-to-global p-value correction as a function of target significance Z. The solid line is from the toy MC approach (and its yellow error band from its statistical limitations), the dotted black line is the upper bound from equation (5.153), and the dotted red line the asymptotic approximation of equation (5.154).

doing so to the accuracy required for 5σ discoveries requires $\mathcal{O}(10^7)$ samples since the rate of up-crossings will be miniscule! Leveraging the asymptotic formulae previously introduced for profile likelihoods, an asymptotic approximation can be taken in which a lower threshold $c_0 \ll c$ is tested using a relatively small toy-MC sample, then extrapolated to the more demanding $q > c$ test. This sampling approach is shown in Figure 5.18(a).

An intuitive feel for what is going on here can be obtained by injecting an asymptotic expression for $\langle N(c) \rangle$ in terms of the chi-squared distribution for $k + 1$ degrees of freedom, in which case equation (5.153) takes the form

$$P(q(\hat{\boldsymbol{\theta}}) > c) \approx P(\chi_k^2 > c) + \mathcal{N} P(\chi_k^2 > c), \tag{5.154}$$

in which $\mathcal{N}$ can can be identified with an effective number of comparisons being performed, despite the $\boldsymbol{\theta}$ space being continuous rather than discretely binned. Using the general asymptotic significance result $Z = \sqrt{q}$, we can identify our threshold test-statistic with a fixed significance $Z_{\text{fix}} = \sqrt{c}$, and the asymptotic trials factor is then

$$T_k \approx 1 + \sqrt{1}\sqrt{2}\mathcal{N}\,Z_{\text{fix}}\frac{\Gamma(k/2)}{\Gamma((k+1)/2)}, \tag{5.155}$$

and hence

$$T_1 \approx 1 + \sqrt{\frac{\pi}{2}}\mathcal{N}Z_{\text{fix}} \tag{5.156}$$

for the important and most common single-parameter case $k = 1$. An example of the resulting linear scaling of the trials factor T_1 with the fixed significance threshold Z is shown compared to the toy-MC approach in Figure 5.18(b).

5.9 Multivariate methods and machine learning

The traditional approach to data analysis in the sciences is to develop domain-specific algorithms and implement them in software by-hand to process recorded data. For example, one might write code to place selections on histogrammed event-level quantities to find signal-rich regions of the data, or to reconstruct tracks or identify jets initiated by b-quarks. However, in physical systems with many variables, or where the underlying physics is poorly understood, such approaches can prove to be difficult or impossible. Furthermore, it is often the case that the sheer volume of data to be processed necessitates the use of solutions that are quicker to evaluate than the available analytic solutions.

In recent years, **machine learning** or **multivariate analysis** (MVA) approaches have become ubiquitous throughout high energy physics, tackling a range of topics across the spectrum from detector-level reconstruction to BSM phenomenology. This includes such problems as processing jet images, tagging top quarks, finding b-jets, reconstructing tracks, classifying Higgs boson events, defining variables for separating supersymmetric events from their SM backgrounds and providing fast interpolations of likelihood functions for global fits of new physics models. In machine learning, a computer learns directly from the data how to solve a particular problem, without needing to be given a specific algorithm that tells it what to do. This approach has become ever more powerful as advances in computing power have allowed users to make much more complex machine-learning systems, especially with the development of 'deep' learning methods in the first decades of the 21st century.

It is important at the outset to distinguish two types of machine-learning problem:

- **Supervised:** In a supervised problem, we have some prior knowledge of a quantity that we are trying to predict, which allows us to train an algorithm on past examples in order to predict future behaviour. For example, we might have simulated some SM events and some supersymmetric events, and we know from the simulation what category ('Standard Model' or 'supersymmetric') each event is taken from. We can then train a classification algorithm to learn how to predict which category events fall into, and define a variable that tells us if any given event is more likely to be supersymmetric or SM in origin. Running the trained algorithm on future events taken from the LHC data will help us define regions of the data that are low in background but potentially high in signal. Supervised learning is typically done in the context of classification or regression.
- **Unsupervised:** In an unsupervised problem, we have no target variable that we wish to predict, and the challenge is instead to infer the natural structure present in the data. Examples include clustering and density estimation. Historically, unsupervised learning has not been utilised much within Large Hadron Collider studies, but this is changing rapidly after the discovery of the Higgs boson which completes the SM. Since we now have no way of telling *a priori* what new physics will look like at the LHC, the discovery of new particles is a matter of either finding anomalies in the dataset, or an overabundance in some region of the data relative to the known background (but which otherwise resembles the background). This is a classic unsupervised problem.

The literature of machine-learning solutions for Large Hadron Collider problems is now growing very rapidly, making a detailed summary beyond the scope of this book. We will, however, briefly describe the two most common machine-learning techniques in use at the LHC.

5.9.1 Boosted decision trees

The **boosted decision tree** (BDT) algorithm is one of the most popular techniques in high energy physics, and is commonly used in analyses that have to separate a signal from a dominant background, but where the available analytic approaches only offer a low signal-to-background ratio. Imagine, for example, that we have a difficult supersymmetric signal to search for, and we wish to define a variable that efficiently separates the signal events from the dominant background events. We can

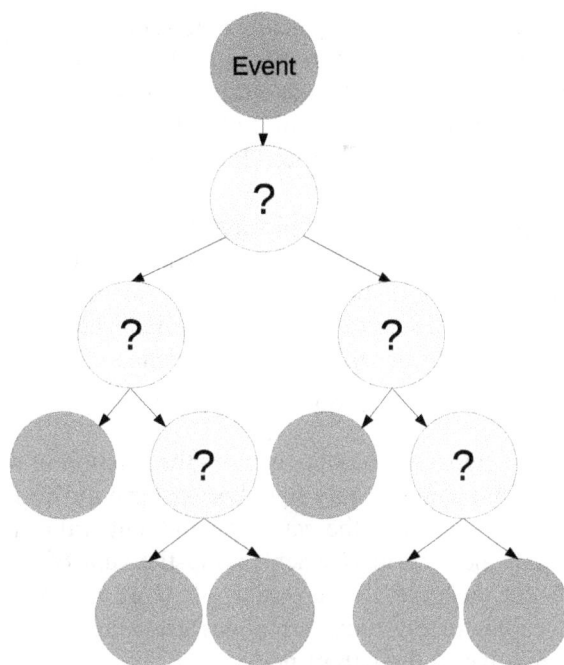

Figure 5.19. Illustration of a decision tree. Nodes are shown as circles, with decision nodes labelled by question marks, and leaf nodes filled in brown. Events passed through the tree will end up at one, and only one, leaf node, which can be used to determine the category of the event. The category (signal or background) is assigned based on which type of event dominates the subsample of events that reach that leaf.

simulate events from the signal and background, thus making this a supervised problem (i.e. the category of each event we train on is known). We will further assume that we are working in a particular final state (defined by the multiplicities and types of the particles we select), and that our analysis can make use of the four-vectors of the final state particles to compute useful functions expected to provide signal–background separation. We will assume in the following that we can define a vector, x, of useful variables for each event. These variables are frequently referred to as the **attributes** or **features** of the event.

The first step towards defining the concept of a boosted decision tree is the **decision tree** itself. A decision tree is a binary-tree structure which performs a series of cuts on attributes to classify the events as either signal- or background-like, organised much like a conventional flow chart (see Figure 5.19). The aim of the tree is ultimately to separate the events into sub-samples which are as pure as possible; that is, some are strongly enriched n signal events, whilst others are enriched in background events. Starting at the top, we first pick an attribute and place a selection on it. Each event will then pass down one of two branches depending on whether its value of the attribute is less than or greater than the cut value, and we end up at another decision node. A selection is then placed on a different attribute, leading to further branches and further nodes. The process continues until some stopping criterion is reached, which is typically placed on either the maximum depth of the decision tree, or the number of events at each node. Once a node cannot be split further, it is a 'terminal' or 'leaf' node, and events at that node are classified as being more signal-like or more background-like depending on the proportion of signal and background events at that node. This classification[27] is performed via an associated node value $c(x)$.

We have not yet specified a recipe for picking the attribute used at each decision node, nor for the cut value placed on the chosen attribute. Based on the logic above, the aim at each internal node of the tree should be to increase the purity of the subsamples that follow the selection at that node. There are many possible procedures, one of which is to use the Gini index, defined as

$$G = S(1 - S) \tag{5.157}$$

where S is the purity of signal in a node, given by the fraction of node events which are signal. For a node that contains only signal or background events, the Gini index is equal to zero. At each node, the variable and cut value are optimised by minimising the overall increase in G when taking the sum of the daughter nodes after applying the cut, relative to the value of G at the parent node. In the calculation, the G values of the daughter nodes are weighted by their respective fractions of events relative to the parent node.

Single decision trees are completely intelligible, since it is obvious why each event x ends up at a particular node and its associated value. The problem is that they rarely end up giving optimal signal-background separation, and they are also typically unstable with respect to statistical fluctuations in the event samples used to train them. A solution is to use a **forest** of trees, which combines a large number of decision trees. The trees may be added one-by-one, allowing information from a previous tree to be used in the definition of the following one. One can also reweight the training events for a particular tree, weighting up those which were misclassified by the previous tree. Each tree is assigned a score w_i during the training process, and the final BDT classifier is constructed from a weighted average over all the trees in the forest,

$$C(x) = \frac{1}{\sum_i w_i} \sum_i w_i c_i(x). \tag{5.158}$$

A potential pitfall of all machine-learning techniques is **overfitting**, which occurs when the algorithm learns features that are unique to the training data set, but which do not occur in the general population from which the training data were drawn. BDTs can be made more robust against overfitting by limiting each tree in the forest to a shallow depth. One can also perform **cross-validation** during training, in which the training sample is split into k equally-sized samples called **folds**, and the training

---

[27] Or regression: BDTs can also be used to parametrise complex functions in the attribute space using the same mechanism. The typical distinction is that classifier trees assign discrete values $c = 1$ for inferred signal events and $c = -1$ for background, while regression of a function f uses continuous values $c(x) \sim f(x)$.

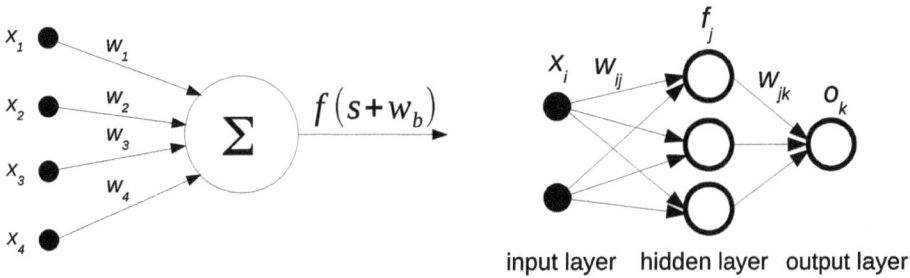

input layer   hidden layer   output layer

Figure 5.20. Core architectural elements of artificial neural networks. (a) A single perceptron, which forms a basic computer model of a neuron. Input variables x_i are fed with weights w_i, and are summed through an activation function which may include a bias parameter. (b) A multi-layer perceptron network, with an input layer of variables x_i, a hidden layer of perceptrons, and an output layer with a single output. The network is fully-connected from layer to layer.

is performed k times. At each iteration, a different fold is chosen as a test data set and is excluded from the training sample. The training is performed using data from the remaining folds, and the trained algorithm is then applied to the test data set to assess its performance. Comparing the performance of each training iteration can be used to evaluate the degree of overfitting.

5.9.2 Artificial neural networks

Many of the advances in machine learning in the present era are being obtained with **artificial neural networks**. The most primitive neural network is the single perceptron shown in Figure 5.20. This takes in some input values x_1 to x_4, evaluates the sum $s = \sum_i w_i x_i$ and then passes this through an activation function $f(s + w_b)$, where w_b is an extra parameter called the **bias**. In the simplest case, this activation function is a step function with $w_b = 0$, meaning that the output is equal to one if $\sum_i w_i x_i > 0$, and is otherwise zero. This mimics the behaviour of the neuron cells found in human and animal brains, that fire if the sum of the electrical signals along the inputs exceeds a certain threshold. The bias parameter allows this threshold to be shifted from zero, and different choices of activation function give rise to different output behaviours. In particular, non-linear activation functions are essential for neural networks to model interesting, non-linear features in generic datasets, and the most popular choices of activation function are the sigmoid ($f(s) = [1 + e^{-s}]^{-1}$), inverse-tanh, and rectified-linear (ReLU) ($f(s) = \max(0, s)$) functions. The latter is piece-wise linear, but non-linear over its entire range.

More complicated artificial neural networks can be built by chaining a bunch of perceptrons together to form layers of neurons, such that the outputs of one layer of neurons become the inputs of the next layer. This gives rise to the feed-forward, multi-layer perceptron (MLP) network shown in Figure 5.21, which in this example has an input layer, a single **hidden layer** and an output layer. At this point, it is helpful to forget any analogy with the human brain, since such a network is almost nothing like how the brain functions in practice. Instead, it is useful to think of an MLP as a 'semi-parametric' function which, under changes of the weights and

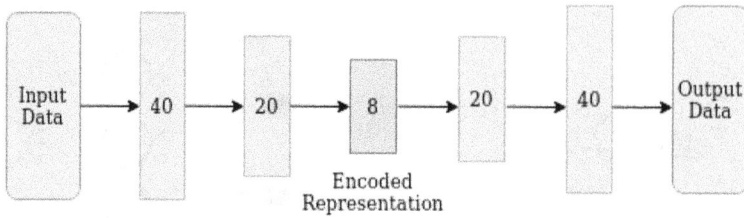

Figure 5.21. Schematic diagram of an autoencoder network. The numbers give the number of neurons in each hidden layer, and the architecture shown has proven to be useful in LHC case studies. Reproduced with permission from van Beekveld *et al* 2020 https://arxiv.org/abs/2010.07940.

biases, allows different output functions to be modelled. Of crucial relevance to supervised learning problems is the existence of a universal Approximation Theorem for MLPs, which states that a single-hidden-layer network containing a sufficient, but finite, number of neurons in the hidden layer can approximate any continuous function to a reasonable accuracy. Note that this requires the activation functions to be suitably non-linear. This makes neural networks an excellent choice for modelling functions whose functional form is unknown, which is exactly the problem under-lying regression and classification problems.

As an example of using a neural network in practice, let us revisit our BDT example in which we wish to train a classifier that can label events as either SM-like or supersymmetry-like. In this case, we have simulated events for both the SM background, and the supersymmetric signal, and we thus have event data where each event contains the category 'SM' or 'SUSY'. We want the network to learn how to tell the category on previous events, before setting it loose on real events from the LHC in an attempt to discover supersymmetry.

The first step is to train the network, much as we had to train our BDT. The input variables x_i will be whatever event features we believe show some discrimination between the background and the signal, potentially including particle four-vector components and useful functions of those components. The output variable is the category that we wish to predict. If we define an MLP architecture (number of hidden layers, number of neurons in each layer, activation functions, etc), we can initialise the weights and biases for the whole network and feed in the input variables for each event. This generates a category prediction for each event which, assuming we randomly initialised the weights and biases, will be wrong for most of the events. We can quantify how good our predictions are by defining a **loss function**, such as the binary cross-entropy

$$L = -\left[y_{\text{true}} \cdot \ln(y_{\text{pred}}) + (1 - y_{\text{true}}) \cdot \ln(1 - y_{\text{pred}}) \right], \tag{5.159}$$

where y_{true} are the true classes of the events, and y_{pred} are the corresponding predictions of the MLP. The higher the quality of our predictions, the smaller the binary cross-entropy. The goal of neural network training now becomes apparent: we must find the values of the weight and bias parameters that minimise the loss function. Having done that, we obtain a network that, for new event data, can be expected to provide an excellent prediction of the category for the events.

Unfortunately, there is no analytic way of determing the correct settings for the weight and bias parameters, and one must rely on numerical optimisation procedures. One choice is **gradient descent**, where the weights are updated as follows:

$$w'_{ij} \Leftarrow w_{ij} - R \cdot \frac{\partial L}{\partial w_{ij}}. \qquad (5.160)$$

Here R is a free parameter known as the learning rate, which controls how large the update steps should be during training. A more refined training strategy is the **Adam optimiser**, which uses a learning rate that is adjusted during the training, rather than being fixed *a priori*. In both cases, the gradients are calculated using a procedure known as **back-propagation of errors**, which expands the gradient and calculates it for each weight independently, starting from the output layer and propagating backward to the input layer. This leads to a substantial reduction in computing costs, which makes the training of large network architectures tractable.

As in the case of boosted decision trees, neural networks are prone to overfitting features in the training data. Cross-validation can again be used to mitigate this, combined with stopping the network training for each iteration when the performance on the test data set starts to get worse (the performance on the training data set will always improve). Another option is to add an extra term of the form

$$L_{\mathrm{L2}} = \lambda \sum_i w_i^2 \qquad (5.161)$$

to the loss function, where λ is a parameter of the algorithm which is smaller than one. The purpose of this term is to penalise situations where the weights w_i become too large, and the method is called L2-regularisation.

A multi-layer perceptron network, though potentially large, is in fact one of the most simple network architectures. Many more can be defined by changing the pattern of connections between layers, and even by adding connections between neuron outputs and earlier layers. A particularly useful innovation for image processing is the use of **convolutional neural networks** which are designed to mimic the vision system of humans and animals. The first few layers detect high level features in the images, whilst the later layers perform standard MLP-like classification on these high-level features. It is thus possible to use neural networks to *automatically* learn features of data that are not obvious analytically, which explains the rapidly expanding literature on neural-network-based jet imaging techniques.

Before we conclude, it is worth stating that neural networks are not limited to supervised applications, but can also be highly useful for unsupervised problems, such as **anomaly detection**. A particular network architecture known as an **autoencoder** is shown in Figure 5.21; this symmetric architecture maps the input variables to a set of output variables that is designed to match the input variables as closely as possible. By training the autoencoder on simulations of SM events, one obtains a system that is able to more-or-less accurately reproduce the input variables for SM events—a **generative network**. When the system is run on new SM events that were not in the training sample, it continues to provide a good prediction of the input variables. When it is run on other types of event, however, it fails to reproduce the input variables, and the discrepancy can be used to determine an anomaly score for the event. Autoencoders can therefore in principle be used to define variables that are highly sensitive to certain types of new physics, without having to know what signal one wants to search for *a priori*—the devil, of course, is in the detail of understanding the impacts of modelling and other uncertainties on the relatively opaque analysis machinery.

Further reading

- There are many excellent books on statistics for high energy physics. Two particular favourites of the authors are 'Statistical Methods in Experimental Physics', F James, World Scientific, and 'Statistical Data Analysis', G Cowan, Oxford Science Publications.
- The statistics chapter of the Particle Data Group 'Review of Particle Physics' (http://pdg.lbl.gov/) provides an excellent and concise review of statistics for HEP applications.
- The original paper on asymptotic formulae for hypothesis testing, formally introducing the Asimov dataset, (*Eur. Phys. J.* C **71** 1554 (2011)) is highly recommended.
- The *Eur. Phys. J.* C 70 525–30 (2010), arXiv:1 005.189 1 paper on approximate calculations of look-elsewhere effect trial factors is a short and enlightening guide to the LEE in general.
- A venerable toolkit for machine learning in high-energy physics is the Toolkit for Multivariate Data Analysis (TMVA), part of the ROOT framework. Recent trends however, have increasingly emphasised non-HEP statistics and machine-learning platforms such as scikit-learn, Keras, TensorFlow and PyTorch. This software landscape is changing rapidly.
- A comprehensive comparison of global optimisation algorithms for particle physics applications can be found in 2021 A comparison of optimisation algorithms for high-dimensional particle and astrophysics applications *JHEP* **05** 108.

Exercises

5.1 A technical design report for a new proposed detector details the expected performance for electrons. The detector is known to give a positive result with a probability of 90% when applied to a particle that is actually an electron. It has a probability of 10% of giving a false positive result when applied to a particle that is not an electron. Assume that only 1% of the particles that will pass through the detector are electrons.

    (a) Calculate the probability that the test will give a positive result.

    (b) Calculate the probability that, given a negative result, a particle is not an electron.

    (c) Calculate the probability that, given a positive result, a particle is an electron.

    (d) Calculate the probability that a particle will be misclassified.

5.2 The Poisson pmf's random variable is the discrete k for fixed λ. Show that it is correctly normalised over all positive k. The $e^{-\lambda}$ term can be taken outside the sum over all k, which can then be identified as the Taylor expansion of e^{λ}. The product is therefore 1, equal to the total probability. A related question is what is the continuous probability distribution over λ, given a fixed (i.e. observed) value of k. Show that the same algebraic expression can be used for this, and is correctly normalised as an integral over $\lambda \in [0, \infty]$.

5.3 Prove that the effective sample size after reweighting is always less than or equal to the unweighted (or uniformly weighted) original. You may find it useful to reduce the problem to the case of two weights, equivalent to adding one more weighted event to an existing set.

5.4 A BSM physics model can in general reduce cross-sections (and hence event yields in search bins) as well as increase them, due to quantum interference. The yield itself is hence not monotonically related to whether an observation is more signal-like or background-like. How does the LLR approach turn this ambiguity into a well-defined ordering criterion for models which can either increase or decrease event yields?

5.5 A random variable x has a Gaussian pdf with mean 70 and standard deviation 30.1.

    (a) What is the probability that $x \leqslant 100$?

    (b) What is the probability that $x > 110$?

    (c) What is the expectation value of x^2 over this pdf, evaluated between 0 and 200?

5.6 A particular detector failure at the LHC is known to follow a Poisson distribution, with an expected rate λ of six failure events per year (and you can assume for this example that the detector is run under the same conditions each year).

    (a) What is the mean number of failure events expected in a year?

    (b) What is the probability of observing no failure events in a particular year?

    (c) What is the probability of observing more than 10 failure events in a particular year?

5.7 Assume that we have measured four variables, and our measured data points are $(x_0, x_1, x_2, x_4) = (3.1, 200.1, 42.6, 7.9)$, $(10.2, 168.3, 39.2, 9.1)$, $(5.7, 192.3, 23.4, 14.2)$, $(19.2, 170.8, 78.2, 131.9)$.

    (a) Calculate the covariance matrix.

    (b) Calculate the correlation matrix.

5.8 Verify the set of equations (5.142). Note that, in taking the derivative with respect to θ_i, k_i is a constant, but λ_i changes when we change θ_i.

5.9 A new experiment is designed to search for a new particle with a mass between 0 and 10 GeV, which should appear as a bump over a smoothly-falling background in the invariant mass distribution of the particle decay products. The background is well-described by the function:

$$m_{\gamma\gamma}^{\text{bkg}} = A(a + bm_{\gamma\gamma} + cm_{\gamma\gamma}^2) \tag{5.161}$$

and the signal is well-described by:

$$m_{\gamma\gamma}^{\text{sig}} = B \times \mathcal{N}(m_{\gamma\gamma}; \mu, \sigma) \tag{5.162}$$

where $\mathcal{N}$ is a Gaussian pdf with mean μ and standard deviation σ. The parameters a, b and c have been independently measured to be $(a, b, c) = (15, -1.2, 0.03)$.

A maximum-likelihood fit is performed, and the best-fit values of the parameters are found to be $(\tilde{A}, \tilde{B}, \tilde{\mu}, \tilde{\sigma}) = (167, 805, 4.63, 1.40)$. The value of $-\log \mathcal{L}$ is 10 at the best-fit point, and 20 at the point $(167, 0, 4.63, 1.40)$. Assuming that Wilks' theorem applies, is the hypothesis with $B = 0$, but the other parameters set to the MLE values, excluded at the 95% CL? What does this mean for the discovery of a new signal?

5.10 Compute the correlation coefficient between variables x and y for the one-to-one association $y = x^2$, with x uniformly distributed between -1 and 1. Compute the entropies of x and y, and their mutual information for the same scenario. Draw a Venn diagram of $H(x)$, $H(y)$, and $I(x, y)$.

5.11 An experimental colleague wants to know how good a variable p_T is at signal/background discrimination between 1 TeV to 4 TeV. If the expected signal-event density is given by $n_S = N_S(p_T/150\text{GeV})^{-4}$ and the expected background is $n_B = N_B \exp\{-p_T/100 \text{ GeV}\}$, what is the mutual information between the S/B discrimination and p_T?

5.12 Show that the 1σ range of the binomial-estimator $\tilde{\epsilon}$ is always in $[0, 1]$, but the 2σ range exceeds it.

Part II

Experimental physics at hadron colliders

IOP Publishing

Practical Collider Physics

Andy Buckley, Christopher White and Martin White

Chapter 6

Detecting and reconstructing particles at hadron colliders

Having described the fundamental constituents of Nature, and how they interact, we have not yet considered the details of the apparatus required to actually *observe* these particles and interactions. Indeed, you may be wondering how we ever worked out that the complex theory described in the previous chapters provides the correct description of Nature, and how we can design and build future experiments that will extend our knowledge of particle theory yet further.

The short answer is that every collider that smashes particles together (or into a fixed target) has at least one **particle detector** that detects the results of those collisions. The detector is composed of various *sub-detectors* that use a variety of different technologies to target specific sorts of object. Typically, a particle collider makes new particles that decay almost instantaneously to Standard Model (SM) particles, and it is these SM **decay products** that are observed in the detector, although we will see some important exceptions to this rule later. By exploiting the known effects caused by different types of particle as they pass through different materials, we can design and build detectors that tell us which particles were produced in a particular collision, and what the energy and momentum of those particles was. For experimentalists, building and testing the operation of detectors (known as 'detector commissioning') is a hugely important part of the job. For theorists, a basic knowledge of how particle detectors operate is no less essential, since it tells us what the limits of our measurements are, what precision is achievable in future experiments for testing new theoretical ideas, and even how to exploit existing detectors in new ways to discover novel interactions beyond those of the SM.

Modern circular collider facilities have a variety of detectors located around the ring. The detectors that we will take most interest in are the ATLAS and CMS detectors of the Large Hadron Collider, which are general-purpose machines for measuring SM processes, and searching for particles beyond those of the SM.

doi:10.1088/978-0-7503-2444-1ch6

ATLAS and CMS are roughly cylindrical, surrounding the beam pipe as shown in Figure 6.1. The **interaction point**, which marks the location where the protons or heavy ions of the LHC actually collide, is in the centre of the cylinder. To understand how these detectors operate, we will first describe the basic approach to measuring the outcome of particle collisions, before moving on to examine the detailed apparatus that allows ATLAS and CMS to perform particle measurements. Throughout, we will refer to the coordinate system that we introduced in Chapter 1.

6.1 Basic approach to particle detection and particle identification

Imagine for a moment that only one particle is passing through a detector from the interaction point. To fully characterise the particle, we would want to know what sort of particle it is, and what its energy and momentum are. Because the particle is moving relativistically, this comes down to knowing what its four-momentum is. The basic principles underlying the ATLAS and CMS detectors are the following:

- The central regions of the detector measure the momentum of charged particles by measuring the tracks that the particles deposit in various media as they pass through. The application of a strong magnetic field bends the trajectory of charged particles, and the radius of curvature of the track is related to the particle momentum. For obvious reasons, the central detectors are called **tracking detectors**.
- The outer regions of the detector are designed to measure the energy of particles, by stopping them and detecting the energy that is released as they stop. Different layers of detector stop different kinds of particle, so that we can get some form of particle identification by looking at where a particle stopped. These energy-measuring detectors are called **calorimeters**.
- Muons are not stopped by the calorimeters, and instead receive further measurements by special **muon detectors** that are located outside the other layers.

We can see this more clearly by cutting a slice through the detector, giving something like Figure 6.2.

Upon leaving the interaction point, our hypothetical particle first traverses the tracking detector, before entering the **electromagnetic calorimeter**, which is designed to stop electrons, positrons and photons. An electron, positron or photon would thus normally stop here, although as one might expect the system is not perfect. Electrons, positrons and photons that escape the electromagnetic calorimeter are

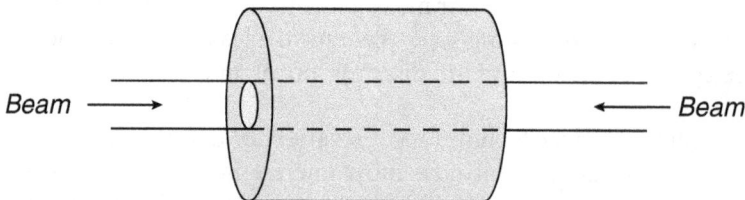

Figure 6.1. Schematic side-on view of a detector in a colliding-beam experiment.

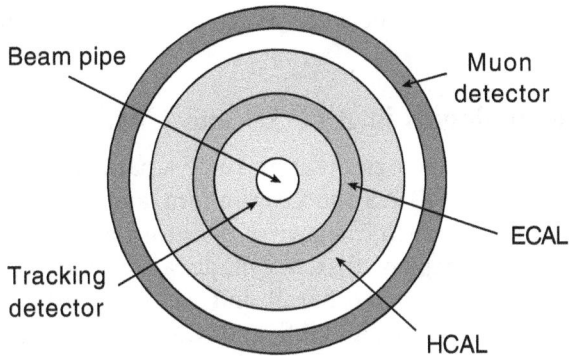

Figure 6.2. Schematic cross-sectional view of a detector in a colliding-beam experiment.

said to **punch-through**, and you will often see 'punch-through' used as a noun to describe the phenomenon of particles escaping beyond their expected final destination. If our initial particle was a quark or gluon, it initiates a jet of particles, which is expected to stop in the next layer of the detector, known as the **hadronic calorimeter** or HCAL. If the particle is a muon, it passes through both calorimeters without interacting much, and leaves further signatures in the outer muon chambers.

The different behaviours of electrons, positrons, photons, jets and muons as they leave the interaction point allows us to tell these particles apart. A photon will not leave a track in the tracker because it is electrically neutral, and it will stop in the electromagnetic calorimeter. An electron or positron will also stop in the electromagnetic calorimeter, but it will also leave a track in the tracking detector. A quark or gluon will initiate a jet that leaves a bunch of tracks close together in the tracker, before depositing most of its energy in the hadronic calorimeter. A muon, meanwhile, will leave tracks in the tracker and the muon chambers, and the careful matching of these tracks allows us to reconstruct the muon four-momentum.

Referring back to the table of SM particles, you may be wondering why we have not yet described the behaviour of neutrinos or tau leptons in the ATLAS and CMS detectors. The answer is that both of these give yet further signatures in the detector. Tau leptons decay promptly to lighter leptons, neutrinos and/or hadrons, and leave characteristic patterns that one may search for specifically in the detector. Neutrinos, meanwhile, are weakly-interacting, and pass through the detector completely unseen. Nevertheless, we are able to at least determine that neutrino production is likely to have occurred by checking the amount of 'missing energy' in an event, as described in Section 6.4.4.

Clearly a detector that is capable of characterising individual particles escaping the interaction point is also capable of characterising multiple particles that escape at the same time. In that case, a key parameter of the detector is the number of readout channels, which is related to the granularity of the detector. If this is not large enough, multiple particles will overlap in the detector components, and we will not be able to identify them as separate particles. As colliders move to higher luminosity, the average number of particles passing through the detector will increase, which necessitates new detector designs with a higher number of readout channels.

6.2 Detector technologies in ATLAS and CMS

Let us now consider the technologies that make up a particle detector. We will pass through each subsystem of ATLAS and CMS in turn, and compare the technologies that were used to implement each category of detector that we defined above. Both detectors feature a **barrel region**, that is cylindrical around the beam pipe, and **forward detectors** that bookend the barrel and measure particles in the large $|\eta|$ region. In physics results from ATLAS and CMS, you will often see object definitions restricted to the barrel region, since this is the region in which the precision of the reconstructed particle four-momenta will be highest. Each detector also comes with a **trigger** system, which is designed to only accept events of sufficient physics interest for later analysis. This is essential, since the very high frequency of LHC collisions means that one cannot record every event, and nor could one search through a massive stack of irrelevant events to find the very rare processes of interest.

6.2.1 Magnet systems

A crucial part of the design of the ATLAS and CMS detectors is the magnet systems that supply the magnetic fields that are so essential for accurate particle tracking. This is exemplified by considering the origins of the ATLAS and CMS acronyms, which derive from *A Toroidal LHC ApparatuS*, and the *Compact Muon Solenoid*, respectively.

The Lorentz force law states that the magnetic force on a particle of charge q is given by

$$\mathcal{F}_M = qv \times \boldsymbol{B}, \tag{6.1}$$

where v is the particle's velocity, and $\boldsymbol{B}$ the magnetic field. This is perpendicular to the particle's direction of motion (as represented by v). If $\boldsymbol{B}$ is also perpendicular to v, the particle will move in a circular trajectory, whose radius r is related to the particle's momentum p by

$$p = Bqr. \tag{6.2}$$

The effect of the applied magnetic field in an LHC detector is thus to cause a charged particle to move in a circle in the tranverse plane. For a particle of fixed momentum p, the radius of curvature is inversely proportional to the strength of the applied magnetic field B. To maximise the precision of the momentum measurement, we want the track to bend by a large amount, which corresponds to a small radius of curvature, which in turn requires us to apply a very large B-field.

ATLAS contains two magnet sub-systems: a central solenoid that surrounds the inner detector, and a toroid system, comprising one barrel toroid and two endcap toroids, that generates a magnetic field for the muon spectrometer. The central solenoid provides a field of 2T (along the beam axis) for the inner detector, and the coil is designed to be as thin as possible, without compromising performance and reliability, in order to reduce the amount of material obscuring the calorimeters. Thus, despite a diameter of 2.5 m and a length of 5.3 m, the thickness is a mere 45 mm. The system is designed to allow accurate measurements of charged particles in the inner detector up to a momentum of 100 GeV. Each of the three toroids consists of eight coils assembled radially around the beam axis, with the 8 'race track' magnets of the barrel toroid mimicked by smaller versions in each of the two endcaps. The immense size of the barrel toroids gives rise to one of the most distinctive visual features of the detector, as the magnets are a staggering 25 m long and 5 m wide. The endcap toroids are a more modest 5 m in length, and are rotated by 22.5° with respect to the barrel toroid coil system in order to provide radial overlap and to improve bending power in the interface regions of the coil systems.

The CMS magnet system differs in a variety of ways from the ATLAS design. Firstly, the central solenoid magnet is stronger (3.8 T), and it encloses the hadronic and electromagnetic calorimeters in addition to the CMS inner detector. This requires a 12 000 tonne solenoid magnet, which is 13 m long, and 6 m in diameter. Rather than having a toroid system for the muon chambers, CMS has a 'return yoke' which is a three-layer, 12-sided iron structure that reaches a maximum diameter of 14 m, and is interleaved with the muon chambers. It contains and guides the field of the solenoid, whilst also acting as a shield for everything except muons and neutrinos.

6.2.2 Tracking detectors

Fundamentally, a tracking detector must register the passage of a charged particle through its medium, via the release and subsequent readout of some sort of signature. By building a detector out of a series of distinct modules, each located at a given spatial position, one can measure a series of 'hits' along the track of a moving particle, which can be used to reconstruct the direction and sign of curvature of the track. This gives us both the particle momentum, and the sign of the particle charge. By using a high granularity of modules near the beam pipe, one can also resolve whether particles were genuinely produced at the interaction point (indicating a very prompt decay of some parent particle), or whether particles travelled a short distance before decaying. The latter feature is expected in the case of jets that were originated by b-quarks, which gives us yet another technique for particle identification.

The ATLAS and CMS tracking detectors have different designs, but share some similarities in their choice of technology. The CMS inner detector is entirely made from modules of silicon, and the first two sub-components of the ATLAS tracker are also made from silicon. Silicon is an example of a semiconductor, in which the there is a small gap (3.6 eV) between the highest filled energy level (the 'valence band') and the next available energy level (the 'conduction band'). A charged particle passing through the semiconductor will release electron–hole pairs which, under the influence of an applied electric field, will drift toward opposite electrodes. However, in pure silicon, there are many more free charge-carriers than those produced by the passage of a charged particle, and the electron–hole pairs quickly recombine. Hence one must find a way to deplete the material of charge carriers before one can usefully apply the technique to measure the position of a particle.

The solution lies in doping. In an 'n-type' semiconductor, donor ions from Group V of the periodic table are added, which introduce energy levels close to the lower end of the conduction band, thus creating a surplus of electrons in the material. In a 'p-type' semiconductor, acceptor ions of group III are added that introduce energy levels close to the top of the valence band, which absorb electrons from the valence band and create a surplus of holes. When these two types of doped semiconductor are brought together, a gradient of electron and hole densities results in the diffuse migration of majority carriers across the junction. The ionised donors now have positive charge, whilst the ionised acceptors acquire negative charge, and the interface region becomes depleted of carriers. There is a potential difference across this 'depletion region', which can be increased through application of an electric field (adding a 'reverse-bias voltage'). This increases the width of the depletion region. Any electron–hole pairs produced by a charged particle passing through the region will drift along the field lines to opposite ends of the junction, and if the p–n junction is made at the surface of a silicon wafer, a prototype silicon detector is obtained. Charge is collected on the surface of the detector, and is amplified before readout.

The density of tracks in ATLAS and CMS is highest nearest the beampipe, since all particles that pass through the full radius of the detector must pass through a much smaller surface area at the first layer of the detector. Accurate hit reconstruction in this region, which is essential for vertex resolution and track reconstruction, thus requires a very high granularity of detector modules nearest the beam pipe. For this reason, both the ATLAS and CMS detectors use silicon pixel detectors for the first few layers of the tracker. The CMS pixel detector has well over 60 million pixels, distributed between layers of a barrel in the middle of the detector, and disks of the endcap at either side. ATLAS also has a barrel and endcap structure, but with far fewer pixels than CMS.

At larger radial distances from the beampipe, the flux of particles drops sufficiently for strip detectors to be a useful replacement for pixel detectors, which brings a substantial cost saving. These comprise modules that contain a series of silicon strips, plus readout chips that each read out a fixed number of strips. One can therefore localise a track hit to at least the location of the strip that fired. The CMS strip detector once again has a barrel and two endcaps, and is equipped with special modules that have two silicon sensors mounted back-to-back, with their strips misaligned at a 100 mrad relative angle. This allows for a 2D measurement of the position of a hit in the plane of the module (rather than a 1D measurement based on which strip fired). The ATLAS strip detector, called the SemiConductor Tracker (SCT) is also arranged into barrel layers and endcap disks, with modules again containing two pairs of wafers that are placed back-to-back to allow 2D hit reconstruction (albeit with a different misalignment angle).

A difference between ATLAS and CMS is that ATLAS has much less silicon. The reason is that ATLAS has another component of the inner detector, which is not replicated in CMS, called the Transition Radiation Tracker (TRT). This is a drift tube system, with a design that incorporates 'straw' detectors. Each of these is a small cylindrical chamber filled with a gas mixture of Xe, CO_2 and O_2, in which the aluminium coated inner wall acts a cathode whilst a central gold-plated tungsten wire acts as an anode. Charged particles passing through ionise the gas, and the resulting ionisation cluster is amplified by a factor of $\approx 2.5 \times 10^4$ whilst drifting through the electric field in the straw. The wires are split in half at the centre and read out at each end, and each channel provides a drift time measurement.

The space between the straws is filled with a polypropylene/polyethylene fibre radiator which increases the number of transition-radiation photons produced in the detector. These are produced when relativistic particles cross a boundary between materials with different dielectric constants, and the threshold above which radiation is produced is dependent on the relativistic factor $\gamma = (1 - v^2/c^2)^{-1/2}$, where v is the particle velocity, and c is the speed of light. The xenon in the straw gas presents a high interaction cross-section to these photons and a signal is produced which has a higher amplitude than the signal arising from minimally ionising particles. There are thus two different categories of signal that one wishes to detect in each straw, and for this reason each channel has two independent thresholds. The lower threshold detects the tracking hits, and the higher threshold is designed for the transition-radiation photons. This higher threshold aids particle identification, as, for example, electrons start producing transition radiation when their momentum is close to 1 GeV, whilst pions start to radiate only when their momentum is close to 100 GeV. At ATLAS, the pion rejection factor is expected to be ≈ 100 for an electron efficiency of 90%.

The ATLAS TRT is intrinsically radiation-hard, and provides a large number of measurements at relatively low cost. A barrel region covers $|\eta| < 0.7$ and the endcap extends coverage to $|\eta| = 2.5$.

6.2.3 Calorimeters

The primary job of the ATLAS and CMS calorimeters is to stop electrons, positrons, photons and jets and measure the energy that was released as they stopped. They also play an important role in determining the position of particles, measuring the missing transverse momentum per event, identifiying particles and selecting events at the trigger level.

High-energy particles entering a calorimeter produce a cascade of secondary particles known as a shower (not to be confused with the parton showers discussed in Section 3.3). The incoming particle interacts via either the electromagnetic or strong interaction to produce new particles of lower energy which react in a similar fashion, producing very large numbers of particles whose energy is deposited and measured. A calorimeter can either be built only from the material that is used to induce the shower, in which case it is called a **homogeneous calorimeter**, or it can use separate materials for inducing the shower and for detecting the energy emitted by particles, in which case it is called a **sampling calorimeter**. A sampling calorimeter thus consists of plates of dense, passive material, alternating with layers of sensitive material. The thickness of the passive layers (in units of the radiation length) determines the number of times the layers can be used, and thus the number of times that a high-energy electron or photon shower is sampled. A thinner passive layer gives a higher sampling frequency, and thus a better resolution. For a fixed containment and signal-collection efficiency, a homogeneous calorimeter will typically give a better energy resolution, because the sampling of the energy in a sampling calorimeter results in a loss of information, and hence in additional statistical fluctuations in the measurement of the shower energy. The ATLAS electromagnetic and hadronic calorimeters and the CMS hadronic calorimeter are all sampling calorimeters, whilst the CMS electromagnetic calorimeter is homogeneous.

A very important property of a calorimeter is the energy resolution, since this will set the precision with which one can reconstruct and manipulate particle four-vectors. The measured energy in a calorimeter will fluctuate around the 'true' energy of the shower, due to effects such as natural fluctuations in the development of the shower, quantum fluctuations in the signal (known as 'photo-electron statistics'), instrumental noise, and, in the case of sampling calorimeters, leakage of the shower from the active detector medium. For both homogeneous and sampling calorimeters, the relative energy resolution is often quoted as

$$\frac{\sigma(E)}{E} = \frac{a}{\sqrt{E}} \oplus \frac{b}{E} \oplus c\%, \tag{6.3}$$

where $\oplus$ denotes addition in quadrature. The a term is known as the **stochastic term**, and it arises from fluctuations in the number of signal generating processes, such as the number of photo-electrons generated in the detector. The b term arises from noise in the readout electronics, plus pile-up. The constant c term arises from imperfections in the calorimeter construction, non-uniformities in the detector response, channel-to-channel intercalibration errors, and losses of components of the shower (either through dead material before and in the detector, or fluctuations in the longitudinal containment of the shower). Good energy resolution at the highest particle energies requires a small constant term, since it will dominate in the limit of large E.

Electromagnetic calorimeters
The interaction with matter for electrons and photons is different from that of hadrons. Electrons and photons penetrate much less deeply than hadrons, and produce narrower showers. Energy loss occurs predominantly via bremsstrahlung[1] for high-energy electrons (which for most materials means energies greater than ≈ 10 MeV), and high-energy photons lose energy via the related process of e^+e^- pair production. The characteristic amount of matter traversed by a particle before undergoing one of these interactions is described by the **radiation length** X_0, and is entirely set by the properties of the material being traversed. The expectation value of the energy of an electron $E(x)$ as a function of the distance x into the material is given by

---

[1] Bremsstrahlung (which translates as 'braking radiation') is produced by the acceleration of a charged particle after deflection by another charged particle.

$$\langle E(x) \rangle = E_0 \exp^{-\frac{x}{X_0}}, \tag{6.4}$$

where E_0 is the incident energy of the electron. We can write a similar equation for the intensity of a photon beam entering the material (where I_0 is the incident intensity):

$$\langle I(x) \rangle = I_0 \exp^{-\frac{7}{9}\frac{x}{X_0}}. \tag{6.5}$$

The CMS electromagnetic (EM) calorimeter is a homogeneous calorimeter made out of ≈ 70000 lead tungstate ($PbWO_4$) crystals. The development of an electromagnetic shower releases scintillation light within the crystals which is then detected to measure the shower energy. Lead tungsten has high density (8.28 g cm^3), a small radiation length ($X_0 = 0.89$ cm), and a small **Molière radius** ($r_M = 2.19$ cm), which describes the radius of the cylinder that contains 90% of the shower. 60000 of the $PbWO_4$ crystals are used in the barrel, which covers the region up to $|\eta| < 1.48$, while separate endcap detectors extend the longitudinal angular coverage up to $|\eta| = 3$. In order to distinguish neutral pions from photons, a special 'preshower' detector is placed in front of the electromagnetic calorimeter endcap between $\eta = 1.653$ and $\eta = 2.6$. This sub-detector contains lead interleaved with silicon strips. The energy resolution for the CMS EM calorimeter barrel reaches 1.1% for a non-converted photon with an energy of 60 GeV. One should note, however, that the constant term varies between 0.3% and 0.5% in the barrel, and between 1% and 1.5% in the endcap.

The ATLAS EM calorimeter is a sampling calorimeter that uses lead and stainless steel to induce showers, and liquid argon as the sampling material. This LAr sampling technique is radiation resistant and combines good energy resolution with other attractive features such as long-term stability of the detector response and relatively easy detector calibration. The geometry of the detector describes a complex accordion shape, chosen to provide complete ϕ symmetry with no azimuthal cracks. The calorimeter has a barrel region (covering the range $|\eta| < 1.475$) and endcap regions (covering the range $1.375 < |\eta| < 3.2$), and the lead thickness in the absorber plates has been optimised as a function of $|\eta|$ to maximise the energy resolution performance. No discussion of the ATLAS ECAL is complete without mention of two points relevant to physics analysis: there is a crack in the coverage at $|\eta| = 1.5$ due to the barrel/endcap transition, and a small crack at $|\eta| = 0$ arising from the fact that the barrel calorimeter is constructed from two identical half-barrels separated by 6 mm at $z = 0$. The ECAL is **non-compensating**; that is, it responds differently to electromagnetic and hadronic showers. This difference is corrected for in the reconstruction software. We have already seen that the inner detector sits between the interaction point and the calorimeters, and the total material seen by an incident particle before it reaches the front face of the calorimeter is $\approx 2.3X_0$ at $\eta = 0$. This amount increases with η because of the angle of the particle trajectory. A presampler, consisting of an active LAr layer of thicknesses 1.1 cm and 0.5 cm in the barrel and endcap regions, respectively, is used to correct for the energy lost by electrons and photons *en route* to the electromagnetic calorimeter in the region $|\eta| < 1.8$. In the transition region between the barrel and endcap, this is supplemented by a scintillator slab, as the amount of obscuring material reaches a localised maximum of $\approx 7X_0$. The energy resolution of the ATLAS EM calorimeter is dominated by sampling fluctuations, and is close to $\approx 0.10 \text{ GeV}^{1/2}/\sqrt{E/1 \text{ GeV}}$ for the entire range of η coverage.

Hadronic calorimeters

Hadrons interact with the nuclei of the calorimeter material via the strong force, and the more complex development of hadronic showers relative to EM showers leads to a number of distinguishing features. Firstly, the fraction of detectable energy from a hadronic shower is lower than that from an EM shower, which leads to an intrinsically worse energy resolution for hadronic species relative to electrons and photons. Secondly, hadronic showers are longer than EM showers, since they are characterised by a nuclear interaction length λ which is typically an order of magnitude greater than X_0. This length is a function of both the energy and type of incoming particle, since it depends on the inelastic cross-section for nuclear scattering. Longitudinal energy deposition profiles have a maximum at:

$$x \approx 0.2 \ln(E_0/1 \text{ GeV}) + 0.7, \tag{6.6}$$

where x is the depth into the material in units of the interaction length λ and E_0 is the energy of the incident particle. The depth required for containment of a fixed fraction of the incident particle energy is also logarithmically dependent on E_0. This ultimately means that hadronic calorimeters have to be larger than EM calorimeters, and the increased width of hadronic showers relative to EM showers means that their granularity is typically coarser than that of EM calorimeters. The deposited energy in a hadronic cascade consists of a prompt EM component due to π^0 production, followed by a slower component due to low-energy hadronic activity. These two different types of energy deposition are usually converted to electrical signals with different efficiencies, the ratio of which is known as the intrinsic *e/h* ratio.

The CMS hadronic calorimeter is a sampling calorimeter divided into four sub-components, with most it composed of alternating layers of brass absorbers and plastic scintillators. The layers form towers of fixed size in (η, ϕ) space. The hadronic calorimeter barrel sits between the electromagnetic calorimeter barrel and the magnet coil, and is sufficiently small that hadronic showers might not be entirely contained within it. For this reason, a hadronic calorimeter outer barrel detector is placed outside the magnet coil, and the two barrel detectors together cover the region up to $|\eta| = 1.4$. Endcap detectors cover the range $1.3 < |\eta| < 3.0$, and a fourth sub-detector extends coverage to large values of $|\eta|$. For pions with an energy of 20 GeV (300 GeV), the energy resolution has been measured to be about 27% (10%).

The ATLAS hadronic calorimeters are also sampling calorimeters, and they cover the range $|\eta| < 4.9$. A tile calorimeter is used for $|\eta| < 1.7$, using iron as the absorber and scintillating tiles as the active material. The system consists of one barrel and two extended barrels, and is longitudinally segmented in three layers. For $\approx 1.5 < |\eta| < 4.9$, LAr calorimeters (similar to those in the ECAL) are used, with the system comprising of a Hadronic End-Cap calorimeter (HEC) extending to $|\eta| < 3.2$ and a high-density forward calorimeter (FCAL) covering the range $3.1 < |\eta| < 4.9$. The FCAL is divided into three sections, one of which uses copper (as does the HEC) and two of which use tungsten for its higher density. The ATLAS hadronic calorimeter, like the ATLAS electromagnetic calorimeter, is non-compensating. The resolution for the full ATLAS calorimeter (i.e. the electromagnetic and hadronic calorimeters together) is $\sigma/E = (48.2 \pm 0.9\%)/\sqrt{E} \oplus (1.8 \pm 0.1\%)$ for $|\eta| = 0.3$ in the barrel region, and $\sigma/E = (55.0 \pm 2.5\%)/\sqrt{E} \oplus (2.2 \pm 0.2\%)$ for $|\eta| = 2.45$ in the endcap region.

6.2.4 Muon chambers

The outermost parts of the ATLAS and CMS detectors are both comprised of muon chambers, whose main functions are to precisely measure the momentum of high-energy muons that are not measured well by the inner detector, and to assist the trigger system in recording muon events.

Precise muon momentum measurements are performed in the ATLAS detector by **monitored drift tubes** (MDTs) over most of the η range, and **cathode strip chambers** (CSCs) at large η and at close proximity to the interaction point. These detectors are all referred to as the 'precision chambers'. Triggering is assisted by the use of **resistive plate chambers** (RPCs) in the barrel, and **thin-gap chambers** (TGCs) in the endcaps. Triggering requires less precise position data, but a response time better than the LHC bunch spacing of 25 ns, hence the use of a separate technology for the purpose. The trigger system covers the pseudo-rapidity range $|\eta| \leqslant 2.4$.

The MDTs are aluminium tubes with central tungsten–rhenium (W–Re) wires, arranged into chambers that contain multilayer pairs of either three or four mono-layers of tubes. The CSCs are fast multiwire proportional chambers, with parallel high-voltage wires strung in a gas volume and closed by conducting planes at 0 V. Precise position measurements along the wires are obtained by determining the centre of gravity of the charge induced on the strips of the conducting plates. Position resolutions of better than 60 μm are able to be obtained, and the high granularity of the system ensures that it copes with the demanding rate and background conditions at high $|\eta|$. Over most of the muon spectrometer, the magnetic field is essentially in the ϕ direction, and thus muons bend in the R–z plane. The MDTs and CSCs provide precision measurements in this plane.

The RPCs of the trigger system are made of a pair of parallel plates separated by a narrow gas gap, with a high electric field applied. Muons passing through release ionisation electrons, and these form an avalanche between the plates whose signal is read out by two sets of strips in orthogonal directions. The measurements are used both to complement the MDT tracking data (the RPC measurement is in an orthogonal direction), and for the level-1 trigger. The TGCs in the endcaps are similar in design to multi-wire proportional chambers, though the anode wire pitch is larger. The muon spectrometer typically gives three precision measurements for $|\eta| < 2.7$, except in crack positions (such as at $|\eta| = 0$). A 20 GeV muon yields a momentum measurement to a precision of 2%, with good acceptance. This rises to 10% as the energy approaches 1 TeV.

The CMS muon system utilises similar technologies to ATLAS. A central part comprised of drift tubes covers the region $|\eta| < 1.2$, whilst a forward part covering $0.9 < |\eta| < 2.4$ uses four layers of cathode strip chambers. Resistive plate chambers cover the region up to $|\eta| < 1.6$. The p_T resolution obtained by the CMS muon system is better than 10% even for muons with $p_T > 2000 \, \text{GeV}$.

6.2.5 ATLAS and CMS upgrades

The detectors described above are the ATLAS and CMS detectors as originally designed and installed for the first two runs of the LHC, which were completed in 2018. For a variety of reasons, these detectors must be upgraded to maintain performance for future runs of the LHC. For example, some detectors are vulnerable to radiation damage, and must be replaced once the performance loss exceeds a critical value. In addition, the LHC is scheduled to run in a high-luminosity mode from 2026, in which collisions will occur between five and ten times more frequently than in earlier runs. This will lead to a significant increase in the flux of particles passing through the detectors, which will then need a finer granularity in order to avoid a loss of resolution due to particles overlapping in the detector components. In the LHC schedule, there are various **long shutdown** periods between future runs in which key detector upgrade and maintenance operations can be carried out.

Upgrade plans for the CMS detector include replacement of the innermost layer of the present pixel detector, using more radiation-tolerant components. The beam pipe is also due to be replaced, in order to allow the edges of future CMS pixel detectors to get even closer to the interaction point. A third generation pixel detector will be installed in the third long shutdown period, and will thus be in place before the high-luminosity LHC commences operation in 2026. At the same time, the existing endcap electromagnetic and hadronic calorimeters will be replaced by a new **high-granularity calorimeter** comprised of 100 layers of hexagonal silicon sensors and plastic scintillator tiles (50 in each endcap). The number of readout channels poses serious problems for software and analysis, and new computational techniques wil need to be employed to process the data. Finally, there is a plan to install 40 large, multi-gas electron multiplier chambers (GEMs) in the inner ring of the first endcap disk, which will be able to measure muons that scatter at an angle of around 10°. GEM chambers are made of a thin polymer foil clad in metal, which is then chemically pierced with 50 to 100 holes per millimetre, and submerged in a gas. Passing muons release electrons from the gas, which drift into the holes before multiplying in a strong electric field and passing to a collection region. Outside of the detector subsystems themselves, the readout electronics are also being substantially upgraded to cope with up to 200 collisions per bunch-crossing.

The ATLAS experiment has similarly ambitious plans to meet future challenges. Some of the endcap muon detectors will be replaced, with the new detectors capable of maintaining low lepton trigger p_T thresholds as the instantaneous luminosity ramps up. The trigger system also requires a hardware upgrade to meet the demands of the high-luminosity LHC, and the inner tracker will eventually be replaced by a new system entirely made of silicon pixel and strip detectors, known as the ITk.

6.3 Triggers

The number of interactions at the LHC is very large indeed, and a majority of these are of little interest to those interested in the frontier of particle physics. Thus, the ATLAS and CMS detectors do not record every interaction but instead are designed to trigger on interesting processes. Limitations on the amount of data that can be stored require an initial bunch-crossing rate of 40 MHz to be reduced to a rate of selected events of 100 Hz, and the challenge is to do this without missing any of the rare new physics processes that motivate the entire experiment. As run conditions evolve at the LHC, further improvements need to be made to the trigger systems to ensure successful data taking.

6.3.1 The ATLAS and CMS triggers

The ATLAS trigger and data-acquisition system (DAQ) is based on three levels of online event selection, with each trigger level refining the decision made at the previous level through the use of additional selection criteria. The process starts with the Level-1 (LVL1) trigger, which is a hardware-based system consisting of a Central Trigger Processor fed by signals that mostly come from dedicated hardware in the calorimeter and muon systems. The system has access to regions of interest in the η–ϕ plane, and identifies objects that resemble muons, electromagnetic clusters, jets or taus. Also available are global sums of the missing transverse energy and the total energy. The time from a proton–proton collision until the availability of the LVL1 trigger decision at the front-end electronics must not exceed 2.5 μs. During this time, information for all detector channels is stored in 'pipeline' memories placed on or close to the detector in harsh radiation environments. The pipeline lengths must be kept as short as possible for reasons of cost and reliability and hence this imposes constraints on the LVL1 latency. One must also consider the time required for signals to leave the detector via cabling.

The LVL1 trigger reduces the event rate from 40 MHz to $\approx$ 100 kHz. Accepted events are then passed to a software-based High Level Trigger (HLT), which uses offline-like reconstruction algorithms on a large farm of processors to reach a refined decision within 300 ms. During Run I of the LHC, the HLT consisted of separate farms, called 'Level 2' and the 'Event Filter'. During the second run of the LHC, these were merged into a single farm to allow better sharing of resources and an overall simplification of the trigger hardware and software. The HLT performs reconstruction in the Level-1 regions of interest, followed by higher-precision reconstruction of full events. Accepted events are written to separate data streams, depending on whether the event is to be used for physics analysis, detector calibration or detector monitoring. Depending on the stream, one does not have to write the full event information, which allows a reduction of the total information flow that can in turn be allowed to increase the total trigger rate.

The CMS trigger utilises two levels. Similar to the ATLAS trigger, the first is a hardware-based Level-1 trigger that selects events containing candidate objects such as detector patterns consistent with a muon, or calorimeter clusters consistent with electrons, photons, taus, jets or total energy (related to $E_{\mathrm{T}}^{\mathrm{miss}}$). Selected events are passed to a software-based high-level trigger via a programmable menu that contains algorithms that utilise the Level-1 candidate objects.

6.3.2 Trigger menus

The various possible trigger outcomes are enshrined in the **trigger menu** for each experiment, which ultimately states what sort of events are going to be recorded by the ATLAS and CMS detectors. Each item in the trigger menu consists of a Level-1 item connected to a high-level trigger chain, connected to a data stream, and the items cover those needed for physics measurements (or searches for new particles), support triggers used for efficiency or performance measurements, alternative online triggers that allow one to use more than one method for triggering on specific objects, backup triggers with tighter selections that can cope with a change in the instantaneous run conditions and calibration triggers. Each physics and combined performance group must fight for their bandwidth in the total event rate of the trigger menu, and it is up to groups to carefully design trigger menus that meet their specific requirements and bandwidth allocation.

6.3.3 Prescaling

The very high rate of events at the LHC necessitates having relatively high thresholds on the Level-1 trigger objects in order to obtain a suitable reduction in the rate of events entering the high-level trigger. As the instantaneous luminosity and centre-of-mass energy of the LHC increase, these thresholds must rise still further to cope with the increased throughput. This has a major impact on physics analyses, since the trigger requirements set the minimum transverse momentum of objects of interest that can be probed by an analysis. For this reason searches for lower mass objects that decay to objects with a relatively small transverse momentum are often better covered by the earlier 8 TeV run of the LHC, than by the later, and much more abundant, 13 TeV data.

For some processes that we wish to trigger on, it is not possible to perform the required analysis with stringent requirements on the relevant object transverse momenta. Examples include detector calibration and monitoring, various studies in the ATLAS and CMS b-physics programmes, and the study of 'minimum-bias events' comprised of both non-diffractive and diffractive inelastic-quantum chromodynamics (QCD) processes, and which must be studied using a very loose trigger selection. To trigger on the relevant events, **trigger prescale** factors are used to reduce the event rate, so as not to saturate the allowed bandwidth. For a prescale factor of N, only one event out of N events which fulfill the trigger requirement will be accepted. Individual prescale factors can be applied to each level of a trigger chain, as required.

6.4 Particle reconstruction

We have now seen how the outcome of an interesting proton–proton collision at the LHC can be observed in the ATLAS and CMS detectors, by producing a range of signatures in different detector subsystems. At this stage, however, we have only learned about the raw signatures that are produced by particles as they traverse the detector. How do we take the electronic signatures of the trackers, calorimeters and muon chambers, and use them to reconstruct particle types, charges and four-momenta? This is accomplished using the ATLAS and CMS **reconstruction** algorithms, and in this section we will explore further exactly how these algorithms identify each sort of physics object. Although the approach is broadly similar between ATLAS and CMS, there are some important differences which arise from the different choices in detector technology.

Typically, particles of each type can be identified by observing where they leave deposits in the various detector subsystems, and it is important to combine information from the different subdetectors to accurately identify particle types. In CMS, and more nascently in ATLAS, this is performed using a **particle-flow** algorithm which takes as input various fundamental 'elements', such as inner detector tracks, electromagnetic calorimeter clusters, hadronic calorimeter clusters and muon chamber tracks. The following different pairings of elements are then considered:

- inner detector track, electromagnetic calorimeter cluster;
- inner detector track, hadronic calorimeter cluster;
- electromagnetic calorimeter cluster, hadronic calorimeter cluster;
- inner detector track, muon track.

If a match is found, the elements involved are grouped into a 'block' which consists of at least two elements (allowing for the possibility of multiple matches between disparate pairs). Particle reconstruction then proceeds by using the blocks to resolve the most distinguishable objects first, then removing all elements associated with that object. For example, muons are the easiest particles to distinguish, because they are the only particle that will leave deposits right up to the outermost regions of the detector. After removing the elements that corresponded to the reconstructed muons, electrons are identified and reconstructed, followed by charged hadrons, photons and neutral hadrons.

Up to the end of Run 2, the ATLAS reconstruction software performed particle identification without using particle-flow methods, but still with a combination of information from all detector subsystems. Particle flow will become the default for Run 3 data processing.

In the following, we give more specific information for each particle species. It is helpful to remember that all reconstructed particles are to some extent defined objects. The electron that one refers to as the outcome of a reconstruction chain is not the same thing as the electron that we refer to as a particle of the SM, being instead a defined collection of detector signatures that is designed to most closely resemble those instigated by a physical electron. This is even clearer in the case of **jets**, which do not formally exist as particles of the SM, rather as a phenomenon that results from the complicated interactions of gluons and quarks. For this reason, one can obtain **fake particles** from detector reconstruction, which are particles that resemble a different object from the one that originated the detector signatures. Examples include electrons faking jets, and charged pions faking electrons. The selections used to define each type of object are carefully designed to reduce these fake contributions as much as possible.

6.4.1 Electrons and photons

Broadly speaking, electrons and photons can be identified by the presence of a shower in the electromagnetic calorimeter, although the showers of electrons and photons cannot be distinguished since brehmsstrahlung causes electrons to emit photons, whilst pair-production causes photons to produce electron–positron pairs. Thankfully, the tracker responds differently to photons and electrons, since photons are electrically neutral and do not leave tracks, whilst electrons and positrons will produce tracks that point to their calorimeter deposits.

ATLAS electron reconstruction starts with the identification of calorimeter **shower seeds** using the **sliding-window algorithm**. This first constructs fixed-size clusters of dimension 3×5 cells, where the size is fixed by the middle electromagnetic calorimeter layer (which is the largest). Deposits within all three layers of the electromagnetic calorimeter within this window are summed to get the total transverse energy deposit within the window. The window position is then moved until the total transverse energy is maximised, after which regions are marked as seeds if their total energy exceeds a threshold that has been chosen carefully to optimise the reconstruction efficiency whilst keeping fake signals and pile-up noise to a minimum.

Seed clusters are then compared with the inner detector tracks, to put them in one of three categories:

- seeds are classified as electrons if they match a track that points to the primary vertex in the event;
- seeds are classified as converted photons if they match a track that points to a secondary vertex;
- seeds are classified as unconverted photons if they do not match any tracks.

Electron and converted photon clusters are rebuilt with a size of 3×7 central layer cells on the barrel, whilst unconverted photons, which typically have narrower showers, are rebuilt with 3×5 central layer cells. In the endcap region, all categories are built using regions of dimension 5×5 cells.

Photons have relatively few processes that can fake them, since the simple criterion of having a set of electromagnetic calorimeter deposits that do not match a track from the primary vertex turns out to be very stringent. Electrons can be faked by more processes, and a particularly prevalent one is a jet faking an electron. Although it is very rare to have a jet which is collimated enough to resemble an electron, and entirely contained within the electromagnetic calorimeter, an enormous number of jets are produced at the LHC, and thus one obtains a reasonable fake rate once one combines these two facts. It is therefore typical to define several different categories of electrons called things like **loose**, **medium** and **tight**, with the exact definitions depending on the current run period and version of the ATLAS reconstruction software. Moving from loose to tight, the electron definitions are set to be more stringent to reduce the fake jet background for electron identification, at the cost of reducing the overall electron reconstruction efficiency. This trade-off between **purity** and **efficiency** is characteristic of all attempts to reconstruct objects from detector signatures.

Electron identification in CMS data proceeds slightly differently through the particle flow algorithm, but the basic approach of matching a track to a set of electromagnetic calorimeter deposits remains the same as ATLAS. Depending on the position of an electron within the detector, and its momentum, it is sometimes better to seed the electron from the tracker rather than from the calorimeter, and it is sometimes the other way around. Very energetic and isolated electrons, where isolated indicates that they are not surrounded by other particles, are seeded using electromagnetic calorimeter information. Clusters with energy deposits greater than a chosen threshold are combined into a supercluster, with the direction of the electron set to the average position of the clusters weighted by the energy deposits (this is called the **barycentre** of the supercluster). The location of the tracker hits is then inferred from this direction.

For non-isolated electrons, this approach is clearly unsatisfactory, since extra particles around the electron will contribute energy to the cells around the electron, and bias the measured barycentre of the supercluster. It is also problematic for electrons with a small transverse momentum which are highly bent by the solenoid magnetic field. Electrons release photons via bremsstrahlung, and for highly curved tracks, these photons are spread over a wide area in the electromagnetic calorimeter, which makes measuring the position of a supercluster difficult. For these two cases, the CMS reconstruction software instead uses all tracks obtained via the tracker as the seeds of future electrons, which are then subjected to a series of quality cuts to try and isolate electron tracks from other tracks.

Photons are identified in CMS by using all of the remaining electromagnetic calorimeter clusters at the end of the particle flow process, at which point none of them should have associated tracks.

6.4.2 Muons

Muons are, in principle, straightforward to reconstruct, given that the only other SM particle that can reach the muon subsystem is a neutrino (which leaves no trace). However, there are subtleties in muon reconstruction which arise from the combination of signatures from different subsystems, and how to cover known gaps in the muon spectrometer acceptance.

Muon reconstruction in the ATLAS data relies primarily on the signals found in the inner detector and the muon subsystem, with occasional assistance from the calorimeters. For muons within $|\eta| < 2.5$, the inner detector provides high-precision measurements of muon positions and momenta. For example, muons with $|\eta| < 1.9$ will typically have three hits in the pixel layers, eight hits in the SCT, and 30 hits in the TRT, which allows one to select muon tracks by placing stringent conditions on both the number of hits in the different subsystems of the inner detector, and the number of silicon layers traversed without a hit. These conditions define a high-quality track which can be accurately extrapolated to the muon spectrometer, in which the three layers of high-precision monitored drift tube detectors can provide six to eight η measurements for a single muon passing through the detector within $|\eta| < 2.7$. A ϕ measurement is obtained from the coarser muon trigger chambers. The hits within each layer of the spectrometer are combined to form local track segments, then these are combined across layers to form a global muon spectrometer track.

At this point, one can define four different types of muon depending on the presence or absence of muon spectrometer and inner detector signatures:

1. Combined muons: in this case, there is both an inner detector and a muon spectrometer track, and there is a good match between the two. The combination of the two tracks yields the best possible precision on the measurement of the muon properties, and these are the only sorts of muon considered in most physics analyses.

2. Segment-tagged muons: in this case, there is an inner detector track, and there are segment tracks in layers of the muon spectrometer which match well with the inner detector track. However, there is no global muon spectrometer track. This might happen in the case of a muon that has a small transverse momentum, or is passing through a region of the detector where only one layer of the muon spectrometer will typically be hit.

3. Standalone muons: in this case, there is no inner detector track, but there is a muon spectrometer track. This typically occurs in the region $2.5 < |\eta| < 2.7$, which is covered by the muon spectrometer, but not by the inner detector.

4. Calorimeter-tagged muons: in this case, there is no muon spectrometer track, but there is an inner detector track. In addition. there is an energy deposit in the calorimeter that matches the inner detector track, and which is consistent with the passage of a minimally-ionising particle. This type of muon has a low purity (since the observed signatures can easily be caused by something else), and the only reason to use calorimeter-tagged muons is in cases where there are known holes in the muon spectrometer.

CMS muon reconstruction occurs first in the particle flow algorithm, and the basic details are similar to ATLAS.

Different categories of CMS muons are defined as follows:

1. Standalone muons: These are formed from hits in the drift tubes, cathode strip chambers or resistive plate chambers of the muon spectrometer, which are combined to form track segments.
2. Tracker muons: All tracks with $p_T > 0.5\,\text{GeV}$ in the inner detector are potential candidates for muon tracks. These are extrapolated to the muon chambers, and if they have at least one match with a track stub, the track becomes a tracker muon track.
3. Global muons: The standalone muon tracks are extrapolated back to the inner detector and, if a match is found, the two tracks are combined to form a global muon. The information from both sub-detectors is then used to obtain the best resolution on the muon track parameters. As with ATLAS, this is the best muon definition, and is used as the standard for most physics analyses.

6.4.3 Jets

We have already seen that when quarks and gluons are produced in the proton–proton collisions of the LHC, they initiate complex **jets**, comprised of collimated sprays of hadrons. We can define jets using the jet algorithms presented in Section 3.6, but these do not yet supply enough information to understand their reconstruction. In the following, we will first describe the ATLAS approach to jet reconstruction (up to the end of Run 2), before turning to CMS.

The main jet-identification algorithm used by the ATLAS reconstruction software is the anti-k_t algorithm with a distance parameter of $R = 0.4$. One can use various objects as input to the algorithm, including inner detector tracks, calorimeter energy deposits, or some combination of the two. Jets constructed from tracks, referred to sensibly as **track-jets** are less sensitive to pile up corrections, since one can only count tracks that point to the primary vertex. However, they have the obvious deficiency that the jet must have passed through the tracker acceptance of $|\eta| < 2.5$ in order to be reconstructed. Thus, it is most common to define jets using the calorimeter energy deposits. The individual cells that contain activity in the calorimeter are first grouped into significant supersets of clusters, bearing in mind the expected noise σ which is calculated as the quadrature sum of the measured electronic and pile-up noise. The clustering algorithm starts from seed cells that have energy deposits greater than 4σ, and iteratively adds neighbouring cells if their energy exceeds 2σ. This is followed by the addition of all adjacent cells. The resulting object is known as a **topo-cluster**. If topo-clusters have multiple local signal maxima with at least four neighbours whose energy is less than the local maximal cell, the topo-cluster is split to avoid overlap between clusters.

It is important to realise that the energy of the topo-cluster is not in fact equal to the energy that was deposited by the hadronic shower, since it has already been stated that the calorimeters respond differently to hadronic and electromagnetic particles. The energy of calorimeter cells is measured at a scale called the **electromagnetic scale** (or EM scale), established using electrons during test beam runs during the development of the detectors. Topo-clusters that are classified as hadronic can have their energy modified by a calibration factor known as the **local cell weighting** (LCW), which is derived using Monte Carlo simulation of single pion events. One can then choose to reconstruct jets from either the EM topo-clusters, or the LCW topo-clusters. The jets then have their energy scale restored to that of simulated truth jets using correction factors known as jet energy scale (JES) calibration factors. Different sets of factors are defined for EM and LCW jets, and the calibration accounts for various processes that may disturb the measured energy of a jet. For example, an origin correction is applied to force the four-momentum of a jet to point to the primary vertex rather than to the centre of the detector, whilst keeping the jet energy constant. Another component must mitigate the known effects of pile-up, that we have previously seen may bloat a jet energy measurement if it is not removed. A separate correction accounts for the difference between the calorimeter energy response and the 'true' jet energy, defined using Monte Carlo simulation, whilst also correcting for any bias in the reconstructed η of a jet that results from transitions between different regions of the calorimeter, or the difference in the granularity of the calorimeter in different locations. One can also apply a correction to reduce the dependence of the jet properties on the flavour of the instigating parto, uing global jet properties such as the fractions of the jet energy measured in the first and third calorimeter layers, the number of tracks associated with the jet, the average p_T-weighted distance in η–ϕ space between the jet axis and the jet tracks, and the number of muon tracks associated with the jet that result from punch-through. Finally, an *in situ* correction can be applied to jets in data, which is calculated as the jet response difference between data and Monte Carlo simulation, using the tranverse momentum balance of a jet and a well-measured reference object.

The calculation of the jet energy scale comes with various associated systematic uncertainties, which must be taken account of in all physics analyses. This is obvious in the case of analyses that require the presence of several jets, in which case the jet energy scale uncertainty can often be a leading source of the total systematic uncertainty on the final results. It is also applicable to analyses that veto the presence of jets, since a jet whose energy is systematically off may end up being below the transverse momentum threshold of the jets that are vetoed in the analysis.

An important additional quantity in the measurement of jets is the **jet energy resolution**, which quantifies, as a function of the detector location, the number of fluctuations in the measurement of a reconstructed jet's energy at a fixed particle-level energy. The jet energy resolution can be carefully measured by studying the asymmetry of dijet events in the detector, where the asymmetry must be generated by misreconstruction within the jet energy resolution.

Early CMS analyses used calorimeter jets, reconstructed from the energy deposits in the calorimeter alone. With improvements in the understanding of the detector, the preferred modern approach is to use **particle-flow jets**, which are obtained by clustering the four-momentum vectors of particle-flow candidates. Particle-flow jet momenta and spatial positions are measured to a much higher precision than for calorimeter jets, as the use of the tracking information alongside the high-granularity electromagnetic calorimeter measurements allows the independent reconstruction of photons and charged hadrons inside a jet, which together constitute roughly 85% of the jet energy. In the reconstruction of the particle-flow jet four-momentum, photons are assumed to be massless, whilst charged hadrons are assigned a mass equal to the charged pion mass. As with ATLAS jet reconstruction, a number of corrections have to be applied to CMS reconstructed jets to bring their energy in line with the 'true' value. Pile-up corrections are determined using a simulated sample of dijet events processed with and without pileup overlay, and are parameterised as a function of the offset energy density ρ, jet area A, jet η and jet p_T (see Section 6.5.1 for further information). A residual correction is applied by studying minimum bias events. A basic correction to the jet response is derived from the difference between MC simulation and data, and is provided as a function of η and p_T. Jets in data receive extra residual corrections that are determined by studying a combination of multijet, $Z(\rightarrow \mu\mu)$ + jet, $Z(\rightarrow ee)$ + jet and γ + jet events. The basic idea in each case is to exploit the transverse momentum balance, at hard-scattering level, between the jet to be calibrated and a reference object. A jet energy scale that differs from unity generates imbalance at the reconstructed level, and one can use this technique to define residual corrections as a function of jet p_T and η. Finally, there are optional jet flavour corrections that are determined from MC simulations. After corrections have been applied, the CMS jet energy resolution is carefully measured for calibrated jets.

6.4.4 Tau leptons

Tau leptons behave very differently from electrons and muons, in that they decay shortly after being produced at the interaction point. This decay is primarily mediated by an off-shell W boson, as shown in Figure 6.3. Roughly 35% of the time, taus decay to an electron or muon, plus neutrinos; 50% of the time, the decay produces one charged hadron, plus neutral hadrons and a neutrino. Finally, 15% of the time, the decay produces three charged hadrons, plus neutral hadrons and a neutrino. Based on these facts, we can make some general remarks about the observation of tau decays in the ATLAS and CMS detectors. Firstly, all of these decay modes feature a neutrino in the final state, and hence it will never be possible to fully reconstruct the energy of the tau. Secondly, leptonic tau decays cannot be

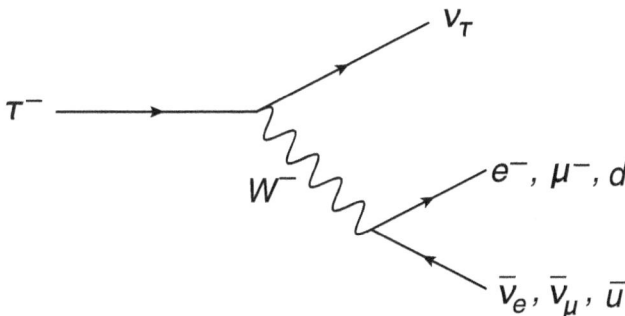

Figure 6.3. Feynman diagram showing the most common decays of the tau lepton.

distinguished from the direct production of electrons or muons at the interaction point, and they thus cannot be identified as tau decays. Thirdly, hadronic tau decays look like special jets, with either one or three tracks (arising from the charged hadron component), and an invariant mass less than the tau mass of 1.8 GeV. This can be compared with a typical jet which has 5–20 tracks and an invariant mass $\approx 0.15\, p_T$, where p_T is the transverse momentum of the jet. These characteristic features make it possible to reconstruct hadronic taus by looking for very narrow jets with a low number of tracks, but it is important to note that the fake rate for tau selection is particularly high. This in turn necessitates tight selections on tau properties that result in a typical reconstruction efficiency of only 40%. Nevertheless, it is important to preserere, since many interesting BSM physics processes are expected to preferentially couple to the third generation of matter, and the identification of tau leptons has already played a very important role in studies of the Higgs boson.

The ATLAS reconstruction software identifies taus by first reconstructing them as jets using the standard anti-k_t $R = 0.4$ jet definition, built from calorimeter topo-clusters. A multivariate analysis is then used to tag taus based on various properties of the jet. This algorithm is subject to continuous research and improvement.

The CMS approach is based on the *hadrons-plus-strips* algorithm which reconstructs the different decays of taus into hadrons. The neutral pions produced in tau decays promptly decay into pairs of photons, which then often convert into electron–positron pairs as they pass through the tracker. These electrons and positrons can acquire a reasonable spatial separation in the η–ϕ plane within the strong magnetic field of the CMS solenoid. To reconstruct the full pion energy, the reconstruction software collects all of the electron and photon candidates within a certain region $\Delta\eta \times \Delta\Phi$ of a candidate jet, with the resulting object referred to as a **strip**. The strip momentum is calculated as the vectorial sum of the momenta of its constituents. The set of charged particles and strips contained in a jet is then used to generate all possible combinations of hadrons for the decay modes h$^\pm$ (where h is a hadron), h$^\pm\pi^0$, h$^\pm\pi^0\pi^0$ and h$^\pm$ h$^\mp$ h$^\pm$, each of which acts as a particular hypothesis for the tau nature of the jet. The compatibility of each of these hypotheses is then assessed by checking that the reconstructed mass of the hadrons in the jet matches resonances that are known to be produced in tau decays, such as the $\rho(770)$ or a$_1$(1260) meson. Jets that fail these requirements can be rejected as taus, as can candidates that have a charge other than ±1, or that have strips outside of a signal cone drawn around the tau, whose radius is typically defined to be $R_{\text{sig}} = (3.0\,\text{GeV})/p_T^h$, where p_T^h is the transverse momentum of the hadronic system. Further selections on the resulting tau candidates can be used to reject fake taus from jets, including isolation requirements and multivariate-based tagging. An MVA discriminant is also used to reject fake taus arising from electrons, whilst fake taus from muons can be rejected by vetoing hadronic tau candidates that have signatures in the muon spectrometer that are close to the tau direction.

All of these techniques so far apply only to taus that are reconstructed as individual particles. In the decay of highly boosted objects, pairs of taus can be emitted very close to each other, such that they overlap in the detector and give rise to a single jet. In this case, jet substructure techniques can be used to improve tau identification, in the same way that they find utility in searches for boosted objects decaying to *b*-quarks.

6.4.5 Missing transverse momentum

The protons travelling along the beam line at the LHC have no component of momentum in the plane transverse to the beam line before they collide. Conservation of momentum thus tells us that the transverse momenta of any particles produced in the collision must sum to zero. This does not mean, however, that the reconstructed transverse momenta will actually sum to zero in any given event. Apart from mis-reconstruction, one expects to see a 'real' imbalance of summed transverse momenta in many SM processes, due to the production of neutrinos that escape the detector without interacting. In the search for BSM physics, our great hope is to see particles of dark matter or weakly interacting species (e.g. WIMPs) that would also contribute to a momentum imbalance by leaving the detector unobserved. For this reason, an accurate measurement of the size of the momentum imbalance in each event is highly sought after by the ATLAS and CMS collaborations, and it explains why a great effort has been expended on ensuring that the detectors enclose as much of the area around the interaction point as possible.

The missing transverse momentum, or $\boldsymbol{p}_{\mathrm{T}}^{\mathrm{miss}} \equiv (p_x^{\mathrm{miss}}, p_y^{\mathrm{miss}})^2$, is a 2D vector that quantifies the momentum imbalance in the transverse plane. Despite not being an energy, the magnitude of the vector is conventionally referred to as 'missing transverse energy' (MET), with symbol $E_{\mathrm{T}}^{\mathrm{miss}}$, and is formed from the components of the vector via:

$$E_{\mathrm{T}}^{\mathrm{miss}} = \sqrt{\left(p_x^{\mathrm{miss}}\right)^2 + \left(p_y^{\mathrm{miss}}\right)^2} \tag{6.7}$$

One can also define the ϕ direction of the missing tranverse momentum as:

$$\phi^{\mathrm{miss}} = \arctan\left(p_y^{\mathrm{miss}}/p_x^{\mathrm{miss}}\right) \tag{6.8}$$

Theorists often think about $\boldsymbol{p}_{\mathrm{T}}^{\mathrm{miss}}$ in the following way. Essentially, it is minus the vectorial sum of the transverse momenta of all reconstructed particles in an event. If this vectorial sum evaluates to zero, there is no missing transverse momentum. If it does not, then there must have been something invisible that was produced in the hard scattering process, and the $\boldsymbol{p}_{\mathrm{T}}^{\mathrm{miss}}$ vector is the sum of the unseen transverse momenta in the event. So an event that makes two neutrinos has a $\boldsymbol{p}_{\mathrm{T}}^{\mathrm{miss}}$ which is the sum of their transverse momenta. Of course, we never see this directly, and can only access it via the sum of visible products, whose transverse vector points in the other direction to balance the event.

Unfortunately, $\boldsymbol{p}_{\mathrm{T}}^{\mathrm{miss}}$ is *much* more complicated than this simple picture would suggest. In reality, measuring $\boldsymbol{p}_{\mathrm{T}}^{\mathrm{miss}}$ involves summing signatures from all detector

---

[2] Note that you will often see this erroneously written as $\boldsymbol{E}_{\mathrm{T}}^{\mathrm{miss}}$, which is a popular convention.

subsystems, and the imperfect resolution on each of these measurements will generate a spurious amount of p_T^{miss} regardless of whether any invisible particles were produced or not. Extra sources of p_T^{miss} include the non-collision backgrounds that we discussed in Section 6.5.2. The connection between p_T^{miss} and exciting new physics makes it critical that every source of spurious p_T^{miss} is well-understood and corrected for if we are to avoid making false discoveries.

A useful starting point for the calculation of p_T^{miss} in the ATLAS reconstruction is the use of terms related to calorimeter deposits. Since muons deposit only a tiny fraction of their energy in the calorimeters, muons reconstructed with the inner detector and muon spectrometer are added to p_T^{miss}:

$$p_T^{miss} = p_T^{miss,calo} + p_T^{miss,\mu}. \tag{6.9}$$

The default prescription for calculating $p_T^{miss,calo}$ is to identify electrons, photons, taus, jets and muons, and any remaining activity that exists in the form of unassigned topo-clusters. For each of these objects, the relevant calibration constants are applied to ensure that effects related to erroneous assumptions about the scale of measured objects do not contribute to the missing transverse momentum measurement. A calorimeter muon term is also calculated by examining the calorimeter regions around the measured muon trajectories, in order to avoid double counting the muon contribution that has already been added separately. This results in

$$p_T^{miss,calo} = p_T^{miss,e} + p_T^{miss,\gamma} + p_T^{miss,\tau} + p_T^{miss,jet} + p_T^{miss,\mu-calo} + p_T^{miss,soft}, \tag{6.10}$$

where $p_T^{miss,soft}$ results from the unassigned topo-clusters. At this point, we have unwittingly over-estimated $p_T^{miss,calo}$ by counting electrons, photons and taus that were also reconstructed as jets. The reconstruction software thus includes an overlap removal step, by associating each observed cluster with only one object, in the order in which they appear in equation (6.10). There are some important subtleties in this procedure, since some objects are reconstructed using a sliding window algorithm in the calorimeter (e.g. electrons), whilst others are defined using topo-clusters (e.g. jets), necessitating a careful matching between the two. The individual objects are also subject to various selections on transverse momenta, quality and/or jet reconstruction parameters, and these are liable to change over time.

The keen reader will note that there are in fact two muon terms that contribute to the missing transverse momentum calculation; one is internal to the calorimeter, whereas the other is external. The external term is set to the sum of the measured transverse momenta for reconstructed muons, and depending on the current definition this may include a mixture of combined, segment-tagged and standalone muons. In this case, it is necessary to be vigilant against biases in the external muon term that can result from the mis-reconstruction of standalone muons. For the internal calorimeter muon term, the prescription differs depending on whether a muon is isolated (within $\Delta R = 0.3$ of a jet) or non-isolated. For isolated muons, a parametrised average deposition in the calorimeter is used to estimate the contribution. For non-isolated muons, the contribution is set to zero, since any deposits left by the muons cannot be identified and will have already been included in previous terms.

The soft term is a little tricker to estimate, since it is not immediately clear how to account for objects that did not reach the calorimeter, or which calibration constants to use for unidentified calorimeter energy deposits. The ATLAS calculation starts by selecting a good quality set of tracks down to a p_T of 400 MeV. Any tracks with a transverse momentum above 100 GeV are ignored. The tracks are extrapolated to the middle layer of the electromgnetic calorimeter, which is selected due to its high granularity, and a check is made as to whether there are any calorimeter deposits with a certain ΔR of each track. If the answer is no, the track is added to an 'eflow' term which counts the contribution from missed objects. If the answer is yes, the track is added and the topo-cluster is discarded, since the tracking resolution is better than the calorimeter resolution at low transverse momentum. A track might in principle match several topo-clusters, in which case the track is added and only the highest-energy topo-cluster is discarded. Finally, there are remaining topo-clusters which have not been discarded by track matching, and these are added to the eflow term.

The performance of the missing transverse momentum reconstruction must be carefully studied, and this can be achieved by comparing data and Monte Carlo simulations for processes such as $Z \to \mu\mu$ + jets and $Z \to ee$ + jets, where the selection of events with lepton invariant masses near the Z peak can produce a clean sample of events with no expected intrinsic p_T^{miss}. Sources of true missing transverse momentum can be studied in $W \to \mu\nu$ + jets and $W \to e\nu$ + jets events. It is important to note that we have described only the current default procedure for the p_T^{miss} calculation, and this must often be studied carefully and re-worked for a particular physics analysis.

In the CMS reconstruction, p_T^{miss} is calculated from the output of the particle flow algorithm, and is computed as the negative of the vectorial sum of the transverse momenta of all particle flow particles. As in the ATLAS reconstruction, this could be affected by the minimum energy thresholds in the calorimeter, inefficiencies in the tracker, and non-linearity of the calorimeter response for hadronic particles. The bias on the p_T^{miss} measurement is reduced by correcting the transverse momentum of the jets to the particle level jet p_T using jet-energy corrections, and propagating the correction into p_T^{miss} via

$$p_T^{miss,corr} = p_T^{miss} - \sum_{jets}\left(p_{T,\,jet}^{corr} - p_{T,jet}\right), \tag{6.11}$$

where $p_{T,\,jet}^{corr}$ represents the corrected values. This is called the *Type-I* correction for p_T^{miss}, and it uses jet-energy scale corrections for all corrected jets with $p_T > 15$ GeV that have less than 90% of their energy deposited in the electromagnetic calorimeter. In addition, if a muon is found within a jet, the muon four-momemtum is subtracted from the jet four-momentum when performing the correction, and is added back in to the corrected object.

In both of the ATLAS and CMS reconstruction procedures, it is essential to apply the event cleaning procedures that are used to remove non-collision backgrounds in order to obtain a reasonable estimate of p_T^{miss}.

6.4.6 Identification and calibration of heavy-flavour jets

Jets initiated by c- and b-quarks produced at the interaction point initiate jets that are composed of hadrons, just like the lighter quarks. However, these **heavy-flavour jets** occupy a unique place in particle reconstruction due to the fact that the primary (weakly-decaying) hadron will typically travel up to a few millimetres before decaying[3].

For b-hadrons, this is due to the strong **CKM-suppression** for third-generation quarks, and the kinematic inability of the b-quark to decay to the top-quark as it would like. In addition, the b-quark **fragmentation function** is relatively hard, meaning the b-hadron formed from the quark before it weakly decays will tend to carry a large fraction—around 80%—of the quark momentum. This generates a large Lorentz boost $\gamma = E/m_{b,c}$ for the b- (and subsequently c-) hadron, and hence an even more distinctive displacement. The c-quark exhibits some of the same features, without CKM-suppression but with the generally long decay time associated with the weak coupling, accentuated by the greater effect of time-dilation on a lighter particle. These macroscopic decay distances, compared to the sub-picometre ones of most hadrons, lie at the core of methods to 'tag' heavy-flavour jets.

The weak decays themselves can be classified either as hadronic or semi-leptonic depending on whether a lepton is absent or present in the final state; an electron or muon is expected in approximately 20% or 10% of jets initiated by a heavy b- or c-hadron, respectively. The larger mass of c- and b-hadrons relative to light-quark hadrons also results in more complex decay chains, with a high multiplicity of tracks expected at the secondary vertex, and with larger track p_Ts relative to the jet axis. These factors provide secondary signatures to the decay-vertex displacement for heavy-flavour jet identification.

These distinct properties of heavy-flavour jets allow them to be identified, with reasonable efficiency, as distinct from jets initiated by lighter quarks or gluons. Naïvely, this might seem pointless, since if the LHC detectors cannot distinguish most of the quark flavours, why does it help to be able to distinguish two of them? The answer, as in the case of tau leptons, is that a number of SM and BSM processes preferentially couple to the third generation of matter, and a knowledge of which jets are c- or b-initiated is vital to searching for supersymmetry, measuring the properties of top quarks, and investigating the nature of the Higgs boson. Measuring the decays of the Higgs boson to bottom and charm quarks is also crucial for cross-referencing Higgs branching ratios with the fermion Yukawa couplings, and hence precision-testing the electroweak part of the Standard Model.

---

[3] Excited heavy-flavour states, such as B^* or D^* vector mesons, de-excite promptly by photon emission, but emission of a much slower W boson is the only way to 'lose' the b or c hadron flavour.

The identification of a jet as a heavy-flavour jet is known as **heavy-flavour tagging**, and the rate at which, for example, a true b-jet is correctly identified is known as the **b-tagging efficiency** for a particular tagging algorithm. As with any tagging technique, there is a trade-off between efficiency and purity, and one must carefully select a working point for the tagging algorithm based on which of these is most important for the analysis at hand.

The geometry of a typical heavy-flavour jet is shown in Figure 6.4. The non-negligible travelled distance of the initial hadron gives rise to a secondary vertex, from which the tracks associated with the heavy-flavour jet will emanate. The granularity of the inner track nearest the interaction point is carefully designed to be able to resolve these secondary vertices. Moreover, if the tracks from the secondary vertex are extrapolated backwards, the point of closest approach of each track to the primary vertex gives a measureable distance called the **impact parameter**, which will clearly differ from the zero values expected from tracks arising from the primary vertex.

Within the ATLAS reconstruction software, there are three main categories of b-tagging algorithm. These use either impact parameter information, inclusive secondary vertex reconstruction techniques, or decay chain reconstruction algorithms. Since these strategies are complementary, they can be combined using a multivariate analysis technique to provide a single b-tagging discriminant. Track information for b-jet candidates is obtained by using a p_T-dependent ΔR matching between tracks and calorimeter jets.

Impact-parameter algorithms take the tracks which are matched to a jet and calculate the signed impact factor significance sig (x):

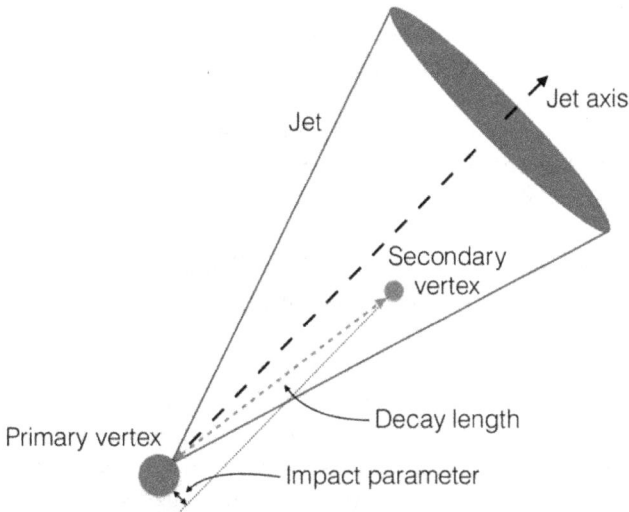

Figure 6.4. The typical geometry of a b-jet. CREDIT: https://cds.cern.ch/record/2229557/files/ATL-PHYS-PROC-2016-193.pdf

$$\text{sig}(x) = \frac{x}{\sigma_x}, \tag{6.12}$$

where x is the measured impact factor for a given track, and σ_x is its uncertainty. The sign is taken as positive if the point of closest approach of a track to the primary vertex is in front of the primary vertex with respect to the jet direction, and negative otherwise. The algorithms then use template functions for the impact factor significance, which have been previously calculated for different jet flavours, and these templates can be calculated in the longitudinal direction along the beamline, in the plane transverse to the beam line or a combination of both. The distributions are much wider for b-jets than for light quark flavour jets, and they can be used to determine the likelihood that a particular candidate jet belongs to a particular category. To obtain the discriminant for each tagger, a ratio of the b-jet and light quark jet likelihoods is used.

The ATLAS secondary-vertex tagger instead tries to reconstruct a displaced secondary vertex, starting from all possible two-track vertices that can be identified from the set of tracks in the event. To improve the performance, the next step is to remove all vertices that are likely to come from photon conversions, hadronic interactions with the inner detector material, or long-lived particles (e.g. K_S, Λ). Next, a secondary vertex is found by iteratively removing outlier tracks until a good candidate is found. One it has been reconstructed, the properties of the associated tracks can be used to define a likelihood ratio for the b-jet and light quark jet hypotheses using quantities such as the energy fraction of the tracks fitted to the vertex with respect to all tracks in the jet, the number of tracks in the vertex, the vertex mass, and the angle between the jet direction and the b-hadron flight direction as estimated from the primary-secondary vertex axis.

A further approach is given by the ATLAS decay chain reconstruction tagger which fits the decay-chain of b-hadrons, by reconstructing a common b-hadron flight direction, along with the position of additional vertices that arise along the direction as the hadron decays to other hadrons. The tracks associated to each reconstructed vertex along the flight direction can be used to discriminate between the b-jet and light quark jet hypotheses, and to calculate the final discriminant.

The value of the tagging discriminant (or their combination) for a candidate jet is known as the **b-jet weight**. Physics analyses can proceed by cutting on this quantity directly, but it is much preferred to follow pre-defined **working points**, which each define a fixed bound on the discriminant output. These working points come with an estimate of the efficiency and purity, plus a prediction of the relevant systematic uncertainties.

The CMS collaboration have developed a variety of b-jet tagging techniques over the years, including a **jet probability tagger** that uses only impact factor information, a **combined secondary-vertex tagger** that uses a wide range of information on reconstructed secondary vertices, and a **combined multivariate-analysis tagger** that combines the discriminator values of various other taggers.

There are two CMS jet-probablity taggers, called the *JP* and *JBP* algorithms. The JP algorithm uses the signed impact-parameter significance of the tracks associated with a jet to obtain its likelihood of originating from the primary vertex. This exploits the fact that the negative impact-parameter significance values of tracks from light flavour jets must arise from within the resolution on the measured track impact-parameter values. Hence, the distribution of the negative impact-parameter significance can be used as a resolution function $\mathcal{R}(s)$ that can be integrated from $-\infty$ to the negative of the absolute track impact parameter significance, $-|\text{IP}|/\sigma$ to get the probability of a track originating from the primary vertex:

$$P_t = \int_{-\infty}^{-|\text{IP}|/\sigma} \mathcal{R}(s)\, ds. \qquad (6.13)$$

Note that this resolution function depends strongly on the quality of the reconstructed track, and the probability it assigns will be lower for a track with lots of missing hits. Nevertheless, tracks that were created by particles from the decay of a displaced particle will have a low value of P_t. The probability that all N tracks in the jet are compatible with the primary vertex is defined as

$$P_j = \Pi \cdot \sum_{n=0}^{N-1} \frac{(-\ln\Pi)^n}{n!}, \qquad (6.14)$$

where $\Pi = \Pi_{i=1}^{N} P_t(i)$. One can then define $-\log(P_j)$ as a b-tagging discriminator, since it gives higher values for b-jets. The JBP algorithm adds a requirement that the four tracks with the highest impact-parameter significance get a higher weight in the calculation of P_j.

The most common heavy-flavour jet definition within the CMS experiment results from the application of the combined secondary-vertex tagger. This uses a neural network to classify jets as b-jets, based on almost 20 input variables taken from the measured jet and track properties. The algorithm is trained on inclusive dijet events, separated into three categories; jets with a reconstructed secondary vertex, jets with no secondary vertex but a set of at least two tracks with a 2D impact-parameter significance value greater than 2 and a combined invariant mass at least 50 MeV away from the K_S^0 mass, and events that pass neither of these sets of selections. In the latter case, only track information can be used in the definition of the tagger.

A typical b-tagging working point chosen for analysis will have an efficiency 77%, with c-jet and light-jet rejection factors of 6 and 130, respectively.

In recent years, both the ATLAS and CMS collaborations have been using refined machine learning methods to improve their taggers. This includes the use of recurrent neural networks that can be applied to an arbitrary sequence of inputs, and the use of deep learning techniques with a larger set of tracks as input data.

Once b-tagging algorithms have been defined, they need to be carefully calibrated by deriving correction factors that make simulated events match the observed data. For any given simulated jet, these corrections depend on both the jet flavour and the kinematics, and there are separate corrections for tagging real b-jets correctly, and for mis-tagging c and light flavour jets as b-jets. In order to derive the correction factors, one needs to take a sample of data events which are known to be rich in *b*-jets, and for which the various contributing processes are well understood. One example is to use leptonically decaying top-pair events, since these should be pure in *b*-jets at the hard-scattering level. The actual event selection applied is to take events with one electron, one muon and two or three jets. Events with two same-flavour leptons can be used to derive data-driven corrections to Z + jets and top-pair samples which dominate the contribution to this final state. The *b*-tagging efficiency is then extracted by performing a 2D maximum likelihood fit as a function of the leading and sub-leading jet p_T, which allows the method to account for correlations between the two jets. Charm jet mis-tag correction factors can be obtained by reconstructing the decay chain of $D^* \rightarrow D_0(\kappa\pi)\pi_s$, since these events are enriched in *c*-jets. After applying a *b*-tagging requirement, the sample of jets will consist of some fraction of true *b*-jets which are correctly selected, and some fraction of *c*-jets which are incorrectly selected. The contamination from real *b*-jets can be estimated by fitting the proper lifetime of the events, which allows one to isolate the *c*-tag efficiency. The light jet tagging efficiency is harder to extract, but it can be estimated by inverting requirements on the signed impact-parameter significance and signed decay length significance of jets to obtain a sample enriched in light quark jets. This sample allows the light jet tagging efficiency to be measured once the contribution of heavy flavour and long-lived particles is accounted for.

Charm tagging is harder to perform than *b*-tagging, due to the fact that the lower mass of *c*-hadrons relative to *b*-hadrons leads to a less displaced secondary vertex, with a lower track multiplicity. Nevertheless, charm tagging is an important topic in its own right within both the ATLAS and CMS collaborations. Within the ATLAS collaboration, charm tagging techniques have been developed using boosted decision trees, with the same set of input variables as are used in *b*-tagging. Two discriminants are used; one that separates *c*-jets from light-jets, and one that separates *c*-jets from *b*-jets. An efficiency of $\approx$ 40% can be achieved for *c*-jets, for *c*-jet and light-jet rejection factors of 4.0 and 20, respectively. The CMS collaboration follows a similar procedure, with a long list of input variables related to displaced tracks, secondary vertices and soft leptons inside the jets. The *c*-jet tagging efficiency can be measured in data by obtaining a sample of events enriched in *c* quarks, for example by selecting events with a W boson produced in association with a *c* quark. This process occurs at leading order mainly through the processes $s + g \rightarrow W^- + c$ and $\bar{s} + g \rightarrow W^+ + \bar{c}$, in which the *c* quark and the W boson have opposite electric charge. The background, meanwhile, comes from $W + q\bar{q}$ events, in which the *c* quark will have the same charge as the W boson half the time, and the opposite charge the other half of the time. After applying a preselection, a high-purity sample of $W + c$ events can be obtained in any variable of interest by subtracting off the same-sign event distribution from the opposite-sign event distribution in that variable. Another approach is to use single leptonic $t\bar{t}$ events, in which one of the W bosons has decayed to quarks. In this case, the decay contains a *c* quark in about 50% of cases. Because of the particular decay chain of the top quark, the energy of up-type quarks produced in W boson decay is larger, on average, than the energy of the down-type quarks produced. This allows the identification of samples of jets enriched and depleted in *c* quarks.

6.5 Rogue signals

So far, we have tacitly assumed an ideal world in which the LHC collides protons, and ATLAS and CMS observe the effects of a single interaction between consitutents of those protons, in the form of measurements of the prompt decay products of any particles produced in the collision. Unfortunately, reality complicates this picture in two distinct ways. The first is that, as we saw in Chapter 3, it is not true that a single hard scattering event between proton constituents is responsible for those parts of the observed events that do actually arise from the proton collisions. As the instananeous luminosity of the LHC is increased, one can expect tens to hundreds of hard scattering events per collision of proton bunches at the LHC, and this leads to pile-up. Secondly, there is no way to guarantee that the particles the ATLAS and CMS detectors observe actually arise from the proton collisions at all, since any sensitive detector will give signatures for *any* particles that happen to pass through it. In this section, we will briefly review the effects of multiple scatters between proton constituents, before describing various **non-collision backgrounds** that can generate rogue signals.

6.5.1 Pile-up

As we saw in Section 1.3, the protons at the LHC collide in bunches, with N_i protons per bunch. The collision of these bunches not only results in proton–proton interactions in which a large transfer of four-momentum occurs, but also in additional collisions within the same bunch-crossing. These latter interactions primarily consist of low-energy QCD processes, and they are referred to as **in-time pile-up** interactions. These are not the whole story, however, since both the ATLAS and CMS detectors are also sensitive to **out-of-time pile-up**, which consists of detector signatures arising from previous and following bunch-crossings with respect to the current triggered bunch-crossing. These contributions grow when the LHC runs in a mode to increase the collision rate by decreasing the bunch spacing. The sum of these distinct contributions is generically referred to simply as **pile-up**.

Pile-up cannot be directly measured, and its main effects are to make it seem like jets have more energy in an event than they actually do, and to produce spurious jets. This in turn will bloat the measurement of the missing transverse-momentum in an event. Thankfully, pile-up effects can be mitigated by exploiting various low-level detector measurements. For example, silicon tracking detectors have a very fast time response, and are thus not typically affected by out-of-time pile-up. They can therefore provide a safe measurement of the number of primary vertices, N_{PV}, in an event, which is highly correlated with the amount of in-time pile-up. For other detectors, such as some calorimeters, the integration time for measured signals is longer than the bunch spacing, and one must develop a careful strategy for pile-up subtraction that takes into account the effects of both in-time and out-of-time pile-up. Here, a helpful strategy is to use the inner detector to identify which charged-particle tracks match the various energy deposits in the calorimeter. This allows the identification of calorimeter deposits that mostly result from non-hard scatter vertices, which can then be rejected as jets.

The actual pile-up correction procedures followed by ATLAS and CMS are detector-specific, and subject to ongoing research. The first step for ATLAS data is to build a pile-up suppression into the topological clustering algorithm that identifies the energy deposits in the calorimeter that are used to build jets. In this procedure, pile-up is treated as noise, and cell energy thresholds are set based on their energy significance relative to the total noise. Raising the assumed pile-up noise value in the reconstruction can be used to suppress the formation of clusters created by pile-up deposits. However, the use of a fixed assumed pile-up noise value does not account for pile-up fluctuations due to different luminosity conditions over a run period, or local and global event pile-up fluctuations. Such effects can be taken account of by attempting to subtract pile-up corrections to a jet on an event-by-event and jet-by-jet basis, by defining the corrected jet transverse momentum:

$$p_T^{corr} = p_T^{jet} - \mathcal{O}^{jet} \tag{6.15}$$

where p_T^{jet} is the naïve measurement of the jet transverse momentum, and $\mathcal{O}^{jet}$ is the offset factor. By defining and measuring a **jet area** A^{jet} for each jet in η–ϕ space (see Section 3.6), along with a pile-up p_T density ρ, a pileup offset can be determined dynamically for each jet via:

$$\mathcal{O}^{jet} = \rho \times A^{jet}. \tag{6.16}$$

For a given event, ρ itself can be estimated as the median of the distribution of the density of many jets, constructed with no minimum p_T threshold:

$$\rho = \text{median} \left\{ \frac{p_{T,i}^{jet}}{A_i^{jet}} \right\}, \tag{6.17}$$

where each jet i has a transverse momentum $p_{T,i}^{jet}$ and area A_i^{jet}. This can be computed from both data and Monte Carlo events.

After this subtraction, there is still a residual pile-up effect that is proportional to the number ($N_{PV} - 1$) of reconstructed pile-up vertices, plus a term proportional to the average number of interaction vertices per LHC bunch-crossing $\langle \mu \rangle$ which accounts for out-of-time pile-up. The final corrected jet transverse momentum is thus given by:

$$p_T^{corr} = p_T^{jet} - \rho \times A^{jet} - \alpha(N_{PV} - 1) - \beta\mu, \tag{6.18}$$

where α and β are free parameters that can be extracted from simulated dijet events, but which are carefully validated through studies of the ATLAS data.

The pile-up subtraction procedure described above removes pile-up contributions to jets from the hard scattering process, and it is also sufficient to remove some spurious jets. However, some of the latter type of jets survive the process, and these are typically a mixture of hard QCD jets arising from a pile-up vertex, and local fluctuations of pile-up activity. Pile-up QCD jets are genuine jets, and can be tagged and rejected using information on the charged-particle tracks associated with the jet (i.e. whether they point to a pile-up vertex rather than the primary vertex). Pile-up jets arising from local fluctuations contain random combinations of particles from multiple pile-up vertices, and these can also be tagged and rejected using tracking information. One such method is to use the **jet-vertex-fraction** (JVF), defined as the fraction of the track transverse momentum contributing to a jet that originates from the hard scatter vertex. More recently, ATLAS has introduced a dedicated **jet-vertex-tagger** (JVT) algorithm.

The CMS collaboration follow a qualitatively similar approach to pile-up subtraction, but with some unique twists. For example, the use of particle flow reconstruction allows CMS to perform charged hadron subtraction to reduce the in-time pile-up contribution from charged particles before jet reconstruction. CMS also applies the jet-area-subtraction method, although the residual correction (the α and β terms in equation (6.18)) is parameterised differently with a specific dependence on the jet p_T. Finally, spurious pile-up jets can be identified using a boosted decision tree discriminant.

It is important to note that these techniques fall down outside of the tracker acceptance, and thus dedicated techniques are needed for pile-up suppression at higher η. CMS again uses a multivariate discriminant, whilst ATLAS uses tracks in the central region to indirectly tag and reject forward pile-up jets that are back-to-back with central pile-up jets. This removes QCD jets, but does not reduce the number of forward jets that instead arise from local pile-up fluctuations.

6.5.2 Non-collision backgrounds

Pile-up constitutes a background for particles from the hard scattering process which, nevertheless, still ultimately originates from the proton–proton collisions themselves. There are, however, other particles that arise from the LHC accelerator complex, that are not directly produced via proton–proton collisions, and these are referred to as **machine-induced backgrounds**. One can also be unlucky, and have a cosmic ray muon pass through the detector at the same time as a physics trigger has fired, leading to a distortion of the observed physics event by additional unrelated detector activity. The sum total of the machine-induced background and the cosmic ray muon background is often referred to as the **non-collision background**.

The first type of beam background arises from interactions between the proton beam, and the various pieces of LHC accelerator apparatus that are located upstream of the detectors. Of particular note are the **collimators**, movable blocks of robust material that close around the beam to clean it of stray particles before they come close to the interaction points. Protons may escape the cleaning process in the primary and secondary collimators, and strike the tertiary collimators that are situated nearest to the ATLAS and CMS detectors. The result is a high-energy electromagnetic or hadronic shower of secondary particles which may eventually propagate into the detectors, despite their heavy shielding, with the particles typically arriving at small radii. This is referred to as the **beam halo** background, and it looks like a relatively collimated flux of particles into the detector from upstream. Further showers can be produced if protons interact with residual gas molecules inside the vacuum pipe which, despite best efforts, will never contain a perfect vacuum. This is referred to as the **beam gas** background, and can be generated either from inelastic interactions with gas molecules in the vacuum pipe near the experiments, or from elastic beam-gas scattering around the ring. There is in fact no clear difference between the beam halo background and the background resulting from elastic beam gas interactions, since the latter will slightly deviate a proton from the beam line, causing it to later hit the tertiary collimator. Inelastic beam gas interactions, meanwhile, will produce a shower of secondary particles that mostly have local effects, with the exception of high-energy muons which can travel large distances and reach the detectors. These muons are typically travelling in the horizontal plane when they arrive at the experiments.

Machine-induced backgrounds cause very funky detector signatures. Results might include fake jets in the calorimeters, which then generate a high missing transverse momentum because the spurious activity will not balance activity from the hard scattering process. Fake tracks might also be produced which, when matched with a calorimeter deposit, may generate a fake electron or jet. Factors that affect the prevalance of machine-induced backgrounds include the beam intensity and energy, the density of gas particles within the beam pipe, the settings of the collimator and the status of the LHC machine optics. It has been possible to simulate machine-induced backgrounds in ATLAS and CMS and obtain reasonable agreement with the observed phenomenology.

The strategy for mitigating machine-induced backgrounds in the LHC detectors first involves carefully studying their properties by, for example, triggering on non-colliding bunches using dedicated detectors placed around ATLAS and CMS. Beam background events can also be identified by the early arrival time of the particles compared to collision products at the edges of the detector. This reveals certain features of beam-induced background events that can be confirmed by detailed simulation. For example, muons from beam–gas interactions near the ATLAS detector will generate particles that are essentially parallel to the beam-line, leaving long continuous tracks in the z direction in the pixel detector. Particles that enter the detector outside of the tracking volume can be studied in the calorimeter and muon subsystems. Fake jets in the calorimeter from beam-induced backgrounds typically have two distinguishing features; their azimuthal distribution is peaked at 0 and π (compared to a flat distributon in ϕ for jets from proton–proton collisions), and the time of the fake jets is typically earlier than that of collision jets. The standard approach for reducing beam-induced background in physics analyses is to apply **jet cleaning** selections that exploit these differences in calorimeter activity. Examples of typical discriminating variables include the electromagnetic energy fraction (defined as the fraction of energy deposited in the electromagnetic calorimeter with respect to the total jet energy), the maximum energy fraction within any single calorimeter layer, the ratio of the scalar sum of the transverse momenta of tracks associated to the jet to the transverse momentum of the jet (which will be higher for collision jets), and timing properties in the calorimeters and muon spectrometer.

The cosmic-ray muon background has also been studied in detail in both ATLAS and CMS. If a cosmic ray muon traverses the whole detector, it will give rise to separate muon tracks reconstructed in the upper and lower hemisphere of the detector, reconstructed as back-to-back muons that do not pass through the interaction point. The two tracks are reconstructed with opposite charge, due to the reversed flight path of the cosmic ray track in the top sectors with respect to the flight path of particles travelling from the interaction point. If the muon time-of-flight through the detector is not fully contained within the trigger window, one might only see a track in one of the hemispheres. Cosmic ray muon tracks can be rejected by placing selections on the impact parameter of muon tracks (which will be substantially larger for cosmic ray muon tracks), and by exploiting topological and timing differences between interaction and cosmic muons in the muon spectrometer. The jet cleaning cuts developed for machine-introduced backgrounds are also efficient for suppressing fake jets from muons that undergo radiative loses in the calorimeters.

It is worth noting that not all signatures searched for in proton–proton collisions at the LHC would involve particles that decay promptly at the interaction point. Some theories give rise to long-lived particles that would *themselves* give a very funky pattern of detector signatures, in which standard assumptions about tracks and calorimeter deposits pointing back to the interaction point break down. In this case, a dedicated re-tracking and re-vertexing for displaced vertices needs to be implemented, and any signatures expected in a candidate theory need to be carefully compared with those expected from non-collision backgrounds.

6.5.3 Detector malfunction

A final background to the measurement of new physics processes is that arising from detector malfunction. For example, the ATLAS detector has approximately 100,000,000 readout channels, and the entire detector cannot be expected to be operational at any given time. Malfunctioning detector components may remove particles entirely from the reconstructed data, bias the four-momentum measurements of particles, and contribute to the missing transverse momentum in an event.

For this reason, both ATLAS and CMS have dedicated data quality systems that identify which data blocks from which LHC runs are safe to use for analysis, rejecting data periods in which substantial malfunctions occurred. Data quality algorithms are run **online** during data-taking itself, usually to generate graphs that shifters watch in the control room as data is taken. This is to allow a swift response if there is a sudden malfunction, with a view to repairing the malfunction on the spot. Similar algorithms are run offline to award a quality indicator to each luminosity block of the LHC data. This is then used to define a **good-run list** that is used by the physics analysis software to automatically choose which stored data files can be safely used for ATLAS and CMS physics analyses. Separate data quality mechanisms are put to use during the development of the ATLAS and CMS simulation and reconstruction software, to automatically flag bad software as the software releases are updated.

Further reading

- An excellent short guide to particle detectors can be found in 'Particle Physics Instrumentation' by I Wingerter-Seez, arXiv:1 804.112 46.
- The ATLAS detector at the time of its installation is described in 'The ATLAS Experiment at the Large Hadron Collider' *JINST* **3** S08003. The CMS detector is similarly described in 'The CMS experiment at the CERN LHC' *JINST* **3** S08004.

Exercises

6.1 Prove that the momentum p of a particle of charge q moving in a magnetic field B is given by $p = Bqr$, where r is the radius of curvature of its circular track.

6.2 Explain why the momentum resolution of an electron in the ATLAS and CMS detectors can be expected to be worse at very high p_T values. Why is it also worse at very low p_T values?

6.3 An electron with an energy of 30 GeV strikes the CMS calorimeter. Estimate the expectation value of its energy once it has moved a distance of 4 cm.

6.4 Match the following descriptions of detector activity to the particle types *electron*, *muon* and *photon*.

  (a) Track in the inner detector, no calorimeter deposits, track in the muon chambers.

  (b) Track in the inner detector, most of the energy deposited in the electromagnetic calorimeter.

  (c) No track in the inner detector, most of the energy deposited in the electromagnetic calorimeter.

6.5 What, if anything, distinguishes an electron from a positron in the inner detector?

6.6 Give two reasons why the granularity of silicon detector modules must be increased in the detector region closest to the beam line.

6.7 You are asked to perform an analysis in a final state with two tau leptons. What fraction of di-tau events can be expected to have two hadronic tau decays in?

6.8 In a certain run period of the LHC, it is expected that there are an average of 24 interactions per bunch-crossing. Assuming that the number of interactions is Poisson-distributed:

  (a) What is the probability that only one interaction occurs in a bunch-crossing?

  (b) What is the probability that ten interactions occur per bunch-crossing?

IOP Publishing

Practical Collider Physics

Andy Buckley, Christopher White and Martin White

Chapter 7

Computing and data processing

The first part of this book has described in some detail the theoretical picture, and corresponding sets of tools and techniques, within which most collider physics phenomenology takes place. The typical approach taken by physics-analysis interpretations is to compare simulated event observables to measured ones, and hence statistically quantify whether a particular model is sufficient or incomplete. This statistical goal, and the need to understand the detector itself, leads to a great need for simulated collider events as well as real ones. While not computationally trivial—in particular, as we shall shortly see, modern high-precision event generation can be very costly—extending these ideas to data analysis within an experiment implies a computational processing system at vast scale.

The conceptual structure of experiment data processing is shown in Figure 7.1. The dominant feature of this diagram is the parallel pair of processing chains for real experimental data as compared to simulated data. Real data is retrieved from the detector readout hardware (after the hardware trigger, if there is one) via the **data-acquisition system** (DAQ), usually passed through a software trigger for more complex discarding of uninteresting events, and the resulting raw signals are operated on by **reconstruction** algorithms. An important distinction is made here between **online processing** such as the triggers, which run 'live' as the event stream arrives from the collider and do not store all of the input events, and **offline processing**, which is performed later on the set of events accepted and stored on disk after the online filtering decisions. In parallel with the event stream, the operational status and alignment information about the beam and subdetector components is entered in a **conditions database**, used to inform the behaviour of the equivalent simulation, and the shared reconstruction algorithms.

A significant caveat to this picture is due for trigger-level analysis, a relatively recent innovation that blurs the distinction between online and offline data-processing tasks. At the time of writing, all LHC experiments have initiated work on trigger-level analysis software, for statistically limited analyses where the rate of

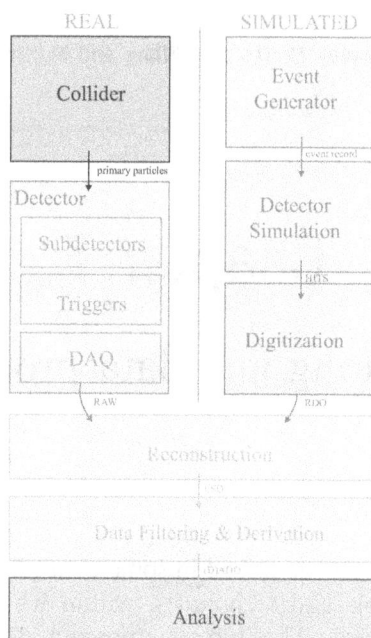

Figure 7.1. Real (LHS) and simulated (RHS) collision data-processing flows in detector experiments.

candidate events is too high for un-prescaled triggering. Much of the pioneering work in this area has been led by the LHCb collaboration, who have the double incentive of a reduced instantaneous luminosity with respect to the other experiments due to intentional beam defocusing, and a core physics programme dependent on rare decays.

Simulated data (sometimes ambiguously known as Monte Carlo (MC) data), by comparison, obviously has no physical detector as a source of events. Instead, a chain of algorithms is assembled to mimic the real-world effects on data in as much detail as is tractable, while taking advantage of opportunities for harmless computational short-cuts whenever possible. In place of a collider, simulated data uses event generator programs, as described in Chapter 3; in place of a physical detector with which the final-state particles interact and deposit energy, a **detector simulation**[1] is used; and in place of the readout electronics of the real detector, simulated data uses a set of **digitization** algorithms. The result of this chain is essentially the same data format as produced by the real detector, which is thereafter processed with the same reconstruction and data-analysis tools as for real data. Simulation, of course, has no online stage: everything is performed at an 'offline' software pace, although this is

---

[1] Detector simulation is often referred to simply as 'simulation'—which can be confusing as *everything* on the right-hand chain of Figure 7.1 is simulated. Another potentially confusing informal terminology is 'Monte Carlo' to refer to event generation, when detector simulation also makes heavy use of MC methods, and indeed in some areas of the field 'MC' is more likely to mean detector simulation. When in doubt, use the full name!

not to say that the simulation chain does not discard unwanted events in a manner equivalent to a trigger.

Were the simulated 'uninteresting' physics and detector effects *perfectly* equivalent to those in the real world, most data analysis would be relatively trivial, but of course in practice there are significant differences, and a great deal of the 'art' and effort of collider data analysis lies in minimising, offsetting, and quantifying such biases.

7.1 Computing logistics

As the cost is so great, and spare capacity in such short supply, errors and duplications in large-scale data processing must be kept to a minimum. The 'full chain' is hence usually decomposed into the components described above, each being maintained and operated by specialist groups within an experimental collaboration. We have already discussed the physics content of quantum field theories and the key elements of their implementation via MC event generator codes in Chapters 2 and 3, and the principles of reconstruction and calibration in Chapter 6; in this chapter we focus on the *logistics* of processing large data volumes through these procedures and the initial phases of data-reduction for analysis, and introduce an overview of the detector-simulation and digitization processes.

In all steps of the computational processing chain, the high-level algorithmic structure is based on an **event loop** design, with initialise and finalise phases each performed once, respectively, at the start and end of each processing job, and an execute phase performed many times—once for each event, in a 'loop' over all the events in a file. This process is typically orchestrated by **data-processing frameworks**, into which physicists inject their own algorithms' code for each phase.

The distribution of such processing jobs is performed in massively parallel executions, either on **local batch farms** of hundreds or thousands of computers, or on the even larger **Worldwide LHC Computing Grid**, or simply 'the Grid', which despite the name does also provide processing capacity for non-LHC particle physics. In these parallel-processing modes, each computational core processes a moderately-sized block of events at a time: the high throughput is due to the 'trivial parallelisation' of independent events.

The Grid capacity devoted to each LHC experiment is around 300k CPU cores and an exabyte ($1 \text{ EB} = 1 \times 10^{18} \text{ B} = 1 \times 10^{9} \text{ GB}$) of data. These computers run 24/7, with processing power distributed internationally in proportion to countries' activity in different particle-physics experiments. Data, both real and simulated, is distributed across these Grid sites so that data-processing and analysis jobs can be sent to the data rather than vice versa, which would be inefficient given the huge data volumes. As collider data is naturally centred at the experimental site, CERN itself is the host of the central 'Tier 0' Grid site which performs primary real-data processing, with several large national facilities providing 'Tier 1' capacity, supported by a larger set of 'Tier 2' clusters. Simulated data, not being tied to the experimental lab, can be initiated anywhere across the Grid, and is duplicated by request to major sites for access and redundancy. Coordination of this **distributed**

data management (DDM) is a major task, comparable to operation of the physical experiment hardware.

7.2 Event generation

As described in detail in Chapter 3, most event generators are built around a core of perturbative field-theory calculations, increasingly at one-loop next-to-leading order (NLO), or even two-loop next-to-next-to-leading order (NNLO). These fixed-order calculations are dressed for realism by a set of more approximate techniques: perturbative 'parton shower' (PS) Markov processes for additional radiative corrections, non-perturbative hadronisation and multiple-scattering models, and a complex suite of particle-decay modelling. The latter can be enhanced through using 'afterburner' programs such as EvtGen, which update pre-generated events to improve specific aspects of the modelling, e.g. heavy-flavour hadron decays.

7.2.1 Event records

As used in experimental collaborations, the fully-exclusive particle-level (as opposed to the pre-hadronization parton-level) simulated events are stored for use as the **MC truth**—the physical, and in-principle exactly measurable events which the rest of the experimental simulation chain attempts to detect (in distorted form) and subsequently reconstruct. Experiment software systems differ, but all store an **event record** for each event, which lists the type (or **particle ID** (PID)) and kinematic properties of the particles produced in that event.

Two main experiment-independent formats, LHE and HepMC are in current use at the MC generator level, the former used for parton-level events and read as input to PS programs and matrix element (ME)/PS matching/merging procedures, and the latter as the native in-memory and data-exchange format for MC generator output events. The on-disk forms of this information are in development, in response to the rise in use of highly parallel computing systems for high-precision partonic event generation, and constraints on LHC data-storage volumes. Both formats encode events as graphs, with interactions at the vertices and particles along the graph edges. Part of an example $t\bar{t}$ event simulated by the Pythia 8 MC generator is shown in Figure 7.2, indicating the great complexity of these simulations.

Whilst LHE records are used to pass partonic events from dedicated ME calculators to the next stage, the format seen by experimental analysers, or in particle-level phenomenology studies, is usually HepMC. The core hard-scattering process from the LHE file is typically rendered into the HepMC graph structure, although not in a standardised way: different event generators may represent the hard-scattering in different forms, e.g. as a single vertex with incoming and outgoing parton legs, or as something more suggestively amplitude-like with internal structure. It is sometimes desirable to propagate the original LHE records along with the more complete MC generator events, mainly for debugging purposes. This is non-trivial, as merging and matching procedures may discard events and reassign their event numbers; the most recent versions of HepMC permit embedding of LHE event data in MC generator event objects for this purpose.

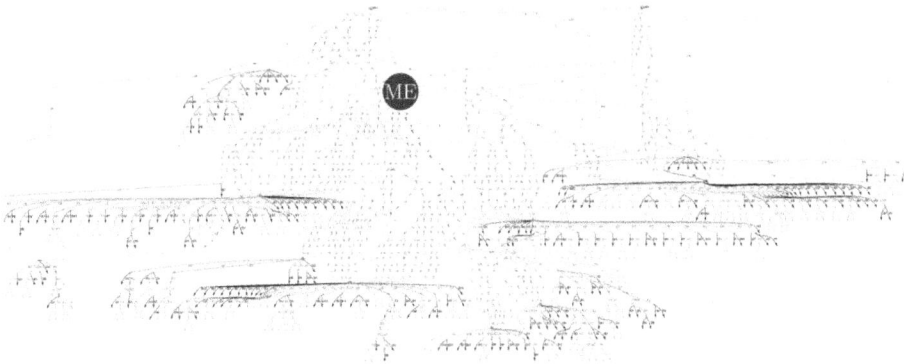

Figure 7.2. Part of a typical HepMC event-record graph for simulation of $t\bar{t}$ production and decay by the Pythia 8 MC generator. The beam particles are at the top of the graph, and simulation proceeds with dashed lines for the pre-hadronisation matrix-element (a single vertex, indicated by the ME mark), PS, and multiple partonic interactions (MPI) model products, and solid lines at the bottom for the final-state particles and hadron decay chains. The complexity of the graph reflects a combination of computational modelling methods and physics, and does *not* fully represent a classical history of event evolution.

While intuitive and reminiscent of Feynman diagrams, the impression given by LHE and HepMC of an unambiguous event history is of course false: every event is produced in a quantum process, with the hard scattering composed of multiple, interfering process amplitudes, perhaps including an entirely different set of internal 'mediator' fields. It is hence impossible to infer production by a particular Feynman graph: at best, the single representation written into an event record will represent the amplitude with the largest non-interference term in the squared ME. The same quantum ambiguity same applies to the PS, which—despite appearances—represents not a physical and time-ordered set of splittings, but the calculational techniques used to effectively resum multiple emissions of quantum chromodynamic (QCD)-interacting particles.

A good rule of thumb is to treat the pre-hadronization 'parton record' as for debugging purposes only, unless deeper interpretation is unavoidable and performed with great care; the post-hadronization particle record, including particle decays, on the other hand, can be happily interpreted as a semi-classical history of independently evolving, non-interacting states. Status codes are attached to particles and vertices, so that they can be unambiguously referred to in event records, and a final unique particle/vertex identifier, the **barcode**, is frequently used to label the origin of an event element in the primary generator or in later processing steps either at the generation or detector-simulation stages[2]. In addition, the type of each particle is referred to using a numerical code called a **PDG ID code**, or **particle ID code**. This is defined in a publication maintained by the Particle Data Group, with specific codes attached to each particle (e.g. 11 and −11 for electrons and positrons, 13 and −13 for muons and antimuons, and so on).

The kinematics and particle IDs of each event, and the internal structure are the headline features of an event record, but several pieces of metadata are additionally

---

[2] Note, however, that this convention is gradually being replaced with separate integer codes for the unique identifier and origin-label facets of MC particle identity.

propagated. Particularly crucial are the MC cross-section estimate, and sets of event weights—we deal in some detail with the latter in the next section. The former may seem strange as an event-wise property, as a cross-section is a physical quantity associated to a process type (and kinematic selection phase-space), rather than to individual events. Although this is correct, an MC generator's estimate of the process cross-section is, like individual events, obtained by sampling from the kinematic phase-space and evaluating the ME at each sampled point: while there is a pre-generation sampling (a.k.a. phase-space integration), the statistical precision on this estimate is improved by the further sampling that occurs during the generation of events. This ever-improving (although in practice, hopefully, rather stable) estimate is hence usually written on an event-by-event basis, with the last event's value being the best estimate, which should be used for physics purposes.

7.2.2 Event weighting and biasing

Not all MC events are equal! There are a variety of reasons that some events may be 'worth' more or less than average and this is reflected in **event weights** to be used when drawing statistical conclusions from e.g. binned histograms filled with property values over a set of events.

In fact, each event can have *multiple* weights, collected in a per-event tuple[3] called the **weight vector**. The first weight in this tuple is called the *nominal*—the default configuration of scale choices, use of a default parton distribution function (PDF), PS variations, inclusion or not of electroweak corrections, etc—and the others are systematic variations on that nominal configuration, or occasionally unphysical trackers of the generation process. In modern event generation, the variation **weight streams** in the weight vector are identified with textual weight names, which identify the origins of each stream. The number of weights can rise as high as 500–1000 per event, bloated in particular by PDF uncertainties: these not only include typically 30–100 variations *per PDF set*, and sometimes more, but the standard convention is to take the envelope of uncertainties between several such leading global-fit sets[4].

All entries in the weight vector, including the nominal, can differ from 1.0. As discussed in Section 5.3.7, a uniform scaling of all weights has no effect on distribution shapes nor on the effective event count N_{eff}—a degree of freedom used by some but not all generators to encode process cross-sections—but non-trivial weights distributions can arise for many reasons:

Imperfect sampling
ME phase-space points, i.e. the kinematic configurations of incoming and outgoing partons on the ME, represented by the symbol Φ, are chosen from a **proposal density** function, $p_{\text{prop}}(\Phi)$, usually an analytic distribution or sum of distributions amenable to the inverse cumulative density function (cdf) sampling technique discussed in

---

[3] In mathematics, a tuple is an ordered list (sequence) of elements.
[4] This motivated the construction of the approximate, pre-combined PDF4LHC PDF set, whose uncertainty bands match those of the total envelope to a reasonable approximation for many purposes.

Section 5.7.1. The proposal distribution will not exactly match the true, far more complicated squared ME function —if we knew this in advance there would be little need for MC techniques—and hence a scaling is required to obtain the physical distribution. This scale-factor is the sampling event weight, given by the ratio of the true ME pdf $1/\sigma \, d\sigma/d\Phi$ to the sampling proposal pdf p_{prop}, evaluated at the phase-space point Φ.

Merging and matching
Some ME/PS merging and matching schemes, such as the CKKW-L treatment which computes weights from the relative probability of an event's partonic configuration to have been produced by a lower-multiplicity matrix element combined with PS emissions. This merging weight is combined with the sampling weight;

Systematics
One can explore the effects of systematic uncertainties such as those mentioned above, by performing systematic variations of elements of the event-generation calculation. Each mode variation is represented by a weight that departs from the nominal weight for each event;

Phase-space biasing
Intentional biasing of the phase-space sampling to achieve a more efficient population of events for the physics application.

The first two of these weight origins are accidental, and in an ideal world of perfect proposal densities and PS algorithms would be equal to 1. But the latter two are quite deliberate. We will address these two types of weight in order.

Accidental weights: sampling inefficiencies and unweighting
Sampling weights from the ME and the ME/PS matching procedure are undesirable: they reduce the statistical power of an event sample of fixed size. For example, a set of weighted events generated directly from the phase-space proposal density will produce physical distributions if filled into histograms using the weights, but if the mismatch between proposal and true pdf is large, then the sampler will have picked many phase-space points that are unrepresentative of the physical process, and they will have correspondingly small weights. Further processing of such low-weight events—through the parton shower, hadronisation, MPI, decays, and perhaps onward into detector simulation and reconstruction—is typically a waste of CPU power, as such events contribute negligibly to assembling a well-converged distribution of events according to the physical cross-section distribution. At NLO and higher computational orders, the problem is made even more intense by the presence of negative weights from the subtraction of loop-term infrared divergences, which actively cancel the contributions of even high-weight samples and further slow the statistical convergence.

Hence, in practice, these weights are usually eliminated by rejection sampling inside the MC generator: weighted events are randomly discarded proportionately to the absolute value of the nominal sampling $\otimes$ matching weight—the absolute value

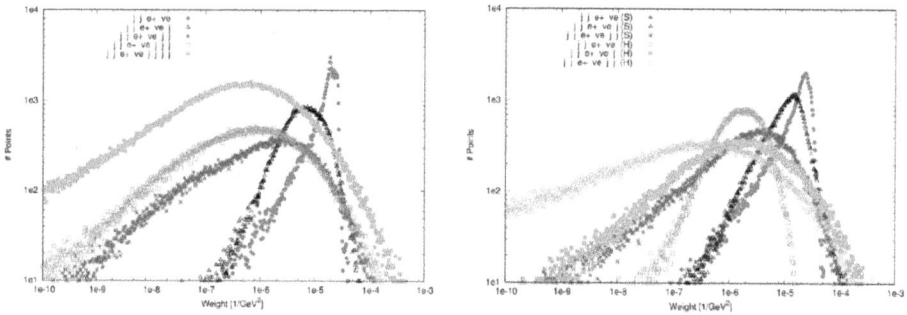

Figure 7.3. Distributions of event weights from the Sherpa event generator in $W(\to e^+\nu_e) + \geqslant 2$ jets events simulated at LO (left) and NLO (right) in the QCD coupling. The wider the weight distribution, the less efficient the unweighting step of the generation. It can be clearly seen that larger numbers of additional jets make the distribution much wider, due to the increasing complexity of the ME as a function of the larger phase-space, and that the intrinsically more expensive NLO calculations (show up to only two additional jets) amplify this problem. CREDIT: plot by Stefan Hoeche/Sherpa Collaboration

retaining the significant negative weights—and the resulting weight vector divided by that nominal absolute-weight to create an unweighted event set which (at least for the nominal stream) converges faster to the physical phase-space distribution. This is referred to as the **unweighting** procedure[5].

Unweighting is a waste of events by some measure—after all, you went to the trouble of sampling the ME and maybe generating a couple of extra parton emissions via ME/PS matching—but it is usually beneficial overall because the punitive expense of downstream processing (detector simulation and reconstruction) is avoided for events not worth the expense. It would be much better, of course, to use a generator with an efficient sampling of the ME and phase-space, but for multi-leg and beyond-tree-level generation this is a tough challenge.

Figure 7.3 shows how additional jets and the inclusion of one-loop terms in MEs both broaden the sampling weight distributions for Sherpa $W(\to e^+\nu_e) + \geqslant 2$ jets events, due to the unavoidable mismatch of current proposal densities to the true phase-space pdf. The deviations of these distributions from ideal delta-function spikes to broad distributions creates two problems. First, low weights lead to poor unweighting efficiencies and hence multiply the CPU cost for generation of unweighted events. This, however, is expensive but not physically fatal; a potentially greater second problem is the tail toward *high* weights—it is possible for the sampler to only discover the true maximum-probability regions of the true matrix element relatively late in a generator run. This creates a problem for unweighting: one cannot rejection-sample events whose sampling weight is outside the [0, 1] range, and so their high weight must be preserved. In bad cases, either from numerical instabilities in the ME code or from genuine high-probability phase-space configurations, this can lead to single events with huge weights which create spikes in observable distributions. There is no magic solution for this, but some heuristics are discussed in Section 7.2.3.

---

[5] In practice, even 'unweighted' events tend to apply a low cutoff so they do not waste time on completely negligible events.

Deliberate weights: systematic variations and kinematic biasing

So much for accidental sources of event weights, which are usually removed by unweighting. What about the weights we induce on purpose? As all non-trivial weight distributions necessarily reduce N_{eff}, this seems counterintuitive, but there are two reasons: mapping systematic variations without making whole alternative event sets, and targetting the sampling of events to converge not to the physical distribution of events in partonic phase-space, but to a biased distribution more efficient for a specific purpose. We address these two use-cases in turn.

Variation weights: The gain of weighting for evaluting systematic variations of MC/theory parameters is clear. These are used primarily to produce uncertainty bands rather than determine nominal predictions, and without weighting would require running the generation several hundred times, at huge CPU and disk-space cost. Such replica runs are indeed currently necessary to estimate the effects of un-reweightable modelling uncertainties such as MPI and hadronisation, but are largely limited to 'core' processes such as V + jets, $t\bar{t}$ and single-top, and a few other electroweak processes of relevance to many analyses as a consequence of their CPU cost. For the many variations such as ME and PS scales and pdf error sets, providing several hundred weighted 'views' of each event's kinematics is far preferable to re-generating statistically independent and often detector-simulated MC samples at unsustainable cost. There are implicit assumptions here: that the systematic variations are not so wildly different to the nominal that their weight stream does not statistically converge (or worse, that phase-space regions completely inaccessible to the nominal should be populated in the variation), and that the degree of inaccuracy in an uncertainty estimate is a less critical issue than inaccuracies in the overall prediction.

Biasing weights: Finally, weight distributions may also be introduced intentionally in the nominal configuration so that the event sample is biased away from the physical distribution, and toward a more convenient one.

The classic example of such intentional kinematic biasing is when obtaining an MC estimate for the shape of a steeply falling distribution—most event-scale variables, such as H_T or leading-jet transverse momentum p_T^{lead}, meet this condition, with a shape falling like $\sim 1/(p_T^{lead})^4$ or similar.

If we were to generate unweighted events naïvely from the MC cross-section distribution, $d\sigma/dp_T^{lead}$, we would end up with a phenomenal number of events at low-p_T^{lead} values (a few tens of GeV) on the left-hand side of our observable plot, and vanishingly few on the right-hand side (typically several hundred GeV to $\mathcal{O}(1\ \text{TeV})$). As the relative statistical uncertainty in each bin scales like $1/\sqrt{N}$, the low-p_T^{lead} event population will be far higher than required to balance statistical precision across the plot, even with use of narrower p_T^{lead} bins at low values, which will be eventually limited by experimental resolution. Generating events from the physical differential cross-section gives the highest N_{eff} per event, but for coverage of such variables it is an inefficient approach, and would require a vast generation campaign—mostly composed of unnecessary low-p_T^{lead} events—to obtain a statistically stable event population in the highest bins. This is bad enough for particle-level samples, and completely unthinkable for those destined to also proceed through detector simulation, reconstruction and analysis.

The simplest way to avoid this curse of falling spectra is to place minimum and maximum scale limits in the matrix-element sampling. The simplest case is in $2 \to 2$ ME processes, since momentum balance means that the kinematic properties of the two final-state partons are identical: the classic approach is to restrict the unique partonic p_T scale to a min–max range, $\hat{p}_T \in [\hat{p}_{T,i}^{min}, \hat{p}_{T,i}^{max})$, for a set of p_T **slices** labelled by the index i. These match up against other such that each $\hat{p}_{T,i}^{max} = \hat{p}_{T,i+1}^{min}$ to cover the whole spectrum range, with the final $\hat{p}_{T,i}^{max} = \infty$. Similar tricks may be performed in any suitable scale variable, including H_T, or the leading (non-unique) $\hat{p}_T$ in multi-leg ME events, and many event generators have a special code implementation called a 'hook' or 'bias' mechanism that allows the user to steer the process. Performing the slicing at ME level avoids the cost of computing PSs and further expensive event dressing for events to be discarded, particularly the dominant population at scales below the current $\hat{p}_{T,i}^{min}$.

The sliced events then need to be stuck back together, which introduces batched event weights, one for each slice i and proportional to the cross-section ratio to the lowest slice, $w_i = \sigma_i/\sigma_1$. As the final particle-level variable (e.g. p_T^{lead}) includes effects from jet-cone size, underlying event, and shower recoils, it will not be strictly limited to the ME-level ranges, but nevertheless this construction sums the slightly over-lapping particle-level distributions correctly to recover the smoothly falling physical distribution from the 'sawtooth' pattern of event-number distribution. The effect has been to turn one huge disparity of naïve event-generation rates into a set of smaller ones, with microcosms of the main problem taking place from the low to high edges within each slice. This is illustrated in the left-hand plot in Figure 7.4.

A smoother, more refined approach to biasing, often called **enhancement**, is to modify the proposal and acceptance functions in the ME sampler to generate

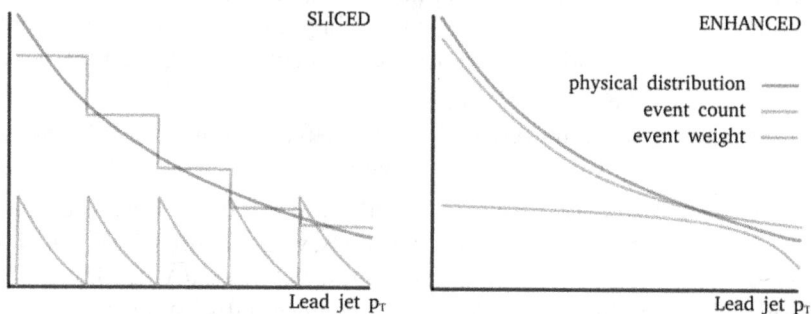

Figure 7.4. Illustrations of event-generation biasing via jet slices (left) and smooth enhancement (right). In both cases, the same physical distribution (the blue line) is obtained, but using different approaches to bias weighting. The slicing method assigns the same weight to all events in each slice, as indicated by the red stepped distribution, with the number of events decreasing steeply within each slice (the grey-line sawtooth pattern). The enhancement method by contract assigns a smoothly varying weight to each event as a function of its lead-jet p_T (the grey line), with an event population kept relatively flat across most of the spectrum, the weight corresponding to the ratio of physical/event-count distributions.

smoothly weighted events whose number distribution does not approach the physical one, i.e.

$$P_{\text{prop}}(\Phi) = P_{\text{phys}}(\Phi) \equiv \frac{1}{\sigma}\frac{d\sigma}{d\Phi} \quad \rightarrow \quad P_{\text{prop}}(\Phi) = P_{\text{enh}}(\Phi) \equiv f(\Phi)\frac{1}{\sigma}\frac{d\sigma}{d\Phi}, \quad (7.1)$$

for **enhancement function** $f(\Phi)$. This can be thought of as an intentional version of the proposal/ME mismatch that causes such problems for high-precision event generation, with the resulting enhancement event weight being $w_{\text{enh}} = 1/f(\Phi)$, the inverse of the intentional bias. This way the population of generated events follows the enhanced $P_{\text{enh}}(\Phi)$ distribution, while the distribution of weights follows the physical $P_{\text{phys}}(\Phi)$.

The canonical example is again the falling $p_{\text{T}}^{\text{lead}}$ spectrum, as shown on the right-hand of Figure 7.4: events are sampled and unweighted according to a biased distribution such as $\tilde{\sigma} = \hat{p}_{\text{T}}^{m}\hat{\sigma}$ for $m \sim 4$, with each event carrying a weight $1/\hat{p}_{\text{T}}^{m}$: high-p_{T} events are hence smoothly generated more frequently than they 'should' be, with weights reflecting the degree to which they should be devalued to recover physical spectra. In this case, with well-chosen m, a reasonable event-sample size, with near-equal event population across the physical spectrum, is possible without the rigmarole of juggling multiple event-slice samples and slice weights. While technically there is a loss of statistical power in such weighting (as in the slicing scheme), what really matters is the shape of the weight distribution within each observable bin: smooth sampling enhancement achieves this near-optimally, without even the residual sawtooth pattern of event statistics across the range of simulated scales. For event generators which do not natively support enhancement functions, the same idea can be used *post hoc* to improve sliced generation, by randomly rejecting generated events proportional to a $(\hat{p}_{\text{T}}/\hat{p}_{\text{T},i}^{\max})^m$ factor: the only difference is that now, rather than weights much less than unity for high-p_{T} events, each slice contains weights $\gg 1$ for events close to its $\hat{p}_{\text{T},i}^{\min}$.

7.2.3 Large-scale event-generation logistics

When deploying large-scale event generation, many logistical issues arise in order to avoid wastage of CPU, or artifacts in the resulting event sets and their variable distributions. To make best use of Grid resources, which are 'high throughput' rather than 'high performance' computing, being composed of large numbers of unexceptional machines, the generation is split into many separate jobs. There is no absolute definition of how large a job should be, but a good rule of thumb is that it should complete in a number of hours, and much less than a 24-hour period. Bookkeeping of the parallel deployment is essential to make sure that different, tracked random-number seeds are used for each job, to avoid event duplication and allow randomly-triggered problems to be reproduced and debugged. These small jobs are typically denominated in factors of 1000 events, for ease of later combination into samples efficiently processable through detector simulation and reconstruction.

For processes with complex MEs, typically with multileg QCD corrections and most critically for NLO samples, deploying a large number of parallel jobs naïvely would be wasteful as all would need to spend a large fraction of their runtime on the initial integration step: sampling the ME phase-space to identify the typical set of dominant kinematic configurations, the maximum-$d\hat{\sigma}/d\Phi$ point, and otherwise mapping out the shape of the proposal density function for event generation. Assuming they all converge on the asymptotically correct mapping, each of the many parallel jobs would have done the same expensive job. In fact, for high-complexity processes this approach would be completely unworkable on the 24-hour nominal timescale of a Grid job: pre-integration runs can take CPU months! So this is an obvious opportunity to save resources by pre-computing the phase-space map and sharing the integration result between many event-generation jobs.

Instead, integration runs are typically computed in a single run on a large computing cluster, using MPI (message-passing interface), multi-machine or multi-thread parallelism to accelerate the sampling. The resulting phase-space map is saved as a **gridpack** file (or collection of such files) which can be distributed with subsequent generation jobs to specify the initial proposal density in place of a new integration. The fact that such a set of parallel jobs relies on the same single MC integration (i.e. the gridpack) means that the cross-section estimates for each job can be highly correlated, and averaging over many such jobs reduces only one statistical component of the cross-section uncertainty. It is hence critical that gridpack pre-integrations be well converged, despite the computational expense.

The assumption that pre-integration does reach a stable approximation of the true ME maximum is not always correct, and it is not infrequent for event generation in huge 'bulk' samples of $\mathcal{O}(100\,\mathrm{M})$ or more events to happen upon points—physically true or numerical anomalies—in the phase-space which are higher-weighted than any encountered in the gridpack integration. Such discoveries invalidate the assumed overestimation of the true ME by the proposal function, and lead to occasional very high-weighted events in the resulting sample, which make themselves known as troublesome **high-weight spikes** in observable distributions. Ironically, such problems become more likely the larger and more CPU-expensive a sample becomes, so 'just generate more' is not a solution... and of course generating less also does not help! A pragmatic approach is hence typically taken to truncate the maximum absolute weight $|w|$, removing the visible spikes, but leaving MC generation coordinators uneasily aware that high-precision generation campaigns are at all times in danger of discovering their own invalidity.

A set of post-processing or 'afterburner' phases are often performed on event samples generated within experimental collaboration frameworks, to further 'improve' their physics content, or finesse their compatibility with later stages. For example, as dedicated high-precision matrix element computations are nowadays used as the input stage to general-purpose generator programs (which formerly computed all MEs internally), dedicated generators are often run on the shower-+hadronisation event record to provide a more precise implementation of e.g. QED radiation corrections or heavy-flavour hadron decays—at least to the extent possible from the semi-classical event graph. Another such variant is the use of **repeated hadronisation** of events whose processing has been frozen at the end of the parton shower evolution, until the primary hadrons (and perhaps specific decay chains) required by a particular analysis have been produced. Post-processing improvements may also be more mundane, such as the removal of spurious event-graph structures irrelevant to physics interpretation but potentially problematic for unwary downstream-processing code.

The final operation applied to generator level samplings is post-generation filtering. We have already encountered one motivation for filtering: the approach used to flatten steeply falling event-scale populations within $\hat{p}_T$ slices was implemented in this way. Filtering on the flavour aspects of events is also important for much physics-analysis interpretation: jet-flavour configurations like $X + c/b$, $X + cc/bb/cb$, for some final-state X, are important because their relative rates according to MC are often a poor match to observations in experimental data, and hence floating their relative normalisations in analysis fits is standard practice. In some important cases, such as the $b\bar{b}$ jet background to $H \rightarrow b\bar{b}$, it also makes sense to simulate a higher number of events than for 'less interesting' flavour combinations. But cleanly separated event samples of these various components cannot be obtained simply by specifying the corresponding MEs in orthogonal process generation runs, as QCD production of c- and b-flavour heavy hadrons in addition to core hard-scattering processes can occur anywhere in the event modelling, from the ME to the PSs and even the multiple partonic interactions machinery. While inefficient, the most physical way to assemble such flavour-component samples is by post-hoc event filtering to separate inclusive event samples into their light- and heavy-flavour subsets. It is naturally harder to assemble large filtered b and c samples than light-parton ones, and the SM samples containing hundreds of millions of high-precision W/Z + HF jet events account for the largest single sink of event-generation CPU for the ATLAS and CMS collaborations. Filtering is also useful to remove critically flawed events which would stymie further processing, and to monitor and flag problematic rates of such issues.

7.2.4 'MC truth' and beyond

The output of the event generation step is a set of data files containing the full event records for each generated event. For some MC event samples, the chain ends here: they are for use in comparisons to detector-corrected measurements only, and the extent of their further processing is reduction of their full 'MC truth' data (noting our earlier caveats about how the 'truth' may not be physically true!) to a more limited **derived analysis data** or 'truth derivations' for data-reduction purposes (to be described in Section 7.5.1). But for most, further processing takes place to allow their interpretation, either as signal or as background processes, on a nearly equal footing with real collider events as seen after event reconstruction. This requires modelling of the interaction of the final state particles in each event with the detector material and data-acquisition electronics.

7.3 Detector simulation and digitization

The role of detector simulation is to transport final-state particles through a model of the detector material and geometry, accounting for both scattering interactions with the material, and curvature in the detector's magnetic field. Both these classes of interaction can scatter the particle and result in the production of secondary particles, which in turn need to be transported. Each material interaction will typically deposit some energy in the detector, which for **sensitive** detector elements stimulates an electronic readout, modelled by digitization algorithms.

The 'gold standard' in detector simulation is the use of so-called **full simulation** tools, in which every particle is individually, and independently, transported though an explicit model of the detailed detector material map, with Monte Carlo (i.e. random-number based) algorithms used to simulate continuous energy loss as well as discrete stochastic processes like the production of secondary particles through electromagnetic and nuclear interactions with the material. This requires a detailed model of the detector geometry and material characteristics, as well as a map of the detector magnetic field in which charged particles curve. These models can be extremely large and complex for a modern general-purpose detector, meaning that 'full-sim' transport of interacting particles is very computationally expensive.

The main full-sim toolkits in collider physics are the Geant and Fluka libraries (and FluGG, which is 'Fluka with Geant geometry code'), both written in C++. As the currently most used transport library, we base our summary here on Geant4. Most experiments will implement their detector simulation as a custom C++ application, building the detector geometry from nested volumes constructed from elementary 3D shapes such as boxes, cylinders and cones, which can be added and subtracted to define **logical volumes** corresponding to coherent detector elements. These logical volumes are linked to material properties used to parametrise interaction processes, as well as being linked to custom signal-handling functions for **sensitive volumes**—those in which energy deposition can trigger a digitization signal.

Considering the complex shapes in e.g. large silicon tracking systems, and convoluted layers of calorimetry, the resulting complete geometry map for an LHC experiment like ATLAS or CMS can occupy $\mathcal{O}(1\ \text{GB})$ in memory. For maintainability, incorporation of varying **run conditions** through the detector lifetime, and because misalignment of coherent subsets of detector components is an important input to evaluating systematic uncertainties on physics-object reconstruction, the in-memory assembly of detector geometries is itself typically a complex application, maintained by the experiment's full-simulation expert team and linked to the conditions database.

7.3.1 From event generation to simulation

The interfacing of event generation to the full-simulation library to first approximation involves simply passing the momentum four-vectors and particle IDs of all generator-stable particles from the generator event record to the transport application. This independently processes each one and returns a list of **hits**, corresponding to energy deposits in logical volumes. A simple expediency is usually applied, to simply delete non-interacting particles rather than transport them through the volume without depositing any energy: this is applied to neutrinos and a customisable list of stable, invisible BSM particles such as the lightest supersymmetric particle[6].

---

[6] And, in Geant4, a special 'Geantino' pseudoparticle which interacts with nothing and exists purely for geometry debugging and timing tests.

One crucial detail does need to be included in the 'copying' operation, and that is accounting for the spread of primary interaction-vertex positions due to the finite widths and lengths of the interacting bunches. The former are usually negligible, on the scale of 5 μm, but at the LHC the latter are substantial and give rise to a beamline **luminous region** of approximately Gaussian probability density in primary interactions, around 20 cm long. This is also called the **beam spot**, but is somewhat larger than the name 'spot' might imply.

As the separation of primary vertices along the z-axis is key for the suppression of pile-up overlays on triggered events, modelling of this longitudinal distribution is essential in simulation, and is achieved easily by sampling from a Gaussian with width and mean positional parameters appropriate to the beam conditions in the run. For MC simulation performed in advance of data-taking runs, with the beam parameters not yet fully known, it may be useful to model using a larger beam-spot than in data, to permit reweighting to match the actual conditions; the statistical wastage through the non-trivial weight distribution means this is not a preferred mode. As MC generator events are all created with primary vertices at the origin, the sampled z position of each event's location can simply be added as an offset to all vertex positions in the generator event record. Secondary modelling can be added for the benefit of precision analyses such as W and Z masses and widths, and forward proton tagging, to account for the fact that the beams have a slight crossing angle at the interaction point, and that there is a $\Delta E/E \sim 1 \times 10^{-5}$ uncertainty in beam energy: both these small effects are propagated into the generator event record via sampling of small transverse and longitudinal momentum boosts, respectively, applied to all event particles.

In practice, the selection of particle inputs to simulation is a little more complex, and it is illustrative to spell this out as the exact treatment of different particle species is of practical importance—particularly in precision measurements where the detector effects are to be corrected for (see Section 11.3). The usual inputs to detector simulation are not *all* final-state particles, but the subset of **primary final-state particles** with macroscopic decay constants τ_0. This subset is most usually chosen to match the condition $c\tau_0 > 10$ mm ~ 3 ps. This definition treats the vast majority of hadrons as decaying 'promptly' within the event generator, without concern about potential detector interaction, while in particular treating the K_S^0 & $\pi^\pm$ mesons and the Λ^0 & $\Sigma^\pm$ baryons as 'generator stable' and in need of decay treatment in the detector simulation library. Their decay particles, while reasonably considered 'final-state', are identified as produced by the transport library and are hence difficult to practically distinguish from secondary particles produced by material interactions. Depending on the definition required by each analysis, this distinct treatment may require some care by the analyser.

As experimental precision and process scales have increased, it has more recently become necessary to also include more promptly decaying particle species, such as b- and c-hadrons in the transport: these may be treated as 'forced decays', in which the more precise simulation of the decay-chain timings and kinematics from the 'in vacuum' event generator are interleaved with B-field bending and material interaction from the detector transport. Special custom treatments are increasingly used for transport of BSM long-lived particles, which may be either electrically charged or neutral, and may interact with detector material for macroscopic distances before decaying within the detector volume and leaving a **displaced vertex** or **emerging jet** experimental signature—a challenge for reconstruction algorithms.

7.3.2 Particle transport

The transport of particles in full simulation involves *stepping* them through the geometry. The fewer steps that can be taken, the more computationally efficient the simulation can be. The operation of a transport code centres on a *stack* (or perhaps several stacks) of particles to be stepped, with the top particle on each stack being the one that is currently undergoing processing. A stepping algorithm decides how far to move that particle, registers the energy deposition as hits in the detector volume, and—depending on which process is determined to have limited the step size—either updates the particle at the top of the stack or replaces/augments it with new particles from the interaction. Several processes compete to limit the step-size, and must be computed for every step:

- distance to boundaries of material volumes—more granular geometries will require more steps;
- particle decays;
- discrete interaction processes that modify kinematics and particle content;
- discontinuities in B-field map;
- tracking cuts to manually stop certain particle species from travelling further than a fixed distance in a particular material.

The baseline trajectory of a particle is calculated assuming a smoothly varying B field, amenable to a standard numerical solver technique such as Runge–Kutta iteration (optimised around the baseline of helical motion in a uniform B field). For the current material, myriad soft interaction processes can be aggregated into **continuous energy-loss** rates specific to particle and material type. The effects of relativistic particle motion on the precession of its spin angular momentum are similarly modelled. Wherever possible, calculations such as remaining proper time to decay are cached or cheaply updated with each step, rather than requiring a full recalculation every time.

The most obvious upper limits on travel distance are the volume boundaries and step-limitation cuts. The latter are easy to deal with, but the former potentially very complex and expensive to compute for complex volumes such as the curved trapezoidal elements of the ATLAS hadronic calorimeter. In Geant4, a sophisticated geometry system is responsible for calculating intersection points to logical volume boundaries: this is done efficiently using a ray-tracing technique called voxelisation which, during initialisation, recursively subdivides the world volume into 3D volume elements or voxels until each 3D bin contains only a small number of volumes. These voxels are then used to optimise the intersection-search algorithm, by reducing the number of potential next volumes to be checked.

Physical processes impose less hard boundaries on step size, as whether an interaction interrupts a step of given length is a probabilistic rather than deterministic question. Modelling stepping limits requires a stochastic approach that considers all possible interactions and decays for the current material. All such interactions have an energy-dependent **mean free path** $\lambda(E)$. For the decay of a relativistic particle, $\lambda(E) = \gamma(E)v(E)\tau_0$, while for material interactions a sum is required over the constituent fractions of molecular or nuclear species characteristic to the volume's material: if a material of density ρ is composed from constituent types i, with mass fractions f_i then the inverse **interaction length** is

$$\frac{1}{\lambda(E)} = \rho \sum_i \frac{f_i \sigma_i(E)}{m_i} = \sum_i n_i \sigma_i(E),$$

where $\sigma_i(E)$ is the energy-dependent interaction cross-section with constituent i. m_i is the constituent's molecular mass, and hence n_i is its number density.

A finite step of size L is naturally written in terms of its effective number of path lengths, $n_\lambda(L) \sim L/\lambda$—actually derived differentially as

$$n_\lambda(L) = \int_0^L dl \frac{dn_\lambda}{dl} = \int_0^L \frac{dl}{\lambda(E(l))}$$

to account for the continuous energy loss. The probability of the particle surviving (i.e. not interacting) over the step is $P(L) = e^{-n_\lambda(L)}$: as this is a simple exponential form, at each step a simple MC method to stochastically generate a number of interaction lengths to be traversed before interaction is to sample a uniform random number $\mathcal{R}$ between 0 and 1 and compute $n_\lambda = -\ln \mathcal{R}$. This can then be inverted to give a physical step size. Applying this algorithm independently to each competing interaction process results in a set of randomly sampled stopping distances, appended to those from the volume boundary and range cut: the process with the smallest such distance is applied, the particle list and kinematics updated, and the algorithm repeats.

As each particle is transported independently of all others (the very good approximation being that the trajectories and energy deposits from one particle do not affect the interactions of another), various strategies have been explored, and continue to be explored, for reduction of particle-transport cost by grouping similar particles or similar material regions together for processing in batch, and for simultaneous transportation of many particles on independent processing threads or ranks. These optimisation activities are highly coupled to the physics goals of the simulation, the detector complexity, and the available computing architectures, and are under continuous development both within the Geant4 collaboration and in experiments.

7.3.3 Discrete interactions

In addition to particle decays, about which we have nothing crucial to add here, there are many kinds of discrete interaction with the detector material which need to be simulated (and the results of the discrete interactions propagated into a detector-level extension of the generator event record). Different codes are structured differently, but roughly speaking these interaction processes fall into the following distinct categories:

Electromagnetic inelastic interactions and scattering: Ultimately, all interactions of particles with a detector are resolved through electromagnetic interactions: indeed, roughly half the CPU time in a full-simulation is usually spent on EM interactions of particles with energies below 10 MeV! Much of this is accounted for via the continuous energy loss parametrisation rather than discrete electromagnetic interactions, but discrete effects are also required. These include

- For charged particles ($e^{\pm}$, $\mu^{\pm}$, and charged hadrons): bremsstrahlung and synchrotron radiation (photon radiation due to particle acceleration in nuclear B fields and detector B field, respectively); ionisation; production of δ-rays (electron emission from atomic shells); and positron annihilation.
- For photons: Compton scattering from atomic electrons, $\gamma \to e^+ e^-$ **photon conversion** (which requires recoiling material to preserve Lorentz symmetry), and the photoelectric effect.

The nature of these processes is to emit additional electrons and photons, which are appended to the particle stack and increase the number of particles to be transported. In dense material, particularly electromagnetic calorimeter volumes which are specifically designed to encourage such interactions, the resulting ballooning of the stack and short step-lengths account for a very large fraction of processing time. This can necessitate computational optimizations, such as aggregating hits to reduce output event size, use of **fast calorimeter simulation** approximations, and use of approximate but more performant electromagnetic modelling.

Electromagnetic multiple scattering is also applied as a correction to the post-step position by the calculation of step corrections from tabulated material properties.

Optical photons

Non-ionising radiation is unimportant in many detectors and hence optical photon handling is omitted from default electromagnetic (EM) transport. Optically sensitive systems such as **Cherenkov ring detectors** do require this additional treatment, to simulate the effects of Rayleigh scattering, refraction, reflection, photon absorption and the Cherenkov effect.

Hadronic interactions

These are the least well-constrained interaction processes, as the plethora of potential nuclear collisions and difficulty of their experimental measurement means they are less empirically determined than EM interactions, and the complexity of the underlying theory precludes full *a priori* calculation. Hadronic interactions are also very sensitive to energy scale, and in detector simulations designed to operate from cosmic ray energies down to nuclear and medical engineering applications, the modelling may need to cover $\mathcal{O}(15)$ orders of magnitude in interaction scale. As a result, a patchwork of hadronic interaction models is used at different energy scales.

Hadronic models are a mixture of data-driven parametrisations and theoretically motivated models. The dominant models in Geant4 are based on nuclear parton-string models, not so different in motivation from the Lund string model we saw used for event generator hadronization modelling in Section 3.4.1.

The main two high-energy hadronic interaction models are the **Quark–Gluon String** and **Fritiof String** models, denoted respectively as QGS and FTF in Geant4. These are valid at

$\gtrsim 20$ GeV and $\gtrsim 5$ GeV respectively but, particularly in hadronic calorimeter material, it is important to simulate the hadronic cascade of increasingly low energies down to much softer interactions, $\mathcal{O}(20$ MeV$)$. These are handled by the Binary Cascade and Bertini Cascade intranuclear cascade models (BIC and BERT), or a nuclear pre-compound model (P). Finally, there is the option of a more detailed high-precision neutron interaction model (HP) down to energies below 20 GeV: this is usually prohibitively expensive in HEP calorimeter simulations and primarily intended for radiation-safety applications.

Our reason to introduce the Geant4 nomenclature here is that many combinations of high, intermediate, and low-scale hadronic physics models are reasonably possible. This scope for variation, with no *a priori* best choice, and with varying CPU costs, leads to many possible choices for a detector simulation, with variation between models giving rise to an additional systematic uncertainty. The choice of a specific set of code elements—chosen from the possible electromagnetic and hadronic interaction models—to perform simulation tasks is called a **physics list**, which is named by concatenating the model names given above.

The most commonly used physics lists are those assembled by framework authors. For LHC experiments, the most important physics lists are QGSP_BERT and FTFP_BERT, i.e. the QGS and FTF high-energy models with pre-compound de-excitation and the Bertini nuclear cascade. Other less common variations include QGSP_BIC and lists with _EMV and _HP suffixes for approximate electromagnetic and high-precision nuclear transport, respectively. These list names will be seen if you work in hadronic reconstruction and calibration (e.g. jets and MET), where they feed through into systematic uncertainty estimates on physics-object performance.

7.3.4 Fast simulation

The intrinsic complexity of full simulation leads to great CPU expense—more CPU was used for MC simulation than any other data processing in the first 10 years of LHC operations, including all stages of real-data event processing. This cost is not sustainable, as it limits the number of events that can be processed, and even by the end of Run 1 (with around 30 fb of data recorded per experiment) limited MC statistics were a limiting uncertainty for many analyses. In such circumstances, a compromise is needed between accuracy and performance, such that the overall scientific power of the experiment can be maximised given the limited computing budget. Prioritising of speed over precision is also relevant to prototyping of physics analyses, and in physics-case studies for designing future collider experiments where no detailed detector simulation or reconstruction software yet exists. Any approach to simulation which trades some accuracy for increased speed is known as a **fast simulation** technique, but within this group there are many different strategies and degrees of trade-off.

Full simulation lies at one end of the spectrum between accuracy and expense, at maximum (but still significantly uncertain) precision but huge CPU cost. At the other end lie parametrised 'smearing' simulations in which physics-object efficiencies and kinematics are derived directly from the MC-truth level via **transfer functions** fitted to full-simulation or *in situ* reference data. The relation of these two methods to one another and to reconstruction is shown in Figure 7.5. The smearing method is

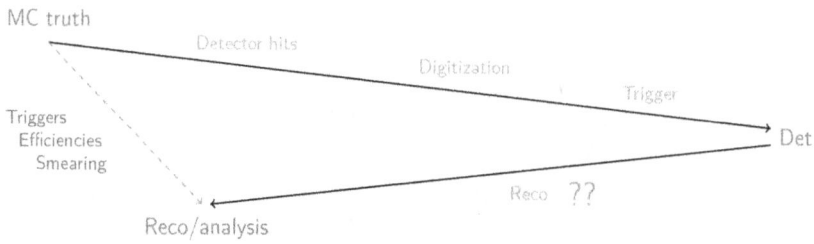

Figure 7.5. The schematic relationship between event-generation 'truth', the increased distortions of that truth induced by detector event-triggering, material interactions, and electronics readout, and the attempt by reconstruction algorithms to regain the true, particle-level picture of what happened. Fast methods for emulation of detector effects + reconstruction algorithms based on transfer functions, known as 'smearing', are feasible because of the small—but crucially important—gap between the truth and reconstructed views of each event.

extremely fast, and in fact acts as a replacement for not simply the simulation but also digitization and reconstruction stages. The compromise, of course, is that these parametrised responses are typically limited to one or two key variables, hence fast-sims lack detailed sensitivity to the specific of each event and its objects. A prominent example is that unless specific whole-event variables are included in the parametrisation, the modelled physics-object reconstruction performance will lack sensitivity to the overall detector occupancy.

In between these extremes lie a variety of semi-fast simulation approaches, in which the full simulation is replaced by an approximation for certain detector systems or particle types. Particular emphasis has been placed on calorimeter fast simulation, given the large fraction of full-simulation CPU spent on high-multiplicity calorimeter showers. A first hybrid approach to solving this problem is to use a **shower library** of pre-simulated calorimeter showers (or 'frozen showers'), in which particles that fall below a certain energy threshold within calorimeter volumes are immediately replaced with a randomly chosen shower 'completion' from the library. This approach is used, for example, in the ATLAS and CMS forward calorimeters, where the high occupancy of the detector leads to a disproportionate CPU cost compared to the physics benefit of that detector system. Even with this replacement, EM calorimetry dominates the CPU cost of simulation, but a more dynamic approach is needed in the physics-critical central region of the detector. This is achieved in ATLAS by the use of parametrised calorimeter-cell responses to the incidence of different particle species over a wide range of energies: longitudinal and lateral shower profiles can be reproduced and tuned to data using a fine grid in $\eta - \phi$, at the cost of losing fluctuations (by use of average responses) and simulation of **punch-through** of incompletely contained hadronic showers into the muon system (where they can register a fake-muon signal). The former, and details such as shower-splitting by wide-angle pion decays, are incorporated by refinements to the naïve algorithm. This approach reduces the calorimeter simulation cost by a factor of 20, and total simulation cost by a factor of 8. More recently, much effort has also been expended on the use of generative neural networks for fast calorimeter simulation, which have the potential to further reduce CPU cost and with better ability to reproduce accurate fluctuations and correlations 'learned' from sets of training events.

After application of calorimeter fast-simulation methods, the CPU cost of simulation is dominated by the tracking detectors, whose complex geometry and requirement of detailed electromagnetic simulation are computationally expensive: further fast-simulation improvements may be made there by use of parametrised physics responses and simplified geometries, achieving up to a factor 100 gain with respect to full simulation for use by physics analyses with less stringent demands on tracking and vertexing accuracy.

Modern LHC detector simulation systems integrate a mix of these methods, swappable by subdetector and particle species and energy regime, similar to the components of Geant4 physics lists, to balance the use of simulation CPU budget between analyses needing large event numbers and those needing maximum accuracy. Specific calibrations may be needed in physics-object reconstruction, to account for differences in response between fast-simulated and full-simulated detector outputs.

7.3.5 Digitization and pile-up

The digitization software is entirely detector-specific, as it converts the simulated energy deposits in sensitive detector volumes (the 'hits') into detector responses called **digits**. These correspond to the electrical signals and timings read out from the detector, including:

- parametrization of each detector's charge-collection behaviour;
- electronics noise (Gaussian, Poisson/binomial for binary outputs, or custom);
- channel-specific details;
- behaviours of the electronic read-out system, leading to raw data objects (RDOs).

Much of this behaviour is run-specific, and captured in the conditions database, documenting the status of the detector for each luminosity block basis (approximately a per-minute status report). Digitization algorithms incorporate the geometry of the hit deposition in addition to the effects listed above, which again can be a CPU-hungry operation depending on the level of detail needed to reproduce key features that impact reconstruction algorithms. As for simulation, this gives an imperative to speed up the digitization step, with **fast digitization** strategies in development to either use fast approximations to the standard digitization functions, or to increase use of trivial parallelisation with corresponding changes to how multiple contributions to digitized outputs are combined.

Digitization is also notable as the point where multiple primary collisions—pile-up—are introduced to the simulation chain. Up to this point, every particle was transported independently through the detector material, meaning that rare signal processes and **minimum-bias** interactions could be generated and simulated completely independently of one another. This is not tenable at the digitization level, as the detector responds to the sum of energy deposits (and their timing) rather than to the individual contributions. Pile-up is simulated by **overlay** of hits from inclusive

QCD samples on top of those from the signal process, and necessarily includes the in-time pile-up and out-of-time pile-up described in Section 6.5.1. The two classes of pile-up influence the detector occupancy and response in different ways: out-of-time pile-up can be reduced by fast detector systems, but others such as calorimeter read-outs may have significant 'dead periods' due to activiation by particles from previous beam-crossings. A correct distribution of pile-up primary vertex positions (particularly for the in-time overlays) is obtained automatically through treatment of the beam-spot distributions in every independent event simulation.

An algorithm to implement pile-up overlay needs to model the bunch structure of the collider by randomly assigning the signal event to a particular bunch-crossing identifier (BCID) S, giving each BCID i a mean number of in-time inelastic interactions μ_i from the run distribution with mean $\langle\mu\rangle$, and finally randomly sampling an actual number of interactions from the Poisson distributions to assign a number of pile-up overlays to each BCID,

$$N_{\mathrm{PU}}^i \sim \mathcal{P}(\mu_i) - \delta_{iS}, \tag{7.2}$$

where the Kronecker-delta term subtracts one pile-up overlay for the bunch-crossing containing the signal interaction (which 'used up' one beam-particle collision). The digitization takes the resulting sequence of timed hit collections surrounding the signal BCID to generate the set of RDOs corresponding to the triggered event. This procedure, of course, produces the pile-up distribution expected before the collider run: **pile-up reweighting** of this to match the actual pile-up $\langle\mu\rangle$ distribution for each data period is standard practice.

A complexity is added by the need for a library of pre-simulated QCD events to be used for the pile-up overlay. With typical $\langle\mu\rangle > 50$, and the pile-up time window for some subdetectors containing $\mathcal{O}(40)$ bunch-crossings, it is easy to require 2000 or more pile-up events to digitize a single signal event. In addition, these must have been simulated with the same beam and detector conditions as the signal event. While **minimum-bias** QCD interactions have lower particle multiplicities than most high-scale signal events, and can be processed using the fast-simulation techniques described earlier, 'wasting' thousands of them per signal event is not tenable. The approach taken is hence to sample soft-QCD events with replacement from a large pre-generated library, as the re-use of generic and relatively uninteresting events in a virtually boundless sequence of random combinations should not bias results. But not all 'minimum bias' events are soft: realistic pile-up simulation also needs to include secondary scatterings with significantly hard activity—from a generator hard-QCD process rather than collective soft-inelastic and diffractive modelling—and these events are sufficiently distinctive that their repetition (albeit at a proportionately low rate) can cause noticable spikes in simulated-event analyses: such events are hence mixed into the pile-up event sampling, but are never replaced.

An alternative approach to pile-up modelling, little used at the LHC (except in heavy-ion events for which minimum-bias modelling is both much poorer and more computationally expensive) but more common at previous colliders is **data overlay**: use of real zero-bias data events rather than simulation to populate the overlay libraries.

This guarantees good physical modelling of the pile-up events and their μ distributions, as the 'events' here are really the full detector activity from a whole zero-bias bunch chain containing thousands of primary interactions, but introduces significant other technical challenges, such as matching of detector alignments between data and simulation. As the outputs from data are by definition already digitized, aspects of the digitization must effectively be inverted to obtain an estimated sum of hits from the pile-up backgrounds, on to which the simulated signal event is added before (re-) digitizing (although in practice the combination is performed more directly). This requires a special detector run-mode for collecting the zero-bias inputs, as calorimeters in particular suppress below-threshold signals in their normal output, but the sum of below-threshold hits from signal and pile-up can be above threshold.

The potentially intimate interaction between pile-up overlay methods and electronic response is illustrated in Figure 7.6, which shows the timing response of cells in the ATLAS LAr calorimeters to cosmic-ray incidences as a function of time from the first hit. The rapid rise of the electronic signal at around 100 ns is followed by a negative sensitivity period around four times longer; the intention of this is that the integral is zero, suppressing out-of-time pile-up effects. The structure of the overlaid pile-up hits hence needs to account not just for spatial distribution and a representative distribution of scattering kinematics, but also the temporal structure in a rather complex way. Systems such as this cause additional complexity for data-overlay, and for fast digitization strategies which attempt to perform the pile-up overlay *after* single-scatter digitization.

Links between the digitization and the MC truth are propagated through digitization from the simulated hits, but now introducing ambiguities as the summed energy deposit in a sensitive detector may be due to several (primary or secondary) particles, including those from pile-up. This ambiguity is maintained through to reconstruction, and in general it is impossible to say that a reconstructed MC physics object 'is' a particular MC truth object, although the higher-energy and better-isolated a direct truth-object is, the greater chance of a clean reconstruction with only a small fraction of its digit inputs traceable to contaminating activity.

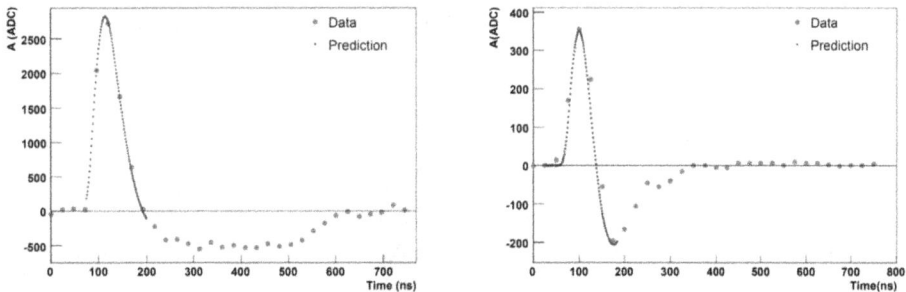

Figure 7.6. Typical timing responses of ATLAS LAr calorimeter cells to cosmic-ray signals, in the hadronic end-cap (HEC, left) and forward calorimeter (FCal, right), showing the use of pulse shaping for pile-up suppression. Plot reproduced with permission from Atlas LAr Collaboration 2009 *J. Phys. Conf. Ser.* **160** 012050. Copyright IOP Publishing. All rights reserved.

7.4 Reconstruction

We have already dealt with the key ideas of physics-object calibration and reconstruction techniques in Chapter 6. The computational implementation of reconstruction repeats the structure and issues of the generation, simulation and digitization steps discussed in the previous sections, with the distinction that it must be run on both collider data and MC-simulated inputs. For MC, reconstruction is typically run in concert with the digitization step, since the intermediate 'raw' data is not of sufficient physics interest to justify the storage requirements. Reconstruction is also re-run more frequently than the earlier simulation steps, to perform so-called **data reprocessing** that accounts for differences in collider and detector conditions compared to expectations, and the continual development of new and improved object calibrations.

Given the complexity of the detector system and the physics objects in need of reconstruction, and the insatiable scientific demand for accuracy and robust uncertainty estimates, computational resources of both speed and storage are critical limitations. Developments in reconstruction are hence focused on improving data throughput, e.g. by making use of novel computational architectures, and in the case of MC simulations, determining whether truth-based shortcuts can be taken without biasing the physics outcomes.

The basic output from the reconstruction step is a very detailed form of event data including both the reconstructed physics objects, and full information of which digitization-level information was used to identify them. This **event summary data** (ESD) is essential for development of reconstruction algorithms, but is more detailed than required by physics analyses. A partially reduced format for physics analysis, the **analysis object data** (AOD) is hence the main output from reconstruction, providing analysis-suitable views of all reconstructed physics objects.

7.5 Analysis-data processing, visualisation, and onwards

Downstream from the reconstruction lies physics analysis—the main motivation of most collider physicists' work. But this is not a single processing stage, nor is it prescribed to the same extent as the centralised data processings that proceed it. It is typically up to each analysis team to make their own decisions about how they prefer to filter, validate, analyse, and plot derived quantities from MC and data at the end of the processing chain.

7.5.1 Data derivations

If every one of the hundreds of analysis teams on an LHC experiment were to individually run over full AOD datasets from the collider and from the dominant 'bulk' MC background simulations, the computational budget for analysis would be immediately overwhelmed. Instead, each analysis group in the ATLAS and CMS collaborations provides a set of 'derivation' datasets, taking the full-sized AOD files from the reconstruction and filtering them into smaller **derived AOD** (DAOD) chunks targeted at particular types of calibration or physics analysis. This refinement can include several approaches:

Skimming

Using a combination of trigger states and reconstructed-object cuts to discard unpromising events;

Slimming

Dropping of unwanted computed quantities at either detector-hit or reconstruction level, should the target analyses have no use for them; and

Thinning

Dropping of containers of physics objects in their entirety.

The data-derivation process can take place in several layers, e.g. construction first of a 'common DAOD' from which physics-group DAODs are derived, and finally are split apart into 'subgroup' DAODs, e.g. those focused on different jet- and lepton-multiplicity samples in W, Z, t, and Higgs studies, or with different MET or H_T requirements for BSM searches. Increasingly, very focused 'mini-AOD'[7] formats have become increasingly used and published by LHC experiments.

Contrary to the impression of strict reduction of information from layer to layer, DAOD production may also compute new variables that will be widely used in their target analyses, e.g. reclustering of large-R jets from small-R ones, whose computation may require inputs from other variable branches that can then be slimmed away. The overall aim of this data reduction chain is to minimise CPU-costly repetition, while retaining an agility within physics groups to provide analysers with the evolving set of variables needed: many physicists in experiment working groups spend time and effort on designing and managing these processes.

7.5.2 Analysis data management and coding

Analysis code is distinct from the rest of the experimental (and modern phenomenology-tool) software ecosystem, in that it is not intended for wide distribution nor extensive re-use. Design and maintainability hence do not merit the same weight for analysis code as when building common frameworks such as those used in simulation or reconstruction, where software engineering concerns and methods are critical for correctness, future-proofing, and scalability.

The priority balance between careful engineering and speed hence tilts toward the latter in the analysis stage, and this freedom can be refreshing when otherwise accustomed to a rather formal and corporatised process. But correctness is still a priority, and is more easily lost with an unstructured and messy process and codebase than a neatly organised and planned one. Even apparently cosmetic aspects like documentation, usability, and standardisation are still important—just in a less formalised way—as the evolution of an analysis is rarely 100% predictable, and when you enter either internal or external peer review, you will discover the value of both flexibility and clear documentation in your analysis code.

Larger analysis teams are more likely to involve code-sharing between team members (including new-joiners, and perhaps also loss of team members over time), and to require coordination of variable naming and script interfaces between different channels which

---

[7] Or micro-, or nano-AODs. We still await pico- and femto-AODs, to current knowledge.

will need to be combined. But even in a single-person analysis, it is guaranteed that the future-you will need to re-run elements of the analysis as bugs are found, collaboration-review processes ask for extra plots and validations, and plot restylings are requested all the way through to final journal approval. In short, plan ahead and take the extra minutes to document and polish your interfaces: it will likely pay off in the long run.

As well as coding style and quality itself, the way in which you stage your data processing is important in balancing fast reprocessing against the flexibility to change definitions. Just as the centralised data-processing systems in experiments sequentially reduce data volumes by the combination of skimming and slimming (and swapping of raw variables for more refined, analysis-specific ones), this process should continue within each analysis. Staged data-reduction not only uses resources more efficiently, but also makes life far easier in the late stages of paper authorship: no-one wants to have to re-run Grid processing of the full event set in order to tweak binning details, or to have to re-run slow parameter fits or samplings to change a colour or line-style in a final histogram: well-designed staging isolates the subsequent, more-frequently iterated, analysis stages from the heavy-duty number crunching of earlier steps.

A typical data-staging scheme at analysis level will look like:

Pre-selection

Grid-based processing of MC and data DAODs resulting in pre-selected event **ntuples**. These are the HEP software name for a set of n variables (i.e. a mathematical n-tuple) per event: effectively a large spreadsheet of n features in columns, against N_{event} rows. Like earlier event data formats, ntuples are rather large files due to their linear scaling[8] with N_{event}—this stage of reduction is likely to reduce many terabytes of DAOD data to 100 GB–1 TB. While modern software chains tend to use 'object persistence' to map composite in-memory code objects to file, in the analysis stage it is more likely that 'flat' ntuples will be used, i.e. each of the n 'branches' in the data tree will be treated as independent. A concrete example is that each component of a four-vector will typically be an independent branch, rather than being combined coherently into a single object: in many cases the first action is to build standard four-vector objects from the less flexible stored components!

Analysis

Running of final analysis-cut optimisation and application, reducing to a smaller set of focused ntuples (most useful for assessing correlations or performing unbinned fits) and/or histograms (in which the per-event information has been lost). This processing is typically achievable using an institutional computing cluster rather than the distributed Grid, and will reduce data volumes to the $\mathcal{O}(\text{GB})$ level.

Numerical results

Use of the final-analysis data in fits or other manipulations achievable on a single or few machines via interactive sessions. This may include optimisations such as rebinning (combination of narrow bins into wider ones with more statistically stable

---

[8] Ntuple formats are usually compressed, using a dynamic compression algorithm such as gzip's DEFLATE, so more predictable (lower-entropy) branches may in fact scale better than linear in N_{event}. Branches with the same value through a file essentially disappear in terms of on-disk memory footprint.

populations), correlation extraction, fits, and detector unfolding. The outputs will be typically 1 MB to 100 MB in size, but ideally unstyled at this point.

Presentation

Transformation of the numerical results data into visually parseable form: sets of 1D or 2D histograms, heatmaps, fit-quality or statistical-limit contours, formatted summary tables, and data formats for long-term preservation, e.g. in the HepData database. This step is likely to be iterated many times both by the analyser alone as they attempt to find the best form, and in response to team and review comments: the choices of colours, line and uncertainty styles, label and legend text, and theory model comparisons involve a mix of physics and aesthetic tastes. Separating it from the vast majority of number-crunching means that this step is quick to run—a sanity-saving analysis life-hack, given how often you're likely to do it.

Unlike the ROOT-based central processing, analysis teams are typically free to use whatever tools and formats suit them best for these internal stages, e.g. text-based CSV or binary HDF5 as favoured by many modern Python-based analysis toolkits. All these points are as valid for phenomenological studies as in experimental collaborations, and indeed as a phenomenologist you will be far more responsible for choosing your tools, documentation, and analysis practices.

Practical tips

While analysis is a myriad-faceted and creative endeavour, and we can no more give a comprehensive guide to performing it than one could explain how to 'do' music or fine art, we end here with a few practical tips for managing analysis code and data processing, to reduce time and repetitive strain:

1. Manage all your code via a version-control system, and commit often. Being able to roll back to previous versions is a great reassurance, and allows you to clean up old scripts rather than archiving them under ad hoc names[9];

2. Use meaningful, systematic and consistent naming: of scripts, data files, directories and subdirectory trees. It really helps if your different classes of data can be identified (both by human and by code) just from the names of the files or ntuple branches, rather than needing a look-up table. Systematic naming and structuring also makes it easier to write secondary code to access them all equivalently, without having to manage a raft of special-case exceptions to your general rule. This includes tedious things like capitalisation: do you really want to have to write out all your branch names by hand because you used pT rather than pt in one of them?

3. Use relative rather than absolute paths in all your scripts, so you can relocate your processing to different systems, e.g. from a departmental cluster to a laptop, or between departments, with minimal pain. At worst, specify the absolute path to your project data in a single place;

---

[9] Enthusiasts can take this a step further: the rise of featureful code-hosting platforms like GitLab makes it now easy to set up continuous integration tests to ensure that changes do not break other parts of the framework, and containerize your code so you can revert not just to a previous collection of source code, but to an immediately runnable previous setup.

4. Stage and filter your data to minimize large-scale reprocessings, and to separate style from statistical content;

5. Document your code, at least for yourself: you won't remember how it works a few months later! Learn to use lightweight methods like comments, docstrings, README files (note that Markdown and RST plain-text formats are often rendered nicely via version-control websites like GitLab and GitHub), and auto-documenting command-line argument systems like argparse and sphinx-argparse;

6. Capture the script commands you use for important processings, by saving them to text files rather than relying on memory or your command-shell history feature;

7. Use project-specific (or even subproject-specific) virtual environments to provide a stable set of code tools that you can return to over months (or years) without fearing that system changes will have broken previously functional analysis code.

8. Archive your interim results and data periodically, e.g. each time that you give a status-update talk or otherwise circulate your results. You could use version control for this (if the files are not too big), or just create well-named archive directories. Make sure to tag the code that produced each results set, too;

9. Always look for ways to improve: voraciously study others' analysis systems and data management practice, and share your own. This is your craft: learn from all sources you can, and hone your skills.

Further reading

- A guide to recent MC generator theory and implementations can be found in the MCnet 2011 'General-purpose event generators for LHC physics' review, *Phys. Rept.* **504** 145–233, arXiv:1 101.259 9.
- The voxelisation process for detector simulation is described in 'Recent Developments and Upgrades to the Geant4 Geometry Modeller', available at https://cds.cern.ch/record/1065741. The GeantWeb pages and Physics Reference Manual, available from https://geant4.web.cern.ch/are an excellent resource.
- Surveys of the practical challenges in scaling MC event generation to higher precisions and luminosities can be found in in CERN-LPCC-2020-002, arXiv:2 004.136 87 and *J. Phys. Conf. Ser.* **1525** 012023 (2020), arXiv:1 908.001 67, and the computational challenges in scaling detector simulation in *Comput. Softw. Big Sci.* **5** 3, arXiv:2 005.009 49 (2021).

Exercises

7.1 A single top-quark is often represented in an event as a chain of $1 \to 1$ 'interactions', reflecting the modification of the ME event first by partonic recoils against the PS and intrinsic k_T of the generator, and by momentum reshuffling as required to put the event on-shell. Eventually, a decay vertex will be reached, for the top decaying to *Wb*. Which top is the 'right' one to pick for physics, and does it depend on the application? How would you design an algorithm to walk the event graph to find your preferred top quark?

7.2 A *b*-quark may be similarly represented in an event record as having several stages of recoil absorption, before entering a hadronisation vertex attached to all colour-charged parts of the event, and emerging as a *b*-hadron. This hadron then undergoes a series of decay steps, e.g. $B^* \to B\gamma$, before undergoing a weak decay that loses its *b*-flavour. Which object would you be most interested in from the perspective of calibrating *b*-tagging performance? Which are you most interested in from the perspective of a physics analyser looking for $H \to b\bar{b}$ events? Are they the same object?

7.3 Photons can be produced at many points in an MC event: in the ME or via ME corrections, in the PS, and in charged-lepton and hadron decays. How would you write an algorithm to determine if a final-state photon was *not* produced in a hadronic decay? How could you distinguish a 'direct photon' from the ME from one produced by the PS? Is that a good idea? Is there a physical difference between ME and PS emissions?

7.4 Calculate the mean transverse displacement of a B^0 meson with mean lifetime $\tau_0 = 1.52$ ps, if it has $p_T = 100$ GeV. Remember to include its Lorentz boost and hence time-dilation factor. How does this displacement compare to the nominal $c\tau_0$ factor? What fraction of such particles produced perpendicular to the beam will decay further than an inner tracker layer at 3 cm?

7.5 How many CPU hours would be needed to generate and simulate detector interactions equivalent to 300 fb for (a) LO 100 mb inelastic minimum-bias events, at 0.01 s and 70 s per event for generation and simulation, respectively? (b) 20 nb NLO $W+b$ events at 200 s and 300 s, respectively? How do these compare to the Grid CPU capacity?

7.6 Simulated pile-up events are generated, simulated, and stored for random overlay on signal MC events. The chosen MC-generator model has too hard a track-p_T spectrum at low scales: for $p_T > 5$ GeV, the MC model shape is roughly $p(p_T) \sim (p_T - 1 \text{ GeV})\exp(-1.5p_T/\text{GeV})$ while the data looks more like $p_T \sim p_T \exp(-1.5p_T/\text{GeV})$. How could rejection sampling be used to create an MC pile-up event store with the data-like distribution in this range? Estimate the efficiency of the sampling.

7.7 To minimize data transfer, analysis jobs for the LHC are dispatched to the Grid sites that store the data, rather than sending the data to the site on which the job is planned to run. Site problems can make data inaccessible if only one copy exists, and popular data can cause bottlenecks at a site, but keeping multiple copies is expensive. How might you distribute the data to the sites to minimize the analysis pain?

IOP Publishing

Practical Collider Physics

Andy Buckley, Christopher White and Martin White

Chapter 8

Data analysis basics

Data analysis in collider physics at general-purpose detectors[1] is typically divided into two categories: 'searches' versus 'measurements'. At first glance the distinction is clear: searches are analyses which look for evidence of the existence of new physics (be that beyond the Standard Model (BSM), rare Standard Model (SM) production and decay channels, or new hadronic states), while measurements characterise physics that is known to exist.

While true, this distinction is not operationally very useful as it tells us nothing about the practical ways in which analyses in the two categories differ. In fact, while measurements are often referred to as '*Standard Model* measurements', a well-performed measurement will be substantially model-independent and may provide broad constraints on hypothetical new physics—a topic to which we will return in Chapter 12. To further emphasise the falseness of the dichotomy, many searches are performed via counting of events that fall into 'signal regions' defined by allowed ranges in reconstructed observables, a method which is not as far from differential cross-section measurements as one might initially expect.

Searches and measurements hence have more in common than separates them, and to reflect this we dedicate the current chapter to the 'basics' of collider data analysis which are to be found in all analyses whether aimed at BSM or SM physics. Data analyses also occur in the development and calibration of object-reconstruction algorithms, and the methods described here should also be found useful by anyone performing such work. In the following chapters we first visit analyses performed at the level of reconstructed objects, and then those corrected for biases introduced by the detector and by reconstruction algorithms. While the first group is the most common mode for searches, and the latter is the modern distinguishing feature of measurements, we find this operational definition more satisfying, and the

---

[1] Life is rather different at specialist collider experiments, in particular flavour-physics experiments.

doi:10.1088/978-0-7503-2444-1ch8

ordering by increasing degree of data processing rather than by complexity of physics model more educationally natural.

The first two actions in any analysis are to select the events of interest, and the relevant **physics objects**—reconstructed electrons, muons, jets or compound objects, etc—within them. These are the subjects of the following two sections, although we take the liberty to present them in the opposite order than they are often summarised in papers, as the object definitions are often important in the event-selection process. The remaining sections of this chapter are concerned with eliminating ambiguities between physics objects so as to map them better to the process under study, basic strategies for estimating and reducing the influence of background processes, and the influence and estimation of systematic uncertainties.

8.1 Data-taking

Fundamentally, without the collider and experiment running, there can be no physics analysis. The experimental data-taking process at a modern collider is operated by teams of specialists, coordinating larger groups of general collaboration members on various categories of operations shifts (from control-room activities to oversight of data-quality and computing). Smooth operation of these aspects is key to a high data-taking efficiency, as indicated in Figure 8.1, and hence more statistical power in eventual physics interpretations.

Data-taking operations typically take place in multi-year **runs**, such as Run 1 of the LHC from 2009–2012, and Run 2 from 2015–2018. The collider is run for approximately half of each year within a run, both to allow for short detector-maintenance periods and to minimise electricity costs, with 24-hour operations during the data-taking periods. Multi-year 'long' shutdowns between the runs permit more invasive maintenance and upgrading, including lengthy processes such as warming and re-cooling the accelerator magnet systems.

As runs naturally divide into data-taking periods between shutdowns, it is common to label these distinctly within the experiment data-processing, particularly

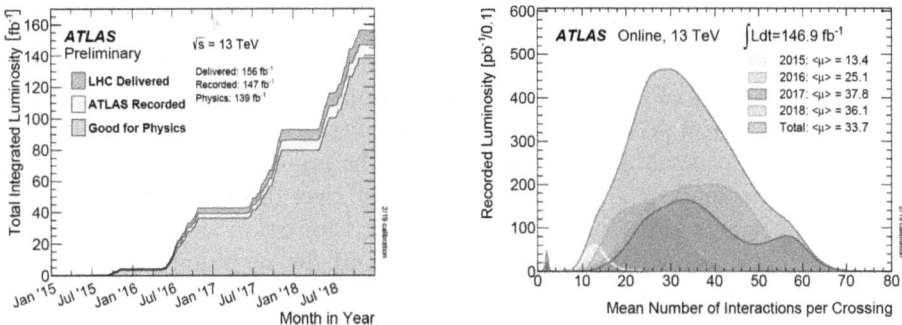

Figure 8.1. Left: integrated luminosity delivered by the LHC, recorded by ATLAS, and passing physics data-quality requirements during LHC Run 2. Right: mean pile-up rate distributions delivered to ATLAS during LHC Run 2. Credit: ATLAS luminosity public plots: https://twiki.cern.ch/twiki/bin/view/AtlasPublic/LuminosityPublicResultsRun2.

when e.g. distinct beam or detector conditions have been used. The mean pile-up rate $\langle \mu \rangle$, i.e. the number of simultaneous pp collisions per bunch-crossing, has been the main driver of distinctions between data periods at the LHC—the reason is clear from the right-hand plot in Figure 8.1, showing the overlapping (and sometimes not so overlapping) pile-up rates in ATLAS Run 2: to avoid major losses of statistical power through MC-event reweighting, several different $\langle \mu \rangle$ distributions were explicitly simulated to match the real data conditions. A frequently used cross-check within data analyses is to subdivide the total dataset into different data-period subsamples with distinct beam conditions, to check for systematic bias from pile-up, occupancy etc.

Within each run, further subdivisions of events are made. The smallest-scale division (other than individual events or bunch-crossings) is the **luminosity block** (or more commonly 'lumiblock') of constant luminosity and stable detector conditions. Typically, these correspond to approximately one minute of data-taking, and hence $\mathcal{O}(1 \times 10^{10})$ bunch-crossings, each of which gets a unique event number within the lumiblock from the collider/trigger clock system. The lumiblock is the smallest set of events that can have a unique set of experimental conditions in the **conditions database**, and be hence declared as suitable or not for various types of data analysis via a **good-run list**—this encodes, for example, whether particular subdetectors were in operation, allowing analyses with less stringent requirements to analyse events for which the detector was in an 'imperfect' configuration.

Between the lumiblock and the run, lie the **data period** (and sometimes sub-period) divisions. Each data period corresponds to a distinct setup of the beam, e.g. the instantaneous luminosity and hence mean number of pile-up interactions per bunch-crossing, or detector. Within each data period, the continual processes of triggering and offline reconstruction (generally known as **reprocessing**, even for the first pass) are performed via on-site computing farms, then distributed to the world-wide computing Grid. It is from here that analysis—either for object calibration or for physics measurement—begins in earnest.

8.2 Object definition and selection

It is difficult to find a unique ideal ordering for the selection of events (covered in the next section) and the definition of the physics objects within them: the selection necessarily makes use of the objects, but in many analyses 'rougher'—and computationally faster—object definitions are used for the first selection stage common to all signal regions or observables (sometimes called the **pre-selection** stage) than in the later parts of the analysis. In this section we brush over this distinction and review common refinements to reconstruction-level objects for use in analyses. But first we take a step back toward the event generation and the quantum event evolution encoded in it, to motivate these extractions from data.

8.2.1 Basic truth-object definitions

In performing a data analysis, our interpretation is almost always conducted via a picture of fundamental processes rooted in Feynman diagrams, with final-state flows of energy and quantum number isolated into a few, well-defined and well-separated partons (including leptons and photons). Notable exceptions to this tend to include studies of non-perturbative effects such as soft-quantum chromodynamics (QCD), collective flows in heavy-ion collisions, or hadron spectroscopy—all cases where single-parton field excitations are not obviously the most appropriate degrees of freedom. On the whole, this picture is an effective cartoon lens through which to approach high-scale interactions, where QCD asymptotic freedom makes the assumption of object independence a workable approximation.

It remains the case, however, that stepping from this hard-process view with a handful of particles in the final state, to realistic collider events with the added complexities of underlying event, hadronisation, and particle decays, introduces a number of unwanted complexities. To reduce realistic events back toward the Feynman cartoon, we can motivate a series of processing techniques on particle-level Monte Carlo (MC), which are analogously applied to data. These are what are often called **truth-object** definitions, as compared to physics-object definitions in general.

Final-state particles and directness

The first is to define what we consider to be a **final-state particle**. On human timescales, the only stable particles are electrons, protons, neutrons, and photons; from the point of a detector, as we mentioned in Section 7.3 during our coverage of detector simulation, a wider class of particles need to be considered 'microscopically stable'. The *de facto* standard currently in use in high energy physics (HEP) is to consider a final-state particle stable if its species mean lifetime $\tau_0 \geqslant 10 \, \mathrm{mm}/c$. Exceptions to this may be introduced in detailed hadronic physics studies, e.g. to account for poor reconstruction of particles with somewhat longer lifetimes, but generally speaking such a definition is used to define the **particle-level** final state.

Of these final-state particles, we have to make an important distinction between the **direct** products of the hard-scattering, and those produced indirectly by subsequent hadron decays. The name **prompt** is often used informally when 'direct' is intended, but this is discouraged due to a better-motivated use of 'promptness' in flavour physics to refer to particles produced at the primary interaction vertex (PV) as opposed to non-prompt particles from displaced vertices: in this context the promptness is better justified as referring to resolveable elapsed time, whereas a majority of indirect particles arise within nuclear length-scales of the PV.

Directness is a tricky business to define, and over time a standard has emerged based on negative logic: in terms of an event graph, we know indirectness when we see it, and should a particle fail to match any of a set of indirectness conditions, we consider it direct. The main distinction we want to draw is between non-strongly-interacting particles (photons, e, μ, τ, neutrinos, and various hypothetical BSM particles) from hadron decays as opposed to the hard scatter—in the hadronisation picture of event evolution, hadron formation occurs at a microscopically distinct timescale, and the absence of quantum interference between final-state amplitudes from these sources means that a hard distinction can be drawn. The clear consequence of this is that *no hadron can be considered direct, nor can any photon or lepton from hadron decays.*

This definition can naturally be iterated, starting from a set of final-state particles. For each particle we start from an assumption that it is direct, and attempt to prove otherwise: if it is an SM hadron it can immediately be labelled as indirect, otherwise its parent particle(s) are asked the same question. Should any of the parents (perhaps through several more layers of recursive questioning) answer that they are indirect, the original particle must also be considered indirect.

The algorithm has to be refined somewhat to reflect non-standardisation in MC event records: while decay chains are semi-classical and can be represented as an unambiguous history of decay branchings of hadrons, leptons, and photons[2], the partonic and hadronization aspects may take many forms depending on the generator code and its hadronization model: string and cluster hadronization are naturally represented in very different ways, and multiple-parton interactions can easily create connections between all colour-charged event structures, meaning that every QCD particle is connected to every other at the hadronization stage. The directness-labelling algorithm naturally fails when attempting to handle such structures, and hence it is standard to block recursion into parton (quark or gluon) parents. This fits with our physical picture where e.g. photon emission from quarks in the hard scattering or its quantum-interfering parton-shower extension is to be considered direct, and distinct from photon emission in hadron decays: in *principle*, a perfect detector could resolve the latter's origin[3]. Photon emissions from direct or indirect leptons also naturally inherit their parent's status.

Jet clustering

Having established a robust set of direct/indirect labels for final-state particles, we are ready to construct definitions of each of our physics objects. We will start with the definition of jets, for which the clustering together of many angularly collimated particles is essential, before moving on to tackle the remaining direct leptons and photons.

The details of jet algorithms have already been visited both in our overviews of theory and of reconstruction methods, illustrating how central jet definitions are to collider analysis at all levels. Note that we ultimately require both a definition of reconstructed jets (that are obtained from detector-level quantities) and truth-level jets, that are somehow to be defined from the event record of our Monte Carlo generators. These definitions are required so that we can answer questions about jet calibration, amongst other quantities. In terms of truth-level definitions, the parameters of the clustering algorithm itself, and of any refinements, should match those used in data. The aspect requiring special care for particle-level MC is which particles to consider as **jet constituents**, i.e. the inputs to the clustering. The key considerations here are *directness* and *visibility*, and the exact definition depends on the experiment and jet reconstruction methods.

---

[2] An awkwardness here is partonically modelled decays, which some event generators write into the hadron-decay stage of the event record: this is discouraged, and in the cases where it remains extant has be to 'fixed' in event generation frameworks as described in Section 7.2.

[3] We will return to the more contentious issue of direct versus 'fragmentation' photons in Chapter 11.

Any experimental jet calibration has to make some assumptions of process independence: MC/data scale-factors must be defined robustly enough that they can be applied in event types very distinct from the processes used in their derivation. This means that direct particles, strongly dependent on the characteristics of the myriad hard-scattering processes, cannot reasonably be included in calibrations.

There is then a free choice of what to do with indirect particles. One can either use a definition of indirect jet constituents that directly maps to what the detector and reconstruction algorithms can resolve, or to something more closely related to the fundamental energy flow cf. the Feynman-diagram cartoon but which requires some extrapolation from detector signatures. Obviously we could include hadrons and their decay products in our truth-jet definition, but an important consideration is what to do about invisible particles from hadron decays, in particular neutrinos from semileptonic hadron decays. These lead to loss of reconstructed momentum, and indeed for this reason b-jets can require distinct energy calibration in precise analyses. In detectors such as ATLAS, whose standard jet reconstruction has been calorimeter-based through LHC Runs 1 and 2, another important source of bias is indirect muons, again particularly in b-jets: these can be reconstructed by the inner tracking and outer muon systems, and hence are naturally included in jets from the particle-flow reconstruction paradigm, but tend to be missed by calorimeter-only jet-finding. Faced with these two challenges, we have a 2×2 set of reasonable combinations for how to deal with them: indirect muons included or rejected, and indirect invisibles included or rejected. All four combinations are plausible, but in practice it is illogical to include indirect neutrinos but to exclude muons.

Finally, we note that current truth-jet definitions at hadron colliders are rarely flexible enough to cover non-standard signatures such as those arising from **long-lived particles** (LLPs). We expect that the resulting signatures of displaced jets and emerging tracks will be incorporated into particle-level jet-constituent definitions via their production-vertex position in future.

Leptons and photons
At particle-level, leptons and photons are relatively straightforward. In both cases, the weakness of electroweak effects means that once emitted they are already close to their full final-state form. Photons in particular may be assumed not to interact any further once emitted, electroweak corrections in their production having been absorbed into the MC emission structure. Neutrinos, of course, behave the same way: in fact, as they cannot be directly resolved at all and only manifest in the event imbalance and hence missing momentum, the word 'lepton' often refers specifically to *charged* leptons, and further (as leptonically decaying taus also cannot usually be resolved) is often used to refer only to electrons and muons.

Little nuance is hence needed in basic truth definitions beyond finding final-state particles with the relevant particle-ID codes for charged leptons or photons[4], but an important refinement is sometimes made to account for quantum electrodynamics

---

[4] Explicitly identifying and using truth-record neutrinos for physics is a distasteful business, but sometimes pragmatic: see chapter 11.

(QED) corrections to final-state charged leptons: **lepton dressing**. This is an approach taken to cluster together the **bare lepton** with nearby photons, noting that in particular electron reconstruction is unlikely to have a resolving power $\Delta R \lesssim 0.1$ due to electromagnetic calorimeter (ECAL) granularity. Algorithms for lepton dressing also, as for truth-jet constituents, differ between experiments: they may be based on clustering algorithms as for jets, or on a simpler, *post hoc* ΔR measure with some care taken to assign each photon to at most one charged lepton. Which photons are used can also be divided by directness, but modern usage almost universally accepts that only direct photons should be used in lepton dressing, contributions from well-understood $\pi^0 \to \gamma\gamma$ decays having been removed in electron calibration. Dressed muons are much closer to their bare counterparts than is the case for electrons, the latter radiating more due to their smaller mass. Lepton dressing is hence a small (and usually ignored) effect in BSM direct-search analyses, but a crucially important one in precision electroweak (EW) analyses such as W, Z, and top-quark mass and width measurements, where, without dressing, the electron and muon channels cannot be combined into a final, flavour-blind result.

8.2.2 Removing physics-object ambiguities

Having now defined our physics objects, we must make a crucial refinement. So far, our definitions permit double-counting of objects due to primary particles which manifest as reconstructed objects in more than one category, or non-hadronic objects that are misreconstructed either directly or indirectly due to hadronic activity (of which there is a surfeit at a hadron collider). The main such cases are:

Electron/photon ambiguity: electrons and photons are both electrpmagnetic (EM)-interacting particles whose reconstruction is largely reliant on signals from the ECAL.

These ambiguities are primarily addressed in e/γ reconstruction by use of track-pointing (and corrections for photon $\to e^+e^-$ conversions) and EM shower-shape information. As none of these reconstruction techniques are perfect, residual uncertainties will need to be propagated.

The most obvious impact at analysis level is that electrons and photons are primarily distinguished using tracking information: shower-shape identification alone is much less performant. For this reason, prompt electron and photon reconstruction is usually limited by analyses to the η acceptance of the tracking detector, rather than the in-principle acceptance of the ECAL.

Jet/photon and jet/electron ambiguities: ECAL signatures are of course also crucial for hadronic jet reconstruction, and if one were to look at the calorimeters alone one would find high-p_T e/γ EM showers meeting the definition of a jet. This has been the case for ATLAS through Runs 1 and 2 of the LHC, in which jets are more-or-less purely calorimetric objects: at analysis stage, electrons and photons *also* appear as jets. By contrast, the CMS experiment from the beginning has been able to use its full set of subdetectors for **particle-flow** object reconstruction.

The reverse mis-indentification is also possible, where a hadron decays electromagnetically or semileptonically to produce real photons or electrons which then leave ECAL signatures. The most important examples of this are the dominant (BR ~ 98%) neutral pion $\pi^0 \rightarrow \gamma\gamma$ decays, and their internal-conversion **Dalitz decay** relatives, $\pi^0 \rightarrow \gamma e^+ e^-$ and $\pi^0 \rightarrow 2e^+ 2e^-$. As for the prompt electron/photon discrimination discussed above, these are considered in e/γ reconstruction, but methods are not perfect and fake electrons and photons may be reconstructed where there is significant hadronic activity.

Jet/muon ambiguities: direct ambiguities between jets and muons are naturally less prevalent than between jets and e/γ, because muons are minimally-ionising particles and pass through the calorimeters with little energy loss: they are unlikely to be misreconstructed as jets. But jets faking muons is still a significant issue, due to leptonic decays of hadrons and **punch-through**. The latter term refers to incomplete containment of a hadronic shower in the calorimeters, such that charged particles 'leak' through the hadronic calorimeter (HCAL) into the muon system and leave a signal used to seed muon reconstruction.

The leptonic decay issue is similar to that for non-prompt electron production, but for muon production there is no contribution from π^0 decay as any muons in that decay would need to be pair-produced, and the pion mass lies below the $2m_\mu \sim 210$ MeV threshold. The main route is hence via the charged pion and kaon weak decays $\{\pi^+, K^+\} \rightarrow \mu^+ \nu_\mu$.[5] While these decays are relatively rare in the detector due to the particles' long lifetimes—both species are usually treated as stable at generator level—there are a *lot* of pions and kaons in a hadron collider environment, and hence the rate of non-prompt muons is significant. Once produced in a hadronic decay, a decay muon will pass through the calorimeter and leave a track/hit in the outer muon system.

Jet/tau ambiguities: As discussed earlier in Section 8.2.1, prompt leptonic taus are almost always considered as part of the prompt electron and muon signal. They are hence not a problem, but hadronic tau-faking is. Due to their 1.777 GeV mass, taus can only be produced (promptly) in high-scale weak interactions, and in heavy-flavour hadron decays, and as such the risk of contamination is strongly biased toward jets faking taus, the reverse process being negligible.

65% of taus decay hadronically to combinations of pions with an odd number of charged pions (often called **prongs**) to conserve charge. Hadronic taus are hence difficult to distinguish from hadronic jets (also dominated by pions), and it is the primary job of tau reconstruction algorithms to disambiguate the two.

---

[5] The electron variants of these decays are subject to **helicity suppression**, a consequence of angular momentum conservation, the perfect left-handedness of the neutrino, and the chirality of weak interactions.

As with other jet/lepton ambiguities, these reconstruction strategies are not perfect, and may err on the side of caution: retaining potential taus unless it is near certain that they are really hadronic jets (by a high prong-multiplicity, for example). Analyses searching for tau leptons therefore need to apply stricter selection cuts on the reconstructed tau candidates, as we saw in Section 6.4.5.

As with the feed-down of real, but indirect, electrons and muons from pion and kaon decays, there is a risk of non-prompt tau contamination in a prompt-tau analysis from D and B meson decays. These primarily occur via the minority (semi)leptonic decays of b-hadrons and—due to their mass-proximity to the tau—the relatively rare pure-leptonic decays of $D^{\pm}$ mesons[6]. While these heavy-flavour hadrons account for a minority of hadron production, the overwhelming rate of QCD hard interactions means that even these small branching ratios to taus can constitute a significant background.

Jet/MET mismeasurement: missing momentum is arguably the most fragile physics-analysis quantity at a collider, being not observed directly but through the imbalance of visible objects and assumptions about what occurs outside the detector acceptance.

As the most structurally complex and hence least-precisely reconstructed hard objects, jet mismeasurement can be a significant source of MET: if the reconstructed energy of a jet is badly wrong, and the rest of the event balances in the transverse plane, a fake MET vector will be generated in the direction of the jet (or opposite to it).

The standard techniques for controlling these errors are **isolation** and **overlap removal**. These are similar, and hence often confused for each other, but they each use a different set of objects. In the case of isolation, the motivation is to insist that non-QCD hard objects be well separated from even diffuse QCD activity to reduce fake rates, while overlap removal is about avoidance of double-counting where two reconstructed hard objects correspond to the same truth object.

Isolation

Isolation is most usually conducted using a ΔR cone around a hard object such as a charged lepton or photon, in which hadronic activity is counted. The cone radius may either be fixed (typically to a moderate size e.g. $\Delta R \sim 0.3$) or be designed to shrink with higher candidate momentum between two fixed limits, reflecting the higher collimation of radiation patterns typical to high-energy particles. The use of cones is not universal: notably ATLAS measurements of hard-photon final states have historically used a square isolation region based on a sum over calorimeter cells in η-ϕ, with a square 4×4 cell region around the photon candidate excluded from the sum.

---

[6] Helicity suppression again means that taus account for the largest of the $D^+ \to \tau^+ \nu_\tau$ branching fractions at BR $= 1.2 \times 10^{-3}$, while the order is reversed in three-body semileptonic decays such as $B^- \to D^0 \tau^- \nu_\tau$.

The typical proxies for hadronic activity used in isolation are
- activated calorimeter clusters or towers, possibly subdivided into ECAL and HCAL terms;
- charged-particle tracks;
- elementary physics objects including charged and neutral hadron proxies reconstructed using particle-flow methods.

The latter two sets of objects will typically be restricted to those associated with the event's primary vertex; this requirement is not possible for raw calorimeter deposits. A typical isolation algorithm will hence involve looking in turn at each lepton/photon candidate and summing the (usually transverse) energies or momenta of clusters or tracks within the given cone radius, excluding those directly associated to the candidate. A decision is then made to accept or reject the candidate according to a maximum threshold of surrounding activity: this may be an absolute or relative threshold for isolation of candidate i from activity in the surrounding patch X_i, e.g.

$$\sum_{j \in X_i} E_{T,j} < E_T^{\max} \qquad \text{or} \tag{8.1}$$

$$\sum_{j \in X_i} p_{T,j} < f^{\max} p_{T,i} \qquad \text{(sometimes called \textbf{gradient isolation})}. \tag{8.2}$$

The thresholds $E_T^{\max}$ and $f^{\max}$ may be functions of the lepton kinematics, most usually parametrised in $\{p_T, \eta\}$. As all combinations of track and calorimeter isolation variables and absolute and relative thresholds contain some orthogonal information, it is not unusual for several such variables to be used concurrently.

A remark is due on these isolation sums, because initial-state effects such as pile-up of final-state particles from multiple coincident primary collisions can necessitate recalibration: even leptons and photons which are well isolated from all other activity in their own event may be scattered with minimum-bias event overlays from the other interactions in the bunch crossing. These effects are exacerbated at hadron colliders by the presence of in-event QCD initial-state processes such as underlying event and initial-state jet production: the former has an effect largely uncorrelated with the hard-scatter objects, much like an extra pile-up collision, while soft initial-state radiation is genuinely part of the hard-scattering process.

An obvious approach to correct for these—in jet energy and mass separately—is to subtract the average amount of such background activity that one would expect in a collision event of this sort. This could be a fixed energy-offset correction in the jet calibration, corresponding to a fixed average p_T **density** in η–ϕ or y–ϕ space, e.g. $\rho = p_T / \Delta\eta\Delta\phi$, multiplied by the geometric area of the jet cone. But this naïve approach immediately hits two problems:
1. Pile-up and underlying event activity can fluctuate greatly from event-to-event: a fixed density will frequently, perhaps usually, be significantly wrong for any given event;
2. The *effective* **jet area** in terms of acceptance of soft, background activity is not really the geometric $A = \pi R^2$, but something more complicated and jet-specific.

Techniques connected to the jet-clustering algorithm can solve both problems, by calculation of both a dynamic area for each jet, and a dynamic median value of ρ.

Dynamic jet areas can either be computed directly by a technique called **ghost association**, or indirectly via a Voronoi tesselation around the clustered constituents. Since ghost-association is illustrative of the importance of perturbative stability in jet clustering, we will focus on that. Recalling our earlier presentation of jet clustering algorithms and measures in Section 3.6, a key feature in modern jet clustering is infrared (IR)-safety: the kinematic results of the clustering should not be affected by the presence of arbitrarily soft particle emissions (or by collinear splittings of hard particles). This feature can be turned on its head by actively adding soft 'ghost' particles to the event, with well-defined directions but with momentum magnitudes many orders of magnitude smaller than real constituents, in the confidence that they will not affect the result. By uniformly distributing $N_{\text{ghost}}^{\text{tot}}$ particles in an η–ϕ rectangle of area $A^{\text{tot}} = 2\pi\Delta\eta$, we can compute an effective jet area

$$A^{\text{jet}} = \frac{N_{\text{ghost}}^{\text{jet}}}{N_{\text{ghost}}^{\text{tot}}} \cdot 2\pi\Delta\eta, \tag{8.3}$$

i.e. the fraction of the total ghost-populated area that is pulled into the jet by the clustering. Such areas can either be computed actively or passively, the labels distinguishing between whether the ghosts are added *en masse* at the beginning of the clustering, or added one at a time at the end: the two definitions are equivalent in the limit of large N_{ghost}, and also equal to the **Voronoi area** based on real jet-constituent areas in the limit of dense events with large numbers of non-ghost jet constituents. As the computational cost of jet clustering grows with the number of jet constituents N, scaling at best as $N \ln N$ using the optimisations available in FastJet, the Voronoi area is faster to compute at the cost of less predictable convergence to the asymptotic value.

Having obtained an area A_j^{jet} for each jet, the jet densities $\rho_j = p_{\text{T}}^j / A_j^{\text{jet}}$ can be computed, and the median density $\rho_{\text{med}}^{\text{jet}} = \text{median}\{\rho_j\}$ obtained. Studies show that in LHC-style events with high pile-up rate μ, this is a good, and event-specific representative of the characteristically lower p_{T} density of minimum-bias jets from pile-up and the underlying event. A jet-specific p_{T} offset $\Delta p_{\text{T}}^j = \rho_{\text{med}}^{\text{jet}} A_j^{\text{jet}}$ can then be used to calibrate the isolation-sum offset. Alternative methods such as CMS' PUPPI pile-up mitigation scheme are rooted in the same ideas, but attempt to infer *per-constituent* pile-up labels by use of particle-flow reconstruction and primary-vertex association.

Both photon and electron radiation patterns (and to a lesser extent muon ones) are more nuanced than this picture suggests. Photon isolation in particular is a theoretically thorny area. We have spoken of using cone isolation methods for both leptons and photons, but in fact the latter are more troublesome than they at first appear. Events can be expected to contain both direct photons (produced in the perturbative partonic process, including QED radiation in parton showers) and

fragmentation photons that are produced non-perturbatively from decays of neutral pions and other light hadrons. In perturbative QCD, this non-perturbative photon production in association with hadrons can be encoded in a set of collinear **fragmentation functions** (see Section 3.4). These describe the momentum relationships between partons and the hadrons and photons they turn into, like parton distributionfunctions (PDFs) in reverse, and as they are collinear they correspond to co-production of hadronic energy precisely collinear to a fragmentation photon[7].

To eliminate fragmentation photons from an object selection, one hence needs to require zero hadronic energy collinear to the photon—but doing so with a finite-sized isolation cone also removes perturbative QCD partons in that region. This is problematic, since banning soft gluons from the collinear region invalidates the Kinoshita–Lee–Nauenberg (KLN) theorem (Section 2.15) needed in perturbative QCD to cancel real and virtual IR divergences. It seems that the cuts in phase-space needed for photon isolation are incompatible with well-behaved perturbative QCD cross-section predictions, since key perturbative and non-perturbative effects live in exactly the same phase-space region. Naïvely relaxing the complete ban does not help either: if we allow a fixed fraction of hadronic energy in the cone, the fragmentation contribution finitely returns.

For this reason, the smooth isolation or **Frixione isolation** scheme was developed, to interpolate between the zero-tolerance and fractional-tolerance approaches. In this method, the requirement is not just that a maximum amount of hadronic activity is allowed in the isolation cone, but that a whole family of isolation criteria need to be considered for all cone radii $0 \leqslant R \leqslant R_0$, where R_0 is the nominal isolation cone size. This criterion can be written in general as

$$\sum_{i \in \{R_{i\gamma} < R\}} E_{Ti} \leqslant \mathcal{X}(R) \quad \text{for all } R < R_0, \tag{8.4}$$

where $R_{i\gamma}$ is the angle between the photon γ and the ith hadron or QCD parton. The radius-dependent energy scale $\mathcal{X}(R)$ can be any function that goes to zero as R does: this ensures that the perfectly collinear region where the fragmentation function contributes is excluded from the sum, while the smooth turn-on of the isolation tolerance preserves the KLN cancellations. In the original paper, it is argued that the form

$$\mathcal{X}(R) \equiv E_\gamma \left(\frac{1 - \cos R}{1 - \cos R_0} \right) \tag{8.5}$$

$$\approx E_\gamma \left(\frac{R}{R_0} \right)^2 \tag{8.6}$$

---

[7] This critique can be applied also to leptons from hadron decays, but the critical difference is the sheer rate of fragmentation-photon production: they are the dominant final-state particle at a hadron collider, largely due to the $\pi^0 \to \gamma\gamma$ decay.

is a good choice based on the leading angular dependences of partonic splitting functions.

The requirement that the Frixione isolation criterion be smoothly applicable all the way down to $R = 0$, however, is incompatible with the finite angular resolutions of real detectors: it cannot be verified whether or not it is satisfied below a certain angular scale. This method hence sees most use in perturbative QCD photon calculations, where it is a standard technique, rather than in experiments where cone isolation algorithms (or simpler, at trigger level) have continued to be used. This has not been such a problem in practice, since hadron-level MC simulations already reduce the sensitivity to IR-safety issues as compared to parton-level calculations, but is less than ideal. As precision QCD studies develop, it seems likely that experimental procedure will evolve to use some form of smooth isolation, either the original Frixione scheme or a less restrictive variant based on the jet-substructure techniques to be discussed in Section 8.4.5.

Overlap removal

Overlap removal (OR) is concerned with eliminating not just misreconstructed prompt physics objects but duplicated ones, due to the sources of ambiguity above, in particular jets masquerading as charged leptons (through ECAL showers and leptons from hadron decays) and *vice versa*. The fact that misidentifications can occur in both directions, and the desire for an unambiguous identity assignment to a given hard object leads to a typical overlap-removal algorithm looking something like the following:

1. For every isolated photon candidate (already disambiguated from isolated electrons by reconstruction), compute the ΔR distance to every jet. If within a radius $\Delta R < R_\gamma$, remove the jet.
2. For every electron candidate, compute the ΔR distance to every jet. If within a radius $\Delta R < R_e$ and the jet has no b-tag, and the electron accounts for greater than a defined fraction of the jet p_T, $p_T^e/p_T^J > f_{eOR}$, remove the jet.
3. For every muon candidate, compute the ΔR distance to every jet. If within a radius $\Delta R < R_\mu$ and the jet has no b-tag, *and* either the jet has fewer than N_{OR} tracks or the muon accounts for greater than a defined fraction of the jet p_T, $p_T^\mu/p_T^J > f_{\mu OR}$, remove the jet.
4. For every tau candidate, compute the ΔR distance to every jet. If within a radius $\Delta R < R_\tau$ and the jet has no b-tag, and the tau has greater than a defined fraction of the jet p_T, $p_T^\tau/p_T^J > f_{\tau OR}$, remove the jet.
5. Finally, for every remaining jet candidate, compute the ΔR distance to every non-jet object (photon, electron, muon and tau). If any is within $\Delta R < R_J$, remove the non-jet.

This example is rather extreme: in practice it's unusual to have an analysis requiring all of high-energy photons, charged leptons, taus, and jets! And for the non-jet objects, the rate of direct backgrounds containing extra electroweak objects is usually sufficiently low that e.g. checks for unexpected hard taus are unnecessary.

The extent of such checks (which necessarily introduce dependences on the reconstruction of more object types, and hence more sources of systematic error) is usually limited to, for example, a requirement that there are no isolated charged-lepton candidates in jet+MET BSM searches, to ensure orthogonality with leptonic searches and ease analysis combination. The complexity of such selection algorithms can create difficulties in re-interpretation of published data, especially if the paper does not describe the algorithm in full detail; as will be discussed in Chapter 12, such issues are a strong argument for experiments to publish a simplified analysis routine in runnable code as a complement to the textual paper and numerical final data.

The logic of the differences between the steps is worth reviewing, however, despite the contrived complexity of this example. First note that some overlap removal will already have been applied via the isolation procedure, but that the exemption of the central region of the isolation cone means that directly overlapping narrow jets—most notably those that *are* the photon, electron, or tau seen from another perspective—remain. Starting with the photon, the removal of an overlapping jet on purely geometric grounds reflects the fact that photon reconstruction will have already applied requirements on tracking, ECAL shower shape, and limited HCAL activity which are very unlikely to be compatible with hadronic jets: here the overlap removal relies on the reconstruction performance.

Moving on to electron overlap removal, the logic is again that if a jet centroid is very close to the electron, i.e. with R_e signficantly less than the jet radius, there is a good chance that the jet *is* the electron. But as non-prompt electrons can also be produced by hadron decays, additional requirements are placed on the tagging state of the jet and the electron energy: these reflect that a lepton with only a fraction of the jet energy is more likely to have been produced via hadronic feed-down or faking by jet constituents, especially if the jet is known to have heavy-flavour content. The same logic applies to the muon–jet overlap removal, but with further refinement of the logic to consider the small track multiplicities associated with minimum-ionising muons. Tau/jet overlap removal, like that for the reconstructed electron–photon ambiguity, naturally builds upon detailed track multiplicity and other inputs to tau reconstruction, as the detector signatures are so closely related.

The final step in the algorithm is to turn the tables and remove any lepton candidates within R_J of a remaining jet. This radius cutoff is likely to be larger than the others, typically equal to the jet radius or larger to ensure there is no geometric overlap between lepton/photon and jet objects. In particular, following the logic above, any lepton candidate within the radius of a b-jet will be discarded.

8.3 Event selection

A key element of any analysis is the number of events (or more generally the sum of event weights) from each fundamental process that are *a priori* interesting.

The aim of the event-selection step is to isolate as far as possible the physical production and decay process (or processes) of interest, by rejecting events likely to be background and accepting those (relatively) likely to be signal. A secondary, logistical motivation is to reduce the volume of data on which to perform the most complex elements of the proposed data analysis.

The number of events of process i selected by an analysis is known as the **event yield**. Its expected value, i.e. the mean of nature's Poisson process of event generation combined with reconstruction and selection algorithms, is given by the product

$$N_i^{\mathrm{exp}} = \sigma_i \epsilon_i A_i L_{\mathrm{int}}, \tag{8.7}$$

where σ_i is the particle-level cross-section, ϵ_i is the **efficiency** due to 'accidental' event losses (e.g. from trigger and reconstruction errors), and A_i is the **acceptance** efficiency for 'intentional' losses due to selection requirements. All these are specific to process i, which may be either considered as signal or background. L_{int} is the integrated luminosity of the collider, i.e. the combination of how intense and how long the relevant data run was. This equation shows the competition between cross-section and luminosity against ϵ and A selection inefficiencies, determining the yields of signal and background processes into the analysis event selection. As an analyser, the cross-sections are given by nature, the available luminosity by the only slightly less omnipotent collider-operations team, and the efficiency by current trigger and object-reconstruction calibrations: the art of event selection is then in engineering the acceptance A to find the best balance between accepting signal and rejecting background event weights.

8.3.1 Event selection as data reduction

The inputs to an analysis, except for the currently special case of trigger-level analyses, will usually be the derived analysis data formats described in Section 7.5.1, which in both MC and data may extend to hundreds of millions of events and many terabytes of data. These data volumes necessitate distributed (Grid) data processing, which is not as nimble as one would like when in the highly iterative process of finalising an analysis and so, in addition to reduction of background contamination, event-selection plays an important pragmatic role in further data reduction, ideally to the scale of a few (tens of) gigabytes that can be efficiently analysed on a small local computing cluster or even a single interactive machine.

All processes have characteristic probability distributions as functions of event kinematics and particle content—often referred to as the **event topology**. The typical approach taken by event selection algorithms is hence to make **selection cuts** on such variables, discarding events which fall outside the cut ranges, so as to maximise the probability of the signal process being selected, and to minimise the probability of selecting events produced by background processes[8]. The definitions of 'signal' and 'background' in this context are entirely dependent on the analysis' aims: SM processes in extreme kinematic regimes may be very interesting to a measurement

---

[8] Note that in general one cannot tell exactly which process produced a particular event, and the question is not even necessarily well-posed: sufficiently similar final states are **irreducible** and may even quantum mechanically interfere such that the event was genuinely produced by a combination of processes. Such is quantum physics.

analysis, but an annoying background in a BSM search. Even 'new', recently discovered processes such as production of Higgs bosons, are considered backgrounds in searches for particles beyond the SM. The adage 'One person's signal is another's background' efficiently captures the essence of this approach.

Even if one does not intend a fully cut-based analysis—see later for some discussion of alternatives—at least a loose, cut-based **pre-selection** will be required to remove events from the most egregiously background-dominated topologies, to reduce the data volume to a locally tractable size.

8.3.2 Event selection variables

Identification of effective selection-cut variables and the optimisation of the cut values are then the first key components of any collider analysis. Typically this process involves a mix of intuition and proliferating histograms of prototype variables to study their signal versus background separation power. There is no fundamental difference between a 'cut variable' and a 'measurement variable': both are observables calculable from an event's physics objects, and so we consider the most common observables in the next section without presuming whether they will be used for event selection or histogrammed as a primary analysis result.

An alternative to the use of cuts is event weighting, whereby disfavoured events are not entirely discarded but instead assigned a probabilistic weight reflecting the confidence that they are associated with a signal rather than background process. Weights have the advantage of avoiding undue wastefulness, by extracting some value from less promising events rather than discarding them altogether.

8.4 Observables

In the following sections we summarise some very common collider physics observables, which are of importance not just in final physics analyses, but also reconstruction and calibration tasks.

8.4.1 Multiplicity variables

Multiplicities, i.e. the numbers of objects of a certain type in an event, act as a joint proxy for the type of process being looked for, and the scale at which it takes place. Most commonly the type of object being counted is a hard-scale proxy for a hard-scattering particle, i.e. a reconstructed jet, lepton, or photon, rather than the raw count of final-state particles which, as discussed in Chapter 3, is highly **IR-unsafe**. A scale cut will be applied for this reason, e.g. only counting leptons with greater than 20 GeV p_T, jets with $p_T > 60$ GeV etc. The multiplicity specific to an object type will typically be denoted in the obvious way as N_{jet}, N_ℓ, N_μ etc.

The process type sets the lowest acceptable multiplicity: for example two jets in a QCD dijet or inclusive jet analysis, two charged leptons (perhaps with a sign requirement) in either a search for leptonic Z boson decays or leptonic WW decays—the two would be primarily distinguished by the presence or otherwise of missing transverse energy—and so on. In the ever-complex case of top quark event selection, which we will return to in more detail in Section 11.2, a combination of minimum numbers of (tagged) b-jets, light jets, and charged leptons will be required: a hadronic top will decay to one b-jet and two (mostly) light jets, and a leptonic top to a b-jet, and charged lepton, and a neutrino (the latter manifesting as missing energy).

8.4.2 Scale variables

In searches for heavy BSM particles, the typical scale of momentum transfer is larger than in SM processes. It is hence common to make cuts requiring key momentum variables to be greater than some threshold, where the *a priori* probability of signal rather than background is higher. There are many variables that indicate the momentum scale of a process, but they can be broken roughly into three groups: single-object scales, few-object scales, and whole-event scales.

The most common choices for single-object scale variables are the transverse momenta of particle proxies such as jets, reconstructed charged leptons, and photons— often ordered such that cuts of decreasing value can be separately placed on the **leading**, i.e. highest transverse momentum, object of each type: for example, the p_Ts of the leading jet, subleading jet etc p_T^1, p_T^2, p_T^3.... The missing transverse momentum is often also treated as a single object, but is more subject to ambiguities of interpretation as of course multiple invisible particles

Few-object scales are often associated with an underlying hypothesis about the particle production mechanism of interest. For example, an analysis looking for a leptonic Z decay (or a higher-mass BSM dilepton resonance) has a clear interest in the invariant masses of reconstructed **same-flavour** dilepton pairs (i.e. ee or $\mu\mu$) after requiring an $N_e \geqslant 2$ OR $N_\mu \geqslant 2$ multiplicity cut. Exactly which pair is best depends on the analysis: if there is ambiguity, one could either choose the leading and subleading lepton pair, or for example the pairing which maximises the invariant mass. One may also choose to place an opposite-charge requirement on the constituents of the pair: a so-called **opposite-sign same-flavour (OSSF)** selection. This requirement is very frequently seen in analyses searching for leptonic decays of high-mass particles, but has an intrinsic trade-off: lepton charge is not perfectly reconstructed, in particular becoming less well-defined as momenta increase and tracks become straighter. Hence, actually applying an opposite-sign requirement may be counterproductive for processes with few expected events, where there is an imperative to preserve signal events above nearly all other factors. We further discuss such trade-offs in Section 8.5.

Similar approaches, but with intrinsically lower resolution and fidelity may be used to find dijet pairs that maximise signal versus background acceptance in hadronic resonance (W, Z, Higgs, or BSM) analyses. The combination of a charged lepton and missing momentum is another interesting case, since not only is the latter ambiguous with regard to single or composite origin, but at hadron colliders is purely a transverse quantity: the missing momentum along the beam direction cannot be inferred as the incoming parton momentum fractions x_1 and x_2 are not known. In this case, a **transverse mass** variable may be usefully defined as

$$m_T^2 = E_T^2 - p_T^2 \tag{8.8}$$

$$= m_1^2 + m_2^2 + 2(E_{T,1} E_{T,2} - \boldsymbol{p}_{T,1} \cdot \boldsymbol{p}_{T,2}) \tag{8.9}$$

$$\approx 2p_T^\ell E_T^{\text{miss}}(1 - \cos(\phi_\ell - \phi_{\text{MET}})), \tag{8.10}$$

where the first line is the general definition in terms of system properties, the second line is a specialisation to a two-object system, and the third a further specialisation to a lepton–neutrino system (as from a leptonic W decay) in which both objects are treated as massless. In the second line, m_i, $E_{T,i}$, and $\boldsymbol{p}_{T,i}$ are respectively the mass, transverse energy, and transverse momentum vector of object i. In the third line the energies are replaced with the lepton p_T and the missing energy, and the dot product reduces to the term involving the azimuthal angles of the lepton and E_T^{miss} vectors, ϕ_ℓ and ϕ_{MET}. While clearly less informative than a fully reconstructed dilepton or dijet mass, m_T is often a useful variable for signal/background discrimination, particularly if the underlying kinematic distributions lend themselves to a clear **Jacobian peak** around $m_T = m_W/2$—see Section 11.2 for more detail.

A useful generalisation of the transverse mass is the **stransverse mass** variable, commonly denoted m_{T2}. This is designed to target final state events with a pair-produced particle that decays semi-invisibly (i.e. leading to two weakly-interacting particles that both contribute to the missing transverse momentum). Imagine that two of the same new particle with mass m_p are produced in an event, and both decay to a massless visible particle V and an invisible particle χ of mass m_χ. The parent particle mass is bounded from below by the transverse mass

$$m_P^2 = m_\chi^2 + 2\left[E_T^V E_T^\chi \cosh(\Delta\eta) - \boldsymbol{p}_T^V \cdot \boldsymbol{p}_T^\chi\right] \geqslant m_T. \tag{8.11}$$

The transverse momentum of the single invisible object from the decay of a given parent particle is of course unknown. Instead, we can construct an observable based on a minimisation over the under-constrained kinematic degrees of freedom associated with the two weakly-interacting particles as

$$m_{T2}^2(M_\chi) = \min_{\boldsymbol{p}_T^{\chi 1} + \boldsymbol{p}_T^{\chi 2} = \boldsymbol{p}_T^{\text{miss}}} \left\{ \max\left[m_T^2(1), m_T^2(2)\right]\right\}. \tag{8.12}$$

This dense formula requires some explanation, particularly since it is a rather ugly looking minimisation over a maximisation. $m_T^2(1)$ is the transverse mass formed using an assumed value of the mass of the invisible particle m_χ, the measured visible transverse momenta of one of the invisible particles $\boldsymbol{p}_T^{V_i}$, and an assumed value for the transverse momentum of one of the invisible particles $\boldsymbol{p}_T^{\chi_1}$:

$$m_T^2(1) = m_\chi^2 + 2\left(E_T^{V_1} E_T^{\chi_1} - \boldsymbol{p}_T^{V_1} \cdot \boldsymbol{p}_T^{\chi_1}\right). \tag{8.13}$$

This represents the transverse mass that is relevant for one of the parent particle decays. We can also make a transverse mass for the other parent particle decay:

$$m_T^2(2) = m_\chi^2 + 2\left(E_T^{V_2} E_T^{\chi_2} - \boldsymbol{p}_T^{V_2} \cdot \boldsymbol{p}_T^{\chi_2}\right), \tag{8.14}$$

where we now take the measured tranverse momentum of the other visible particle, and assume some value for the transverse momentum of the other invisible particle. For an assumed value of m_χ, called the **test mass**, and an assumed set of components for the two invisible transverse momenta, we can calculate these two invariant masses and work out which is bigger. We can then vary the assumed values of the components of the invisible transverse momenta (subject to the constraint that the two transverse momentum vectors sum to the observed missing tranverse momentum), and find the values of the components that minimise the maximum of the two transverse masses. This is typically done by numerical optimisation. This then gives us a value of m_{T2} which is unique for that event, based on the assumed value of m_χ. Different choices of m_χ lead to different definitions of the m_{T2} variable.

It can be shown that, with the correct choice of test mass, the m_{T2} distribution over all events is bounded from above at the parent mass m_p. In addition, if we choose $m_\chi = 0$, the m_{T2} distribution is bounded from above by $(m_p^2 - m_\chi^2)/m_p$. Note that different types of m_{T2} variable have been defined in the literature for complex decay processes, based on which of the visible particles are included in the transverse mass calculations that enter the formula.

Composite objects built from a subset of the event are not limited to two-object systems: obvious counterexamples are top-quark decays $t \to bW(\to \{\ell^+\nu_\ell, q\bar{q}\})$, three-body decays, or multiboson decays such as $H \to ZZ \to \{4\ell, q\bar{q} + 2\ell,...\}$. In these cases, three- or four-body composite systems are natural proxies for the original decaying object (and indeed we will return to the interesting challenges of top-quark reconstruction in Sections 11.2.2 and 11.2.3). Note, however, that neither of these examples is a *fundamental* $1 \to 2$ process, but instead compositions of $1 \to 2$ Feynman rules. This is a consequence of renormalizability in the SM: one cannot have more than two fermions connected to a vertex while maintaining dimensionless coupling constants. The only renormalizable four-point vertices in the SM are $g \to 3g$ (but the massless gluon does not set a natural scale for a three-jet mass) and triboson production: the latter of course are themselves reconstructed from final-state dilepton pairs. While off-shell effects break this cartoon picture, it retains enough truth for placing mass constraints recursively on pairings to be a useful reduction strategy for better selecting such signals from their combinatoric backgrounds.

Other properties of the few-object composite may also be useful, such as its p_T: in the lowest-order Feynman amplitudes for simple processes such as $pp \to V$ (for any decaying resonance V), the resonance p_T is zero: it is only through (mostly QCD) radiative corrections that a finite transverse momentum appears. If wanting to either select or eliminate such processes, the momentum properties of the composite may be very useful.

The final set of composite scale variables is those which represent all, or nearly all, of the event. As such, they are built from all or most of the elementary physics objects. The most common such object is H_T, the scalar sum of p_T over all visible 'hard' physics objects:

$$H_T = \sum_{\text{objects}} |p_T|. \tag{8.15}$$

Sometimes the set of objects used in an H_T computation is restricted to just a sum over jets, but most often in analyses without leptons. What counts as a 'hard enough' object to contribute is also analysis-dependent, with minimum-p_T and maximum-(pseudo)rapidity requirements being common. Regardless of these details, the common motivation is to construct an upper estimate of the hard-process energy scale, in which every object contributes constructively.

The logical further development of this idea, mostly seen in BSM search analyses, is the inclusion of MET in the scalar sum along with the visible objects:

$$m_{\text{eff}} = \sum_{\text{jets}} |p_T| + \sum_{\text{leptons}} |p_T| + E_T^{\text{miss}}. \tag{8.16}$$

The m_{eff} symbol suggests that this variable is the 'effective mass' of the event, but it only represents a mass in an illustrative sense—it is more an upper limit on the scale of momentum transfer to have taken place in the hard process.

8.4.3 Angular variables

In addition to energy, momentum and mass scale variables, we can define variables that are related to the geometries of events. This can be very important in isolating processes of interest. Take, for example, a process like $pp \to jj\gamma$ (or replace the γ with any other QCD singlet): geometrically this encompasses all sorts of configurations, of which some distinctive examples are (i) a 'Mercedes' topology with roughly equal (transverse) angles between each of the final-state objects, (ii) a 'dijet-like' config-uration in which the jets are more or less back-to-back and the photon is much closer in direction to one jet than the other, and (iii) a 'jets versus photon' system where the two jets are close together and are collectively recoiling against the photon.

At fixed (leading) order, there is no deep difference between these configurations (although they will have different differential cross-sections $d\hat{\sigma}/d\Phi$ in the kinematic phase-space Φ), but for realistic collider final-states, close-together configurations are more susceptible to e.g. QCD and QED resummation corrections, and perhaps also non-perturbative QCD effects. They hence may induce more model-dependence into an analysis; depending on whether the point is to measure and constrain that relatively soft physics, or to minimize its impact, the analysis may choose to place cuts that favour or exclude some of these angular topologies from the event selection.

At hadron colliders, the most obvious angular variable is the azimuthal angular difference $\Delta\phi_{ij} \equiv \phi_j - \phi_i$ between two objects i and j. This apparently trivial variable can often lead to bugs in analysis code if not approached carefully: the periodicity in ϕ means that one must be careful when taking the difference between azimuthal angles located on opposite sides of the arbitrary $\phi = 0$ direction. Computationally, this is usually done by mapping the angle into the acute range $|\Delta\phi| \leqslant \pi$, and if using a signed angle re-inserting a -1 factor if j defines the anticlockwise rather than clockwise edge of the acute angle.

A common use of azimuthal separation is in selecting events with large MET: as missing momentum can readily be faked by jet mismeasurement, requiring that the transverse MET vector point away from any single jet via a $|\Delta\phi|$ cut is an effective 'cleaning' strategy to improve sample purity.

A subtle issue when working with $\Delta\phi$ (or raw azimuthal angle ϕ) values is that they are naturally bounded in $[0, \pi)$ or $[0, 2\pi)$ (or $[-\pi/2, \pi/2)$ or $[-\pi, \pi)$), which involve irrational numbers. One cannot book a computer histogram with range ends exactly on irrational values, and close-enough approximations to avoid artefacts or incomplete coverage[9] are inconvenient to report. A good solution is to report such angles in rationally-bounded forms like $\Delta\phi/\pi$, or e.g. $\cos\phi$ where that is well-motivated by the physics.

The periodic nature of azimuthal angles can also be troublesome when e.g. supplying kinematics to a multivariate 'machine learning' toolkit, where it is often implicit that distances between values are computed linearly, but the points at opposite ends of ϕ and signed-$\Delta\phi$ ranges are infinitesimally close.

In addition to the transverse angular variable ϕ, it is also useful to consider longitudinal angular information, in the form of the rapidity or pseudorapidity. Despite the usual problem of the unknown hadron collider beam-parton boost, use of the longitudinal component is important in event selection for two reasons. Firstly, the angular acceptance of a real detector relative to the beam-line is never perfect: the acceptance of tracking detectors for charged-lepton identification and jet flavour-tagging, and the high-resolution parts of calorimeters, are typically limited to $|\eta| < 3$. This imperfection of the detection apparatus needs to be acknowledged in both search and measurement analysis types. Secondly, asymmetric event rapidity distributions tend to be associated with low-scale event types without very high-energy or high-mass objects, as they indicate a combination of large and small incoming parton x_1 and x_2 values: one cannot easily produce events with characteristic scales greater than a few tens of GeV if only one incoming parton is supplying the energy. This is the reason that the LHCb detector, primarily for the detailed study of few-GeV hadron production and decay, is a one-armed spectrometer at high-η, while the general-purpose detectors, more optimised for high-scale measurements and searches, are symmetric and concentrate most subdetector precision around $\eta = 0$.

In typical collider analyses, therefore, a cut to require key objects to have a small $|\eta|$ or $|y|$ value ~1 or 2 is very common as a proxy for event scale, as well as ensuring availability of the highest-resolution detector components. Rapidity differences are also meaningful, particularly as collider pseudorapidities and rapidities are by design invariant under changes of the unknown beam boost. A particularly striking motivation for rapidity-difference cuts in event selection is the study of colour-singlet fusion processes, in which there is no exchange of colour quantum numbers between the primary incoming partons. In this case, as there is no colour dipole to radiate, the **central region** (low-$|\eta|$) of the detector is expected to be relatively devoid of activity, the leading-order cartoon picture being spoiled only by subleading radiative and non-perturbative (e.g. multiple partonic interaction) physics. Searches

---

[9] It is not uncommon to see a ϕ plot that finishes at 3.2, with a systematically low final bin as the $[\pi, 3.2)$ part cannot be filled but still contributes to the normalisation.

for electroweak vector-boson fusion, $pp \rightarrow Vjj$, for example, are characterised by two **forward jets** at high absolute rapidity, and a central $V = \{W, Z, \gamma\}$, with a **rapidity gap** of low QCD activity between the two jets. The size of the gap, $\Delta\eta_{ij} \equiv \eta_j - \eta_i$ (and obvious variants for Δy_{ij}, and their unsigned versions) is a useful cut to exclude backgrounds from the more common processes with central QCD activity.

Finally, we consider the combination of longitudinal and transverse angular variables into a two-dimensional angular difference: $\Delta R_{ij} \equiv \sqrt{\Delta y_{ij}^2 + \Delta\phi_{ij}^2}$ and its η-based equivalent[10]. By incorporating information from both directions in the rapidity-azimuth plane, ΔR is in principle a more informative variable than either $\Delta\phi$ or $\Delta\eta$ alone. But by being distributed in a two-dimensional variable space, one must be aware that its distribution has a non-trivial phase-space component: perfectly uniform ϕ and y/η distributions nonetheless have a peak at $\Delta R \sim \pi$. In practice, ΔR is used more as an **isolation** and **overlap-removal** variable than an event-selection one, with separate, unconflated $\Delta\phi$ and $\Delta\eta/\Delta y$ cuts being used for event selection.

8.4.4 Event shape variables

Event shapes are an important class of variables motivated by the desire to characterise the general flows of energy and particles in various directions via a small number of variables. By construction, such low-dimensional quantities cannot capture the full complexity of an event, but this is entirely the point: they reduce it to the leading terms of physical interest, cf. a **multipole expansion** or Taylor series.

The most basic such variables emerged from lepton collider experiments such as those at the SLC and LEP colliders, and were particularly important (along with the development of jet-clustering algorithms) in enabling quantitative comparisons between analytic and MC-based QCD theory calculations and experimental observations. The variables characterise event shapes in quite generic ways, e.g. how spherical, collimated, or oblate an event is.

Later developments in event-structure observables, motivated by the increased prevalence of events with multiple QCD jets or high-momentum EW resonances, have extended the basic observables to reflect the increased topological complexity. These correspond conceptually to higher-order terms in our multipole expansion. In nearly all cases, the event-level shape observables described here have intra-jet equivalents to be introduced in the following section on **jet substructure**.

The historical role of event shapes as a meeting point of experiment and phenomenology means that event shapes have a careful relationship with QCD singularities, in particular the IR and collinear safety discussed in Chapter 2.

---

[10] Unfortunately there is no standard notation to discriminate between the two definitions, and despite arguments for the primacy of 'true' rapidity y, the pseudorapidity's direct identification with the angular position of detector elements means that $\Delta R_{ij}^{\eta} \equiv \sqrt{\Delta\eta_{ij}^2 + \Delta\phi_{ij}^2}$ is widely found in use under the name ΔR.

Precision measurements of IR-safe event shapes were, and remain, key to the measurement of the running of the strong coupling α_S.

Thinking again of event structure by analogy with terms in a multipole expansion, the zeroth term in the expansion is the average 'displacement' of the entire event, i.e. the **momentum balance** (or p_T balance at a hadron collider),

$$p^{\text{tot}} = \left| \sum_i \mathbf{p}_i \right| \quad \text{or} \quad p_T^{\text{tot}} = \left| \sum_i \mathbf{p}_{T,i} \right|. \tag{8.17}$$

If the sum here runs over all particles in a perfectly measured event devoid of invisibles, the balance is trivially zero by momentum conservation; in reality this delta-function will acquire some width due to acceptance, energy scale calibration, and resolution effects. This corresponds to the missing energy/missing p_T physics object, which plays a key role in distinguishing between fully resolved events such as QCD scattering or $Z \to q\bar{q}, \ell\ell$, against SM or BSM sources of missing energy ranging from direct neutrinos to WIMPs, dark-sector particles etc.

In analyses, it is also not infrequent, to use a hard-object version of event balance, e.g. the sum of jets and/or leptons to distinguish between processes in which particle combinations should be symmetric and those expected to be asymmetric. An example of this is the use of event **hemispheres** in lepton colliders, or azimuthal event partitionings at hadron colliders.

The next set of variables characterise the degree of collimation or spherical symmetry of the energy flows. We describe these as defined at lepton colliders, where the unambiguous centre-of-mass frame allows all three spatial directions to be considered equivalent (up to acceptance effects around the beam direction, to be corrected): hadron-collider versions may be defined easily by setting all z components to zero and hence reducing the dimensionality into the transverse plane. The 'classic' event shapes of this type are the **thrust** and **sphericity** (or spherocity[11]). We begin with the thrust:

$$T = \max_{|\mathbf{n}|=1} \frac{\sum_i |\mathbf{n} \cdot \mathbf{p}_i|}{\sum_i |\mathbf{p}_i|}, \tag{8.18}$$

in which the unit three-vector $\mathbf{n}$ is oriented to maximise the sum of momentum-projection magnitudes over all particles i. A neat algorithm exists to find exact solutions from combinations of event-constituent vectors but this is nonetheless expensive, scaling as $\mathcal{O}(n^3)$ in the number of constituents. T is constrained to lie between 1/2 and 1: a large value corresponds to 'pencil-like' events in which a large fraction of energy flow goes forward and backward along the thrust axis, and the 1/2 limit to perfectly isotropic events. As QCD jet production is dominated by a lowest-order two-parton final state with the two partons perfectly balanced at fixed order, with three-jet configurations naïvely suppressed by a factor of α_S, thrust distributions

---

[11] Some authors draw a nomenclature distinction between the full and transverse forms, but the literature is not unified in this respect.

$d\sigma/dT$ tend to rise from low values at $T \sim 1/2$ to a peak close to 1.[12] In collider physics we feel most comfortable with distributions that have a peak on the left-hand side of the plot, and hence $1 - T$ is often used in place of the thrust, frequently with the same name even though large values of $1 - T$ are not intuitively 'thrusty'. The plane transverse to the thrust vector $\boldsymbol{n}$ may be analysed in the same way, giving an orthogonal **thrust major** axis and finally the **thrust minor** vector perpendicular to the other two. The thrust projections along these axes, respectively denoted M and m, are also useful event-shape variables, often combined into the single **oblateness** variable, $O = M - m$.

Our second such variable characterises more or less the same shape information, but in a different and more convenient way. The **generalised sphericity tensor** is defined in terms of the three-momentum vector components p^a (for $a = 1, 2, 3$), as

$$S_r^{ab} = \frac{\sum_i p_i^a p_i^b |p|_i^{r-2}}{\sum_i p_i^r},$$ (8.19)

where $a, b \in \{1, 2, 3\}$ are spatial axis indices, and again the sum i runs over all event constituents (which may be individual particles, or objects like jets). The r parameter is free: the classic sphericity used $r = 2$, resulting in an IR-unsafe observable that could not be calculated at fixed-order and was subject to significant corrections in low-order resummation: this motivated much further phenomenological study of the thrust. Nowadays $r = 1$ is more likely to be used: this choice renders the sphericity tensor IR-safe.

As for the three thrust variables T, M, m, the sphericity tensor can be reduced to a set of principle components, this time by simple matrix diagonalisation rather than complicated combinatoric optimisation. This gives a thrust-like set of orthogonal event-alignment vectors, each with a corresponding eigenvalue. Several conventional sphericity components are defined in terms of these eigenvalues $\lambda_1 \geqslant \lambda_2 \geqslant \lambda_3$ as

$$S_r = 3/2(\lambda_2 + \lambda_3) \qquad S_2 \equiv \textbf{sphericity}, \ 0 \leqslant S_2 \leqslant 1 \qquad (8.20)$$

$$A_r = 3/2\lambda_3 \qquad A_2 \equiv \textbf{aplanarity}, \ 0 \leqslant A_2 \leqslant \tfrac{1}{2} \qquad (8.21)$$

$$C_r = 3(\lambda_1\lambda_2 + \lambda_1\lambda_3 + \lambda_2\lambda_3) \quad C_1 \equiv \textbf{C-parameter}, \ 0 \leqslant C_1 \leqslant 1 \qquad (8.22)$$

$$D_r = 27\lambda_1\lambda_2\lambda_3 \qquad D_1 \equiv \textbf{D-parameter}, \ 0 \leqslant D_1 \leqslant 1. \qquad (8.23)$$

The sphericity variables S_r interpolate between pencil-like and isotropic events, and the aplanarities A_r play a role similar to the subleading thrust measures (major/minor or oblateness) in characterising the activity perpendicular to the main event axis and particularly out of the main event plane. Pleasingly for intuition, events

---

[12] In practice, resummation of multiple emissions gives enhancements by kinematic-logarithm factors which complicate this simple picture of power-suppressed contributions. They also shift the peak from a divergence at exactly 1 to a finite peak slightly below. These corrections notwithstanding, the fixed-order expansion is a good conceptual starting point.

with larger values of sphericities are more spherical, and those with smaller values of aplanarity have their activity closer to exclusively occuring in a plane. The aplanarity variable in particular is frequently used in full three-dimensional form in LHC BSM direct search analyses, particularly computed from jets, as a distinguishing variable between QCD-like and BSM decay-chain kinematics.

This brings us to the final set of event-shape variables, which arose as multi-jet physics and jet substructure techniques rose in prominence during the LHC era. Many variables can be found in the literature, but one of particular prominence is the **N-jettiness** variable—as a precursor to its more famous direct relative, the N-subjettiness (which we will encounter shortly). N-jettiness, originally motivated by QCD resummation calculations for classifying Higgs-candidate events by their jet multiplicity is defined as

$$\tau_N^d = \sum_i \min\{d_a(p_i),\ d_b(p_i),\ d_1(p_i),\ \cdots,\ d_N(p_i)\}, \tag{8.24}$$

where the i sum runs over a set of elementary event constituents with four-momenta p_i, and the $d_{a,b}(p_i)$ and $d_k(p_i)$ functions are distance measures between p_i and four-vectors representing the beams (a, b) and a set of $k \in 1...N$ jet axes, respectively. These jet axes may be defined using several schemes and the distances may be any IR-safe measure (i.e. linear in $|p_i|$), but a not inaccurate characterisation of the normal behaviour is to pick the set of N jet axes that minimise τ_N. The astute reader will note a joint echo of hadron-collider inclusive jet clustering (in the unique assignment of particles to jets via a distance measure, including beam distances) and of the thrust event shape (in the choice of event axes by minimising the shape measure). The same properties that made thrust an IR-safe shape variable with low-values corresponding to two-jet events make the N-jettiness a powerful variable for identifying events with N distinct jets, although the nomenclature here takes a sadly unintuitive turn, with highly n-jetty events acquiring a *low* τ_N score.

We will return to and extend the concept of dynamic N-jet measures in the following section. In conclusion regarding whole-event shapes, we should also mention that there are many more such ways to characterise event energy flows, from literal Fourier/wavelet expansions to machine-learning methods such as energy-flow and **particle-flow networks** (the latter based on an again familiar-looking distance measure). And again, these ideas can largely be applied to the internal structure of jets as well as to whole events—a topic which in recent years has seen prodigious activity on both the phenomenological and experimental sides, and which we now address.

8.4.5 Jet (sub)structure variables

Jets are an event microcosm, and as such many of the event shape observables just described also apply in some form to jets. Use of the internal structure of jets can be a powerful phenomenological technique for rejecting QCD backgrounds in the search for **boosted** heavy resonances (either SM or BSM) whose decay products are collimated by their parent's high momentum into a single jet cone. Jets with large radius parameters $R \sim 1.0$—historically called 'fat jets'—are typically used for such studies, applying a boosting heuristic that decay products from a heavy resonance of mass M undergoing two-body decay will be captured within a single large-R jet cone if its (transverse) momentum is greater than twice the resonance mass:

$$p_T^{\text{jet}} \gtrsim p_T^{\text{boost}} = \frac{M}{2R}.$$

(8.25)

Jet structure observables can also be important for the labelling of normal-width light jets (without b- or c-tags) as being dominated by quark or gluon amplitudes. These are usually referred to as **quark jets** or **gluon jets**, respectively, although of course there is no perfect one-to-one correspondence: if nothing else, hadronic jets are colour singlets and quarks and gluons obviously are not. In fact, it is the difference in non-zero colour charge of quarks and gluons that enables experimental distinction between their jets: initiating gluons carry the adjoint colour charge $C_A = 3$ while quarks carry $C_F = 4/3$, and for negligible quark masses it is the magnitude of this charge that dominates the amount of QCD radiation. Gluon jets hence radiate more than quark jets, and tend to have more constituents and to be 'wider' in η–ϕ.

We start, therefore, by defining a set of simple jet-structure properties: the constituent multiplicity, width and mass. The multiplicity and mass are familiar. The former is the number of clustered final-state particles (or frequently, for better accuracy via tracking, the number of *charged* particles) in the jet, and the latter is the invariant mass of the four-vector sum of constituent momenta,

$$M^2 = \left(\sum_i E_i \right)^2 - \left(\sum_i p_i \right)^2$$

(8.26)

where the i index as usual runs over the jet constituents. The **jet width** is less familiar: it is defined as the p_T-weighted mean ΔR distance of jet constituents from the jet centroid:

$$W = \frac{\sum_i \Delta R^i p_T^i}{\sum_i p_T^i}.$$

(8.27)

As the width is linear in the constituent momenta, it is an IR-safe observable, with the intuitive interpretation that larger widths correspond to jets whose energy flow is spread more widely around the centroid.

In fact, all these measures are examples of a more general class of observables, **jet angularities**. The generalised angularity is defined as

$$\lambda_\beta^\kappa = \sum_i z_i^\kappa \theta_i^\beta$$

(8.28)

$$\Rightarrow = \sum_i \left(\frac{p_{T,i}}{p_T^{\text{jet}}} \right)^\kappa \left(\frac{\Delta R_i}{R} \right)^\beta,$$

(8.29)

where as usual the sum is over jet constituents. Here the z_i variable is a fraction of a jet energy measure and θ_i is an angular variable: various definitions exist in the literature, but here they are made concrete in the second line as the ith constituent's p_T fraction with respect to the containing jet, and the ratio of the constituent ΔR angle from the jet centroid to the jet radius parameter R. These angularities are parametrised by the κ and β parameters, whose variation defines the whole 2D family of angularities, shown in Figure 8.2. The $\kappa = 1$ line is particularly important, as it defines the IRC-safe subfamily of angularities: on this line, $\beta = 1$ is the jet width and $\beta = 2$ is monotonic to jet mass; off it, $(\kappa, \beta) = (0, 0)$ is the IR-unsafe constituent multiplicity. κ values greater than unity place more emphasis on high-p_T constituents, much like the generalised exponent in the usual jet-clustering measure, and large β values place more emphasis on wide-angle radiation: angularities are hence a natural basis for characterising QCD radiation around its soft and collinear singularities.

In fact, angularities are so closely associated with QCD phenomenology that a leading technique in jet physics is directly motivated by them. The problem with fat jets is that they contain extra gunk from pile-up and soft radiation that pollutes the clean substructure that we wish to find. Before we can use fat jets, we must therefore subject them to a process known as **jet grooming**, which aims to remove as much of the gunk as possible. The study of angularities has led to the **soft-drop** method, one of many approaches for removing 'inconvenient' QCD radiation from a jet and exposing its hardest components, either from hard QCD splittings or heavy-particle decays (see also 'filtering', 'trimming', '(modified) mass-drop', and various other

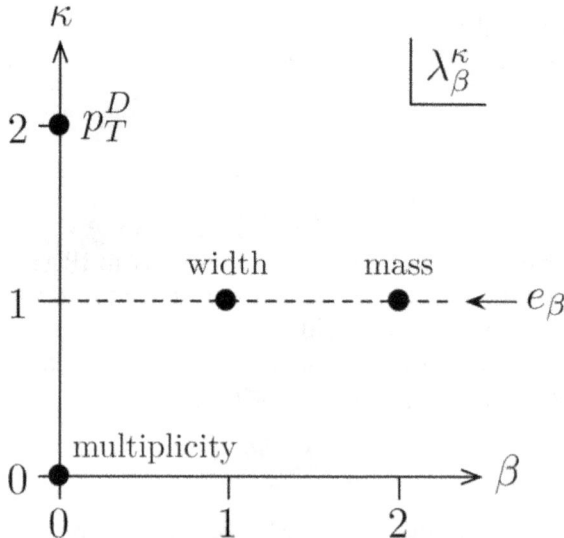

Figure 8.2. Identification of 'classic' jet structure variables with points in the generalised-angularity (κ, β) plane, highlighting the IR-safe subset of variables along the $\kappa = 1$ line. Diagram reproduced with permisson from Larkoski A, Thaler J and Waalewijn W 2014 *JHEP* **11** 129, arXiv:1408.3122.

methods you will find in the literature and the FastJet plugin collection). In soft-drop, as with other methods, the key idea is to step backward through the jet cluster sequence, first undoing the last clustering, then the second-last, and so on. At each declustering step, the two branches with transverse momenta $p_{T,1}$ & $p_{T,2}$, and angular separation ΔR_{12} are used to compute the inequality

$$\frac{\min\{p_{T,1}, p_{T,2}\}}{p_{T,1} + p_{T,2}} > z_{\text{cut}}\left(\frac{\Delta R_{12}}{R}\right)^{\beta}. \tag{8.30}$$

This is the **soft-drop condition**. If it evaluates as true, the softer branch is sufficiently hard and narrow and is retained; if it evaluates false, the softer branch is dropped from the cluster sequence and the grooming procedure continues to the next declustering. Note the role of the z_{cut} variable, which sets the minimum p_T fraction needed for a branch to be retained, and the β exponent (cf. angularities) which makes it *easier* for collinear branches to survive the cut: this is why it is a *soft*-drop procedure: it removes low-energy contaminants while protecting the collinear QCD singularity. Soft-drop is a powerful (and from a theory perspective analytically amenable) technique for cleaning residual pile-up, underlying event, and soft QCD radiation from jets, to more clearly expose their key structures.

We now proceed to jet-shape variables designed for sensitivity to the 'pronginess' of a jet's internal energy flows. Here we find two classes of observable, which as it happens bear close relation to the N-jettiness and sphericity event shapes. The first really *is* N-jettiness, but limited to the subset of final-state particles clustered into the jet: the N-**subjettiness**, denoted $\tau_N^{(\beta)}$. The β term again represents an angular exponent, and is often dropped, assuming $\beta = 1$. As before, it is a variable which takes values which asymptotically approach zero as the energy flows within the jet approach perfect coincidence with a set of N subjet axes. Previously we did not specify the IR-safe definition of distance to be used, as different measures are possible, but for N-subjettiness we do not have the complication of beam distances, and for the distance of constituent four-vector p_i to in-jet axis $\hat{n}_k$ it is rare to use a definition other than

$$d_k^{(\beta)}(p_i) = \frac{p_{T,i}\Delta R(p_i, \hat{n}_k)^{\beta}}{\sum_i p_{T,i} R^{\beta}}, \tag{8.31}$$

where R is the original jet-clustering radius, giving an explicit form for first a fixed-axis N-subjettiness $\tilde{\tau}_N^{(\beta)}$ and finally the operational form in which the axes themselves are optimised to minimise the measure:

$$\tilde{\tau}_N^{(\beta)} = \frac{1}{\sum_i p_{T,i} R^{\beta}} \sum_i \min\{\Delta R(p_i, \hat{n}_1)^{\beta}, \cdots, \Delta R(p_i, \hat{n}_N)^{\beta}\} \tag{8.32}$$

$$\Rightarrow \tau_N^d = \min_{\{\hat{n}_k\}} \tilde{\tau}_N^{(\beta)}. \tag{8.33}$$

While directly useful in their own right, these observables acquire additional power when used—in an echo of the likelihood-ratio discriminant between hypotheses—in the form of an N-**subjettiness ratio** between different hypothesised N,

$$\tau_{NM}^{(\beta)} = \tau_N^{(\beta)}/\tau_M^{(\beta)}. \tag{8.34}$$

Most commonly, the distinction sought is between adjacent N and M, i.e. $\tau_{21}^{(\beta)}$ or $\tau_{32}^{(\beta)}$. These are, respectively, used in attempts to distinguish truly two-prong jets with small $\tau_2^{(\beta)}$ from less structured QCD jets which only achieve small values by chance, and to discriminate three-prong decay like those of hadronic top-quarks from QCD and hadronic W and Z decays. Use of soft-drop jet grooming can further improve this resolving power. The N-subjettiness ratio is a powerful technique but problematic phenomenologically, as fixed-order perturbative QCD calculations diverge when the $\tau_{N-1}^{(\beta)}$ denominator approaches zero: they are hence not IR-safe. They are, however, safely calculable with all-orders resummation, an important practical relaxation of the IR-safety concept called **Sudakov safety**.

Our final class of jet substructure variables is **energy correlation functions** (ECFs), which as it happens capture some of the features of sphericity to complement N-subjettiness's extension of the thrust concept. The core definition is that of the generalised N-point ECF,

$$\mathrm{ECF}(N, \beta) = \sum_{i_1<i_2<\cdots<i_N\in J}\left(\prod_{a=1}^{N} p_{\mathrm{T},i_a}\right)\left(\prod_{b=1}^{N-1}\prod_{c=b+1}^{N}\Delta R(i_b, i_c)\right)^{\beta}, \tag{8.35}$$

where once again all the i terms are jet constituent indices, and β plays a role in emphasising or de-emphasising wide-angle emissions. $\mathrm{ECF}(N, \beta) = 0$ if the jet has fewer than N particles. Being linear in the p_Ts of the constituents, ECFs are IRC-safe, and by comparison with $\tau_N^{(\beta)}$, there are now no axes to optimise: the angular sensitivity is incorporated through the angular differences between every ordered pair of constituents.

This form, while general, is rather difficult to parse. As with N-subjettinesses, we are mainly interested in rather low values of N, to probe sensitivity to few-particle splittings and decays. The more comprehensible explicit forms for $N = 1\ldots4$ are

$$\mathrm{ECF}(1, \beta) = \sum_i E_i \tag{8.36}$$

$$\mathrm{ECF}(2, \beta) = \sum_{i<j} E_i E_j \theta_{ij}^{\beta} \tag{8.37}$$

$$\mathrm{ECF}(3, \beta) = \sum_{i<j<k} E_i E_j E_k (\theta_{ij}\theta_{ik}\theta_{jk})^{\beta} \tag{8.38}$$

$$\mathrm{ECF}(4, \beta) = \sum_{i<j<k<l} E_i E_j E_k E_l (\theta_{ij}\theta_{ik}\theta_{il}\theta_{jk}\theta_{jl}\theta_{kl})^{\beta}. \tag{8.39}$$

The key feature of ECFs is that if there are only N significant energy flows (i.e. subjets), $\text{ECF}(N + 1, \beta) \ll \text{ECF}(N, \beta)$, and this rapid dependence on the order of the correlator means that again we can use adjacent-order ratios as discriminants:

$$r_N^{(\beta)} = \frac{\text{ECF}(N + 1, \beta)}{\text{ECF}(N, \beta)} \tag{8.40}$$

behaves much like N-subjettiness $\tau_N^{(\beta)}$, going to small values for jets with N subjets. Having drawn this analogy between $r_N^{(\beta)}$ and $\tau_N^{(\beta)}$, it is natural to ask whether we can go further and also use ECFs to build discriminants like $\tau_{N,N-1}^{(\beta)}$: indeed we can, via the double-ratio

$$C_N^{(\beta)} = \frac{r_N^{(\beta)}}{r_{N-1}^{(\beta)}} = \frac{\text{ECF}(N + 1, \beta)\,\text{ECF}(N - 1, \beta)}{\text{ECF}(N, \beta)^2}. \tag{8.41}$$

The C notation is a nod to the fact that these ECF double-ratios are not just superficially similar to sphericity (a two-point correlator of three-momentum components), but are an exact generalisation of the C-parameter defined earlier in terms of sphericity eigenvalues. As with $\tau_{N,N-1}^{(\beta)}$, small values of $C_N^{(\beta)}$ indicate that a jet contains N subjets. $C_2^{(1)}$ in particular is a provably optimal discriminant of boosted two-prong decays against QCD backgrounds and is used in boosted-jet searches, as is the related D_2 variable for three-prong versus two-prong discrimination:

$$D_2^{(\beta)} = \frac{r_3^{(\beta)}}{r_2^{(\beta)}} \cdot \frac{\text{ECF}(1, \beta)^2}{\text{ECF}(2, \beta)} \tag{8.42}$$

$$= \frac{\text{ECF}(3, \beta)\,\text{ECF}(1, \beta)^3}{\text{ECF}(2, \beta)^3}. \tag{8.43}$$

Further generalisations introduce an extended family of N, M and U correlator combinations, but these are less used. Energy correlators are amenable to analytic QCD calculations, and have shown themselves to be a powerful general family of observables for extracting analysing power from the inner structure of hadronic jets.

8.5 Performance optimisation

The basic job of an LHC data analysis is to study or search for some particular signal process amongst other background processes, which raises the obvious question of how well optimised a given analysis is. It turns out that there is no single 'correct' answer for quantifying this, since it is dependent on the overall goal of the analysis. Specific examples include:

- In a search analysis for a specific model, your ideal metric may be the exclusion reach of 95%-confidence limit contours in the model-parameter space (e.g. masses and/or couplings)[13];

---

[13] But beware over-optimizing for features of unrealistic straw-man models, such as BSM simplified models: see Section 10.1.2.

- In a more generic model, the metric could be the *expected* limit maximisation across the broad familiy of such models—perhaps, but not necessarily, expressible as the volume within a higher-dimensional set of exclusion contours;
- In a single-number measurement, the aim is typically to minimise the total measurement uncertainty;
- In a multi-bin differential measurement, the metric may fall into a nebulous combination of maximising resolution (i.e. minimising bin sizes), and minimising uncertainty (perhaps by increasing bin sizes). The competition between these two can motivate use of different bin-sizes, using smaller ones where event yields and observable resolutions permit. Measurements may also be motivated by discriminating between models, in which case a search-like metric of optimising discrimination/limit-setting power may be appropriate.

The exact balance of expected signal and background yields to maximise analysis performance will depend on these various priorities, and in turn on the signal- and background-process cross-sections, the integrated luminosity (since absolute yield values will enter the Poisson likelihood and discriminant measures presented in Chapter 5), and on the nature of the systematic uncertainties which (through nuisance parameters) dilute the statistical power of the results. Unfortunately, in most situations one cannot reasonably put the whole analysis chain—running over perhaps hundreds of millions of events, estimating hundreds of uncertainties, and performing expensive likelihood fits—into a numerical optimisation code to 'automatically' find the optimal event-selection cuts: we must usually work instead using heuristics and judgement.

8.5.1 Summary estimators of significance

A useful class of quantities for assessing analysis performance (or the expected performance) is that of **significance estimators**. It is always useful to have simple estimators to hand, and the simplest available are functions comparing the expected signal yield S to the expected background yield B. These yields can be estimated in the first instance by running the analysis selection algorithms on nominal-configuration MC event samples. Which form these summary functions take depends on whether the analysis is likely to be **statistics-dominated** or **systematics-dominated**, i.e. whether the uncertainties that limit the experimental power stem from the statistical fluctuations inherent to small event counts, or the limits in resolution and calibration of the apparatus (including software).

In the former case, where statistical 'jitter' is the dominant uncertainty, we recall the simple result that the variance of a Poisson distribution is equal to its mean, λ, and hence its standard deviation is given by $\sqrt{\lambda}$. In a search for a new signal, we will only be able to conclusively declare that it is present if we measure a total event yield $S + B$ that is significantly higher than would be expected from background B alone. 'Significantly', here, means that the excess above expectation $(S + B) - B = S$ must be larger than several background standard deviations. The measure $S/\sqrt{B}$ is hence a useful heuristic in judging whether significant is statistically achievable in principle, before assessing the tricker effects of systematic mismeasurement (or in the fortunate limit that they are negligible). Considering also the statistical uncertainty in S, and combining them in quadrature, the statistic $S/\sqrt{S+B}$ is similarly useful.

If the statistics are expected to be large enough that the background yield should be stable and easily allow statistical resolution of S, we move instead to the systematics-dominated regime. Here a full evaluation of the *a priori* uncertainty (before likelihood profiling), and hence expected significance, will require very expensive runs over systematic variations of the object reconstruction, and over the set of theory uncertainties associated with the MC event-generation. But as systematic uncertainties are fixed relative errors, which unlike statistical relative errors do not reduce with accumulating luminosity, a shortcut is to look at the ratios $S/\epsilon B$ or $S/(\epsilon_S S + \epsilon_B B) \sim S/\epsilon(S+B)$, where ϵ is a representative total relative systematic uncertainty, and $\epsilon_{S,B}$ a pair of such uncertainties specific to signal or background event characteristics.

A less simple approach is to use the result given in equation (5.145), that for most variants of the log likelihood ratio q_μ, the median expected significance of a statistical test based on that LLR is the square-root of its Asimov-dataset value, i.e. the value obtained when the data are exactly equal to the expected values. For a single Poisson-distributed event count k, the Asimov LLR is

$$t_A = 2\ln\left(\frac{\mathcal{P}(k = s + b; \lambda = s + b)}{\mathcal{P}(k = s + b; \lambda = b)}\right) \tag{8.44}$$

$$= 2\left[(s + b)\ln\left(\frac{s + b}{b}\right) - s\right], \tag{8.45}$$

and so the Asimov expected median significance of the single bin is

$$Z_A = \sqrt{2\left[(s + b)\ln\left(\frac{s + b}{b}\right) - s\right]}. \tag{8.46}$$

A more detailed treatment also permits inclusion of estimated systematic uncertainties on the background, σ_b, giving the final estimator

$$Z_A^{\text{syst}} = \left[2\left((s + b)\ln\left[\frac{(s + b)(b + \sigma_b^2)}{b^2 + (s + b)\sigma_b^2}\right] - \frac{b^2}{\sigma_b^2}\ln\left[1 + \frac{s\,\sigma_b^2}{b(b + \sigma_b^2)}\right]\right)\right]^{1/2}. \tag{8.47}$$

Given the explicit probabilistic picture used here, these expressions can be generalised to multi-bin likelihood-ratio tests by summing the LLR contributions from each bin to form a composite LLR before taking the square-root.

8.5.2 Selection performance and working points

Often when developing an event selection procedure, cut-based or otherwise, the general framework of cuts or weightings is clear, but the exact values to use are not.

The same situation is also natural in object reconstruction The space of these indefinite values hence contains a set of **working points** for the algorithm, many of which might be reasonable. Again, we need a quantitative metric for how to pick the optimal working point for the analysis' purposes. When developing a new algorithm, it may also be interesting to consider how the overall performance of the algorithm compares to others, across the whole space of reasonable working points.

The key quantities to summarise selection performance are again functions of event counts from signal and background, but now including the pre-selection numbers from signal and background as well as the post-selection ones: we denote these as S_{tot}, B_{tot} and S_{sel}, B_{sel}, respectively.

A first pair of important quantities are already quite familiar: the signal and background **efficiencies**. The signal efficiency, also known as the **true-positive rate** (TPR) is

$$\epsilon_S = S_{\text{sel}}/S_{\text{tot}}, \tag{8.48}$$

and the background efficiency, also known as **false-positive rate** (FPR) or **fake-rate**, is

$$\epsilon_S = B_{\text{sel}}/B_{\text{tot}}. \tag{8.49}$$

'Positive' in this nomenclature means that an event was selected, and 'true' or 'false' whether it *should* have been.

The second kind of quantity important for characterising selection performance is the **purity**,

$$\pi = \frac{S_{\text{sel}}}{S_{\text{sel}} + B_{\text{sel}}}, \tag{8.50}$$

i.e. a measure of the extent to which the selected sample is contaminated by background events which we would rather have discarded. There is a natural tension between efficiency and purity: one can trivially achieve 100% efficiency by simply accepting every event; equally, full purity can be approached by tightening cuts until virtually no events pass. The space of interesting selection working points lies between these two extremes.

A useful tool for visualising the space of working points that produce these signal and background yields, and hence efficiency and purity metrics, is the **ROC curve**— the name derived from the quasi-military jargon 'receiver operating characteristic' and reflecting its origins in wartime operations research. ROC curves come in several formulations, but the classic (and arguably easiest to remember) form is as a scatter plot of working points in the 2D space of TPR versus FPR. Very often, the set of working points plotted in the ROC are drawn from variation of a single operating parameter and can hence be joined by lines to give the classic 'curve'. If the TPR is plotted on the horizontal axis, and the FPR on the vertical, a typical curve will tend to curve from bottom-left to top-right below the diagonal, reflecting the trade-offs between high acceptance and high purity. Examples of such ROC curves are shown in Figure 8.3. The general aim in optimising the choice of working point on this curve is to pick a point with large TPR and small FPR, i.e. as far to the

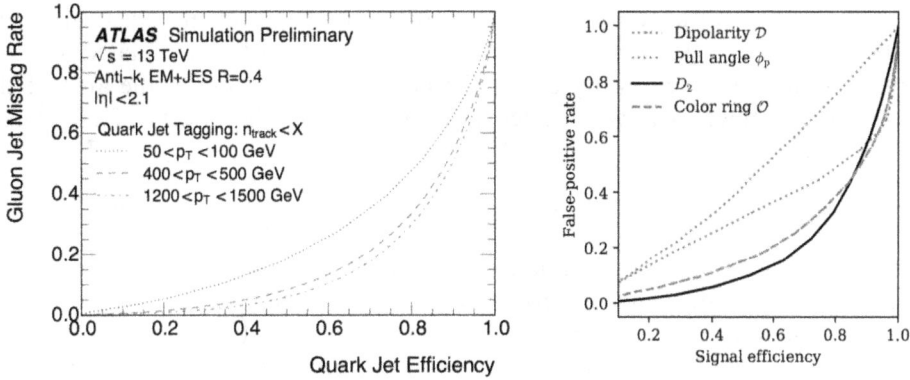

Figure 8.3. Example ROC curves, both in the TPR versus FPR format described in the text. Left: ATLAS 'tagging' of quark-like and gluon-like jet categories using charged-particle multiplicity cuts in various jet-p_T bins. Credit: ATL-PHYS-PUB-2017-009 Right: Discrimination of colour-singlet boosted $H \to b\bar{b}$ decay from the dominant QCD background using a variety of jet substructure observables. Reproduced with permission from Buckley *et al* 2020 *SciPost Phys.* **9** 026, arXiv:2006.10480.

right-hand side and toward the bottom of the plot as possible. As it is unlikely that both aims can be met simultaneously, the exact 'best' working point within this region requires a quantitative metric beyond informal 'eyeballing'. Such a metric will depend on the requirements of the analysis, and the relative rates of its signal and background event types.

An oft-cited measure of ROC curve performance is the **area under the curve** or AUC. This is defined as you might expect, for the TPR/FPR axis layout we have described. A larger area under this curve indicates a more performant selection algorithm in general. But note that AUC does not characterise a single working point, but the whole method: as such, it often does not answer the question of most interest, namely 'What specific settings are best for *my* analysis?' What is needed is usually one operating point well-suited to the application, at which point the rest of the curve is irrelevant.

And indeed the ROC curve cannot fully answer that question, for the simple reason that it deals entirely in efficiencies for signal and background selection individually, losing the crucial context of the relative sizes of pre-selection signal and background, S_{tot} and B_{tot}. As it happens, plotting the signal efficiency against purity does give a handy visual indication of a relevant metric, since the rectangle between the origin and any working point on the ROC curve is

$$\pi \times \epsilon = \frac{S}{S + B} \times \frac{S}{S_{\text{tot}}} \tag{8.51}$$

$$= \frac{S^2}{S + B} \times \frac{1}{S_{\text{tot}}} \tag{8.52}$$

$$\propto (S/\sqrt{S + B})^2, \tag{8.53}$$

our summary significance from earlier, in the case of dominant statistical uncertainties. This gives a useful rule of thumb for picking a statistics-dominated working point from an efficiency–purity curve: the point which maximises the area of the subtended rectangle is *a priori* interesting. Of course, the effects of systematic uncertainties must also be considered in a final optimisation.

8.6 Estimating backgrounds and uncertainties

No analysis data or object selection is perfect: there will always be contributions at some level from independent processes that you weren't targetting, and mis-identification or mismeasurement of your physics objects. The two topics are related in the sense that both can contribute to systematic offsets in measured observables. Depending on the type of analysis you are performing, you may wish to either correct for these defects or simply to represent them accurately. In this section we will cover a series of basic methods to estimate the effects of backgrounds and measurement errors, and to propagate and minimise their impact on final detector-level results.

8.6.1 Backgrounds

The simplest way to estimate backgrounds is to simulate them using MC event generation and simulation. The MC generator, or perhaps a state-of-the-art parton-level calculation, will provide an estimate of the total cross-section, and the event kinematics, detector simulation, reconstruction, and analysis algorithms will furnish the efficiency and acceptance terms.

But of course this presupposes that the calculations, the phenomenological parts of event generation, the simulation geometry and material interactions, digitization and pile-up are all modelled accurately for the phase-space being selected: while the technology is very impressive, this is a huge demand. It is also an unacceptable assumption for an experimental measurement: other than processes whose expected upper cross-section estimate is orders of magnitude below the experimental resolution, MC estimates must be challenged and refined using *in situ* data as much as possible. Even then, a combination of theory and experimental uncertainties will be associated with the background contributions: we will consider these in the next section.

Reducible and irreducible backgrounds
Background processes fall into two main categories: **reducible** and **irreducible backgrounds**, respectively, defined as those which can or cannot be removed by use of extra event/object-selection cuts.

One category of reducible backgrounds is 'fakes', where an elementary physics object is mis-identified: jets can fake electrons and photons, light jets can fake hadronic taus or b-jets, and so on. These are reducible in principle by either better reconstruction algorithms (a long-term project) or by using tighter working points in e/γ reconstruction, b-tagging, and isolation—at the usual cost of losing some signal efficiency while increasing sample purity.

A more complex type of fake is the **combinatoric background**, which occurs in the reconstruction of multi-body decays such as top quarks or exclusive hadron decays. For example, a hadronic top-quark will decay to a b-jet and at least two other, usually light-quark, jets, but hadron collider events are awash with QCD (also mostly-light) jets: it is easy to choose a wrong combination of jets which looks reasonable, but does not correspond to the set of jets involved in the hard interaction and decay. As hard-process scales increase, the potential for such errors can also increase as the phase-space for extra QCD jet activity opens up. Combinatoric backgrounds also occur between non-jet elements of events, such as pairing leptons with the correct b-jets in dileptonic $t\bar{t}$ production, or finding correct lepton pairs in diboson $ZZ \rightarrow 4\ell$ events. These problems are reducible in principle by use of further cuts in the assembly of the parent system, for example:

- invariant-mass requirements in the object selection: an SM top-quark decay passes through a W resonance, and as such the majority of correct light-jet pairs will have an invariant mass close to the W pole mass $\sim$80.4 GeV. Similar restrictions on lepton-pair invariant masses are standard in Z reconstruction, and may be either a double-sided **window cut** requiring $m_{\text{low}} < m_{\text{inv}} < m_{\text{high}}$, or a single-sided one: the choice depends on the dominant source of contamination, and the usual trade-off between purity and efficiency. The size of the window is limited by the experimental resolution on m_{inv}.

  Note that these cuts bias the selection, perhaps unacceptably. In diboson events in particular, for example $H \rightarrow ZZ^* \rightarrow 4\ell$, the fully resonant mode with both Z bosons on-shell may be only part of the desired signal process.

- lepton flavour and charge requirements: if attempting reconstruction of a rather degenerate system such as dileptonic $t\bar{t}$, $ZZ \rightarrow 4\ell$, $WW \rightarrow 2\ell 2\nu$ etc, and unable to make tight invariant mass cuts, there is in principle power in the choice of lepton flavours, and use of charge identification. The former necessarily involves a loss of events, but with the benefit of great purity: in $t\bar{t}$ the same-flavour $\{ee, \mu\mu\}$+jets+MET events are discarded to give less ambiguous $e\mu$+jets+MET configurations; $ZZ \rightarrow 4\ell$ is knocked down to $2ee2\mu\mu$, and so on.

  If losing half your events isn't an option, identifying the lepton charges can help in highly degenerate circumstances: there are far fewer pairings in an event with known $2e^+2e^-$ than in a $4e$ view of the same. But the kicker is that charge identification has its own inefficiency and mistag problems: another source of uncertainty to be investigated and propagated. The charge-identification error rate increases with lepton energy, as it is reliant on track curvature in the detector B-field, and high-p_T tracks asymptotically approach a charge-agnostic straight line.

- MET cuts to reject backgrounds from e.g. W.

- angular cuts between physics objects, e.g. to select back-to-back systems compatible with a two-body production process, and reject combinations of kinematically uncorrelated objects.

We will return to top quark reconstruction in Chapter 11.

When assessing background contributions to an analysis, all which can enter the event selection need to be considered, whether due to equivalent final-states, fakes, or combinatorics. The most important source of fakes at hadron colliders is the plethora of light-jet backgrounds: any target final-state containing a charged lepton (particularly electrons), photon, tau, or b-jet should be considered under the transformation of replacing that object (or each combination of such objects) with light-jet fakes. In the case of b-jets, there can also be significant contamination from c-jets and taus.

For example, if reconstructing the $Z(\to \ell^+\ell^-)H(\to b\bar{b})$ final-state, one also needs to consider contributions from:

- $Z(\to \ell\ell) + \geqslant 2$ jets, including pure light-jet, c–light, cc, bc, and bb jet flavours from various sources, with acceptance into the event selection according to the product of cross-section and b-(mis)tag rates for the chosen tagger working point;

- $W(\to \ell\nu) + \geqslant 2$ jets, as above but with a single direct lepton, the other being faked by either a jet misidentification or an indirect lepton. With 'tight' (high purity) lepton-identification definitions, this process should have a low acceptance efficiency;

- Inclusive $t\bar{t}$, with the dileptonic channel entering mainly via the same-flavour $ee/\mu\mu$ top decays, semileptonic via a lepton-faking jet, and all-hadronic $t\bar{t}$ via two electron fakes;

- Single top, via a lepton fake and either a real or mistagged b-jet;

- Diboson configurations: $Z(\to \ell\ell)Z(\to b\bar{b})$, $W(\to \ell\nu)Z(\to b\bar{b})$ with a lepton fake, and $W(\to \ell\nu)W(\to \ell\nu) + \geqslant 2$ jets are the most obvious combinations, but many more diboson feed-in modes are possible when mistagging and fakes are considered. Diboson cross-sections are themselves small—around 10 pb to 100 pb—and fake/mistag rates are usually well below 10%, so for most signal processes these effects are small, but in e.g. a Higgs analysis they are not negligible.

- Multijet QCD backgrounds via real and mistagged b-jets, and lepton fakes. This source of background almost always needs to be considered, even if expected to be small by back-of-envelope calculations with fake and mistag rates, because the cross-sections are so high, and the MC modelling unreliable.

In this case, the top and W channels are reducible to some extent by MET cuts (the missing momentum being low in the signal process, as compared to the backgrounds with direct neutrinos), and the others are amenable to some extent by use of tighter b-tag and lepton working points. But the combination of several tight cuts can destroy the signal efficiency, usually below what is acceptable: given large incoming luminosity, many analyses would rather populate more bins or explore further p_T-distribution tails than discard signal events. So backgrounds are to some extent a fact of life: how, then, to improve on their MC estimates?

Data-driven background estimation

'Data-driven' estimation of unknown or semi-known analysis elements is a heavily overused phrase in HEP literature, and covers a wide range of techniques, some making significantly more use of data than others. In general, though, the idea is to improve or replace an uncertain estimate, e.g. from MC modelling, with a better one informed by data in a phase-space directly relevant to that of the analysis. The usual approach is to change some analysis cuts in a way that produces an event selection dominated by the background of interest, and to interpolate or extrapolate measurements taken in those inverted selections into the signal phase-space.

The most intuitive example of this method is an analysis where the signal process is expected to be localised in a scale variable—most usually the invariant mass, m, of a dijet, diphoton, or dilepton pair, or similar. If one has confidence that the signal will be contained within a mass window $m \in [m_{min}, m_{max}]$, the data in the complementary **sideband** regions $m < m_{min}$ and $m > m_{max}$ can be used to obtain estimates of the background normalisation and shape. As differential cross-sections typically fall fast with increasing scale, simply using a constant average background cross-section in the signal window is not sufficient: the usual approach is to fit a functional form $d\tilde{\sigma}/dm$ as a function of m to the data outside the mass window of interest.

This approach is compatible with the common **blinded analysis** approach, where to avoid (even involuntary) biasing and 'cherry-picking' of features that might only result from statistical-fluctuations, the expected signal-region is masked out of all analysis plots until very late in the development and review process. The sidebands must also be chosen carefully to ensure that they are in a 'well-behaved' region of phase-space: a good example of this is the Drell–Yan dilepton mass spectrum, which at low masses contains narrowly-peaked contributions from hadrons such as the closed-charm and bottom ($J/)\psi$ and Υ states: these peaks are clearly not representative of the continuum Z/γ^* Drell–Yan background and so must also be masked out of sideband fits.

Exactly what function to use for sideband interpolation is something of an artform, and different areas of physics from inclusive jet measurements to $H \to \gamma\gamma$ have evolved quite complex refinements over time. A balance needs to be struck between a 'physical' degree of smoothness, without going so far as to hide either detector or physical glitches in the background spectrum, and with a reasonable degree of both functional and statistical uncertainty in the fit propagated to the resulting background estimate in the mass window. Generic approaches to parametrisation have become common in recent years, in particular the use of **Gaussian-process smoothing**, which effectively explores a discretised space of all functional forms with a frequency cutoff given via a covariance matrix. Gaussian-process and neural-network smoothings extend neatly to higher-dimensional sidebands, although they are no magic bullet: the resulting fits and their variations must be cross-checked, as must all such implementation details.

A second major form of background estimation is the use of **discrete sidebands**, most notably inversion of binary cuts such as tagging status. The usefulness of this is obvious in the case of a single binary cut, such as requiring that one jet carries a b-tag or that a lepton is isolated: inverting the cut to require *no b*-tag/isolation gives a strong indication of the background rate and shape of light-jet events. But how should one extrapolate the discrete sideband into the signal region? A neat way to do this is to combine several such cuts, and to use the scalings between different background regions to estimate a transfer function into the signal selection, as illustrated in Figure 8.4: this is variously known as the **ABCD method**, which is a subset of a more general technique known as the **matrix method**.

Let us start with the classic 2D sideband case, e.g. an attempt to reconstruct a $2 \times$ *b*-jet state, with contamination from mis-tagged *c*- and light-jets. There are then two possible tag states (B and $\mathcal{B}$) for each of the two candidate jets, giving a total of $2^2 = 4$ configurations: three sidebands and one signal region, which we label as $A, B, C, D = \mathcal{B}\mathcal{B}, B\mathcal{B}, \mathcal{B}B \;\&\; BB$, respectively. Let us assume that other background contributions are not significant in these regions, or have been separately estimated and subtracted: we are only concerned with mistag rates. Starting from the 'least signal-y' sideband A, we can estimate a rate of first-jet mistag as $f_1 = N_B/N_A$, where N_I is the number of events in region I, and a similar rate of second-jet mistag

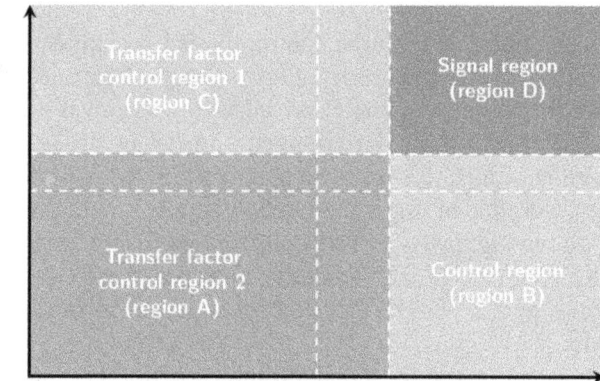

2nd large-R jet					
1t1b	J (7.6%)	K (21%)	L (42%)	S	
0t1b	B (2.2%)	D (5.8%)	H (13%)	N (47%)	
1t0b	E (0.7%)	F (2.4%)	G (6.4%)	M (30%)	
0t0b	A (0.2%)	C (0.8%)	I (2.2%)	O (11%)	
	0t0b	1t0b	0t1b	1t1b	

Leading large-R jet

Figure 8.4. Matrix-method sidebands for background estimation in the 2D 'ABCD' configuration, with 'buffer' zones between regions for continuous variables (top), and the 16-region A–S configuration used by ATLAS' all-hadronic boosted top-quark measurement (bottom). Bottom reproduced with permission from ATLAS Collaboration 2008 *Phys. Rev.* D **98** 012003, arXiv:1801.02052. Copyright IOP Publishing. All rights reserved

as $f_2 = N_C/N_A$. An estimate of the background contamination of the two-tag signal region through mistagging is then

$$\tilde{N}_D = f_1 f_2 N_A = N_A \cdot \frac{N_B \times N_C}{N_A^2} = \frac{N_B \times N_C}{N_A} \quad (8.54)$$

$$\text{or } D \approx \frac{B \times C}{A}, \quad (8.55)$$

where in the last term we switch to a simpler notation $N_I \to I$. Note that here, the 'path' taken from the fully-light A region to the D signal region can either be $A \to B \to D \Rightarrow f_1 f_2$ or $A \to C \to D \Rightarrow f_2 f_1$, which are multiplicatively equivalent. This method of extrapolating background estimates from fully background-dominated selections into signal-sensitive ones can be extended further by use of more binary cuts. An interesting, if extreme, example can be found in ATLAS' Run 2 all-hadronic boosted $t\bar{t}$ measurement, in which two large-R jets are used as top-quark candidates. The signal region requires both such jets to possess both a b-tag and a top-tag—the latter being a label derived from jet-substructure observables and designed to favour jets with a clear SM-top-like hadronic decay structure. With two tags on each jet, there are now a total of $2^4 = 16$ regions, 15 of which are sidebands to the 4-tag signal region. Labelling the regions (and their estimated background-event counts) from no-tag A to 4-tag S, as shown in Figure 8.4, gives a naïve ABCD-like estimator between the first and fourth matrix indices of $S \approx (J \times O)/A$, but this assumes an absence of correlation between tagging rates on leading and subleading jets. These correlations can also be estimated, via alternative paths through the first three row and columns of the matrix (containing only untagged or incompletely tagged jets, and hence background-dominated), resulting in a refined four-dimensional matrix estimator of the background in region S,

$$S = \frac{J \times O}{A} \cdot \frac{D/B}{C/A} \cdot \frac{G/I}{E/A} \cdot \frac{F/E}{C/A} \cdot \frac{H/I}{B/A} \quad (8.56)$$

$$= \frac{J \times O}{A} \cdot \frac{H \times F \times D \times G \times A^4}{(B \times E \times C \times I)^2}. \quad (8.57)$$

Note that the correlation-term numerator consists of the incompletely-tagged regions in the 2×2 submatrix of the second and third matrix columns, and the denominator from those regions in the first column or row, i.e. with at least one fully untagged candidate jet.

8.6.2 Uncertainties

The experimentalist Ken Peach noted that 'the value you measure is given by Nature, but the error bar is all yours'. As this suggests, the place where an experimentalist gets to show their skill and ingenuity is in assessing and reducing the impact of uncertainties. This is a balance: the smaller the error bar (or more generally, the smaller the covariance entries, or the more tightly localised the likelihood or posterior pdf), the more accurate and powerful the measurement, but above all it must be a *fair* representation of the measurement uncertainty. Underestimation of error is a worse crime than conservative overestimation, as the former can actively mislead the whole field while the latter simply reduces the impact of your measurement.

As ingenuity is key to exploration and constraint of uncertainties, we certainly cannot give a comprehensive guide here. But in the following we will illustrate a few key techniques for handling both statistical and systematic uncertainties. As for backgrounds, which are themselves a source of uncertainty, central elements will be the use of:

- nuisance parameters (such as the background rates) to encode a space of reasonable statistical and systematic variations;
- and sidebands or similar **control regions** (CRs) to act as a constraint on the otherwise unfettered nuisances.

Taking the Poisson likelihood from equations (5.131) and (5.132) as a canonical measure for data compatibility over a set of signal-region or histogram-bin populations and model expectations, we can extend it easily to include a set of control-region measurements:

$$\mathcal{L}(k; \theta = \{\phi, \nu\}) = \prod_i \mathcal{P}(k_i; b_i(\nu) + s_i(\phi, \nu)) \cdot \prod_j \mathcal{P}(m_j; b_j(\nu)) \qquad (8.58)$$

$$= \prod_i \frac{e^{-(b_i + s_i)}(b_i + s_i)^{k_i}}{k_i!} \cdot \prod_j \frac{e^{-b_j} b_j^{m_j}}{m_j!}. \qquad (8.59)$$

where $m = \{m_j\}$ is the set of control-region observed counts. As denoted here, it is usual for CRs to be implemented such that they are sensitive to non-signal processes only: this can be explicit backgrounds whose significant contamination of the signal region needs to be constrained using CR measurements (cf. the ABCD examples above) or any process which can be used to constrain estimates of object mismeasurement.

Of course, there are already estimates of mismeasurement for most objects in the form of calibration uncertainties: a lepton or jet energy scale correction, for example, will come with estimates of its uncertainty—usually in the form of a pair of 'up and down' variations on the scale parameter, around the **nominal configuration**. These calibrations are necessarily generic, and the aim of analysis-specific signal regions is to constrain *more tightly* than the *a priori* calibration uncertainty. It would be computationally excessive—although ideal in principle—to have to re-perform a full set of generic object-calibrations within an analysis, e.g. including unrelated photon–jet and dileptonic $t\bar{t}$ event samples and a maximal set of reconstruction nuisances in every analysis to constrain jet and MET reconstruction and b-tagging in a fully correlated way. Instead, the set of calibration up/down systematic variations are propagated through the reconstruction, data-reduction, and analysis software chains to produce sets of data and MC **yield templates**, which can then be interpolated between using nuisance parameters. Whether linear or more complex interpolation is used, and whether a prior (most commonly a unit Gaussian with the variation templates at the 1σ points) is applied is an issue specific to each analysis. Given this framework, a profile-likelihood fit (or Bayesian analysis) can be performed, providing estimates of not just the parameters of interest, ϕ, but also the nuisances ν, the confidence/credibility ranges on each, and the correlations between them.

Statistical uncertainties

Statistical uncertainties are in fact rather easily handled, either by use of discrete probability distributions, or by estimators derived in their continuum limit.

If passing the measurements to a Poisson-based likelihood calculation or similar, the pdf (or rather probability mass function (pmf)) 'automatically' contains the statistical variance. You should not have any problem with discreteness in the observed count: the number of observed events passing cuts and falling into bins or signal/control regions is always an integer. The same is not generally true of MC predictions: these contain weights from MC biasing, from scaling the effective integrated luminosity of their sample to match the observed luminosity, and from pile-up reweighting and other calibration scale-factors. In most cases, this is fine: there is no discreteness requirement on the Poisson mean, λ, and that is what MC estimates are mostly used for. When using weighted MC samples as pseudodata (e.g. to create an Asimov dataset), it can become necessary to round their continuous values to integers, creating a small, but unconcerning **non-closure** in checks to test that the MC parameters can be re-obtained in a circular fit.

If converting instead to a continuous variable, for example by normalising, a Gaussian estimator for the statistical error can be obtained via the variance, standard deviation and standard error on the mean given earlier in Section 5.3.

Statistical uncertainties, unlike most systematic ones, are also amenable to improvements by widening or merging bins: the **statistical jitter** of bin values (or profile means) and random empty bins visible in overbinned plots are a consequence of a bin's relative statistical error scaling like $1/\sqrt{N_{fills}}$, and this can be ameliorated by making underpopulated bins wider. While it is *simplest* with available tools to 'book' histograms with uniform bin widths, and various heuristics exist to pick optimal uniform bin widths, a more optimal strategy is to apply a binning motivated by the expected distribution density—this way, each bin will be roughly equally populated, with an expected relative uncertainty matched to the analysis requirements. A simple and common approach to variable binning, in particular for rapidly falling distributions such as the jet p_T spectra shown in Figure 8.5 is **log-spaced binning**, with $N_{bin} + 1$ bin edges uniformly spaced in the logarithm of the x-axis variable.

Such binning optimisations should be tempered by sanity: it is far easier to report and interpret 'round-numbered' bin-edge positions in tables and numeric data publications, so some human rounding to e.g. the nearest 10 GeV is usually wise following the numerical optimisation. The same imperative applies to binnings in irrational-number ranges, most notably azimuthal angles: if reporting e.g. binned $\Delta\phi$ values, expressing the angle either in degrees or in fractions of π is far preferable to having to record cumbersome and inaccurate bin edges like 0, 0.31415, 0.62832, ...!

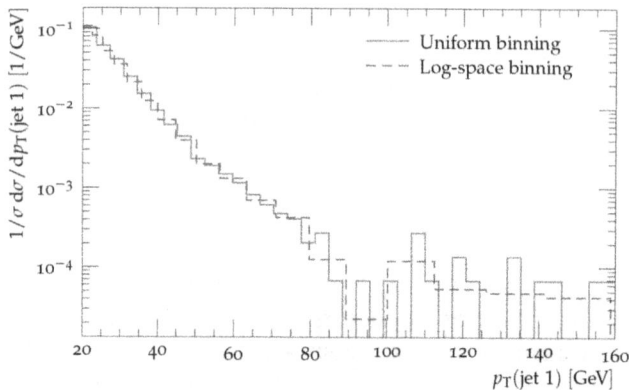

Figure 8.5. Illustration of uniform versus log-spaced binning for a leading-jet p_T distribution, filled from the same MC events. The increasing bin widths in the log-spaced version help to compensate for the decreasing statistical populations (while binning as small as detector resolution permits at low-p_T, with fewer bins), while statistical jitter and empty-bin gaps are evident in the uniformly binned variant.

Bootstrapping: Estimating means and variances is one thing, obtainable from the set of events populating the analysis bins, but a tougher challenge is estimating the correlations between those bins. The reason is that the bin values in any measurement are the best estimate of physics quantities, and to be that best estimate the maximum available number of events needs to be used. But correlations are computed by comparing multiple datasets and seeing how their elements co-vary: if we only have a single, all-events dataset, we have no ensemble to compute the correlations from. How can we get an impression of the statistical distribution from which our universe drew our single observed dataset, without wasting our precious number of recorded events? The answer is found in a neat technique called **bootstrapping**[14].

In the simplest form of bootstrapping, we compute our observable N_{replica} times using sampling of values from our available dataset with replacement, i.e. it is allowed to sample the same event or entry more than once. This results in several replica datasets, one from each sampling, which fluctuate relative to one another in and provide the ensemble from which we can compute covariances and so on. Note here that there is no assumption that the bin values are event sums: they can have been transformed, scaled, divided, or normalised, and will still statistically fluctuate in a representative way. The power of the bootstrap method is that it can accurately compute the correlated statistical uncertainties between these complex quantities, such as the negative correlations that must be introduced by a normalisation procedure—if the integral is fixed and one bin fluctuates upwards, all others must on average fluctuate down to compensate. The bootstrap technique also accurately

---

[14] From the idiomatic phrase 'pull yourself up by your bootstraps', nowadays meaning to improve oneself without external help. There is also a degree of cynicism embedded: try as you might, pulling up on your shoes will not actually help to lift you across any obstacle, much as the accuracy of bootstrapping is limited by the actual observation of only one dataset.

reflects the impact of resolution effects on spreading events stochastically between bins, as the sampling with replacement is effectively random-sampling from not only the physical distribution, but the experimental resolutions it is convolved with.

This is nice, but performing 100, 1000, or many more random-input iterations of an analysis that is fundamentally using the same set of events is painfully inefficient. Fortunately, an efficient refinement is possible using weighted histograms: this is the **Poisson bootstrap**. In this procedure we perform no sampling with replacement, but instead book N_{replica} copies of each result histogram within the analysis routine: the same concept as the many with-replacement replicas of the analysis run, but now there is only a single execution. For each event in the execute loop, we sample a N_{replica}-tuple of bootstrap event weights from the unit Poisson, $\mathcal{P}(1)$: depending on the replica number, each observed event may be considered to have an unchanged value of 1 event, or to have fluctuated down to be worth 0 events, or to have statistically fluctuated up to represent 2, 3, or more events seen in alternative histories of our experiment. We then fill our N_{replica}-stacks of histograms as normal—typically we use a code interface which pretends to be a single histogram, but dispatches the fill value to the internal replicas with appropriate unit-Poisson weights. The correlations can then again be calculated, with the same statistical properties.

The bootstrap is an amazing and simple technique to implement. Once familiar with it, you will find uses everywhere.

Systematic uncertainties

Systematic uncertainties propagate into an analysis from every ingredient that is imperfectly known, other than the statistical limitations of the data. Sources of systematic uncertainty occur all the way down the processing chain from event generation, to collider and detector simulation, to reconstruction and calibration.

This definition is necessarily vague, because some such uncertainties may result from statistical limitations either in calibration data or in MC samples: nevertheless, if it can be rendered into the form of a pdf on a nuisance parameter, it can be treated as a systematic effect. For ease of application, it is usual for these sources of uncertainty to each be decomposed into a maximally orthogonal set of **elementary nuisances**, for example by the covariance-diagonalisation procedure used to derive parton density Hessian error-sets. Each elementary nuisance is hence *a priori* uncorrelated, but as they are likely to influence bin populations and hence data-compatibility in structured ways, correlations between systematic nuisances and across measured bin-values are likely to be induced by any fitting procedure: the corresponding **pre-fit** and **post-fit** distributions and errors are often seen in analysis documentation, with the post-fit version being the best estimate of the true calibration values and uncertainties specific to the analysis phase-space.

Examples of sources of systematic uncertainty include (but are certainly not limited to):

Physics modelling: despite best efforts by many theorists and experimentalists, MC event generation remains imperfect, particularly for processes subject to large radiative QCD corrections (notably Higgs-production), loop effects (the $t\bar{t}\,p_{\text{T}}$ distribution and related quantities have been found to be difficult to model at LHC energies, and appear to require differential NNLO + resummation calculations for a good description of data), and EW corrections (in particular in the increasingly important high-p_T tails of many processes).

The uncertainties in MC modelling are often explicit in the calculation, e.g. the technically unphysical factorization and renormalization scales (μ_F and μ_R) that reflect truncation of the fixed-order perturbative expansion; in the merging and matching of fixed-order and parton-shower radiation modelling, cf. the resummation and merging (μ_Q and μ_J) scales and Powheg HDAMP parameter; or in the degrees of freedom of parton density fits, expressed as PDF error sets.

Parton density functions characterise best estimates on the momentum and flavour structure of the colliding beam-particles, usually obtained via global PDF fits performed by international phenomenology collaborations. The best-known of these are the NNPDF group, the CT (formerly CTEQ) group, and the MSHT (formerly MRST, MSTW, MMHT, and several other names) group, each of which uses largely the same datasets from ep and fixed-target experiments, with varying admixtures of Tevatron and LHC data. Other often-used PDF fits include the fixed-flavour ABM sets, the ep-specific HERAPDF sets, and fits from the LHC experiments and the xFitter collaboration.

A PDF's general functional form (by construction very free in the case of NNPDF), detailed data-selection, the fit-quality metric, and the uncertainty treatment lead to different PDF central fits and uncertainty bands: these are captured in the members of each **PDF set**, where the central or nominal fit is the first member in the set, and the uncertainty structure is encoded via a following structure of error-members that represent either diagonalised Hessian variations on the nominal, bootstrapped equivalent fits for CoI construction, or two-point up/down systematic variations on a single nuisance parameter such as α_S. The LHAPDF software library both collects the main published sets and provides computational tools for their use.

A typical PDF error set has 40 or 100 entries, but it is normal to combine several sets from different fitting groups to minimise parametrisation bias. As the 68% CL bands from different global-fit sets (e.g. the latest NNPDF, CT, and MSHT NNLO fits) often do not overlap, it was agreed by the LHC community to use a standard PDF uncertainty estimate—the **PDF4LHC procedure**. This procedure simply computes the total enclosing envelope of CL bands for the major sets. The PDF weighting of MC events is typically used to evaluate these uncertainty bands, either by explicitly computing the constituent bands and taking the envelope or by use of summary PDF4LHC Hessian error sets which produce a very similar result more conveniently.

Scale uncertainties are nowadays mostly evaluated by variations about a nominal value, usually setting $\mu_F = \mu_R$ to a characteristic dynamic scale μ_{nom}, related to an event-specific quantity such as the H_T or e.g. $\mu_{nom}^2 \sim M_X^2 + \sum_{jets} p_{T,j}^2$, where X is the heavy resonance dominating the hard-scattering amplitudes. The use of dynamic scales including QCD activity avoids historical problems with fixed scales like $Q_{fix} \sim M_X$ in kinematic regions where the momentum transfer into QCD jets is comparable to M_X: in such cases, fixed-order calculation could be dogged by spurious negative cross-sections induced by a poor choice of the unphysical, and technically unknowable, scales. The normal process of uncertainty estimation around this nominal is to use a **seven-point scale variation**, where μ_F and μ_R are varied up and down by factors of (usually) two, in all combinations other than the anticorrelated $(\mu_F, \mu_R) = (2\mu_{nom}, \frac{1}{2}\mu_{nom})$ or $(\frac{1}{2}\mu_{nom}, 2\mu_{nom})$, which destabilise calculations by introducing artificially large logarithms via the mismatch between the two scales. This point-wise scale-dependence scheme is not foolproof, but has become standard as it is much more scalable than detailed study of smooth variations in the (μ_F, μ_R) space, and in most cases produces similar results.

Once the parton shower and hadronisation stage is entered, the number of parameters inflates enormously due to the increasingly less robust physical foundations of the modelling: taken naïvely, one could easily proliferate $\mathcal{O}(50)$, or perhaps even 100 elementary nuisances from this stage, requiring control regions from across the whole range of LHC physics from soft-QCD onward. Thankfully, as for physics-object calibrations, representative groupings of leading MC generator uncertainties are typically gathered by MC tuning groups into a more usable set of <10 tuning nuisances. Still, as for PDF sets, it is often considered insufficient to rely on a single MC generator programme, and so 'two-point' discrete variations such as interfacing the same hard-scattering events to both the Pythia and Herwig generators are often used in an attempt to cover the reasonable range of predicted shower and hadronisation behaviours.

It is common for calculations to be combined, most commonly using an NLO or multi-leg event generator to make explicit events and hence obtain shape estimates in observable distributions, but with the process cross-section normalization estimate taken from a higher-order analytic calculation. The picture we hence need to work with is of several distinct signal and background processes able to significantly enter the analysis event selection, each with a set of $\mathcal{O}(10)$ scale uncertainties, 40 to 300 PDF uncertainties, and 5 to 10 parton-shower and similar uncertainties. The naïve modelling systematic space is hence $\mathcal{O}(1000)$-dimensional—too large to approach directly, an issue which we will return to shortly.

MC statistics: As well as their intrinsic physical uncertainties, modelling via MC generation implies a finite number of events, and many analyses find themselves significantly affected to some extent by the statistical limitations of their MC event samples. We should note, for clarity, that MC statistics are not the same as the real-world signal/background statistics: the effective integrated luminosity L_{int}^{eff} of MC samples does not have to be the same for every process, nor do they need to match the integrated luminosity of the collider: it is the effective bin count N_{eff} (cf. equation (5.76)) rather than the scaled expected yield that defines the statistical stability of a yield estimate.

Analysis elements particularly affected by MC statistics are the tails of scale distributions and multi-differential distributions (the curse of dimensionality affects the population of naïve binning just as it does the coverage of naïve sampling algorithms). Two-dimensional constructs such as covariance and correlation matrices, and the response matrices used in detector unfolding (see Section 11.3) are included in these susceptible quantities. While these limitations can to some degree be ameliorated by use of biased event generation as described in Section 7.2.2, it is hard to cover all relevant corners of phase space.

These uncertainties are typically treated as systematics in MC-driven background estimation, rather than statistical uncertainties in the data *per se*, although their transformation into systematic form does require the machinery we have established for systematic uncertainties. Via these procedures, every bin of every distribution will have at least one degree of statistical MC uncertainty: should any multi-dimensional objects be involved, the number of associated nuisances can be astronomical. Again, we shall shortly return to the need for reducing the nuisance dimensionality.

Collider luminosity and pile-up: Moving to the actual experimental apparatus, the first source of uncertainty is not the detector but the collider. The rate of collisions is determined primarily by the instantaneous collision luminosity L, itself determined by the combination of beam-current and beam optics specific to each collider run and lumiblock. Luminosity is calibrated to the level of $\mathcal{O}(1\%)$ by dedicated calibration runs using **Van der Meer scans** in which the transverse positions of the two colliding beams are scanned through each other and the resulting inclusive interactions measured: this acts as a measurement 'floor' on the accuracy achievable in measurement of absolute cross-sections, as the instantaneous luminosity uncertainty propagates into the integrated luminosity used in equation (8.7), and hence a 100% correlated uncertainty across all measured cross-section bins.

Pile-up modelling is related to luminosity and collider conditions, being a result of beam optics. Monitoring of beam conditions leads to the estimated μ distribution of Poisson mean rate of inelastic collisions, but further sources of uncertainty result from the different (not yet precision-calculable) models for minimum-bias QCD interactions, and the impacts of in-time and out-of-time pile-up modelling in the digitization procedure.

Detector simulation: Even the very CPU-intensive full-simulation frameworks suffer from significant uncertainties, although they are constrained as far as possible by the use of reference data from e.g. calorimeter material test-beam data during the development and tuning of the full-simulation framework hadronic physics and multiple-scattering models. These uncertainties are expressed mainly through the use of the pre-assembled physics lists. Detector-element misalignments, within known tolerances, are similarly taken into account by special detector-simulation configurations.

As assessing these uncertainties requires multiple expensive runs of the full-simulation machinery, it is not computationally feasible for every physics-process MC sample to be simulated in several different ways: instead, these variations are explored in dedicated calibration samples for use by physics-object reconstruction teams, and are folded into the reconstruction systematics we explore below.

For the various flavours of fast-simulation, these uncertainties are compounded by the approximations made in search of processing-speed gains. These are estimated, and the underlying full-simulation uncertainties propagated, by the teams managing simulation strategies within each experiment: the additional fast-simulation uncertainties may be exposed directly to analysers, in addition to the fundamental modelling uncertainties encapsulated within physics-object calibrations.

The beam-spot modelling is typically left unvaried in simulation uncertainty estimation, except in forward-physics experiments such as AFP, ALFA, and TOTEM which operate 'Roman Pot' silicon detectors for forward-proton tagging close to the beam, several hundred metres downstream from their main-detector interation points. These are naturally more sensitive to the beam conditions than analyses using the main detector apparatus at relatively high angles to the beamline.

Energy/mass/momentum reconstruction: All reconstructed physics objects suffer from biases and resolution in their momentum-parameter estimates. At hadron colliders these are normally expressed as usual via transverse variables, e.g. jet, lepton, or photon p_Ts, or missing transverse momentum.

These uncertainties can be divided into directional resolutions and miscalibrations, and scale uncertainties. The latter tend to dominate reconstruction systematics, given the importance of scalar p_Ts in deriving useful physics-cut and observable quantities such as H_T and m_{eff}, but angular uncertainties are significant for high-resolution objects such as tracks, and in e.g. $\Delta\phi$ cuts between identified physics objects, such as separating jets from each other and the E_T^{miss} vector.

Different object types have different scale-uncertainty characteristics, most notably driven by the difference between calorimeter-driven objects (jets, electrons, photons, and MET) and tracking-driven ones (charged hadrons, muons, and vertexing). Track momentum uncertainties grow at high scales due to straight-line error: it is hard to accurately measure very high momenta via tracks, as their curvature scales like $1/p_T$ and hence the uncertainty on the p_T estimation explodes as the curvature becomes small. By contrast, calorimeter relative uncertainties decrease with increasing scale due to sampling stability and the physical increased collimation of high-p_T jets. Jet calibration requires scale corrections to be applied to account for different hadron/lepton responses in the calorimeters, as discussed in Chapter 6: these propagate into the systematics as **jet energy scale** and **jet mass scale** uncertainties.

MET, being dominated by calorimeter terms, tends to also stabilise in events with a large amount of hard-scattering activity, as the visible-event balance is then dominated by collimated objects with low calorimetric uncertainty, and uncertainties from sampling and soft-term calibration become subleading.

This is why missing momentum uncertainties tend to scale as $\sim 1/\sqrt{\sum E_\mathrm{T}}$. As MET is computed from the sum of calibrated hard objects, plus the soft-term contributions, MET systematics are correlated with the uncertainties of all reconstructed types contributing to the analysis event-selection.

Lepton, photon, tau, and jet-flavour identification: all these physics-object types are identified through relatively complex algorithms—often based on machine-learning methods—operating on a multi-variable set of reconstruction inputs. These include ECAL shower-shape discriminators for e/γ discrimination, counting of charged-particle 'prongs' and other jet-shape variables for tau discrimination from hadronic jets, and a large variety of impact-parameter and decay-chain characteristics in the case of heavy-flavour tagging. The cuts implicit in these algorithms are dependent on the modelling used in their training (for example, the calorimeter response and heavy-hadron production and decays), and on other reconstruction uncertainties feeding into their data-driven calibration. These lead to effective systematic uncertainties on the efficiency and mis-ID rates assumed in the interpretation of the reconstructed events.

Practically, systematic variations of this sort tend to enter the final stages of a physics analysis in the form of **templates**—histograms representing the alternative histograms obtained using variations of simulation and reconstruction behaviour by '1σ' shifts away from the nominal behaviour one elementary nuisance at a time. The effects of changes are often decomposed into independent shape uncertainty and normalisation uncertainty components, especially for systematics from MC modelling, where the expected shapes and normalisations of background processes often come from different sources.

It is usual for these variations to be provided as a pair of templates $h_\pm$ for each systematic nuisance ν_i, corresponding to $\nu_i = \pm 1$ deviations from the nominal configuration $\boldsymbol{\nu} = \{0, 0, \dots, 0\}$ with its corresponding template histogram h_0. Provided the deviations are small enough, the three template copies of each histogram can then be interpolated as a function of each ν_i, permitting a histogram estimator $\tilde{h}$ to be constructed from the templates for any point in the nuisance-parameter space, e.g. the piecewise linear

$$\tilde{h}(\boldsymbol{\nu}) = h_0 + \sum_i \begin{array}{ll} \nu_i(h_+ - h_0) & \text{for} \quad \nu_i \geqslant 0 \\ -\nu_i(h_- - h_0) & \text{for} \quad \nu_i < 0 \end{array} \tag{8.60}$$

or one of many smoother forms.

The exception to this approach is discrete changes such as a change of event generator, for which only two-point templates (and possibly a set of them, for more than two discrete model choices) are possible. Such two-point variations may be decomposed into several interpolation components, each with its own nuisance-parameter, to check that no significant bias or unphysicality is introduced by linearly interpolating all bins between the discrete templates in a fully correlated way.

Constraining systematics. The set of interpolated templates for each elementary nuisance parameter allow a smooth parametrisation of the likelihood function, with changes in a nuisance propagating in very non-trivial ways into the expected values of measurement bins or regions, and thus changing the probability of having seen the observed data. In a Bayesian treatment, explicit priors can be placed on the nuisances—most commonly unit Gaussian priors centred on the nominal $\nu_i = 0$, but sometimes more nuanced (e.g. non-negative priors on background cross-section expectations, or flatter priors between *a priori* equally valid MC models).

Some systematic effects will need to be coordinated in further processing between processes and observables—jet energy scales or lepton isolation behaviours should not be expected to be different between event types, or between plots of different object properties from the same events—but others which appear superficially linked are in fact distinct. This again particularly applies to MC modelling systematics: calculation scale modification factors such as those applied to μ_R, μ_F etc do not have a universal value to be discovered and constrained, and hence are expected to be different for each process type, but should in general be the same between observables. As ever, there will be room for debate.

Representative confidence or credibility bands may then be extracted as described earlier in frequentist or Bayesian approaches, respectively, by scanning the nuisance-parameter space in either profiling or marginalising approaches (i.e. respectively optimising or integrating over nuisance parameters). Representative values of the extracted measurement (e.g. cross-section or signal-yield estimate) can be obtained within these by either a maximum-likelihood/Bayesian estimate, or the modes or medians of the marginal distributions in the parameters of interest.

The widths of nuisance-parameter distributions propagate into these POI intervals, and limit the power of the analysis. Use either of likelihood-ratio profiling or Bayesian posterior probabilities in combination with control regions enables these nuisance-parameter pdfs to be tightened up, within the assumption that the systematic-uncertainty model is sufficiently general to include a good approximation to reality, and hence improve analysis precision.

This full statistical machinery is not always necessary: systematic variations may be able to be constrained separately from fitting, e.g. by control regions, and in the case of MC generation sometimes by matching of process elements to higher-order calculations. These constraints are usually performed in specialist groups and bundled with standard working points of the algorithms for relatively unspecialist use by analysers. In the analysis itself, further constraints can be derived by, for example:

Lumiblock partitioning: splitting the dataset by run conditions, to check for consistency of extracted events between, e.g. different pile-up or instantaneous-luminosity conditions;

Comparison of decay channels: as SM-lepton mass effects should be negligible in nearly all cases at high-energy colliders, deviations between results extracted via muon- and electron-based decay channels of the same hard-process are most likely to be due to scale, ID, or isolation systematics. Similar effects can apply when considering decays that go through both W and Z decays: not only are there various e and μ combinations, but also the number of charged leptons can be different, inducing different sets of systematic uncertainties in each channel. Separating and comparing the channels can flag problems and permit uncertainty constraints;

Cut loosening/inversion: as for background estimation, techniques such as tag-requirement inversion (cf. the ABCD/matrix method), and cut and working-point loosening provide control regions for uncertainty constraints;

Studying isolation variables: jet backgrounds are often key, and are naïvely poorly constrained due to the relative unreliability of QCD-multijet event MC modelling. Explicit study of lepton/photon isolation spectra without the isolation cut, combined with loosening of various cuts, can provide information on how much residual contamination can enter from misidentification of non-jet physics objects.

Standard candles: if the analysis itself is not aiming to measure a fundamental SM parameter, such as an EW coupling or Higgs/W/Z/t mass, the prior knowledge from earlier experiments of what these values 'should' be (assuming they are known to better precision than available via the current analysis) permits constraint of systematics biasing the observables away from their standard values.

Pruning systematic uncertainties: practically, fits and scans on systematics suffer from the usual curse of dimensionality, and may be extremely slow to process, or fail to converge at all: it is trivially easy to acquire a parameter space dominated by tens or hundreds of systematic nuisance parameters, but comprehensiveness can render an analysis computationally impractical. Ironically, in statistical fits, the largest problems can arise from the least important parameters, as they create flat directions in the parameter space which confuse optimization and sampling algorithms.

The solution to both problems is to **prune** the large list of naïve nuisance parameters to focus only on those which significantly impact the final results. This can be done approximately, but usually sufficiently, directly on the reconstruction-level histogram templates, e.g. retaining only those nuisance parameters whose 1-σ variations make at least one bin value shift by a relative or absolute amount (or a conservative combination of both).

Further reading

- Jet clustering algorithms, the pros and cons of various distance measures, and some simple are discussed in the classic pre-LHC *Towards Jetography* paper, Salam G P *Eur. Phys. J.* C **67** 637–86 (2010), arXiv:0906.1833.
- Some interesting discussions on the distinctions between rapidity and pseudor-apidity, and between missing momentum versus energy are discussed in the short preprint paper Gallicchio J and Chien Y-T Quit using pseudorapidity, transverse energy, and massless constituents (2018), arXiv:1802.05356.

- The Frixione isolation scheme is described in Frixione S Isolated photons in perturbative QCD *Phys. Lett.* B **429** 369–74 (1998). An isolation scheme based on jet substructure techniques can be found in Hall E and Thaler J Photon isolation and jet substructure *JHEP* **09** 164 (2018).
- The *N*-jettiness variable was first defined in '*N*-Jettiness: An Inclusive Event Shape to Veto Jets' by Stewart *et al Phys. Rev. Lett.* **105** 092002 (2020). Generalised jet angularities and ECFs are discussed in detail in *JHEP* **11** 129 (2014), arXiv:1408.3122, and in *JHEP* **06** 108 (2013), arXiv:1305.0007, respectively.
- An excellent but unfortunately unpublished guide to 'Background Estimation with the ABCD Method' has been written by Will Buttinger and can currently be found at https://twiki.cern.ch/twiki/pub/Main/ABCDMethod/ABCDGuide_draft18Oct18.pdf or via a Web search.

Exercises

8.1 Show that at $\eta = 0$, the pseudorapidity variable behaves as a longitudinal equivalent to the azimuthal angle ϕ.

8.2 Given a conical jet of radius R and a smooth background activity of p_T density ρ, what are the biases Δp_T and Δm on the jet's p_T and mass, respectively?

8.3 Derive the phase-space of the ΔR observable, i.e. for pairs of points randomly distributed on the η–ϕ plane, what is the functional form of the resulting probability density $p(\Delta R)$, and why does it peak at $\Delta R = \pi$? (As an extension, how would you expect finite η-acceptance to modify the naïve distribution?)

8.4 An analysis pre-selection yields a pair of 1D signal and background histograms in an observable x, given in the table to the right. The final selection is to be defined by a requirement $x > x_{cut}$, where x_{cut} needs to be optimised.

x	S	B
0	0	0
1	2	2
2	4	3
3	3	8
4	2	6
5	2	2

Compute the ROC curve corresponding to these histograms and the final selection cut. By eye, in what range of x would you expect the best working point to be found? Which point gives the better a) S/B and b) $S/\sqrt{B}$ scores? If typical systematic uncertainties are <10%, which of these is likely to be the more appropriate metric for these data?

8.5 In a results histogram with two bins, the error bars for each bin can be constructed either by finding a confidence (or credibility) interval on the marginal likelihood (or posterior) distribution for each bin, or by projecting the equivalent 2D interval's contour on to each bin-value, b_i. If the likelihood function $\mathcal{L}(b_1, b_2; \boldsymbol{D})$ is a two-dimensional Gaussian distribution, will the marginal or global construction give larger error bars? *[Hint: think about the role of the tails in the 2D distribution when integrating or projecting.]*

8.6 Inclusively simulated W production has a 1×10^{-5} selection efficiency as the dominant background process in a BSM search requiring high-p_T leptons. If the W cross-section is 44 nb, what is the expected background yield from 300/fb? Roughly, what is the lowest cross-section for a 20% efficient signal process that can be excluded at 95% CL? How many background events would need to be naïvely simulated to perform this analysis, and how can this be made more tractable than naïvely simulating all W-boson production and decay events?

8.7 Show that equation (8.47) reduces to equation (8.46) in the case of small background uncertainties.

8.8 A signal process is calculated to produce an expected yield of 55 events after analysis cuts, and a mean background yield of 500 events. If there are no uncertainties on the background prediction, what is the expected significance of the measurement? If the only uncertainty on the background-yield estimate comes from MC statistics, how small does the relative statistical uncertainty need to be to achieve 95% CL significance? (You may want to make a plot.) How many background events need to be generated to achieve this, compared with the central expectation of 500?

8.9 Given the suggestion of equal-population as an optimal binning strategy, how many bins can be used on a 10 000-event sample if 2% statistical precision is desired in each bin? How will the position of the free bin edges be related to the cumulative density function (cdf) of the expected distribution, and hence how is the problem of binning related to inverse-cdf sampling?

8.10 Suggest good binning schemes in a) lepton azimuthal angle ϕ, b) $\Delta\phi$ between the two leading jets in dijet events, c) leading-jet p_T in inclusive-jet events, d) dilepton mass from 70 GeV to 110 GeV, and e) $H \to b\bar{b}$ di-b-jet mass in the range $m_{b\bar{b}} \in 100$ GeV to 150 GeV.

IOP Publishing

Practical Collider Physics

Andy Buckley, Christopher White and Martin White

Chapter 9

Resonance searches

In Chapter 4 we noted the various ways in which beyond-Standard Model physics might manifest itself at the LHC. In particular, if a new particle decays purely to visible Standard Model (SM) decay products, one can observe a *resonance* in the invariant mass distribution of its decay products. Such resonance searches, or 'bump hunts', have a rich history, including the discovery of the Z and Higgs bosons, and the previous discovery of various mesons that nowadays act as important backgrounds for new physics searches. In this chapter, we will explain the origin and behaviour of resonances in detail, using a particular example of a diphoton resonance search. Although the basic techniques will generalise to other final states, we will also comment on the differences between the various final states at the LHC, and why some are much easier to explore than others. Along the way, we will explain various aspects of LHC data analysis that are also useful for non-resonant behaviour.

Let us begin by clarifying exactly what we mean by a *resonance*. Imagine that some new particle A decays to two visible particles d_1 and d_2, by which we mean particles that are produced near the interaction point (with b-quarks and c quarks possibly having a small displacement from the interaction vertex), and which are fully reconstructed in the detector. In this case, we can measure a four-momentum for each of the decay products, which we will call $P^\mu_{(d_1)}$ and $P^\mu_{(d_2)}$, and we will call the four-momentum of the parent particle $P^\mu_{(A)}$. If we are in the rest frame of the A particle, conservation of four-momentum tells us that

$$P^\mu_{(A)} = P^\mu_{(d_1)} + P^\mu_{(d_2)}. \tag{9.1}$$

If we square both sides, we simply get the rest mass of the particle A on the left-hand side (in natural units), and we get the invariant mass of the two daughter four-momenta on the right-hand side. Since this is a Lorentz scalar, this is true in any frame, including that of the LHC detectors, and this gives us a convenient recipe for searching for A particles at the LHC.

Imagine that our theory of the A particle predicts that it can decay to two photons. In that case, we can simply take all of the proton collision events that produced pairs of photons, add the two photon four-momenta together and square the result to get the invariant mass. Events in which the two photons were produced by an A particle will have an invariant mass close to the A mass. The reason that it is *close to* and not *precisely equal* to the A mass is due to that fact that quantum mechanics allows A particles to be produced with a four-momentum that slightly violates the relativistic invariant mass formula, which broadens the invariant mass distribution of the decay products. In addition, the four-momenta of the decay products are not reconstructed with 100% accuracy, which leads to a further broadening of the invariant mass peak. Both of these effects will be dealt with in more detail below.

By selecting all events with two photons in, we do not only select events that contain an A particle decaying to two photons. The SM can give us pairs of photons by other processes (including those with two real photons, and with one or two fake photons), and these give a smooth continuum of invariant masses with no peaking structure. Thus, discovering the A particle requires seeing a bump in the diphoton invariant mass spectrum that is large enough to be clearly visible above the background. The height of the A bump is related to the product of the A production cross-section and its branching ratio to photons, and the shape of the bump may also be distorted by interference between the SM background and the A signal. The definition of *clearly visible* must also take account of the statistical and systematic uncertainties on the SM background, which might conspire to make normal SM processes appear bump-like in a limited region of the invariant mass spectrum. As with most topics in physics, hunting for bumps starts off simply, but rapidly becomes a very complicated business indeed.

9.1 Types of resonance

Before working through a detailed example of an experimental resonance search, it is useful to perform a quick survey of the overall landscape of resonances. In this book, we will focus on two-body resonances, in which there are two visible particles produced when the new particle decays[1]. Much of what we say can be generalised to the case of an n-body resonance. Even in the case of the two-body decay $A \to d_1 d_2$, there are several possible options to consider:

- d_1 and d_2 might be electrons, muons, photons or jets, whose four-momenta are directly reconstructed.
- One or both of d_1 and d_2 might be SM particles that decay immediately to other SM particles, for example top quarks, W bosons or Z bosons. In this case, the four-momenta of the decay products of d_1 and d_2 can be summed to produce a feature in an invariant mass distribution, but the final state has more than two particles, despite the fact that A decays to only two things.
- One or both of d_1 and d_2 might be non-SM particles which decay immediately to visible SM particles, in which case one could search for both the A particle itself, and the new particle decay products. The final states in this case would

---

[1] By 'visible', we mean either that the particle is directly reconstructed, or it decays to particles that are directly reconstructed. We therefore do not count direct production of neutrinos.

also have more than two SM particles in, and could be arbitrarily complex depending on the nature of the BSM physics involved.

In the following, we focus on the first two of these options, on the basis that the techniques we cover could be generalised to the more complicated third case. Typically, LHC searches for two-body resonances target cases where the new particle decays to pairs of the same sort of object, for example a pair of muons (which may have opposite sign), an electron–positron pair, a pair of photons, a pair of b-jets, and so on. Examples of theoretical models that would produce this behaviour include models with extra Z-like bosons (commonly called Z' bosons), and models with extra Higgs bosons. However, there is no reason not to search for all possible pairs of SM particle in the final state, since at least one well-motivated extension of the SM can be found for each case. Different flavour lepton pairs can be produced by the decay of super-symmetric particles in the case that R-parity is violated. Exotic particles called leptoquarks are expected to decay to a quark or gluon along with a lepton. If there are extra W-like bosons in Nature, these might decay to different flavours of quark. Excited quarks or excited leptons might decay to a photon, W boson, Z boson or Higgs boson, plus a lepton, whilst excited quarks would decay to a quark or gluon along with a boson. Vector-like top and bottom quarks can decay to a top quark plus gauge boson, Higgs boson, quark or gluon. Extra Higgs bosons, meanwhile can produce a variety of final states, including a Z or W boson plus a photon and another Higgs boson. For researchers on ATLAS and CMS, finding options that remain uncovered by existing searches is an excellent way to build and sustain a career.

It might be the case that a resonance is produced that is moving at high speed in the frame of the ATLAS or CMS detector (frequently called the *lab frame*). For example, the resonance might recoil from a jet that is radiated off an incoming parton in the LHC collision process. In this case, there is a sizable Lorentz boost between the lab frame and the rest frame of the particle. Thus, the four-momenta of the decay products, which are well separated in the particle rest frame, may appear collimated in the lab frame, to the extent that they cease to be reconstructed as distinct objects. We then say that the resonance is *boosted*, and we can perform a special type of analysis to try and find the resonance, using the jet substructure techniques introduced in Section 8.4.5.

9.2 Anatomy of a diphoton resonance search

9.2.1 Event selection

To expand on our basic explanation of resonance searches, let us now walk through a particular analysis in detail. We will choose a diphoton resonance search, and will summarise a variety of techniques that are useful for background characterisation and signal modelling. These are taken from a particular study by the ATLAS experiment that was carried out in thrilling circumstances after a previous search had revealed tentative evidence for a discovery. The analysis sadly ruled out the existence of a new particle, but not before the publication of over 100 theoretical papers interpreting the original result!

Photons make nice objects for a resonance search, because they are reconstructed with high precision, and the rate of diphoton production in the SM is relatively low which reduces the background. Nevertheless, there are many details to get right, the first of which concerns the event selection. It is typical to choose photons that satisfy *tight* selection criteria on the shower shapes in the electromagnetic calorimeter, which helps to reduce backgrounds that produce fake photons. In addition, the analysis can be restricted to the region where the calorimeter behaviour is best, which in ATLAS means focussing on the region $|\eta| < 2.37$, excluding the transition region $1.37 < |\eta| < 1.52$ between the barrel and endcap calorimeters. Finally, one can impose isolation criteria on the photons, using combinations of tracker and calorimeter information.

We should note again at this point that we have passed an important milestone in the journey from theoretical to experimental physics. What a theorist calls a *photon* is a gauge boson of the SM. What an experimentalist refers to as a photon carries the same name as that particle of the SM, but is in fact something quite different; it is an object that results from applying a set of careful definitions to a series of detector signals. Changing those definitions would change the observed number of events at the LHC, the properties of those events, and the relative proportion of different contributions to the set of events with two photons.

Having decided what to call a photon, we must now decide how to choose the events that have two photons in. As we learnt in Chapter 6, the only events that are recorded by the LHC detectors are those that pass some *trigger* condition, and there is an extensive trigger menu that contains various conditions on the recorded events. The ATLAS search utilised a trigger that required events to have one photon with a tansverse energy $E_T > 35$ GeV, plus another photon with $E_T > 25$ GeV, where the photon definition at the trigger level differs from the analysis in requiring a relatively loose selection on the photon shower shapes. In theory, a trigger should pass no events that do not satisfy this criterion, and pass all events that do satisfy the criterion. In practice, however, the efficiency of events passing a trigger *turns-on* somewhere near the E_T threshold, and then plateaus at some final efficiency which is lower than 100%. The selection of the events for analysis is usually required to be more stringent than the trigger condition, to ensure that the trigger is maximally efficient for the final selected events. Measuring the efficiency of the trigger then becomes an important part of the experimental analysis.

In addition to the diphoton selections, there are standard *data quality* selections that are designed to reject data from run periods where a sizable proportion of the detector was malfunctioning. It is also necessary to apply a minimum selection on the diphoton invariant mass ($m_{\gamma\gamma} > 150$ GeV), since the effect of the photon E_T selections is to sculpt the low $m_{\gamma\gamma}$ region, and distort the expected shape. This leads to an important general consideration for resonance searches: more stringent trigger requirements will lead to a greater region of the invariant mass that can no longer be searched for resonances unless alternative techniques are used.

9.2.2 Modelling the diphoton background

We have now made sufficient progress to define both the concept of a photon, and the concept of an event with two photons in. Having made those selections, we can now try and discover resonances by looking at the diphoton invariant mass distribution of the ATLAS or CMS events that pass those selections, and comparing to the distribution expected if there are no new particles in Nature beyond those of the SM. Searching for a signal of any kind involves first developing a detailed *model* of the background that is as accurate and precise as possible, and in our present case that means predicting the shape of the diphoton mass distribution that arises from each particular background contribution, and also determining the relative contribution of each background source. This model will eventually be fitted to the LHC data along with a model of the resonant signal we are searching for, allowing us to extract the relative normalisation of the signal and the background. If the resulting signal normalisation is a small enough number, we can conclude that there is no evidence for a signal. If it is large, we have a potential discovery on our hands.

For the newcomer, it is by no means obvious how to model the background, and the following questions may have already occurred to you:

- *How can we identify which processes in the SM can produce two real photons?* A general answer is that we can work out which Feynman diagrams involve incoming gluons and quarks and end in two photons, and we can then try and work out which diagrams will contribute most (crudely based on the number of vertex factors in the diagram, and some rough knowledge of the relative size of the parton distribution function for the incoming state). Whilst possible, this is largely unnecessary, since the large number of analyses at ATLAS and CMS means that almost all final states have already been investigated. Thus, looking at papers that cover similar final states to that of interest will already tell you what the dominant background contributions will be (particularly if you are careful to refer to papers in which the particles in the final state have similar typical energy and momenta as those you are interested in). Those who have spent decades analysing collider data can usually rattle off the dominant SM background processes in a given final state off the top of their head, a feat which never ceases to terrify students.
- *Is there any contribution from backgrounds that have a slightly different final state than the one of interest?* The answer here is always 'yes', and you can again refer to previous work to get a quick estimate of the processes of interest. For diphoton searches, you need to consider backgrounds that can produce jets that fake photons, since the small probability of a jet faking a photon is compensated for by the vastly higher production cross-section for dijet events at the LHC. For other final states, you might have to consider either fake particles, or real particles that accidentally fall out of the acceptance for the analysis (by either passing through a detector region that is outside of the accepted volume, or by having transverse momenta that fall below the threshold for selection). Neglecting these possible contributions is a frequent source of error for theoretical papers proposing new searches at the LHC.
- *Having identified a background, how do we predict the invariant mass distribution expected from that background?* There are several possible answers to this problem, and techniques for doing this are constantly evolving. As we saw in Chapter 8, one answer is to find the best Monte Carlo generator available for

your particular SM process, and run events from that through a detector simulation to get an invariant mass distribution. This may require a separate calculation of the production cross-section for the overall normalisation (for example, if the latter results from a higher-order calculation than that available in the Monte Carlo generator). We also saw in Chapter 8 that there are a variety of *data-driven* techniques for estimating backgrounds directly from the LHC data, and these are usually preferred when they are available. We will spend a great deal of this section exploring these techniques and presenting concrete examples that might prove useful in future work. It is common for different members of an analysis team to try different methods in order to check their consistency. The model with the smallest sytematic uncertainty should then be selected as the nominal method, with the others remaining as cross-checks.

If we examine recent searches for diphoton resonances, we find out quickly that there is a *reducible* background that arises from SM events in which one or both of the reconstructed photon candidates is a fake (e.g. a jet that mimics a photon), and an *irreducible* background in which an SM process has produced two real photons. The irreducible background is comprised of three main processes:

- **The Born process:** $qq \rightarrow \gamma\gamma$ which is of $O(e^2)$ in the electromagnetic coupling constant e.
- **The box process:** $gg \rightarrow \gamma\gamma$ which is of $O(e^2 g_s^2)$, in the electromagnetic and strong coupling constants e and g_s. Note that, although this suggests that the process is suppressed relative to the Born process, this is more than compensated for by the gluon parton distribution function dominating over that of the quarks.
- **The bremsstrahlung process:** $qg \rightarrow q\gamma\gamma$ which is of $O(e^2 g_s)$.

Photons can also be produced by jet fragmentation, in which case they are still deemed to be part of the irreducible background (note that this is distinct from a jet faking a photon). Thus, the production of a photon plus a jet, or the production of multiple jets with one or more jets producing photons, are both considered to be part of the irreducible background.

Two distinct approaches to modelling these backgrounds can be pursued, with different strategies required depending on how far into the tail of the diphoton invariant mass distribution the search aims to delve. For a search that aims to reach very high invariant mass values, where few events remain, the background can be modelled using a combination of Monte Carlo simulation and data-driven methods. For a search that does not aim as high in the invariant mass, it is possible to fit a functional form to the background distribution to describe its shape, then use that in the final comparison with the observed LHC data. We will deal with each of these approaches in turn.

Background estimation for searches extending to large invariant mass values
In the first approach, the shape of the invariant mass distribution for the irreducible background can be determined by Monte Carlo simulation, and it is important to

take the best generators available in terms of quantum chromodynamics (QCD) order, and also to compare multiple generators to determine the accuracy of the model. In the most recent analysis, two different next-to-leading order in QCD generators were used, with both found to give a consistent shape. Systematic uncertainties on the prediction can be investigated by simulating new event samples in which the parton distribution functions are varied within their uncertainties (or switched to a different set), and samples in which the factorisation, renormalisation and fragmentation scales are varied.

It then remains to find the shapes of the photon+jet and dijet backgrounds, which together form the reducible background. In this case, a *data-driven* technique is used, in which *control samples* are obtained in the LHC data by selecting events with different photon definitions. For example, one can select events in which the photon with the smallest transverse momentum fails the tight selection criteria (whilst still passing looser criteria), whilst the photon with the highest transverse momentum passes the tight selection criteria. These are referred to as γj events, since they are enriched in events where the photon candidate with the smallest p_T actually results from a jet. One can also define an analogous control sample for $j\gamma$ events, plus a control region where both photons fail the tight selection criteria, which is enriched in jj events. The shape of the invariant mass distribution in these samples is then fitted with a function of the form

$$f(x) = p_0 \times x^{p_1 + p_2 \log(x)} \times \left(1 - \frac{1}{1 + e^{(x-p_3)/p_4}}\right) \qquad (9.2)$$

where $x = m_{\gamma\gamma}/\sqrt{S}$ and p_i are free parameters. This may appear to be a very random choice of function, but typically functions of varying complexity will be tried, before the simplest function that gives a good answer is selected. The uncertainty in the shapes determined from the control samples can be investigated by varying the loose photon definitions that are used to select the photon candidates that fail the tight criteria.

Having identified the shapes of the different contributions to the background, it still remains to determine their relative proportion. The composition of the total diphoton background can be determined using a variety of data-driven techniques, and we here give two recent examples for inspiration. It is important to realise that there is no strictly correct answer, but instead different members of an analysis team will typically try different methods, leaving the one that ended up having the smallest systematic uncertainty to be selected as the preferred option. The lesser methods can be then be used to assess the accuracy of the nominal method.

- **The 2 × 2 sidebands method:** In this method, control regions of the data are defined by taking the normal signal selection, but playing with part of the definition of a photon, similar to the techniques that we first described in Section 8.6.1. For each photon candidate, one can define different photon selections as: a) the regular signal selection; b) the isolation requirement is not met instead of being met; c) part of the tight identification requirement is not met instead of being met (relating to electromagnetic (EM) shower shapes in the calorimeter); and d) both the isolation and tight requirements are not met.

This gives rise to 16 different regions of the data in which two photons appear, but there are differing selections on the photon candidates (e.g. one photon matches (a), the other matches (a); one photon matches (a), the other matches (b), and so on). Now assume that there is some number of diphoton events $N_{\gamma\gamma}$ with two true photons; some number of events $N_{\gamma j}$ with one photon and a jet, where the photon has the highest transverse momentum; some number of events $N_{j\gamma}$ with one photon and a jet, where the jet has the highest transverse momentum; and some number of events N_{jj} in which the photon candidates both result from jets. In addition, we can write the proportion of true photons that pass the isolation requirement as ϵ_γ^{isol}, which is usually referred to as the (isolation) selection *efficiency*, and we can write a similar efficiency for passing the tight selection requirement. We can also define quantities that give the proportion of jets that pass the isolation and tight selection requirements, and these are conventionally referred to as *fake rates*. It is assumed that the isolation and tight selection requirements are uncorrelated for background events. The number of observed events in each region can then be written as a series of functions of $N_{\gamma\gamma}$, $N_{\gamma j}$, $N_{j\gamma}$, N_{jj}, the efficiencies, the fake rates, and a correlation factor for the isolation of jet pairs. This gives a system of 16 equations, each of which gives the observed yield in a particular region. Assuming that the efficiencies for true photons have been measured previously (which is usually true), the system of equations can be solved numerically using a χ^2 minimisation method with the observed yields and true efficiencies supplied as inputs. The method allows the simultaneous extraction of $N_{\gamma\gamma}$, $N_{\gamma j}$, $N_{j\gamma}$, N_{jj}, plus the correlation factor and the fake efficiencies, which ultimately determines the background composition in the signal region. The method can be applied in bins of the diphoton invariant mass, which allows the shape of the diphoton invariant mass distribution from each contribution to be estimated separately.

- **The matrix method:** In this method, diphoton events where both photons satisfy the tight selection requirement are classified into four categories depending on which photons satisfy the isolation requirement: both photons (PP), only the photon with the highest transverse momentum (PF), only the photon with the lowest transverse momentum (FP), or neither photon (FF). One can then write a matrix equation that relates the number of observed events in data in each of these regions to $N_{\gamma\gamma}$, $N_{\gamma j}$, $N_{j\gamma}$ and N_{jj}, where the elements of the matrix all involve factors of the isolation efficiencies for photons and jets. Once these have been measured in a separate analysis, the sample composition can then be obtained by inverting the matrix equation. As with the 2×2 sideband method, the matrix method can be applied in bins of kinematic variables, and thus can be used to determine the shape of the invariant mass distribution arising from each contribution to the data.

Once the background composition has been measured, the template shapes for the irreducible and irreducible background can be summed with their relevant proportions to determine the shape of the total diphoton background. The overall normalisation of this total background is left as a free parameter when the shape is compared to the LHC data using the statistical test described in Section 9.2.4. Uncertainties in the measured proportion of each background contribution can be considered as systematic uncertainties on the shape of the total background, in addition to the shape uncertainties that are unique to each contribution.

Background estimation for searches extending to moderate invariant mass values

When a resonance search is only conducted up to moderate values of the invariant mass, there are still plenty of events left in the tail of the distribution beyond the search region. In this case, one can fit a functional form to data to describe the background shape, which is ultimately fit simultaneously with a signal model to extract the relative normalisation of signal and background. The range over which the function is fitted to the data extends below the resonance search region, and also above the search region.

A family of functional forms is chosen, which has been found to be useful in previous analyses. For ATLAS analyses, the shape of the diphoton invariant mass distribution is assumed to be given by

$$f_{(k)}(x; a_k, b) = N(1 - x^{1/3})^b x^{\sum_{j=0}^{k} a_j (\log x)^j}, \qquad (9.3)$$

where $x = m_{\gamma\gamma}/\sqrt{s}$, a_k and b are free parameters, and N is a normalisation factor. For any particular choice of k, the function has $k + 2$ free parameters, and the choice of k sets the precise functional form that will be used with the LHC data.

How can we choose the best value of k? The answer is to first obtain a sample of typical background events by generating a large sample of mock diphoton data using Monte Carlo generation, with photons smeared by the detector resolution. Control samples defined in real LHC data are used to supply estimates of the dijet and photon + jet backgrounds, using a similar method to that described above, and these are fitted with a functional form to provide smooth templates. The total background can then be estimated as the sum of these templates. For different choices of k, one can then fit this 'background-only pseudo-data' with equation (9.3) for different choices of k, including also a template for the signal model which we will describe in the next section. If such a fit were to reveal the presence of signal, we can say with confidence that such a signal is spurious, since we fitted a sample that only contains background events! This then gives us a way to select k: any choice of k whose resulting functional form readily generates spurious signals should not be used to fit real LHC data, since we will falsely conclude that we have a signal when there is not one in the data.

In practice, k is chosen by performing signal-plus-background fits for different values of k, for different signal masses, and requiring the eventual choice of functional form to give a spurious signal of less than some nominal figure over the full investigated mass range (usually of the order of 20%, but increased for low mass searches). Several functional forms will pass this test, and the one with the smallest number of degrees of freedom is selected for the final analysis. One can also perform hypothesis tests of pairs of functional forms of varying complexity. Once a functional form is chosen, the spurious signal analysis can be used to assign an uncertainty to the background model.

9.2.3 Modelling the diphoton signal

In addition to modelling the diphoton background, we also need to model the contribution to the invariant mass distribution that results from our resonant new physics. One option for this is simply to generate Monte Carlo events for our chosen signal model before passing them through a detailed simulation of the detector. Such an approach is certainly accurate, but it is also relatively inflexible. Ideally, we would like a model that, for a given set of parameters, gives us the expected shape of the signal analytically, thus allowing us to test many signal points without detailed simulation. Such an analytic model can be tested on a limited number of fully-simulated models, before being employed in the final statistical comparison with the observed data.

An analytic model can be developed by convolving a function that describes the detector resolution with a function that gives the invariant mass distribution (also called the *lineshape*) at particle-level (i.e. particles that are reconstructed with 100% precision).

Detector resolution
Previous experience with diphoton resonance searches had shown that the ATLAS detector resolution could be described well by a **double-sided crystal ball** (DSCB) function, given by

$$
N \times
\begin{cases}
e^{-t^2/2} & \text{if } -\alpha_{\text{low}} \leqslant t \leqslant \alpha_{\text{high}} \\[2em]
\dfrac{e^{-\alpha_{\text{low}}^2/2}}{\left[\dfrac{\alpha_{\text{low}}}{n_{\text{low}}} \left(\dfrac{n_{\text{low}}}{\alpha_{\text{low}}} - \alpha_{\text{low}} - t \right) \right]^{n_{\text{low}}}} & \text{if } t < -\alpha_{\text{low}} \\[3em]
\dfrac{e^{-\alpha_{\text{high}}^2/2}}{\left[\dfrac{\alpha_{\text{high}}}{n_{\text{high}}} \left(\dfrac{n_{\text{high}}}{\alpha_{\text{high}}} - \alpha_{\text{high}} + t \right) \right]^{n_{\text{high}}}} & \text{if } t > \alpha_{\text{high}}
\end{cases}
\tag{9.4}
$$

This function is Gaussian within some range α_{low} to α_{high}, outside of which it follows a power law. The Gaussian is given in terms of $t = (m_{\gamma\gamma} - \mu_{\text{CB}})/\sigma_{\text{CB}}$ (with a peak position μ_{CB} and width σ_{CB}), and α_{low} and α_{high} are given in units of t. n_{low} and n_{high} are the exponents of the power law on the low and high mass sides of the Gaussian, respectively. Finally, N is a parameter that sets the overall normalisation. Note that this function describes the effects of the finite detector resolution on the diphoton invariant mass, which is really an effective description of what results from the slight mismeasurement of each photon four momentum.

There is unfortunately no way to know the parameters of this function *a priori*. Furthermore, the parameters of the DSCB function vary with the mass of the resonance, since resolution effects on each photon depend on the transverse momentum of the photon, which in turn depends on the mass of the resonance. One must therefore fully simulate signal samples for different resonance masses, and extract preferred values for the DSCB parameters at each mass. A parametric function can then be fitted to the extracted DSCB parameters, which provides our desired analytic method for determining the detector resolution for a given resonance mass. n_{low} and n_{high} are typically treated as constants, whilst polynomials of first or second order are used for the other parameters. The parameterisation can then be verified by taking a simulated sample whose resonance mass was not used to derive the DSCB parameterisation, and testing how closely the DSCB function from the parameterisation matches a new fit to the DSCB parameters.

Signal lineshape

The signal lineshape for a given model can be calculated by multiplying the analytic differential cross-section for resonance production and decay to photons by a function that parameterises the incoming parton luminosity. We will develop one example in this section, but the basic techniques would easily generalise to other processes.

Consider a spin zero resonance being produced by incoming gluons, which then decays to photons, as shown in Figure 9.1. Such a resonance is called an *s*-channel resonance. Assuming some new physics creates couplings between the scalar, gluons and photons at energies much higher than the LHC, we can use the effective field theory techniques of Chapter 4 to write a Lagrangian that is valid at LHC energies:

$$\mathcal{L}_S = c_3 \frac{g_s^2}{\Lambda} G_{\mu\nu}^a G^{\mu\nu a} S + c_2 \frac{g^2}{\Lambda} W_{\mu\nu}^i W^{\mu\nu i} S + c_1 \frac{g'^2}{\Lambda} B_{\mu\nu} B^{\mu\nu} S, \qquad (9.5)$$

where S is the new scalar field, and $G_{\mu\nu}^a$, $W_{\mu\nu}^i$ and $B_{\mu\nu}$ are the gluon and electroweak field strength tensors. These terms are the gauge-invariant terms of lowest mass dimension that can couple the SM gauge fields to the new scalar. The g_i factors are the various coupling constants of the SM, and the $\{c_i\}$ are free parameters that adjust the normalisation of each term. Λ is an additional constant (see the exercises). Note that the Lagrangian is written before electroweak symmetry breaking, so we see the $W_{\mu\nu}^i$ and $B_{\mu\nu}$ fields rather than the gauge boson fields; the coupling to photons will emerge from the second two terms after electroweak symmetry breaking.

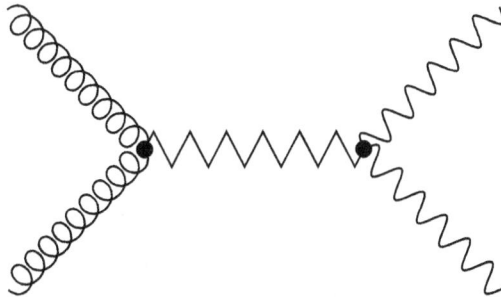

Figure 9.1. A scalar particle produced in the *s*-channel.

For an s-channel resonance, the differential cross-section (in terms of the Mandelstam variable s) for the $gg \to \gamma\gamma$ process is given by

$$d\sigma_{\text{EFT}} \propto \int ds \frac{1}{s} \frac{|a_{gg \to S}(s)|^2 |a_{S \to \gamma\gamma}(s)|^2}{(s - M_X^2)^2 + (M_X \Gamma_X)^2} \tag{9.6}$$

where we can recognise the relativistic Breit–Wigner distribution[2]

$$f_{\text{BW}} = \frac{1}{(\hat{s} - M_X^2)^2 + (M_X \Gamma_X)^2}. \tag{9.7}$$

The numerator of this expression must be calculated for our specific effective field theory, and it can be shown to be[3]

$$|a_{gg \to S}(s)|^2 |a_{S \to \gamma\gamma}(s)|^2 = \frac{e^2 g_s^2 a^2 b^2 s^4}{256\pi}, \tag{9.8}$$

where e and g_s are the electromagnetic and strong coupling constants, and a and b are given by

$$a = \frac{-4(c_1 + c_2)}{\Lambda} \qquad b = \frac{4c_3}{\Lambda}. \tag{9.9}$$

The non-trivial form of a as a function of the coefficients in the original Lagrangian results from the fact that we wrote the $SU(2)$ and $U(1)$ field strength tensors before electroweak symmetry breaking. The s^4 dependence results from the fact that both vertices of the s-channel resonance diagram involve couplings between S and the derivative of SM fields (since the field strength tensors involve the derivatives of fields, rather than the field themselves).

Putting these together, we obtain

$$\frac{d\sigma_{\text{EFT}}}{ds} \propto \frac{s^4}{s} \times f_{\text{BW}} = s^3 \times f_{\text{BW}} \tag{9.10}$$

Let us now clarify what s is. In any one collision, it is the fixed centre-of-mass energy, and our current form for the differential cross-section assumes a flat s distribution. In fact, LHC collisions will produce s values with a certain weight given by the gluon luminosity $\mathcal{L}_{gg}$ (for the simple case of gluon fusion), and we also know that $m_{\gamma\gamma}^2 = s$ (hence $ds = 2m_{\gamma\gamma} dm_{\gamma\gamma}$). We can thus finally write

$$\frac{d\sigma_{\text{EFT}}}{dm_{\gamma\gamma}} \propto \mathcal{L}_{gg} \times m_{\gamma\gamma}(m_{\gamma\gamma})^6 \times f_{\text{BW}} = \mathcal{L}_{gg} \times m_{\gamma\gamma}^7 \times f_{\text{BW}}. \tag{9.11}$$

---

[2] Note that this is given up to a normalisation factor that can be neglected in the present discussion.

[3] At this point, theorists might be confident that this is a relatively straightforward calculation in quantum field theory, but experimentalists may feel the onset of terror. In fact, such calculations can now be performed by software codes for matrix element calculations, which substantially lowers the bar for experimentalists, whilst ensuring that theorists can correct mistakes in their algebra.

For a given set of parton distribution functions, the gluon luminosity can be plotted as a function of q^2 (the momentum exchange in the scattering process), and a functional form can be fitted to the distribution. Past experience has shown that the function is well described by

$$\mathcal{L}_{gg}(q^2) \propto \left(1 - \left(\frac{q}{13\ 000}\right)^{1/3}\right)^{10.334} \left(\frac{q}{13\ 000}\right)^{-2.8}, \tag{9.12}$$

where q is expressed in GeV.

Equations (9.11) and (9.12) together give us our analytic model for the signal contribution to the diphoton invariant mass distribution. As usual, it should be tested against fully-simulated Monte Carlo events for the model of interest to check that it does indeed provide an accurate description of the invariant mass distribution expected in LHC data. In doing this comparison, it is important to check that the form of the Breit–Wigner distribution that was chosen for our differential cross-section matches that used by the Monte Carlo generator, since different generators use slightly different forms for the Breit–Wigner function.

9.2.4 Statistical approach for extracting signal significance

We now have our signal and background models, and it only remains to compare them to the data to determine whether we have any evidence for physics beyond the SM. For a particular resonance hypothesis of a given mass m_X and width Γ_x, this is accomplished by performing a maximum likelihood fit of the signal and background models to the observed $m_{\gamma\gamma}$ distribution. Assuming that we call our signal model $f_S(m_{\gamma\gamma})$, and our background model $f_b(m_{\gamma\gamma})$, the predicted $m_{\gamma\gamma}$ distribution is given by

$$N_S f_S(m_{\gamma\gamma}) + N_b f_b(m_{\gamma\gamma}), \tag{9.13}$$

with N_S and N_b giving the number of signal and background events, and $f_S(m_{\gamma\gamma})$ and $f_b(m_{\gamma\gamma})$ assumed to be normalised to unit area. For a given set of the parameters of this model (which includes N_S and N_b along with the parameters of the functional forms f_S and f_b), one has a prediction for the yield in each bin of the diphoton invariant mass, which can be compared to the observed data using a suitable likelihood. Uncertainties in the signal and background shapes are added to the fit as nuisance parameters, with either Gaussian or log-normal penalty terms added to the likelihood, and fits are performed for a variety of candidate masses. Each fit assumes that only one signal component could be present in the data, and the whole invariant mass distribution (above some lower limit) is used for comparison with the signal and background models.

For each signal hypothesis (m_X, Γ_x), one can calculate a local p-value for the compatibility of the data with the background-only hypothesis, using the hypothesis testing framework introduced in Section 5.8.1. This can be based on the test statistic

$$q_0(m_X, \Gamma_x) = -2 \log \frac{L(0, m_X, \Gamma_x, \hat{\hat{v}})}{L(\hat{\sigma}, m_X, \Gamma_x, \hat{v})}. \tag{9.14}$$

There is a lot to unpack in this equation. σ is the signal yield, and ν are the nuisance parameters. Parameters with a hat are chosen to unconditionally maximise the likelihood L, whilst parameters with a double hat are chosen to maximise the likelihood in a background-only fit. Hence, the numerator of this quantity expresses the likelihood of the background-only hypothesis, in which the nuisance parameters are chosen to maximise the likelihood. The denominator is the likelihood of the signal hypothesis where both the signal yield and the nuisance parameters are chosen to maximise the likelihood.

One method for presenting the results of the search is that presented in Figure 9.2, in which the local p-value for the compatibility of the background-only hypothesis is shown in the plane of the assumed mass and width of the resonance. As explained in Chapter 5, local p-values are susceptible to the look-elsewhere effect, in which the fact that we have looked in many regions of the data can be expected to have produced a locally significant p value at least somewhere. One way around this is to estimate a global significance by generating a large number of pseudo-experiments that assume the background-only hypothesis. For each experiment, you perform a maximum-likelihood fit with the width, mass and normalisation of the resonance signal model as free parameters (constrained to be within the search range). The local value of each pseudo-experiment is computed, and the global significance is then estimated by comparing the minimum local p-value observed in data to the distribution derived from the pseudo-experiments.

An alternative approach to presenting results is to fix the assumed width of the resonance to some assumed value, then display the upper limit, at 95% CL, on the resonance production cross-section times branching ratio to photons. An example is shown in Figure 9.3, where the limit has been calculated using the CL_s technique

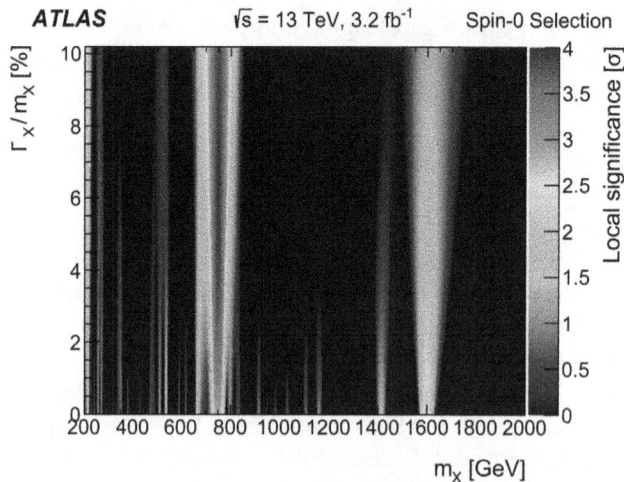

Figure 9.2. An example presentation of resonance search results, consisting of the local p-values for the compatibility of the data with the background-only hypothesis, in the plane of the assumed mass and width of the resonance. Reproduced with permission from *JHEP* **09** 001 (2016).

introduced in Section 5.7.1. Evidence for a signal in this presentation manifests itself as a difference between the expected and observed CL_s limits.

9.2.5 Interference effects

In our discussion so far, we have assumed that the effect of a signal in the LHC data is simply to produce diphoton events that add an extra contribution to the diphoton invariant mass distribution, without any change to the original background distribution. This assumption is encapsulated in equation (9.13). Recall, however, that the way to calculate the rate of scattering for, e.g. two incoming gluons at the LHC to two outgoing photons, is to sum the amplitudes corresponding to each possible Feynman diagram, and only then to take the square of the amplitude. This sum is not in general equal to the sum of the squares of the individual amplitudes. Rather, it may contain cross-terms that constitute *quantum interference* between each of the interactions that contributes to the total amplitude. Equation (9.13) therefore amounts to neglecting interference effects completely.

Unfortunately, neglecting interference effects is a pragmatic compromise rather than a well-motivated assumption. The details of signal-background interference are highly dependent on the new physics that gives rise to the resonance, which is unknown *a priori*. To give only one example, a scalar being produced via gluon-fusion through a fermion loop interferes with existing gluon-induced SM processes, with the precise effects depending on both the mass and width of the scalar, and the masses of any fermions that can appear in the loop. Different models will generate different patterns of interference, and the ultimate effect is to change the production rate of the resonance, whilst distorting the invariant mass distribution of the decay products in such a way as to change the apparent mass of the scalar resonance. The potential presence of these effects complicates both the interpretation of a new discovery in a resonance search, and the presentation of null results in the form of

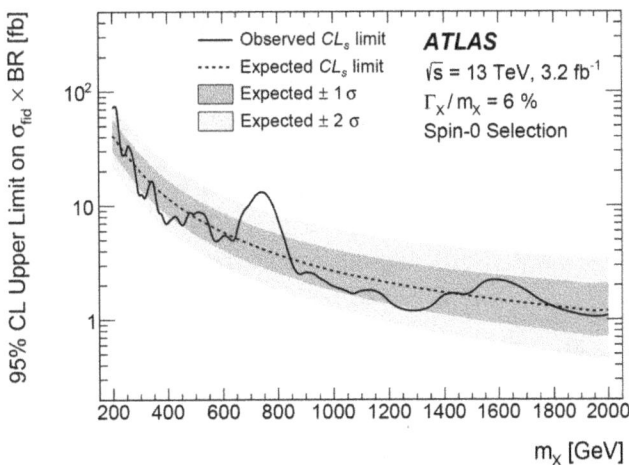

Figure 9.3. An example of an upper limit on the resonance production cross-section times branching ratio to photons. Reproduced with permission from *JHEP* **09** 001 (2016).

cross-section times branching ratio limits, which are not well-defined in the case of interference.

A model-specific way to deal with interference effects in resonance searches is to perform a full Monte Carlo simulation for specific model parameter points that include interference, and state whether these points are excluded or not by the statistical analysis in the presentation of the results. This could be presented in the form of an exclusion limit in a plane of model parameters, similar to the semi-invisible searches for new physics presented in Chapter 10. A model-independent approach has recently been proposed in which the shape of the distribution in data is modelled by the usual background template functional form, plus a suitable sum of even and odd functions that can span the space of possible interference effects, thus obviating the need for a precise model of the new physics. A positive result could then be cast into different models by reproducing the approach of the template function analysis.

9.3 Jet resonance searches

We have now completed our detailed review of a specific resonance search, using the relatively clean diphoton final state. We could in principle repeat this process for every final state of interest, but will choose instead to focus on a particular final state that commands lots of attention from experimentalists due to the large number of models that are expected to produce it. We will examine *jet* resonance searches and, at the outset, it is useful to think about what is similar to a diphoton search, and what is different.

- If a resonance decays to two jets that are reconstructed as separate objects, then the search involves looking for a bump-like feature in the dijet invariant mass distribution of two jet events. This is directly analogous to the diphoton case. However, it might be the case that a resonance decays to SM particles that themselves decay to produce jets. The boost of the final state partons in the laboratory frame causes the resulting jets to overlap in the detector, and one must apply the **jet substructure** techniques of Section 8.4.5 to search for the resonance.
- Diphoton resonance searches are assisted by the relatively low cross-section of diphoton production at the LHC, which helps in two ways. Firstly, it means that the background to the resonance signal is quite small, which means that the cross-section times branching ratio to photons that can be probed is relatively small. Secondly, it means that triggering on, and storing, diphoton events is straightforward, since there are not too many of them to record even if the transverse momentum of the photons is relatively low. For dijet events, both of these points change dramatically for the worse. Practically every event at the LHC is a dijet event (or more generally an **inclusive jet event**), with everything else being a tiny correction. This makes the background for resonance searches huge. It also means, of course, that a jet trigger is only feasible if we raise the transverse momentum requirement of the jet (or jets) which, as we learned above, sculpts the low mass region of the dijet invariant mass distribution in such a way that we cannot safely perform

resonance searches in that region. Typical dijet resonance searches therefore concentrate on the high mass tail of the dijet invariant mass distribution, with dedicated search techniques required to probe the low mass region (e.g. looking for a resonance that recoils against a photon emitted by an incoming parton).

In the following, we sketch a recent CMS dijet resonance search to give a sense of the techniques that are applicable to dijet searches in general. The first step is to decide on a trigger strategy, and in our chosen example this involved using two triggers, and accepting events if they passed either one of them. This amounts to taking a logical OR of the two sets of trigger conditions. The first trigger simply required a single jet with $p_T > 550$ GeV, where this threshold can be expected to change with the operating conditions of the LHC over its history (whilst remaining fairly stable within a given run period). The second trigger required $H_T > 1050$ GeV, where H_T is the scalar sum of the jet p_T for all jets in an event with $|\eta| < 3.0$. H_T is a useful concept for event analysis at the LHC, although its definition often changes depending on what one is searching for in an analysis. Having passed the trigger requirement, events were selected for further analysis if the dijet invariant mass m_{jj} was greater than 1.5 TeV for jets with an $|\eta|$ separation of less than 1.1, or if $m_{jj} > 1.5$ TeV for $1.1 < |\Delta\eta| < 2.6$. The jets used for this selection were specially-defined 'wide jets', in which jets reconstructed with the anti-k_T clustering algorithm with a distance parameter of 0.4 were merged based on their spatial separation. In the wide-jet algorithm, the two leading jets in the event are used as seeds, and the four-vectors of all other jets are added to the nearest leading jet if $\Delta R = \sqrt{(\Delta\eta)^2 + (\Delta\phi)^2} < 1.1$, eventually forming two wide jets which are used to define the dijet system for the calculation of m_{jj}. This procedure mops up gluon radiation from nearby the final state partons, which improves the dijet mass resolution. If one were to instead increase the distance parameter used in the original definition of the jets, this would add extra pile-up and initial state radiation contributions that are less welcome. The signal region is defined by requiring the two wide jets to satisfy $|\Delta\eta| < 1.1$. This is because the background contribution arising from t-channel dijet events has an angular distribution which is roughly proportional to $1/[1 - \tanh(|\Delta\eta|/2)]^2$, which peaks at large values of $|\Delta\eta < 1.1|^4$.

The dominant background for the dijet search is the production of two or more jets via QCD processes, which is dominated by the t-channel parton exchange referred to above. The CMS analysis utilised two methods for modelling the dijet background, the first of which was to fit the invariant mass spectrum with the functional form

$$\frac{d\sigma}{dm_{jj}} = \frac{a_0(1 - x)^{a_1}}{x^{a_2 + a_3 \ln(x)}} \tag{9.15}$$

where the a_i are free parameters, and $x = m_{jj}/\sqrt{s}$. The second method, which was found to reduce the systematic uncertainties, involved deriving the background

---

[4] Scattering enthusiasts will recognise this as the same angular distribution as Rutherford scattering!

model for the signal region using two control regions in the data. It is thus another example of a data-driven method. The first control region CR_{middle} requires the difference in η between the two wide jets to satisfy $1.1 < |\Delta\eta| < 1.5$, whilst the second control region CR_{high} requires $1.5 < |\Delta\eta| < 2.6$.

To get the shape of the dijet invariant mass distribution in the signal region, the number of events observed in data in each bin of the m_{jj} distribution in CR_{high} is multiplied by an appropriate 'transfer factor'. For a given bin, the transfer factor is defined using a leading-order Monte Carlo simulation of QCD dijet production, coupled to a suitable simulation of the CMS detector. The predicted number of events in a given bin i of the m_{jj} distribution in the signal region, N_i^{pred} is given by:

$$N_i^{pred} = K \times \frac{N_i^{SR,sim}}{N_i^{CR_{high},sim}} \qquad (9.16)$$

where $N_i^{SR,sim}$ and $N_i^{CR_{high}}$ are the simulated yields in the corresponding bins of the m_{jj} distributions in the SR and CR_{high} region. K is a factor which corrects for the absence of higher-order QCD and EW effects in the Monte Carlo generator, in addition to experimental systematic effects. This itself can be derived from data using the second control region CR_{middle}, in which the wide jets have a more similar η separation to the signal region than those in CR_{high}. First, we define $R = N_i^{CR_{middle}}/N_i^{CR_{high}}$, which can be evaluated using either data events or background events, and which expresses the ratio of the number of events in a given bin i of the m_{jj} distribution in the CR_{middle} region with the number in the corresponding bin in the CR_{high} region. We then define K as

$$K = \frac{R^{data}}{R^{sim}} = a + b \times (m_{jj}/\sqrt{s})^4 \qquad (9.17)$$

where a and b are free parameters. In other words, we assume that the factor which corrects the simulation to the data can be parameterised by the expression on the right, and we must extract a and b from a fit to the data. a and b are in fact included as nuisance parameters in the final comparison of the resonance signal and background models to the observed CMS data.

Signal models are harder to generate than for the diphoton case, since the width of the expected lineshape depends on whether the new resonance decays to a pair of quarks (or a quark and an anti-quark), a quark and a gluon, or a pair of gluons. The simplest way to obtain signal predictions for the dijet invariant mass distribution is to perform a Monte Carlo simulation of suitable physics models, from which one finds that the predicted lineshapes for narrow resonances have Gaussian cores from the jet energy resolution, and tails towards lower mass values from QCD radiation.

Having defined the signal and background models, the extraction of limits on the cross-section times branching ratio to jets as a function of the resonance mass can follow a similar treatment to that defined in the diphoton case. Note that, in the case of the data-driven background estimate, the fit of the signal and background model to data can be performed simultaneously in both the signal and CR_{high} regions, thus accounting for possible signal contamination of the CR_{high} region.

Further reading

- An excellent recent summary of resonance searches at the LHC is given in arXiv:1907.06659, which includes a summary of the theoretical origin of different resonances, plus the results of searches at the second run of the LHC.
- The ATLAS diphoton search that we worked through is based on *JHEP* **09** 001 (2016), arXiv:1606.03833. An excellent resource for understanding this search in detail is Yee Chinn Yap's PhD thesis, available at http://cds.cern.ch/record/2252531/files/?ln=el.
- A model-independent approach for incorporating interference effects in resonance searches can be found in Frixione *et al* 2020 Model-independent approach for incorporating interference effects in collider searches for new resonances *Eur. Phys. J.* C **80** 1174 (2020).
- Our dijet resonance search example is based on CMS-PAS-EXO-17-026.

Exercises

9.1 Draw the Feynman diagrams for the three main SM processes that produce two photons in the final state.

9.2 Write the matrix equation that expresses the number of events in each region defined by the matrix method in terms of the quantities $N_{\gamma\gamma}$, $N_{\gamma j}$, $N_{j\gamma}$, and N_{jj}.

9.3 Explain what the normalisation constant Λ in equation (9.5) represents, and why it is necessary.

9.4 By browsing the literature or otherwise, write down the main backgrounds for a *WW* resonance search in the fully-hadronic final state.

9.5 As explained in the text, a common deficiency of dijet resonance searches is that jet triggers often require a large jet transverse momentum for the leading jet in each event.

    (a) Explain why this makes it impossible to search for resonances at low invariant mass.

    (b) How might initial state radiation be used to circumvent this problem?

9.6 Plots such as that shown in Figure 9.3 often lead to confusion in the high energy physics community.

    (a) Which feature of Figure 9.3 indicates the possible existence of a signal in the data?

    (b) What is the likely mass of the resonance if this signal were true?

    (c) Would the same conclusion automatically apply to a resonance with a relative width of 10%? Give a reason for your answer.

IOP Publishing

Practical Collider Physics

Andy Buckley, Christopher White and Martin White

Chapter 10

Semi-invisible particle searches

We have now described how searches for new particles at the LHC can be performed if we see all of the decay products of the particles. Things become more complicated if the new particles decay *semi-invisibly*, which will generically occur if one or more of the decay products are weakly-interacting, and hence exit the detector without trace. This will happen if the new particles decay to one or more neutrinos of the Standard Model (SM), or if they decay to one or more new particles that are themselves unable to interact with the detector. It is important to note that we are using 'weakly-interacting' in a colloquial sense, meaning that the particle might interact through a new force that has a small coupling constant, rather than necessarily interacting through the weak interaction of the SM.

In this chapter, we will describe techniques for dealing with semi-invisible particle decays in LHC searches, and we will see that a variety of strategies can be employed to infer the presence of new invisible particles, and even to measure their masses and couplings in some cases. Semi-invisible particle decays are a generic feature of any model that predicts a dark matter candidate in the form of a weakly-interacting massive particle, and are a particularly important feature of supersymmetry searches.

10.1 General approach for semi-invisible particle searches

In the case of a resonance search, the new particle we are searching for can usually be treated as a generic state with an unknown mass and coupling to the particles in the final state, and we do not need to worry about the precise physical theory that gives rise to the new particle when we are actually performing the search[1]. In contrast, however, the term 'semi-invisible particle search' covers a huge range of possible physics models, each of which might have a very different phenomenology.

---

[1] An exception would occur if we wanted to unambiguously model interference between the hypothetical signal and the background, which is rarely done.

Faced with this mass of models, we should immediately ask if there is anything that can be said in general about models that include semi-invisible particle decays. Thankfully, there is: all events with invisible particles in will contain missing transverse energy from the escaping particles. We can therefore frame semi-invisible particle searches from the outset as searches in the final state 'missing transverse energy plus something else'. We may then start to classify what the 'something else' might be, as follows.

It might be that our model predicts a single new weakly-interacting particle, that interacts with quarks or gluons such that it can be made directly in proton–proton collisions. For example, the simplest approach to dark matter is to hypothesise that there is a single new particle responsible for it, and that its interactions with SM fields can be described either by an effective field theory with a contact interaction, or by a theory with an explicit mediator between the SM fields and the new dark matter particle. In such models, the new dark matter particle will typically be produced in pairs, and if *only* a pair of dark matter particles is produced, there is nothing for the detector to trigger on in the event. Instead, one must search for events in which a parton was radiated from the initial state, which gives rise to events with a single jet, plus missing transverse energy. Other possibilities include events with a single photon, Higgs boson or weak gauge boson plus missing transverse energy.

It might instead be the case that there is a plethora of new particles in Nature, and the particles that are produced by the proton–proton collisions are not the invisible particles themselves. A classic example is that of supersymmetry: here one would expect to produce the coloured sparticles at a much higher rate than the weakly-coupled sparticles (if their masses are similar), and hence one should generically produce squarks and gluinos. These could then decay to supersymmetric particles which can themselves decay further, producing what is known as a **cascade decay**, culminating in the release of various SM decay products, plus one lightest neutralino for each sparticle that was produced. The lightest neutralinos generate anomalous missing transverse energy, whilst the SM decay products can give rise to high multiplicity final states that are very rare in the SM.

There is a very important difference between these two types of scenario. In the second case, as we shall see below, it turns out that we can form various invariant masses from the SM decay products, and these allow us to infer something about the masses of the particles involved in the cascade decay chain. In the case where we directly produce the weakly-interacting particle, it is much harder to access the properties of the particle. Nevertheless, how we actually go about *discovering* either of these scenarios looks remarkably similar, and we can break the process down into the following steps:

1. We choose a particular final state where we expect the signal to be highly visible. This can either be due to a large combined cross-section times branching ratio to that particular final state in the signal model, or due to the total SM background being very small in that final state relative to the signal process.

2. We simulate the relevant SM backgrounds by running suitable Monte Carlo generators, before passing the results through a simulation of the relevant LHC detector. We also do the same thing for the signal model. In the case where the signal model has a number of free parameters, we choose some **benchmark models** (i.e. specific points in the parameter space), and simulate those.

3. For each benchmark point, we try and find kinematic variables (i.e. functions of the four-vectors of the SM particles in each event, plus the missing transverse momentum), that differentiate between the signal and background events. In other words, we try and find functions whose histograms over the event sample would look very different for the signal and background. We then define **signal regions** by placing selections on these variables to reject as much of the background as possible, without removing too much of the signal. The selections (or 'cuts') can be optimised using the performance optimisation techniques described in Section 8.5.

4. We then develop accurate models of the different SM backgrounds in the signal regions, assuming that the results of our Monte Carlo simulations may not be perfectly accurate. This builds on the approach outlined in Section 8.6.

5. We compare the predicted background yields in the signal regions with the observed yields in the LHC data, and perform a statistical procedure to determine if the SM-only hypothesis is still viable. We may also determine if any of the simulated signal benchmark models are excluded at the 95% confidence level.

Each of these short descriptions hides myriad complexities, and it should not surprise you to learn that the precise details of how to implement these steps depend on the model being searched for. We will expand our knowledge of each of these steps in turn, using examples from past ATLAS and CMS searches.

10.1.1 Choosing a final state

In order to search for a beyond-Standard Model (BSM) physics theory at the LHC, the theory must be mature enough that we know the new particles of the theory, how they will decay (which may be a function of the parameters of the model), and what the kinematics of those decay products are. Assuming that this is the case for a hypothetical model of interest, our first job is to identify final states of the form 'missing transverse momentum plus something' in which our signal should eventually be visible over the background after we make some selections on kinematic variables. At this point, it may not be obvious why this is necessary. For example, for models that introduce many new particles, with many different decay modes, why not look in all final states simultaneously in an attempt to find evidence of anomalous events? The answer is that the dominant SM backgrounds differ wildly in each final state, and each of those backgrounds must be modelled using dedicated techniques that do not easily transfer between final states in all cases. A second reason to break up searches into different final states is political: it becomes much easier to organise efforts across hundreds of researchers if their work can be grouped into similar signatures, allowing methods to be easily shared amongst search teams that are looking for similar final states. Although there are techniques on the horizon that would be more ambitious in their combination of information across final states (e.g. using unsupervised machine learning), we will assume at this point that the dominant paradigm will persist for some years.

Having decided that we need to select a particular final state, how do we go about choosing it? In models with multiple particles, we can proceed by identifying the particles that have the highest production cross-section, and then choosing the final state that results from the highest decay branching ratio for those particular particles. A classic example is given by supersymmetry, in which the LHC production cross-section is dominated by squark and gluino production if these sparticles are light in mass. The simplest squark decay produces a jet and a lightest neutralino, with more complex decays featuring more jets and/or leptons. The simplest gluino decay produces two jets and a lightest neutralino, with more complex decays also producing more jets and/or leptons. On average, we would expect to see jets more often than leptons in these events. One can therefore assume that there will be a high combined branching ratio for producing events with a few jets plus missing transverse energy, and there will be reasonably high branching ratios for events containing several jets plus missing transvrse energy. The flagship searches for supersymmetry at the ATLAS and CMS experiments thus focus on final states with two jets plus missing energy, three jets plus missing energy, and so on.

Having identified our final state, we must also consider the potential SM backgrounds to see if our search is really going to be viable. Given that the LHC search programme has been developed for well over a decade now, the simplest way to identify the backgrounds in a given final state is to read existing LHC search papers that target a similar signature. Nevertheless, we can provide some general principles for identifying backgrounds if you are either going beyond the existing literature, or do not want to assume existing knowledge.

First, we can look up the production cross-sections of a large range of SM processes, such as those shown in Figure 4.1 of Chapter 4. Next, we have to work out which of those processes can possibly mimic the final state of our hypothetical signal process, and their relative production cross-sections multiplied by the relevant branching ratios for producing particular final states then give us a first idea of the relative proportion of each of these backgrounds. It must be stressed, however, that this is by no means the final answer. The imposition of kinematic selections will have different effects on each of these background contributions, so that the dominant background in the final analysis may not be that with the highest production cross-section. It can also happen that SM processes that do not produce the final state of interest at the level of their lowest-order Feynman diagram can still form part of the background for a search if some of their particles fall out of the acceptance of the final state (e.g. due to some particles having a low transverse momentum), or if initial state radiation increases the object multiplicity.

In Figure 4.1, the total LHC production cross-section is many orders of magnitude higher than the highest individual cross-section shown, which is that of

W boson production. The difference between these two is dominated by multijet production instigated by quantum chromodynamics (QCD) processes. Taking our two jets plus missing energy supersymmetry example, therefore, we expect the overwhelmingly dominant background to be multijet production, at least at the outset. We should also expect contributions from W and Z boson events, where the bosons are produced in association with jets. Looking up the possible decays of W bosons, we find that it can decay to a tau lepton and a neutrino some of the time, which gives us missing energy and a potential jet (depending on how the tau decays). We might also miss a light lepton from a leptonic W boson decay, which means that W boson events where the W boson decays to an electron or muon plus a neutrino will form an additional background. Z bosons, meanwhile, can decay to two neutrinos, giving missing energy plus jets that were emitted as initial state radiation in the event. Other backgrounds result from top quark pair production, single top quark production, and diboson production (i.e. pairs of W and/or Z bosons), with some of these only being sizable in our search regions that require multiple jets plus missing energy (rather than two jets plus missing energy). These SM processes vary in their capability of producing real missing energy, and also in the typical p_T values of the jets they will produce, which means that the final relative proportion of each of these backgrounds will depend critically on the kinematic selections that we later apply to try and extract the signal. It is this fact that makes the search for squark pair production in the two jets plus missing energy final state feasible at all, since a modest missing energy cut plus tight selections on the transverse momenta of the two jets reduce the dijet background considerably.

To take another example, imagine that we are looking for stop quark pair production, having failed to see the light quark partners and gluinos for over a decade. The simplest way that each stop quark can decay is to a top quark and a lightest neutralino, assuming that the mass difference between the stop quark and the lightest neutralino is larger than the top mass. The SM process that most resembles this is top quark pair production, which differs from the signal process only by the absence of two lightest neutralinos in addition to the top quarks. Since these lightest neutralinos are invisible to the detector, this manifests itself in a modification of the expected missing transverse momentum of the events. Each top quark decays to a b quark and a W boson, and we can therefore define three different final states depending on how the W bosons decay (a 0 lepton final state, a 1 lepton final state, and a 2 lepton final state, each of which also includes missing transverse momentum). Each final state receives further contributions to the background from single top production, W and Z boson production, diboson production and multijet production. The large cross-section of top quark pair production relative to stop quark production for even moderately heavy stop quarks makes this a very challenging search in general.

A much simpler example of choosing a final state results from our dark matter simplified model example given above, in which case we expect a single jet and missing energy, a single photon and missing energy, and so on. Here the model itself does not allow for a terribly complicated range of options.

10.1.2 Simulating backgrounds and benchmark signal models

Once we have chosen a final state and identified the dominant SM background processes, our next job is to generate Monte Carlo events for each relevant background process, and apply either a fast or full simulation of the ATLAS or CMS detector. For theorists wanting to develop and benchmark a new search method, it is common to use leading-order Monte Carlo generators, and fast simulations, on the basis that a proof of principle does not require a fully-accurate answer. Examples of where next-to-leading order (NLO) generation might be necessary include dark matter simplified models (where NLO effects have been shown to modify the missing energy distribution substantially), and any process where initial state radiation is particularly important. For experimentalists within a collaboration, the use of NLO generators, or at least leading order plus explicit jet radiation at the matrix element level, is mandatory in order to have any hope of reproducing the complexities of the LHC data. Both of the ATLAS and CMS collaborations employ a dedicated Monte Carlo team to generate samples for all SM backgrounds of interest. It is also standard to normalise the Monte Carlo yields using higher-order production cross-sections where these are available. In rare cases, you may need to extend this recipe with new background samples if your particular search is probing a region of the LHC data that does not strongly overlap with previous analyses.

Generating Monte Carlo samples for SM backgrounds is made easier by the fact that we know the SM processes that are likely to contribute to a given final state. Signal generation, on the other hand, is made much more complicated by the fact that we do not know *a priori* exactly what our signal scenario will look like. Almost all theories that we wish to search for at the LHC come with a number of free parameters, which we can either view as the masses and branching ratios of the new particles, or the fundamental parameters of the Lagrangian density of the theory that allow us to calculate those masses and branching ratios. It is rare that we have any knowledge of the values of these parameters in advance of our search, and yet we need to simulate benchmark models on which to develop and test our analysis approach. Various approaches have been developed to help us choose the benchmark models, none of which are entirely satisfactory.

To illustrate the problem, consider a dark matter simplified model, in which a hypothesised dark matter particle is able to interact with SM fields through a mediator particle. This is arguably the simplest concrete way of explaining dark matter, but it is already deceptively complex. For example, what is the spin of the dark matter particle, and what is the spin of the mediator? Both will affect the LHC signatures. Having decided on the spins, we still have to choose the mass of the dark matter particle, m_{DM}, the mass of the mediator particle, m_{med}, the coupling strength between the dark matter particles and the mediator, g_{DM}, and finally the coupling strengths between the mediator and SM quarks. The last of these could be utterly generic in principle, with separate parameters for the coupling to each quark species. For spin-1 mediators, it is generally assumed that there is a single coupling to quarks g_q, whilst for spin-0 mediators it is generally assumed that the coupling is of the same form as the SM Yukawa interactions, with a single coupling strength modifier g_q that scales all of the Yukawa couplings. Both of these amount to an extra assumption about the model.

We can see immediately that our naïvely simple dark matter model has a *minimum* of four parameters (m_{DM}, m_{med}, g_{DM} and g_q), and we also effectively have separate models for each combination of assumed particle spins. The classic way of generating benchmark models in this case is to treat each spin combination as a separate model, fix the couplings to some assumed values (e.g. $g_{DM} = 1$, $g_q = 0.25$), then generate different models in the m_{DM}-m_{med} plane. These models can be used to train the analysis optimisation, and also to present the final results in terms of an exclusion limit in that plane.

If even a simple model ends up looking complicated, how do we deal with a complicated model? The Minimal Supersymmetric Standard Model provides an excellent example, since it contains at least 24 Lagrangian parameters that can affect the LHC phenomenology, and we have relatively little knowledge about the allowed ranges of these parameters. If one instead views the model in terms of the masses and branching ratios of the new particles, there is an infinite array of options, each of which has dramatic effects on the LHC observables. The current standard approach to dealing with the problem is to use another kind of **simplified model**, where we assume that only particular supersymmetric partners will be produced, with fixed decays to specific particles. An example is shown in Figure 10.1, wherein one assumes that only charginos and neutralinos will be produced at the LHC, in specific combinations, and they will only decay to an electroweak gauge boson plus a lightest neutralino. It is typically assumed that the masses of the $\tilde{\chi}_2^0$ and $\tilde{\chi}_1^{\pm}$ particles are equal, and one can then generate different models in the $m_{\tilde{\chi}_2^0} - m_{\tilde{\chi}_1^0}$ plane on which to train analyses and present results. Each of these assumptions about masses and couplings has consequences, and we must also assume what the dominant content of the neutralinos and charginos is given that they are admixtures of different particles (this assumption will affect the production cross-section). By taking many different simplified models and training analyses on all of them, it is hoped that the LHC search programme will be able to discover supersymmetric particles no matter how Nature has chosen their masses and couplings. The reality, however, is that

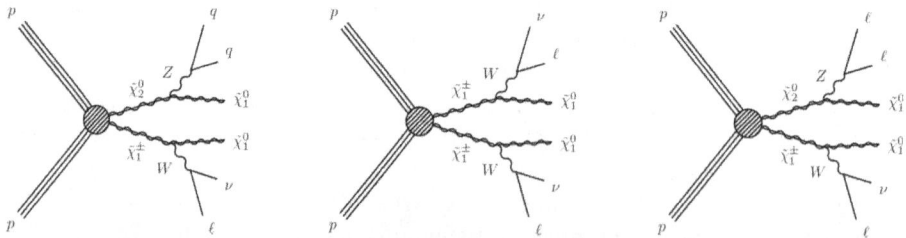

Figure 10.1. An example of a simplified model used for the optimisation of ATLAS and CMS supersymmetry searches. It is assumed that neutralinos and charginos are pair-produced, each decaying only to an electroweak gauge boson and a lightest neutralino. This results in final states with leptons, jets and missing energy.

over-optimisation on such models removes sensitivity to any models that differ from the underlying assumptions, and we must always strive to invent better search techniques that are less model-dependent.

10.1.3 Defining signal regions

Assuming that we have decided on our strategy for generating signal models, and we have a complete set of Monte Carlo samples to work with, we must now decide how to go about finding signal regions that are best-suited to the discovery of these benchmark points. This amounts to finding a set of selections on kinematic variables that removes as much of the background as possible, whilst retaining much of the signal. This optimisation must naïvely be done on each of our benchmark models separately, although we can expect points that are close in parameter space to be covered by similar signal regions. We thus might find that a few signal regions are sufficient to cover a large selection of our benchmark models.

The first part of any signal region selection is the **pre-selection** that selects the particular final state we are interested in. This works in a very similar fashion to the diphoton resonance search that we detailed in Section 9.2.1. We apply multiplicity cuts to select the final state objects, but must also carefully impose lower bounds on the p_T of the objects to ensure that we are on the plateau of the trigger that originally selected the events. Indeed, choosing which triggers to use is itself an important part of the search strategy, and it is often the case that a logical OR of various triggers will be more efficient than a simpler choice. Within the ATLAS and CMS collaborations, the pre-selection will also typically include so-called **event cleaning** cuts that are designed to remove non-collision backgrounds, or known detector issues. As a general rule, the pre-selection cuts should be as loose as possible, so that the starting point for our more aggressive optimisation is a selection that retains potential sensitivity to as many parameter points in our theory as possible.

Once the pre-selection cuts have been chosen, we are in a position to define our signal regions. All of the many hundreds of LHC searches for disparate phenomena within specific final states can be summarised by the following recipe: *Construct functions of the four-vector components of the final state objects in each event that provide maximum discrimination between the signal of interest and the dominant background processes, apply selections on one or more of these functions, and look in the regions of the data where the background is expected to be small.* In order to proceed, then, we need to know what sorts of functions to use for specific types of signal. Many of the variables that we have already met (e.g. in Section 8.3.2) are very useful for semi-invisible particle searches, such as:

E_T^{miss}: Models with extra invisible particles in, or even anomalous production of SM neutrinos, would be expected to give a different distribution of E_T^{miss} than that generated by the SM backgrounds within a specific final state. For heavy supersymmetric particles decaying to lightest neutralinos, the E_T^{miss} distribution is considerably broader than the backgrounds, making E_T^{miss} a very effective discriminant.

m_T: The transverse mass is useful for understanding W boson backgrounds (see the next section), and also for designing signal regions that target W boson production from BSM particles.

m_{T2}: The **stransverse mass** was originally designed for supersymmetric analyses, but is generally applicable to any case where a pair-produced object decays semi-invisibly. Recall that, if we assume a parent particle of mass m_p decays to a visible particle and an invisible one of mass m_χ, m_{T2} has a maximum value given by $m_{T2}(m_\chi) = m_P$ in the case that it is calculated using the right mass m_χ. If we assume instead, through lack of *a priori* knowledge, that the invisible particle is massless, m_{T2} instead has a maximum value of $m_{T2}(0) = \frac{m_P^2 - m_\chi^2}{m_P}$. A histogram of the m_{T2} distribution thus has a different endpoint for signal events than it has for background events, which can provide excellent discrimination between the two in some cases.

H_T: In supersymmetric analyses, it is very common to define some variant of H_T, as the scalar sum of the p_T of most of the interesting objects in the event. The details are analysis-dependent, but the fact that the variable is correlated with the energy scale of the hard process means that it has a broader distribution for events in which heavy BSM particles were produced, relative to SM backgrounds.

m_{eff}: Much like H_T, the scalar sum of the p_T of the interesting objects in the event, plus the E_T^{miss}, has a broader distribution for signal events than for background events, in the case where the signal events involve production of much heavier particles than we encounter in the SM. It is a very common variable in supersymmetric analyses in particular.

There is an entire industry devoted to inventing additional new variables for semi-invisible particle searches, and a few of the most popular are:

- $\frac{E_T^{miss}}{\sqrt{H_T}}$ and $\frac{E_T^{miss}}{m_{eff}}$: Both of these variables are attempts to normalise the E_T^{miss} to some meaningful scale of the total hadronic activity in the event (assuming that H_T and m_{eff} are defined using jets only). They often provide extra discrimination between multijet backgrounds and signal processes, beyond the use of E_T^{miss}, H_T and m_{eff} alone. When to use either is usually a case of trial-and-error, but it has been found that $\frac{E_T^{miss}}{\sqrt{H_T}}$ outperforms $\frac{E_T^{miss}}{m_{eff}}$ for final states with a low number of jets.

- $\Delta\phi(\text{obj}, p_T^{miss})$: This gives the polar angle between the direction of an object, and p_T^{miss}. In SM multijet events, p_T^{miss} often arises from jet mismeasurements, in which case the jets with the highest transverse momenta in the events are often closely aligned with the direction of p_T^{miss}. In searches for new physics, it is thus common to place a lower bound on $\Delta\phi(\text{jet}, p_T^{miss})$ for some number of the highest-p_T jets.

- α_T: The α_T variable is designed to characterise how close an event is to being a dijet event. This is particularly useful in semi-invisible searches in purely hadronic final states, since the dijet background is so large. First, all jets in the event are combined into two pseudo-jets in such a way that the difference in the E_T of the two jets is minimised. Then α_T is defined as:

$$\alpha_T = \frac{E_T^2}{\sqrt{2p_T^1 p_T^2 (1 - \cos \phi_{12})}}, \tag{10.1}$$

where E_T^i is the transverse energy of the i'th pseudojet, p_T^i is the transverse momentum of the i'th pseudojet, and ϕ_{12} is the polar angle between the two pseudojets. For a perfectly measured dijet event, $\alpha_T = 0.5$, and it is less than 0.5 if the jets are mismeasured. Events with $\alpha_T > 0.5$ either have genuine missing transverse energy (and are thus likely to be signal-like), or they are multijet events where some jets have fallen below the p_T threshold for the analysis. In practice, the α_T distribution for multijet events is very steeply-falling at $\alpha_T = 0.5$.

- m_{ll}: For events with two opposite-sign, same-flavour leptons in, it is straightforward to reduce the SM Z boson background by placing a selection on the dilepton invariant mass that excludes the Z peak.

- m_{CT}: This is another example of a variable whose distribution has a well-defined endpoint that is a function of the masses of the particles that we are observing. Imagine that a particular new particle δ can decay to two particles α and ν, where α is invisible, and ν is visible (i.e. a SM particle that interacts with the ATLAS and CMS detectors). The visible particles ν_1 and ν_2 are assumed to have masses $m(\nu_1)$ and $m(\nu_2)$, and we measure their four-momentum in some frame to be $p(\nu_1)$ and $p(\nu_2)$. Their invariant mass, formed from $p(\nu_1) + p(\nu_2)$ is invariant under any Lorentz boost that is applied to the two visible particle four-momenta simultaneously. However, we can also construct quantities which are invariant under equal and opposite boosts of the two visible particle four-momenta. That is, when we boost one particle by a particular boost, and we apply the equal and opposite boost to the other particle. It turns out that the following variable is invariant under these 'back-to-back' or 'contra-linear' boosts:

$$m_C^2(\nu_1, \nu_2) = m^2(\nu_1) + m^2(\nu_2) + 2[E(\nu_1)E(\nu_2) + p(\nu_1) \cdot p(\nu_2)] \tag{10.2}$$

This is very similar to the normal invariant mass formula, but with a plus sign instead of a minus sign on the three-momenta term. To give an example of where a contra-linear boost might be useful, imagine that the two particles ν_1 and ν_2 are produced by separate decays of δ particles that are pair-produced in an event. We will call these two particles δ_1 and δ_2, where the subscript is merely intended to distinguish the fact that two particles were produced. Then if we start in the $\delta_1\delta_2$ centre-of-mass frame, contra-linear boosts would be the boosts that would be required to put us in the δ_1 mass frame and the δ_2 mass frame. Then this formula would be applied after the boosts by taking the ν_1 four-momentum from the δ_1 rest frame, and taking the ν_2 four-momentum from the δ_2 rest frame. m_C would be useful if the $\delta_1\delta_2$ rest frame was the same as the laboratory frame, such as at an electron-positron collider, in which case it allows us to obtain a useful invariant of the produced system. At hadron colliders, however, we have an unknown boost along the beam direction that arises from the differing proton momentum fractions of the colliding partons in the event. The obvious solution is to refer only to quantities in the transverse plane, which are longitudinally boost-invariant. This finally gives us the **contransverse mass**:

$$m_{\mathrm{CT}}^2(v_1, v_2) = m^2(v_1) + m^2(v_2) + 2[E_{\mathrm{T}}(v_1)E_{\mathrm{T}}(v_2) + \boldsymbol{p}_{\mathrm{T}}(v_1) \cdot \boldsymbol{p}_{\mathrm{T}}(v_2)] \quad (10.3)$$

In the special case that $m(v_1) = m(v_2) = 0$ and there is no upstream momentum boost of the δ particles (from e.g. initial state radiation), the maximum value of m_{CT} is given by:

$$m_{\mathrm{CT}}^{\max} = \frac{m^2(\delta) - m^2(\alpha)}{m(\delta)} \quad (10.4)$$

Various formulations exist in the literature for correcting m_{CT} to be more robust under the effects of initial state radiation.

- In models such as supersymmetry, particles can undergo very complicated cascade decay chains, such as the squark decay

$$\tilde{q} \to \tilde{\chi}_2^0 \to q\ell^{\pm}\tilde{\ell}^{\mp} \to q\ell^{\pm}\ell^{\mp}\tilde{\chi}_1^0, \quad (10.5)$$

where q denotes a quark, ℓ denotes a lepton, and various supersymmetric particles feature in the decay chain. There are three SM decay products that result from this decay, and pair production of squarks would give us six products in total. It can be shown that the distributions of the invariant masses of different combinations of these decay products have kinematic endpoints given by functions of the sparticle masses in the cascade decay chain. If we define $\tilde{\psi} = m_{\tilde{\chi}_2^0}^2$, $\tilde{q} = m_{\tilde{q}}^2$, $\tilde{\ell} = m_{\tilde{\ell}}^2$ and $\tilde{\chi} = m_{\tilde{\chi}_1^0}^2$, we obtain the following endpoints:

$$(m_{\ell\ell}^2)^{\mathrm{edge}} = \frac{(\tilde{\psi} - \tilde{\ell})(\tilde{\ell} - \tilde{\chi})}{\tilde{\ell}}; \quad (10.6)$$

$$(m_{\ell\ell q}^2)^{\mathrm{edge}} = \begin{cases} \max\left[\dfrac{(\tilde{q} - \tilde{\psi})(\tilde{\psi} - \tilde{\chi})}{\tilde{\psi}}, \dfrac{(\tilde{q} - \tilde{\ell})(\tilde{\ell} - \tilde{\chi})}{\tilde{\ell}}, \dfrac{(\tilde{q}\tilde{\ell} - \tilde{\psi}\tilde{\chi})(\tilde{\psi} - \tilde{\ell})}{\tilde{\psi}\tilde{\ell}}\right]; \\ \text{except when } \tilde{\ell}^2 < \tilde{q}\tilde{\chi} < \tilde{\psi}^2 \text{ and } \tilde{\psi}^2\tilde{\chi} < \tilde{q}\tilde{\ell}^2 \\ \text{where one must use } (m_{\tilde{q}} - m_{\tilde{\chi}_1^0})^2 \end{cases} \quad (10.7)$$

$$(m_{\ell\ell q}^2)_{\max}^{\mathrm{edge}} = \max\left[\frac{(\tilde{q} - \tilde{\psi})(\tilde{\psi} - \tilde{\ell})}{\tilde{\psi}}, \frac{(\tilde{q} - \tilde{\psi})(\tilde{\ell} - \tilde{\chi})}{\tilde{\ell}}\right]; \quad (10.8)$$

$$(m_{\ell\ell q}^2)_{\min}^{\text{edge}} = \max\left[\frac{(\tilde{q}-\tilde{\psi})(\tilde{\psi}-\tilde{\ell})}{\tilde{\psi}}, \frac{(\tilde{q}-\tilde{\psi})(\tilde{\ell}-\tilde{\chi})}{2\tilde{\ell}-\tilde{\chi}}\right]. \tag{10.9}$$

Were we to discover supersymmetry at the LHC, these endpoints could be used to measure the masses of the sparticles (up to an overall mass scale). However, it is often very useful to repurpose these formula for SM cascade decay processes that involve successive two-body decays (such as top quark decay), where we can use them to localise a particular SM background in a variable that remains unconstrained for the signal process we are looking for.

- Recursive-jigsaw variables: The problem with many of the variables above is that they are highly correlated with each other. Thus, any analysis that uses several of them is using the same information multiple times, whilst ignoring other information that we might have about the events in a particular final state. One can think of the choice of variables in an analysis as a **dimensional reduction** of the original event information, and the aim should be to retain as much of the information in the original four-vectors of the final state objects as possible.

A modern approach is to use a basis of variables called 'recursive-jigsaw variables', which are uncorrelated by construction. In the **recursive-jigsaw reconstruction** (RJR) method, we attempt to guess an approximation to the rest frame of each of the intermediate particle states in an event. A natural basis of kinematic variables is then constructed by evaluating the energy and momenta of different objects in these reference frames. Although this guess of the rest frames will not be correct in every event, the defined variables are well-behaved approximations to the rest-frame behaviour *on average*, and this turns out to be very powerful for LHC semi-invisible particle searches.

The starting point for the RJR method is to assume a particular *decay tree*, or decay topology, for the events we are interested in. This is chosen to match the particular signal process that we are looking for. For example, imagine that we are looking for a case where two particles P_a and P_b are produced at the LHC, with each subsequently decaying to a visible system V_i and an invisible system I_i, for $i \in a, b$. We can visualise the decay tree as in the first panel of Figure 10.2, and it is typical of gluino pair production and squark pair production with subsequent decay to quarks and lightest neutralinos. The next step is to define a procedure for taking the final state objects in each event, and apportioning them to the visible systems V_a and V_b. For example, in a jets plus missing energy search, we must decide which jets are going to be assumed to form V_a, and which jets are going to be assumed to form V_b. Since we cannot know exactly which jets came from the decay of which new particle in an event, we have to come up with a rule which will get the right answer on average. A useful approach is to take the collection of the jet four-momenta in an event, $V \equiv p_i$ and their four-vector sum p_V, and then calculate each of the p_i in the rest frame of p_V (the V rest frame). Different partitions of the jets are then considered, such that each jet can only appear in either V_a or V_b, but all jets must be put into one of the two groups. For each partition, the sum of

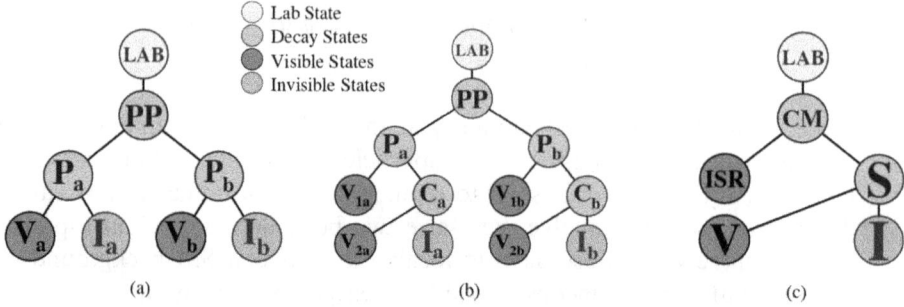

Figure 10.2. Examples of decay trees from a recursive-jigsaw analysis of sparticle production. Reproduced with permission from *Phys. Rev.* D 97 112001 (2018).

four-momenta in the V frame $p_{V_i} = \sum_{j=1}^{N_i} p_j$ is calculated for each group, where $i = a, b$, and there are N_i jets in group i. The combination that maximises the sum of the momentum of the two groups $|p_{V_a} + p_{V_b}|$ is chosen. This choice implicitly defines an axis in the V frame which is equivalent to the thrust axis of the jets, and the masses of each four-momentum sum $M_{V_i} = \sqrt{p_{V_i}^2}$ are simultaneously minimised.

The remaining unknowns in each event are associated with the two invisible systems, each of which represents a four-vector sum of the invisible particles produced by the decay of each new particle that was produced in the hard-scattering process. We do not know the masses of these particles, nor their longitudinal momenta, and we do not know exactly how they sum to form the single E_T^{miss} vector that was reconstructed for the event. The RJR algorithm determines these unknowns by iteratively minimising the intermediate particle masses that appear in the decay tree. This allows a guess of the rest frame to be chosen for each particle in the tree, and one can construct variables in each of these rest frames simply by boosting the observed momenta in the event to the frame, and constructing variables as functions of the boosted four-vectors. Generically, one can construct variables that have units of mass (equivalent to energy and to momentum in natural units where $c = 1$), and variables that are dimensionless, such as angles between objects. This gives a rich basis of largely uncorrelated variables that have proven to be very powerful in searches for semi-invisible particles. RJR can get complicated very quickly, but there is a standard library called RestFrames that can be used to define decay trees and their associated variables.

To demonstrate how we use this arsenal in practice, let us consider a particular example of a semi-invisible particle search. We will take the flagship, 0-lepton search for supersymmetry performed by the ATLAS experiment in 2014. Table 10.1 lists the various kinematic selections for four of the signal regions in the analysis. These signal regions are grouped first of all by the jet multiplicity, and then within each multiplicity by the relative strength of the kinematic selections ('l' means loose,

Table 10.1. Definition of the signal regions used in the 2014 flagship ATLAS supersymmetric search for squark and gluino pair production. Credit: *JHEP* **09** 176 (2014).

Requirement	Signal Region			
	2jl	2jm	2jt	3j
E_T^{miss} [GeV] >	160			
$p_T(j_1)$ [GeV] >	130			
$p_T(j_2)$ [GeV] >	60			
$p_T(j_3)$ [GeV] >	—	60		
$\Delta\phi(\text{jet}_{1,2,(3)}, \boldsymbol{p}_T^{miss})_{min}$ >	0.4			
$E_T^{miss}/\sqrt{H_T}$ [GeV$^{1/2}$] >	8	15	—	
$E_T^{miss}/m_{eff}(N_j)$ >	—	0.3		
m_{eff}(incl.) [GeV] >	800	1200	1600	2200

'm' means medium and 't' means tight). In all cases, events are vetoed if they contain electrons or muons with $p_T > 10$ GeV.

We can understand the logic of the analysis by working through the table row-by-row, and rationalising where each specific selection comes from. The selections in the first two rows are necessary to ensure that the events are taken from the region of the data where the trigger is fully-efficient. A detailed reading of the original analysis shows that events were required to have passed a 'jet-met' trigger, which requires a trigger-level jet with $p_T > 80$ GeV, and GeV. The trigger is fully efficient for offline selections of $E_T^{miss} > 160$ GeV and $p_T(j_1) > 130$ GeV, where it is assumed in the table that the jets are arranged in order of decreasing GeV. Note that the use of this trigger has already introduced some model-dependence to the analysis, since the analysis is already blind to any signal that would generate a smaller typical value of the the E_T^{miss}, or leading jets with lower p_T. Next, we see various selections on the p_T values of the other jets in the event, depending on how many are selected in the final state for each signal region. These will have been tuned carefully on the simulated benchmark signal and background models, to provide extra background rejection without substantially reducing the number of expected signal events.

The fifth selection in table 10.1 is on the polar angle between the leading jets and the $\boldsymbol{p}_T^{miss}$ vector. This significantly reduces the multijet background, for which the jets are frequently well-aligned with $\boldsymbol{p}_T^{miss}$, since the $\boldsymbol{p}_T^{miss}$ arises from jet mismeasurement. Each of the jets in the signal regions must have $\Delta\phi(\text{jet}_{1,2,(3)}, \boldsymbol{p}_T^{miss})_{min} > 0.4$.

Finally, we see selections on various quantities that provide most of the discrimination between squark and gluino production and the SM backgrounds. $E_T^{miss}/\sqrt{H_T}$ is the main discriminant variable for squark pair production, which is expected to populate the two-jet signal regions. For gluino pair or squark-gluino production, $E_T^{miss}/m_{eff}(N_j)$ was found to be more effective in Monte Carlo studies, where the m_{eff} definition for each signal region depends on the number of jets selected for that region. There is an additional selection on an inclusive definition of m_{eff} that is the same in all signal regions. This is found to enhance the significance for squark and gluino production through its sensitivity to the overall mass scale of the particles produced in the hard scattering process of an event.

Although every analysis at the LHC is different, and some are much more complex than this example, all searches will resemble this one in terms of the basic approach. When designing a search, one should start from the minimum set of selections consistent with the trigger conditions, then proceed to develop selections that are designed to reduce each dominant reducible background, whilst tuning the precise selection cuts on the Monte Carlo samples available.

10.2 Developing accurate background models

The act of choosing signal regions is equivalent to identifying the regions of the data where we think a major discovery might be lurking. In principle, the only remaining step is to look in those regions of the data and see if the number of events we count is consistent with the SM expectation or not. The total expected number of SM events is the sum of the individual background contributions, and we already have Monte Carlo samples available for these backgrounds. Thus, a naïve approach to particle searches at the LHC would be to simply compare our total Monte Carlo background yield in the signal regions with the observed number of events in LHC data.

Unfortunately, this simple approach is very naïve indeed, and would have disastrous consequences in many cases. The simple fact is that the Monte Carlo yields alone are not accurate enough to fully model the LHC data, which is sometimes due to deficiencies of the Monte Carlo generators themselves, and sometimes due to deficiencies of the detector simulations used to model the ATLAS and CMS detectors. It is therefore necessary to develop refined background models for each of the SM processes that we expect to contribute to the signal regions, and only then is it safe to look at the LHC data and decide whether we have made a new discovery. Until we are satisfied that our background models are adequate, we must absolutely refrain from looking in the signal regions, which is referred to as a blind analysis strategy.

For many backgrounds, the Monte Carlo samples accurately model the shape of the kinematic variables that are used to define the signal regions, but the overall normalisation might be off. This is typical of backgrounds where the final state objects from the hard-process match the final state used in the signal region definition (ie. the processes do not enter the signal acceptance due to the presence of fake leptons, for example). In this case, the standard way to refine our Monte Carlo model for a particular background process is to define a region of the data, called a **control region** that we expect to be dominated by that background. We can then define a normalisation factor that corrects for any discrepancy between the observed data and the Monte Carlo in the control region. This is called a **data-driven background estimate**, and the yield in the signal region is given by

$$N(\text{SR, scaled}) = N(\text{SR, unscaled}) \times \frac{N(\text{CR, obs})}{N(\text{CR, unscaled})}, \qquad (10.10)$$

where $N(\text{CR, obs})$ is the observed number of events in the control region, $N(\text{SR, unscaled})$ is the Monte Carlo yield in the signal region, and $N(\text{CR, unscaled})$ is the Monte Carlo yield in the control region. The normalisation factor can also be applied to histograms of kinematic variables, by scaling the whole histogram by the constant normalisation factor. This formula makes it clear that we are correcting the Monte Carlo yield for the background in the signal region by the ratio of the observed and Monte Carlo yields in the control region, and is the simplest way to picture what is going on. The dominant alternative in the literature, however, is to regroup the terms to give

$$N(\text{SR, scaled}) = N(\text{CR, obs}) \times \left[\frac{N(\text{SR, unscaled})}{N(\text{CR, unscaled})} \right], \qquad (10.11)$$

where the quantity in square brackets is called the **transfer factor**, and it is common to refer to it as a quantity that extrapolates the observed control region yield to the signal region. In reality, transfer factors are not usually defined by hand, but are floated in a combined likelihood fit that compares the Monte Carlo background yields with the observed yields in all signal and control regions simultaneously. This allows for a consistent normalisation of each background process across all defined regions, since the likelihood fit must balance the contribution of all processes in all regions to match the observed data. The fit derives both the central value for each transfer factor, and an associated uncertainty.

Choosing good control regions for SM background processes is an art in itself, and arguably constitutes most of the effort in performing an analysis within an experimental collaboration. As a general rule, one should aim to make a control region for each dedicated signal region, also trying to ensure that the kinematic selections in each control region are as similar as possible to their corresponding signal region. This minimises systematic uncertainties on the transfer factor that result from extrapolating from one region in 'kinematic variable space' to a wildly different region. However, if the kinematic selections for a control region hardly differ from those of its signal region, then we should expect it to be dominated by the signal rather than the background process we are targetting. This is called **signal contamination**, and it is very dangerous for the following reason. If a control region had an excess related to the presence of a signal, our data-driven background procedure would increase the normalisation of the relevant background, and 'calibrate away' any observed excess in the signal region. To guard against this, it is common to test all control region definitions with the generated signal Monte

---

[2] The fact that we do not know exactly what any potential signal at the LHC will look like, making any attempt to test signal contamination doomed to failure, is usually politely ignored. At least we can say that the particular signal we are optimising on does not contaminate the control regions. Any more general solution would require a model-independent analysis technique.

Table 10.2. Examples of control regions used in the 2014 flagship ATLAS supersymmetric search for squark and gluino pair production. Credit: *JHEP* **09** 176 (2014).

CR	SR background	CR process	CR selection
CRW	$W(\to l\nu)$+jets	$W(\to l\nu)$+jets	1 $e^{\pm}$ or $\mu^{\pm}$, 30 GeV $< m_{\mathrm{T}}(l, E_{\mathrm{T}}^{\mathrm{miss}}) < 100$ GeV, b-veto
CRT	$t\bar{t}$ and single-t	$t\bar{t} \to b\bar{b}qq'l\nu$	1 $e^{\pm}$ or $\mu^{\pm}$, 30 GeV $< m_{\mathrm{T}}(l, E_{\mathrm{T}}^{\mathrm{miss}}) < 100$ GeV, b-tag

Carlo samples to assess the level of signal contamination, tweaking the control region selections as necessary until the contamination is reduced[2].

To illustrate effective control region design, table 10.2 contains two of the control region definitions from the supersymmetry search that we referred to above. These regions were defined for each of the signal regions in the analysis, and in each case the selection on the inclusive m_{eff} variable was retained from the signal region definition. The other selections were modified to define regions dominated in particular backgrounds. The first row gives details of a control region for the W+jets background, which involves selecting events with 1 electron or muon in, imposing a b-tag veto (since b-jets are expected only rarely in W+jets events), plus restricting the transverse mass in events to lie between 30 GeV and 100 GeV. This selects a sample that is overwhelmingly dominated by W+jets events where the W boson has decayed leptonically, creating real missing transverse energy from the escaping neutrino. Furthermore, the transverse mass is well known to be concentrated in this range for leptonic W boson events. The dominant W+jets contribution in the signal regions instead arises from events where the W boson decays to a tau neutrino and a hadronically-decaying tau lepton, which gives events with jets plus missing transverse energy. To use the control region events to model this, we can simply pretend that the electron or muon in our control sample is a jet with the same four-momentum. This is a very neat trick for defining the control region, since the 1 lepton selection makes it very distinct from the signal region, but with similar kinematics for the final state objects. The second row of table 10.2 contains selections for a control region that selects both top pair and single top events. Once again, single lepton events are selected, and are used to model the fully hadronic top decay background by pretending that the lepton is a jet with the same four-momentum. The b-veto requirement of the W control region is changed to a b-tag, which preferentially selects events with a top quark (since top quarks will decay to a b quark and a W boson).

Control regions give us the means of refining our Monte Carlo estimates of each background contribution, but ideally we should have a way of testing that our background models accurately reproduce the LHC data before we look in the signal region. Otherwise, we will remain unsure that any discrepancy observed in the signal region actually results from new physics. The trick is to use **validation regions** which are distinct from both the signal and control regions, whilst being carefully chosen to reduce signal contamination. Defining validation regions is often difficult, since any insight we had on how to enrich a given region with particular background events has usually already been used in defining the relevant control region. For the control regions in table 10.2, various validation regions were defined. The first took events from the W and top control regions and reapplied the $\Delta\phi(\text{jet}_{1,2,(3)}, p_T^{\text{miss}})_{\text{min}}$ and $E_T^{\text{miss}}/\sqrt{H_T}$ or $E_T^{\text{miss}}/m_{\text{eff}}(N_j)$ selections that were missing from the control region definitions. This makes the validation region much more similar to the signal region, whilst still remaining different due to the 1 lepton selection. The lepton was then either treated as a jet with the same four-momentum or, in a separate set of validation regions, treated as contributing to the E_T^{miss}. Extra validation regions were obtained by taking regions in which at least one hadronically decaying τ lepton was reconstructed, with a separate b-tag requirement used to separate the W from the top validation region. When designing an analysis, it is worth doing a thorough literature review of analyses in a similar final state in order to learn strategies for effective validation region design.

So far, we have assumed that the Monte Carlo samples we obtained at the start of our analysis process provides an accurate description of the shape of each background contribution, and the only correction we need to apply is to the overall normalisation. For some backgrounds, however, this is woefully inadequate. The classic example is that of the multijet background, for pretty much any analysis that requires a moderate missing transverse energy cut. It requires a vast amount of computing time to generate enough multijet events to populate the typical signal and control regions we encounter in these searches, to the extent that reliable Monte Carlo background models are not available. Thankfully, we already have a much better generator of multijet events, which is the Large Hadron Collider itself! Since, to first order, every event at the LHC is a multijet event, with everything else being a tiny correction, we can instead define a purely data-driven approach to modelling the multijet background that removes the need for Monte Carlo generators at all. For the ATLAS supersymmetry analysis, a **jet-smearing** method was used, defined as follows:

(1) 'Seed' jet events are first defined by taking events that pass a variety of single jet triggers with different p_T thresholds, and the selection $E_T^{\text{miss}}/\sqrt{E_T^{\text{sum}}} < 0.6 \text{ GeV}^{1/2}$ is applied, where E_T^{sum} is the scalar sum of the transverse energy measured in the calorimeters. This selection ensures that the events contain well-measured jets.

(2) A **jet response function** is defined using Monte Carlo simulations of multijet events, by comparing the reconstructed energy of jets with the true energy. This function quantifies the probability of a fluctuation of the measured p_T of a jet, and takes into account both the effects of jet mismeasurement and the contributions from neutrinos and muons that are produced from the decay of jet constituents of heavy flavour. To do this, 'truth' jets reconstructed from generator-level particles are matched to detector-level jets within $\Delta R < 0.1$ in multi-jet samples. The four-momenta of any generator-level neutrinos in the truth jet cone are added to the four-momentum of the truth jet. The response is then given as the ratio of the reconstructed jet energy to the generator-level jet energy.

(3) Jets in the seed events are convoluted with the response function to generate pseudo-data events. These are compared to real multijet events in a

dedicated analysis, and the response function is adjusted until the agreement between the pseudo-data and real data is optimised.

(4) The seed jet events are convoluted with the final jet response function that emerged from step (3). This provides a final sample of pseudo-multijet events.

Once this method is complete, the pseudo-multijet events can be treated the same as a Monte Carlo sample. A dedicated multijet control region was defined for the ATLAS supersymmetry analysis, in which the signal region requirements on $\Delta\phi(\text{jet}_{1,2,(3)}, \boldsymbol{p}_{\text{T}}^{\text{miss}})_{\text{min}}$ and $E_{\text{T}}^{\text{miss}}/\sqrt{H_{\text{T}}}$ or $E_{\text{T}}^{\text{miss}}/m_{\text{eff}}(N_{\text{j}})$ were inverted. The second of these requirements could then be reinstated to define a corresponding validation region.

In other analyses, fully data-driven background estimation methods are typically used to measure backgrounds that contain fake objects, such as fake leptons or fake photons. Both ABCD and matrix method approaches are common in the literature.

We are now at the point where we can state the number of events expected in the signal region for each of the SM background processes. However, it is equally important to attach an uncertainty to each contribution, which combines both the **statistical uncertainty** related to the finite number of Monte Carlo events generated, and a long list of **systematic uncertainties**, a detailed inventory of which was provided in Section 8.6.2. As we saw there, for systematic uncertainties, an experimentalist typically has to run their analysis code several times, each time using a particular variation of systematic quantities within their allowed ranges such as the jet energy, resolution and mass scales, the resolution and energy scale for leptons and photons (plus their reconstruction efficiencies), the scale of missing energy contributions, the efficiency for tagging b jets, the trigger efficiencies and the overall luminosity. Theoretical systematic uncertainties can be defined by, for example, varying cross-sections within their allowed uncertainties, comparing different Monte Carlo generator yields, varying the QCD renormalisation and factorisation scales for a Monte Carlo generator, and varying the parton distribution functions. These variations define a series of predicted yields in each control, validation and signal region, which can be used as inputs to likelihood fits. One can also define bin-by-bin uncertainties for kinematic distributions in each region, before using these to define systematic uncertainty bands on histograms of the variables that were used for the particle search.

10.3 Comparing the observed LHC data with background models

Having now defined accurate background models with uncertainties for each SM background contribution, it really is the case that we can look in the LHC data and see if the observed signal region yields are consistent with the SM expectation. In practice, this requires performing a statistical procedure that is able to quantify the strength of agreement with the data, and we must therefore rely on the hypothesis testing methods from Chapter 5. Frequentist methods are the standard approach for semi-invisible particle searches, and we must therefore define a likelihood that encapsulates the comparison of our background models with the observed data, before maximising that likelihood and defining confidence intervals on its free parameters.

A typical likelihood for an LHC semi-invisible particle search includes some parameters of interest (e.g. the assumed supersymmetric particle masses, or the rate of a generic signal process), the normalisation factors for the background processes, and a nuisance parameter θ_i for each systematic uncertainty i. The systematic parameters can be defined such that $\theta_i = \pm 1$ corresponds to the $\pm 1\sigma$ variations in the systematic uncertainties, whilst $\theta_i = 0$ corresponds to the nominal yield. One can then construct a total likelihood that is a product of Poisson distributions of event counts in the signal and control regions, and of additional distributions that implement the systematic uncertainties. This will be given by a function $\mathcal{L}(n, \theta^0 | \mu, b, \theta)$, where n is a vector containing the observed event yields in all signal and control regions, b contains the predicted total background yields in all signal and control regions, and μ contains the parameters of interest. θ^0 contains the central values of the systematic parameters (which, like n should be considered as observed quantities), whilst θ contains the parameters of the systematic uncertainty distributions. Assuming for simplicity that we have only one signal region, a suitable choice for $\mathcal{L}$ is

$$\mathcal{L}(n, \theta^0 | \mu, b, \theta) = \mathcal{P}(n_S | \lambda_S(\mu, b, \theta)) \times \prod_{i \in CR} \mathcal{P}(n_i | \lambda_i(\mu, b, \theta)) \times C_{\text{syst}}(\theta^0, \theta), \quad (10.12)$$

where n_S is the observed yield of events in the signal region, n_i is the yield of events in the i'th control region, and $C_{\text{syst}}(\theta^0, \theta)$ is an assumed distribution for the systematic uncertainties. It is common to take this to be a multidimensional Gaussian. The expected number of events for each Poisson distribution is given by a function $\lambda_j(\mu, b, \theta)$, which contains the details of the transfer factors between the control regions and the signal region, and between control regions. For example, imagine that we had a single parameter of interest μ which gave the strength of a hypothetical signal in units of the nominal model prediction. We could then write

$$\lambda_S(\mu, b, \theta) = \mu \cdot C_{\text{SR}\rightarrow\text{SR}}(\theta) \cdot s + \sum_j C_{j\text{R}\rightarrow\text{SR}}(\theta) \cdot b_j, \quad (10.13)$$

and

$$\lambda_i(\mu, b, \theta) = \mu \cdot C_{\text{SR}\rightarrow i\text{R}}(\theta) \cdot s + \sum_j C_{j\text{R}\rightarrow i\text{R}}(\theta) \cdot b_j, \quad (10.14)$$

where the index j runs over the control regions. The predicted number of signal events for our nominal model is given by s, and the predicted yields in each control region are given by b_j. C is the matrix of transfer factors, whose diagonal terms $C_{\text{SR}\rightarrow\text{SR}}$ are equal to unity by construction, and whose off-diagonal elements contain the transfer functions between the different regions (control regions and signal region) defined in the analysis.

Hypothesis testing can then be performed as in Chapter 5 by defining the profile log-likelihood test statistic,

$$q_\mu = -2\log\left(\frac{\mathcal{L}(\mu, \hat{\hat{\boldsymbol{\theta}}})}{\mathcal{L}(\hat{\mu}, \hat{\boldsymbol{\theta}})}\right).$$ (10.15)

In fact, it is typical to use three different sorts of likelihood fit in the analysis of semi-invisible particle search data. The first is a **background-only fit** which seeks to determine how compatible the SM background expectation is with the observed event yield in each signal region. In this case, the likelihood fit is performed without using the observed yields in the signal regions as constraints, but only the yields in the corresponding control regions. It is then assumed that signal events from new physics do not contribute in the signal regions. The significance of any excess over the background that is observed in the signal regions can then be quantified by calculating the probability (as a one-sided p-value p_0) that the signal region event yield obtained in a single hypothetical background-only experiment is greater than that observed in the real LHC data. This background-only fit is also used to estimate the background yields in the validation regions, which gives a check on the validity of the SM background measurements before the signal yields are unblinded.

If no excess is observed, a second likelihood fit, called a **model-independent fit**, can be used to set 'model-independent' upper limits on the number of BSM signal events that can contribute to each signal region. These limits do not presuppose any particular new physics model, but are instead related to the fact that the observed yields, the predicted background yields, and the total uncertainties on the background yields set a limit on the total number of non-SM events that could be present whilst still remaining consistent with the observed data. It is common to use the $\mathrm{CL_s}$ prescription of Section 5.8.1 for deriving the limits, and to state them at the 95% confidence level, and it is important to run Monte Carlo pseudo-experiments rather than relying on asymptotic formulae to ensure that the limits have the correct coverage properties. When normalised by the integrated luminosity of the data sample, these limits can be interpreted as upper limits on the visible cross-section of a new physics model (defined as the product of the production cross-section for the new physics model with the acceptance and reconstruction efficiency). The model-independent fit differs from the background-only fit only in the fact that the number of events observed in each signal region is added as an input to the fit, and a single signal strength parameter μ is added as a parameter of interest. It is assumed that signal contamination of the control regions does not occur, since there is no way of determining the strength of the signal contamination in the absence of a specific physics theory.

A third class of likelihood fit uses actual model information, typically utilising the Monte Carlo samples for the particular model of interest. For example, in the supersymmetric case described above, one could perform a likelihood fit that uses a specific supersymmetric benchmark point to set an upper limit on the signal cross-section *for that particular model*. This proceeds in the same way as the model-independent fit, except that the Monte Carlo yield in each region can be scaled by the signal strength parameter. This in turn gives us a way to model the signal

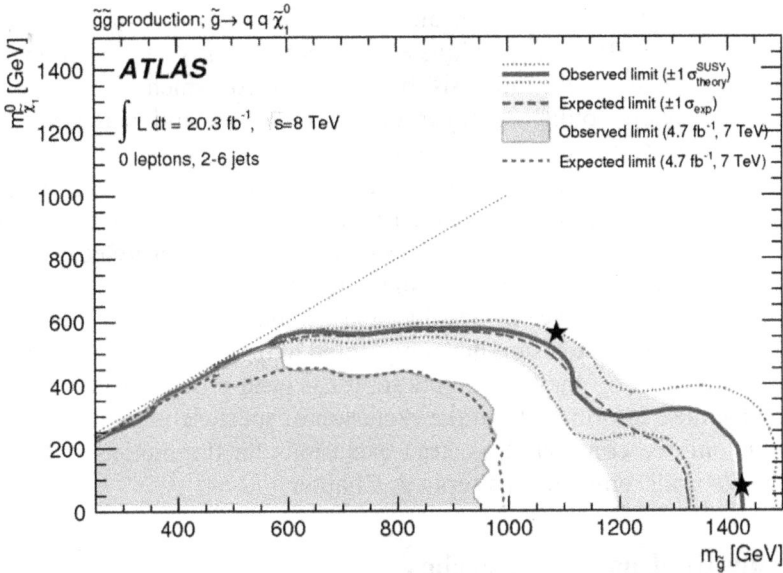

Figure 10.3. An example of a simplified model limit for a supersymmetry search, generated by performing separate hypothesis tests on benchmark Monte Carlo models in the parameter plane. Reproduced with permission from *JHEP* **09** 176 (2014).

contamination of the control regions, and take into account the experimental and theoretical uncertainties on the supersymmetric production cross-section and kinematic distributions, and the effect of correlations between the signal and background systematic uncertainties. Repeating this for different benchmark models leads to **exclusion plots** such as that shown in Figure 10.3, in which model points have been generated for a simplified model of gluino pair production, with all gluinos decaying to two quarks and a lightest neutralino. Different models are then given by different choices of the gluino mass and the lightest neutralino mass. The **observed limit** is calculated from the observed signal region yields for both the nominal signal cross-section, and the red line delineates the region for which points are excluded at the 95% confidence level (points below the line are excluded). This line can be determined by, for example, interpolating CL_s values in the plane. The uncertainty band on the red line is obtained by varying the signal cross-sections by the renormalisation and factorisation scale and parton distriibution function uncertainties. The **expected limit** is calculated by setting the nominal event yield in each signal region to the corresponding mean expected background, whilst the yellow uncertainty band on it shows how the expected limit would change due to $\pm 1\sigma$ variations in the experimental uncertainties. A major discrepancy between the observed and expected limits (which has not occurred in this case) might indicate evidence for an excess consistent with the region of the parameter space where the deviation is observed. In plots like these, it is common to use the signal region with the best expected significance to determine the exclusion at each point in the plane, rather than combining signal regions with an appropriate treatment of the correlations between them. Note also that previous experimental results have been included to show the advantage of the analysis being presented which, in this particular example, is largely due to the higher centre-of-mass energy and increased integrated luminosity relative to the previous analysis.

For supersymmetric searches in particular, plots like that shown in Figure 10.3 are ubiquitous, and there are even summary plots that attempt to collapse all of the observed limits into one figure. All such plots should be interpreted with extreme caution, however. Each of the 2D simplified model planes is a vanishingly thin slice of the large dimensional space of possible sparticle masses and couplings, and there is no guarantee that the exclusion limits presented in the simplified model planes will persist as you wander off the planes. Indeed, it has been shown repeatedly that these limits provide false information on the exclusion of sparticle mass ranges, and they should thus only be considered as strict exclusions on the models that are being presented. We shall return to this topic in Chapter 12.

10.4 Long-lived particle searches

The preceding sections have all dealt with the case that the new particles produced at the LHC decay promptly, such that their decay products emerge from the interaction point. This implies that all tracks in the event point back to this interaction point, within the normal bounds expected within the SM. However, many theories predict that particles will be metastable on detector timescales, for a variety of reasons. There might be a symmetry of a new physics model that forbids a particular new particle decay, but it might only approximately hold so that it manifests itself as an increased particle lifetime. Alternatively, a new particle might have small couplings to the lighter states that it can decay to, or there might a phase space suppression due to, for example, a small mass difference between the particle and its decay products. It is common to use $c\tau$ to characterise the behaviour of long-lived particles, where τ is the proper lifetime of the particle. Thus, $c\tau$ is the proper decay length of the particle, which can be mapped to how far the particle will typically travel in the ATLAS or CMS detectors in the laboratory frame before decaying. We can distinguish different qualitative kinds of long-lived particle signature based on both the value of $c\tau$, and the couplings of the new particle with SM particles.

The first, and simplest, case occurs if $c\tau$ is large enough that the particle leaves the detector without decaying, and its properties are such that it does not interact on its way out. In this case, the phenomenology resembles prompt production, and we can simply apply the semi-invisible search techniques that we developed above[3].

---

[3] Note that this tells us that we can never use the LHC alone to distinguish a completely stable weakly-interacting particle from a meta-stable one, which ultimately means that the LHC alone cannot unambiguously discover a dark matter candidate. This is independent of the need to correlate LHC observations with astrophysical measurements, which is the only means of testing whether a particle produced at the LHC can provide the dark matter that we see in our universe.

More interesting behaviour occurs if $c\tau$ leads to a particle decay *within the detector itself*. What happens then depends on the precise properties of what we now call a **long-lived particle**. If the particle interacts electromagnetically, it may leave a track in the inner detector, which has unusual ionisation and propagation properties. If it does not, we may instead see localised deposits of energy inside the calorimeters, without associated tracks. Particles that interact with either of the calorimeters, and for which $c\tau$ is high enough that they reach them, may stop in the calorimeters. They might then decay in a different bunch-crossing, which would lead to very confusing and unusual events. Another set of options results from particles whose $c\tau$ value means that they typically decay within the inner detector, with possible signatures including displaced vertices in the inner detector, or disappearing, appearing or kinked tracks. Longer-lived particles might instead give odd track signatures in the muon spectrometer.

Long-lived particle searches are without doubt the hardest analyses to perform, since they often require a complete rethink of how object reconstruction works. In addition, the strange signatures of long-lived particles can frequently resemble noise, pile-up or mis-reconstruction of standard detector objects, none of which are typically well-modelled by Monte Carlo simulations. Understanding the background to a long-lived particle search thus often requires an encyclopedic knowledge of possible detector anomalies, and a willingness to implement and drive changes in the reconstruction software.

Further reading

- The ATLAS supersymmetry search that we used as an example is Search for squarks and gluinos with the ATLAS detector in final states with jets and missing transverse momentum using $\sqrt{s} = 8TeV$ proton-proton collision data *JHEP* **09** 176 (2014). We have also made use of an updated analysis that used recursive jigsaw reconstruction, detailed in *Phys. Rev.* D **97** 112001 (2018).
- The RestFrames package can be found at restframes.com, and more information on recursive jigsaw reconstruction can be found in *Phys. Rev.* D **96** 11 112007 (2017).
- A useful resource on the likelihood treatment for ATLAS semi-invisible particle searches is Baak M *et al*, HistFitter software framework for statistical data analysis *Eur. Phys. J.* C **75** 153 (2015).
- We have barely had space to scratch the surface of long-lived particle searches. A comprehensive recent review be can found in arXiv:1903.04497.

Exercises

10.1 A search is performed for smuon pair production at the LHC, in which each smuon is expected to decay to a muon and a lightest neutralino.
  (a) If the lightest neutralino is assumed to be massless, what is the endpoint of the m_{T2} distribution for the supersymmetric signal process?

   (b) A dominant background is expected to arise from WW production, with both W bosons decaying leptonically. What is the endpoint of the m_{T2} distribution for this process, calculated assuming that the invisible particle is massless?

   (c) What condition on the smuon and lightest neutralino masses results from imposing the constraint that the m_{T2} distribution for the signal process must exceed the endpoint of the m_{T2} distribution for the background process?

10.2 It is proposed to search for semi-invisible decaying new particles that are expected to produce a final state with three charged leptons plus missing transverse energy from escaping new particles. The model further predicts that at least two of the charged leptons will have the same flavour and opposite charge. Determine the dominant SM backgrounds for the analysis.

10.3 Prove that m_C is invariant under contra-linear boosts.

10.4 A top quark can decay via the cascade decay chain $t \to bW \to bl\nu$, where l is a charged lepton.

   (a) What is the maximum value of the invariant mass m_{bl}?

   (b) How might this be used in a search for a new particle that also produces a b quark and a charged lepton via subsequent two-body decays, assuming that the invisible particle in the new particle decay has a non-zero mass?

10.5 Simplified models are popular in searches for supersymmetric particles.

   (a) Explain what a simplified model is in the context of a supersymmetric particle search.

   (b) Explain how simplified models are used in analysis *optimisation*. What is the main deficiency of this approach?

   (c) Explain how simplified models are used in the presentation of analysis results. What is the main deficiency of this approach?

10.6 In a particular LHC search, the number of observed events in a control region is 107, the predicted number of Monte Carlo background events in the control region is 91.2, and the predicted number of Monte Carlo events in the signal region is 10.7.

   (a) Explain why the predicted Monte Carlo yields are not integers.

   (b) What is the transfer factor for extrapolating events from the control region to the signal region?

10.7 Assume that the stop quark always decays to a top quark and a neutralino. Explain how a 0 lepton, 1 lepton or 2 lepton final state can arise from stop pair production and decay.

IOP Publishing

Practical Collider Physics

Andy Buckley, Christopher White and Martin White

Chapter 11

High-precision measurements

Now that we have explored the two basic types of direct particle search at the LHC, let us continue on to consider the topic of measurements. We will primarily distinguish such measurements from the analyses considered in the previous chapters by their attempts to correct for biases introduced by the detection process, and/or to fully reconstruct decayed resonances. These are not absolute conditions, and there are undoubtedly searches which contain these elements, as there are measurements which do not—but for our purposes here, it is a useful line to draw.

Analyses as described so far in this book operate in terms of detector-level observables: fundamentally, the numbers of collider events found to have reconstructed properties that fall into binned ranges or categories. Statistically there is no uncertainty on these event counts: a fixed integer number of events will be found in each reconstruction-level bin, and the observed yields can (via the use of Monte Carlo (MC) and detector simulation tools) be used to constrain the masses, couplings or signal-strengths of new-physics models. For such analyses, what matters most is that the expected significance of the analysis to a particular choice of new physics model is maximised. The limiting factors on such analyses arise from the statistical uncertainties due to the finite sample size, and systematic uncertainties from defects both in the model predictivity and the detector's resolutions and biases.

By contrast, a 'precision measurement' analysis is motivated to make the best possible estimation of how the observed events were really distributed, usually without hypothesis testing of any particular model in mind. Most notably, the imperfect detector has no place in such estimates: we want to know what would have been observed not at ATLAS or at CMS specifically, but by anyone in possession of a detector with known and correctable reconstruction inefficiencies and biases. In a sense, what we want to know is what would have been observed by an experiment with a *perfect* detector. As the integrated luminosity is also specific to the experiment, we typically also divide our best estimate of this out of the inferred event yields (and propagate its uncertainty) so that the analysis target is now a set of total cross-sections $\sigma \sim N/\epsilon A L_{\mathrm{int}}$, differential cross-sections of the form $\mathrm{d}\sigma/\mathrm{d}X$ in some variable X, or combinations of them into intrinsic, more-or-less directly observable properties of nature such as cross-section ratios σ_A/σ_B, or asymmetries $(\sigma_A - \sigma_B)/(\sigma_A + \sigma_B)$.

In a sense, this changes nothing: we have merely shifted the target of our statistical inference from fundamental model parameters to the detector-independent values of observables. But the focus on physically meaningful observables, and the specific nature of inverting detector effects affect the nature of the analysis. It may seem that by insisting on idealised observables as an intermediate point between reconstruction-level observations and constraining model parameters, we are necessarily weakening the power of our model-specific inferences. This has historically been true, but it has traditionally been due to defects in the data-publishing process rather than fundamental limitations. With sufficiently complete publication of the inference model and measurement uncertainties, and use of a complete enough observable set, detector-independent measurements can have as much model-constraining power as their more direct, reconstruction-level counterparts described in Chapters 9 and 10.

In this chapter we will consider specific methods for **unfolding** detector effects, for explicit reconstruction of decayed particle resonances, and for optimal combination of several semi-independent measurements of the same observable, and in Chapter 12 we consider again how best to publish measurements of all sorts for *post hoc* re-interpretation against arbitrary new physics models.

11.1 Fiducial definitions, volumes and cross-sections

We have dealt so far in reconstruction-level quantities: event counts and the resulting effective cross-sections in terms of physics objects as seen by a particular detector experiment. Using the machinery to be discussed in Section 11.3, we could correct for these detector effects to obtain instead a detector-independent estimate at **particle level**. We could even go further and correct for the annoying effects of low-scale quantum chromodynamics (QCD), subtracting away the pesky underlying event, hadronisation, and so on, and eventually arrive at a partonic full-phase-space cross-section. Doing so introduces various modelling artifacts, however, and so the central object for the extraction of detector-independent observables is not a partonic full-phase-space cross-section, but a **fiducial definition** of the measurement. The origins of this rather baroque term are unimportant and indeed confusing, so we instead focus on it as currently understood: it gives the observables as they would be measured by a perfect version of the real detector. That is, a detector without reconstruction inefficiencies or biases. This idealised experiment becomes the target of the data analysis, the aim being to map the uncertainty-free set of real observations on to a probability-density map of their corresponding fiducial observables.

It is crucial, however, that we do not idealise too far: if our detector has only finite geometric acceptance, or the trigger and reconstruction performance is unusably low below e.g. some p_{T} threshold, we should not extrapolate our perfect detector to have perfect coverage where the real one had effectively none. This picture corresponds to a closed region of phase-space highly overlapping with the performant regions of the real detector, within which physics object performance is perfect: this is the **fiducial volume**. Its power is that it defines a detector-independent class of measurement in which the assigned probabilities are dominated by the observations and understanding of the contributing experiment, rather than any specific physics model. By contrast, any extrapolation outside the fiducial volume is necessarily dependent on assumptions about things that could not be directly observed and in most cases will either be highly model-dependent or highly uncertain. The motivating principle of fiducial measurements, above all others, is not to extrapolate, and hence to make the measurement as precise and model-independent as possible, based on what the experiment was actually able to observe[1]. Theory predictions made more inclusively than the detector could have seen, or at parton level, require the application of separately estimated cut efficiencies and/or **non-perturbative corrections** for comparison to a fiducial measurement.

Use of the fiducial volume in model interpretations from detector-independent observables hence minimises the risk of contamination by such problematic regions. Fiducial volumes restricted beyond the essential level imposed by detector limitations can also be useful, as they permit comparison (and combination) of several experiments' measurements of the same fiducial observables within a commonly accessible phase-space.

The exact fiducial volume is analysis-specific: an analysis measuring only inclusive jets naturally has access to a larger fiducial volume than one using the same detector which needs restricted-acceptance lepton or jet flavour-tag reconstruction. In practice, fiducial observables are implemented in terms of event analysis on simulated collider events via an MC event generator, such that the fiducial volume is expressed via the fraction of inclusive events to pass the analysis selection cuts. As these particle-level events can be passed through the detector simulation and reconstruction chain, and a closely related reconstruction-level analysis performed, a one-to-one mapping of simulated events can be obtained, allowing at least in principle the construction of an arbitrarily complete mapping between fiducial and reconstruction-level observables. Practical attempts to obtain and apply such a mapping, and hence fiducial observables as seen in real collider data, lie at the heart of the detector-correction methods described in Section 11.3.

11.1.1 Fiducial cross-sections

A fiducial cross-section, then, restricts to event configurations observable by the detector, the so-called **fiducial volume**. The most obvious such restriction is geometric acceptance: a fiducial volume cannot include events that fall into uninstrumented regions and cannot be reconstructed based on final-state information. Less obvious restrictions are p_T thresholds, below which accurate reconstruction cannot be performed, and requirements on the primary-interaction vertex position.

---

[1] As the host of venerable UK game show *Catchphrase* used to urge his contestants, 'Say what you see'.

Within this basic restriction to reconstructibility, fiducial cross-sections can be 'total' or differential, the distinction being whether they are a single-number interaction rate for observable events, or if that category is more finely subdivided into (approximations to) smooth shape variations in observables. The spirit, if not necessarily the formal definition, of fiducial differential cross-sections is that the independent variable X in a differential cross-section $d\sigma/dX$ should be one based on final-state observable properties. This excludes parton-level process types, including separation of a particular final state by partonic production channel, e.g. gluon fusion versus $q\bar{q}$ Higgs production modes separated out in simplified template cross-section (STXS) studies, as will be discussed in Chapter 12.

As exclusive event generation with matched parton showers, hadronisation, and other effects 'dressing' the core phenomenological QCD matrix elements provides the most complete description of events currently available, hence the particle-level analysis of such events provides the best available estimation of the **fiducial acceptance** of the event and object selections for a particular analysis. Fully-exclusive generation, however, is not the state of the art in terms of inclusive cross-section calculations, and it is standard to combine this best-estimate of the acceptance with a high-precision cross-section estimate, σ_{tot}, which results from either a multi-loop fixed-order calculation or a resummed one. The **fiducial cross-section** from theory is then

$$\sigma_{\text{fid}} = \sigma_{\text{tot}} \frac{\sum\limits_{i \in \text{acc}} w_i}{\sum\limits_{i \in \text{all}} w_i}, \tag{11.1}$$

where the $\sum_i w_i$ terms are the sums of event weights in the showered MC sample, either in total or those accepted by the analysis for the denominator or numerator, respectively. The same applies to differential fiducial cross-sections $d\sigma_{\text{fid}}/dX$ by considering the acceptance restricted to a single bin. As a histogram bin already by definition contains the sum of weights accepted into its fill-range, the differential cross-section is usually obtained by simply dividing by the incoming sum of MC weights $\sum_{i \in \text{all}} w_i$ (giving a unit normalisation for the histogram), and multiplying by the high-precision total cross-section σ_{tot} of choice.

11.1.2 Fiducial definitions

In practice, there is flexibility in the definition of the fiducial volume and observables, as a few examples will illustrate:

- The conception of the fiducial volume as being based on the post-hadronisation final state means that, while temptingly convenient, it is not allowed to directly access electroweak resonances inside the MC event record. This can feel pointlessly inconvenient: why force reconstruction of a leptonic Z from the final-state leptons and photons, when it is already pre-encoded in the MC record? But the fiducial approach here saves us from problems that occur when taking the event record too literally. For example, a resonance may not have a well-defined physical momentum, or even a well-defined rest frame, and beyond leading order there may not even be such a thing. Working backward from the final state therefore gives more physically unambiguous observables than working forward through the steps of current calculation techniques. And once QCD-charged objects are involved, as is always the case in top-quark decays, event-record literalism introduces a myriad of problems as every parton may be related to every other: parentage and history are not really questions we should ask about partons.

• The intuitive position is that fiducial observables have to be constructed from the set of particles considered stable by the event generator. But this introduces as many questions as it answers, as the stability of a given particle species is configurable in most generators.

A *de facto* standard of $c\tau_0 > 10$ mm is often assumed, as a nominal macroscopic-displacement that would allow 'easy' reconstruction of any shorter-lived parent, and which happens to fall in a gap between the lifetimes of known hadrons. But several longer-lived hadrons are *difficult* to reconstruct specifically because their lifetimes result in partial tracker signals: the accuracy of a measurement may be significantly increased by excluding them from the fiducial definition and hence avoiding extrapolations.

Another consideration arises from unstable-hadron reconstruction: if the 'real' truth-level final state has enough information to reconstruct its parent decay chains, there is little point in going through that explicit process using perfectly accurate input momenta and particle-identification, when the semi-classical hadron-decay event record can be used directly. This is a convenient application of the fiducial philosophy that 'if it *could* have been obtained experimentally, you can use it'. This approach is universally taken when performing fiducial truth-jet flavour-labelling: the tagging b and c hadrons are not explicitly reconstructed, but rather looked up directly in the unstable-hadron record, and matched to the jet (by ΔR comparison, or via ghost-association in the clustering process).

Connected to the idea of in-principle decay-chain reconstruction is **directness** or **promptness**, mentioned earlier in Section 8.2.1. If decay chains can in principle be reconstructed, then particles *not* from hadron decays can be experimentally identified as coming 'directly' from the hard-scattering process, modulo perhaps electroweak corrections such as photon emission. With care, this directness labelling of final-state particles enables simpler fiducial definitions—provided the in-principle identification is substantiated with something concrete, such as the use of lepton isolation methods.

• As we are concerned with minimising extrapolations, and including the geometric and other detector-acceptance limits in the fiducial definition, there is a natural question of how accurately these need to be reported. Do detector 'cracks'—uninstrumented regions, particularly between calorimeter barrel and endcap sections—need to be included in the truth definitions? There is a judgement call here: when we fill in histograms with contributions from regions *beyond* the detector's reach, e.g. into very low p_T or high $|\eta|$, we are undoubtedly extrapolating; when we fill in contributions from regions *between* measured ones, we are arguably interpolating instead—a far less egregious crime, and a process which can be reasonably data-driven rather than subject to significant modelling whims. In measurements, as opposed to reconstruction-level MC analyses, cracks are often ignored in the fiducial definition, and hence corrected for in the measured data. Similar details, such as the efficiency turn-off for jet-flavour tagging as the jet moves out of the tracker acceptance, are often also skipped or dealt with approximately, in the expectation that the type of assumption made is not a crucial one for the downstream physics-interpretation purpose.

• Another area for careful consideration is neutrinos: realistically, the reconstruction efficiency in any real detector is so low that neutrinos are treated as missing energy experimentally, not even explicitly passed to detector-simulation codes. But they are physical, decohered particles, and *in principle* one could integrate an efficient neutrino detector into a collider experiment—in a ludicrous way such as embedding the experiment in highly-instrumented Antarctic ice or perhaps a neutron star, but nonetheless not physically invalid. Should this mean that we can treat neutrinos as stable particles and directly access their kinematic and flavour properties?

Fiducial definitions need to consider pragmatism at least as much as principle. The fact that a fantasy detector could in-principle measure a particle does not mean that the real one can, and to infer definite properties from a (vanishingly) low-efficiency reconstruction is to introduce extrapolation and hence MC-dependence: the very thing the fiducial scheme is intended to avoid. So in general neutrinos (and hypothetical beyond the SM (BSM) invisibles) cannot be used directly: their experimental defining property of being reconstructed not by what is seen, but by what is not, makes them particularly vulnerable to modelling and extrapolation biases. Hence, when reconstructing missing (transverse) momentum, it is preferable to make the fiducial definition using reconstructed objects, either simple or composite, and with an acceptance cut.

But in specific situations, where the MET would be dominated by a single—usually direct—invisible particle, the simplicity of directly identifying the target truth particle may be a stronger argument than fiducial-principle purity. If there is minimal bias in the reconstruction process used, why not jump directly to the relevant truth-particle quantity. If life is not so simple, as is often the case in busy environments such as top-quark decay, a truth-particle analogue of the experimental reconstruction is probably safer.

• For many measurement analyses involving charged-lepton final states, the focus is primarily on the event kinematics, particularly the lepton system's relation to accompanying QCD activity, and whether the particular lepton in a selected event is an electron or muon is a sub-leading effect. In this picture—while noting that at the time of writing, hints of lepton non-universality are the most prominent observed deviations from the SM—a convenient approach for MC model comparisons is to combine the electron and muon channels into a single, generic 'lepton' measurement. At the partonic level, this is trivial, since for high-energy collisions both the electron and muon can be safely treated as effectively massless, but in fiducial definitions some extra work is needed to account for the higher-mass muon's suppression of QED radiation relative to the electron. Without doing so, 'bare' final-state muons would have more energy than equivalently produced electrons, since the muons radiate less. The optimal balance is one that minimises model-sensitive extrapolations from reconstruction to truth levels, and is achieved using the lepton-dressing recipe described in Section 8.2.1.

- We earlier discussed the reconstruction-level design of isolation methods, to separate direct non-QCD particles from their indirect cousins produced in non-perturbative fragmentation processes. These methods can be more or less directly encoded in a fiducial definition as well, for example summing the hadronic transverse energy in a cone around a bare or dressed lepton and applying the absolute or relative isolation criteria. But fiducial definitions do not have to exactly match the reconstruction procedure: they just have to achieve a similar enough result. As we already discussed, a particle's *directness* is a property which in principle could be obtained from forensic final-state reconstruction of hadron decays. Depending on the isolation procedure used, it may even be acceptable to shortcut the isolation method at least partially, by use of the directness property.

11.2 Complex object reconstruction

Particularly in precision electroweak physics analyses (and in equally precise b-physics studies involving hadron decay chains), it is not enough to observe generic final-state distributions such as p_T sums, unbiased jet-clusterings, and so on, but we must try and reconstruct particular particles from their decay products. The principle is trivial in the case of fully-visible final states such as $X \to \ell\ell$ or $Y \to jj$—simply identify the final-state physics objects and add their four-momenta—although in reality combinatorics and (particularly jet) resolutions are still a major challenge. But in event topologies involving invisible particles, the challenge is an order of magnitude greater.

In the previous chapters we encountered the recursive-jigsaw method and the m_{T2} (and higher-order) methods for partially reconstructing systems where the details of the unstable resonance are not known; here, we add to this collection methods for reconstructing systems where the details are known, or at least we believe them to be. Chief amongst these are reconstructions of events with a single leptonic W, and with either one or two leptonic top-quarks.

11.2.1 Leptonic W reconstruction

Due to worse momentum resolutions for hadronic jets than for charged leptons, inclusive-W and W + jets event types are most precisely measured via the leptonic W-decay modes, $W^- \to \ell\bar{\nu}$ and its charge-conjugate. Compared to the equivalent Z decays, however, these have the obvious difficulty that the final-state neutrino cannot be seen directly: the missing momentum must be used as a proxy. This cunning plan has several deficiencies:

1. As the initial-state colliding-parton momenta are not known (cf. parton distribution functions (PDFs)), the longitudinal component of the missing momentum cannot be known even in principle with respect to the partonic scattering rest frame. Even more crucially, since the missing momentum components are estimated via negative sums of visible momenta, and the dominant longitudinal momentum flows are in the uninstrumented area close to the beam-pipe, even the lab-frame missing longitudinal momentum cannot be measured with any accuracy. The best-available proxy is the purely transverse missing-momentum vector, p_T^{miss}.

2. The neutrino cannot be distinguished from any other invisible particle. While reconstruction of the system would be almost entirely unaffected by a different near-massless invisible object (or objects) in its place,[2] physics processes with massive invisibles contributing to p_T^{miss} + jets event signatures will be misreconstructed.

If one does not need to reconstruct the W completely, these limitations are not a major problem as much relevant information is captured in differential distributions such as those of the charged-lepton transverse momentum, p_T^ℓ, and the transverse mass,

$$m_T = \sqrt{2p_T^\ell p_T^{\text{miss}}(1 - \cos \Delta\phi)}, \tag{11.2}$$

(which uses nearly all the measured kinematic information, other than the charged-lepton's longitudinal component). Neglecting spin effects, the motion of the W and assuming a two-body decay, the transverse momentum of the lepton is

$$p_T^\ell = \frac{1}{2}M_W \sin \theta, \tag{11.3}$$

allowing the differential cross-section in the lepton p_T to be written in terms of the polar angle θ as

$$\frac{d\sigma}{dp_T^\ell} = \frac{d\sigma}{d\cos\theta}\frac{d\cos\theta}{dp_T^\ell} \tag{11.4}$$

$$= \frac{d\sigma}{d\cos\theta}\left|\frac{d\cos\theta}{d\sin\theta}\right|\frac{2}{M_W}. \tag{11.5}$$

The **Jacobian factor** $|d\cos\theta/d\sin\theta|$ is easily computed using the usual trigonometric identities, and re-substituting equation (11.3) gives the resulting rest-frame p_T^ℓ lineshape

---

[2] Only 'almost' entirely, since non-SM spin effects could also manifest in the final state, via angular distributions.

$$\frac{d\sigma}{dp_T^\ell} = \frac{d\sigma}{d\cos\theta}\left(\frac{2p_T^\ell}{M_W}\right)\frac{1}{\sqrt{(M_W/2)^2 - (p_T^\ell)^2}}. \tag{11.6}$$

This expression clearly has a singularity at $p_T^\ell = M_W/2$, which due to resolution effects, the motion of the W, and off-shell effects (cf. the W's Lorentzian mass lineshape) is rendered as a finite peak with a sharp cutoff to the distribution. Similar kinematic endpoints appear in other distributions such as m_T, and can be used to constrain the system subject to different systematic uncertainties: examples are shown in Figure 11.1. Note that we have here considered the W mass, but this analysis is relevant for any two-body decay of a heavy object to a charged lepton and a light invisible particle, and hence the model assumptions can be kept minimal.

For complete W reconstruction, however, stronger assumptions must be made in order to solve for the unknown missing-object momentum components: its longitudinal momentum, p_z^{miss}, and mass, $\mathcal{m}$. The obvious assumptions to make, given the depth to which the electroweak model was tested at LEP and other lepton colliders, is to assume SM W decay kinematics. These manifest via the missing-object constraints $(p_x, p_y) = (p_x^{miss}, p_y^{miss})$, i.e. a single invisible object, and $\mathcal{m} = 0$ as for a neutrino (within the ATLAS and CMS experimental resolution). The single remaining unknown is hence p_z^{miss}. To fix this, we again assume the SM and use the W mass as a standard candle: to the extent the W can be treated as on-shell, the sum of the fully-reconstructed lepton and the partially-reconstructed neutrino four-vectors should have an invariant mass equal to the W. We can hence express a first-order estimate of the neutrino p_z, assuming massless leptons, via the constraint

$$M_W^2 \approx (E_\nu + E_\ell)^2 - (\boldsymbol{p}_\nu + \boldsymbol{p}_\ell)^2 \tag{11.7}$$

$$\approx (|\boldsymbol{p}_\nu| + |\boldsymbol{p}_\ell|)^2 - (\boldsymbol{p}_{T,\nu} + \boldsymbol{p}_{T,\ell})^2 - (p_{z,\nu} + p_{z,\ell})^2 \tag{11.8}$$

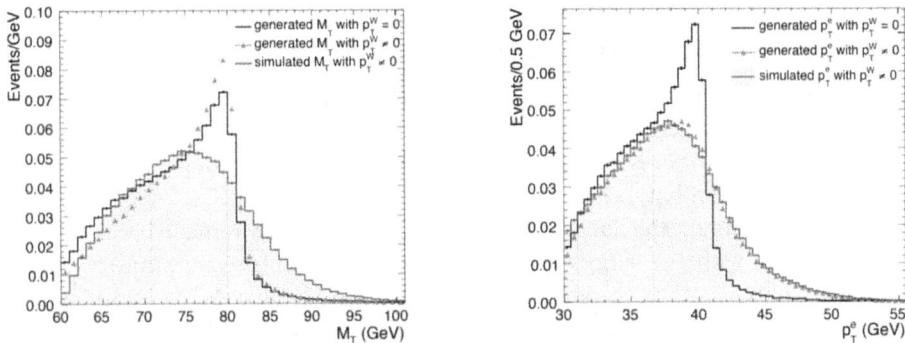

Figure 11.1. The ℓ + MET transverse mass spectrum (left) and electron p_T distribution (right) in $W \to e\nu$ decay. The black and red lines show the distribution with $p_T(W) = 0$ and a realistic W p_T spectrum, respectively, and the shaded region shows the distributions after including detector and reconstruction effects. Reproduced from Sarah Malik, Precision measurement of the mass and width of the W Boson at CDF FERMILAB-THESIS-2009-59TRN: US1004113 (2009).

$$= 2\Big[|p_\nu||p_\ell| - \boldsymbol{p}_{\mathrm{T},\nu} \cdot \boldsymbol{p}_{\mathrm{T},\ell} - |p_{z,\nu}||p_{z,\ell}| \Big]. \tag{11.9}$$

This form contains two dependences on p_z^ν: one explicit, and the other hidden inside $|p_\nu|$. Expanding, we can rearrange into a quadratic formula with the resulting neutrino longitudinal-momentum estimate

$$\tilde{p}_{z,\nu} = \frac{p_{z,\ell} Q^2 \pm \sqrt{p_{z,\ell}^2 Q^4 - p_{\mathrm{T},\ell}^2\left(|p_\ell|^2 |p_{\mathrm{T},\nu}|^2 - Q^4\right)}}{p_{\mathrm{T},\ell}^2}, \tag{11.10}$$

where $Q^2 = \boldsymbol{p}_{\mathrm{T},\nu} \cdot \boldsymbol{p}_{\mathrm{T},\ell} + M_W^2/2$, and we use the experimental $p_{\mathrm{T}}^{\mathrm{miss}}$ vector in place of $\boldsymbol{p}_{\mathrm{T},\nu}$ and $|p_{\mathrm{T},\nu}|$.

It is important to note that the quadratic leads to an ambiguity in the $p_{z,\nu}$ assignment, as there are in general two values of the longitudinal momentum that will satisfy the mass constraint. Nevertheless, it is a very useful technique for re-acquiring an almost unambiguous event reconstruction, under the assumption of SM kinematics. Of course, this approach is unviable in analyses wishing to *measure* the W mass!

This picture is rather idealised, though. In practice, the fact that $p_{\mathrm{T}}^{\mathrm{miss}}$ includes contributions other than the direct neutrino, imperfect reconstruction resolutions, and the finite W resonance width spoil the exactness of the solution, and result in imaginary components appearing in the quadratic $p_{z,\nu}$ estimates. These can in fact be used as heuristics to break the quadratic degeneracy, such as choosing the solution with the smallest imaginary components, or which requires the smallest modification of lepton and jet kinematics (allowed to vary on the scale of their resolutions) to achieve a real-valued solution. Such fits are typically performed via a kinematic likelihood expression, with resolution-scaled Gaussian penalty terms on the kinematic variations. For variety, we will introduce these concepts in the following section, where the same ambiguities enter the reconstruction of single-leptonic top-quark events.

11.2.2 Single leptonic top-quark reconstruction

Top quarks occupy a special position in the SM in terms of reconstruction: they are the only colour-charged electroweak (EW)-scale resonance, and have near-unity Yukawa and Cabibbo–Kobayashi–Maskawa (CKM) couplings. These factors introduce several conflicting issues:

Yukawa coupling: the large and 'natural' top Yukawa means it has strong interactions with the Higgs field, making the top sector comparable to direct Higgs analyses as a promising venue for BSM physics to manifest. The possibility of such modifications means that assuming SM kinematics in top-quark reconstruction is particularly risky.

Colour-charge: tops radiate gluons in their production, propagation and decay, and are dominantly produced via QCD interactions sensitive to parton distribution functions (particularly the high-x gluon and any b-quark content generated by Dokshitzer–Gribov–Lipatov–Altarelli–Parisi (DGLAP) evolution). The necessity for jet production in top-quark decay is a challenge for precision measurement, as reconstructed jet energy and mass resolutions are significantly less precise than for charged leptons (cf. the other EW-scale resonances, W and Z, which do have completely leptonic decay routes.)

CKM coupling: $|V_{tb}| \sim 1$, and the large phase-space available for its relatively light decay products, means that the top decays before hadronising: our closest possible view of a bare quark. But both QCD and EW corrections to top-quark kinematics have been found to be critically important for the accuracy of MC/data model comparisons.

These factors introduce problems for our preferred precision-analysis method, the fiducial definition: the field has still not worked out a standard approach which addresses all these issues. In this section, we will explore the methods available for precision top-quark reconstruction with a single leptonic top, and in the next section the even more awkward case with two. These methods provide useful techniques at the reconstruction-level regardless of the level to be corrected to.

Our workhorses for top-quark physics analysis will be leptonic tops—those whose decay-daughter W decays leptonically, as shown in Figure 11.2(a). Naïvely, fully hadronic top-quark decays are more attractive as they have a fully reconstructible final-state without missing energy. But two factors put paid to this hope: firstly, either singly or in a $t\bar{t}$ pair, hadronic tops have few distinguishing signatures from the overwhelming QCD multi-jet background; and secondly, the resolutions on masses and other composite observables obtained from jets are much poorer than are required for e.g. a top-quark property measurement. Some published studies have been made using jet-substructure techniques to reconstruct high-momentum tops, reducing the jet background, but for now the momentum resolution remains restrictive. By contrast, events with at least one leptonic top have a direct charged lepton (immediately reducing the background by orders of magnitude, since a high-scale weak process must have been involved somewhere), significant p_T^{miss}, and better resolution with which composite-object reconstruction techniques can be attempted. The downside is the ambiguities introduced by the invisible neutrinos: at

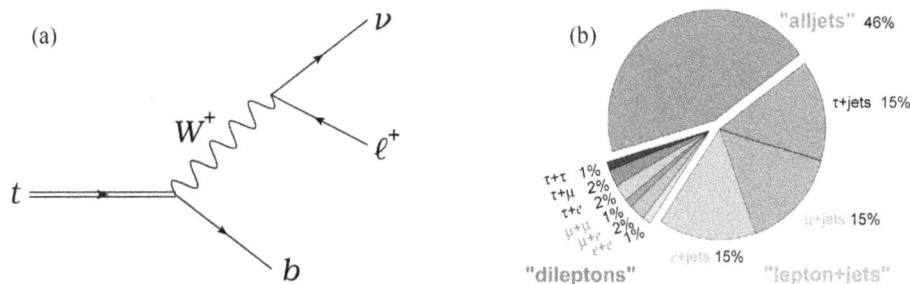

Figure 11.2. (a) Leading-order Feynman diagram for leptonic top-quark decay, via the 99.8%-dominant $t \rightarrow b$ CKM channel. (b) Top-quark pair branching fractions. Credit: diagram courtesy of the DØ Collaboration.

least one per leptonic top, and two in the case of tau leptons (one from the tau production and one from its leptonic decay to e or μ).

The classes of top-pair decays producing a charged lepton (including taus, which themselves decay hadronically $\sim$65% of the time, and leptonically 35% of the time) are shown in Figure 11.2(b). In addition to these we can add leptonic decays from single-top production, although these have much smaller cross-sections due to being induced via weak interactions; both the $t\bar{t}$ and single-top production mechanisms are shown in Figure 11.3. From these fractions we see that roughly 38% of top-pair events have exactly one direct electron or muon in their final state, the **semileptonic** $t\bar{t}$ topology; and $\sim$6% in the **dileptonic** $t\bar{t}$ mode with two direct e/μ. Given the natural trade-offs between resolutions, yields, and the ambiguities of missing momentum,

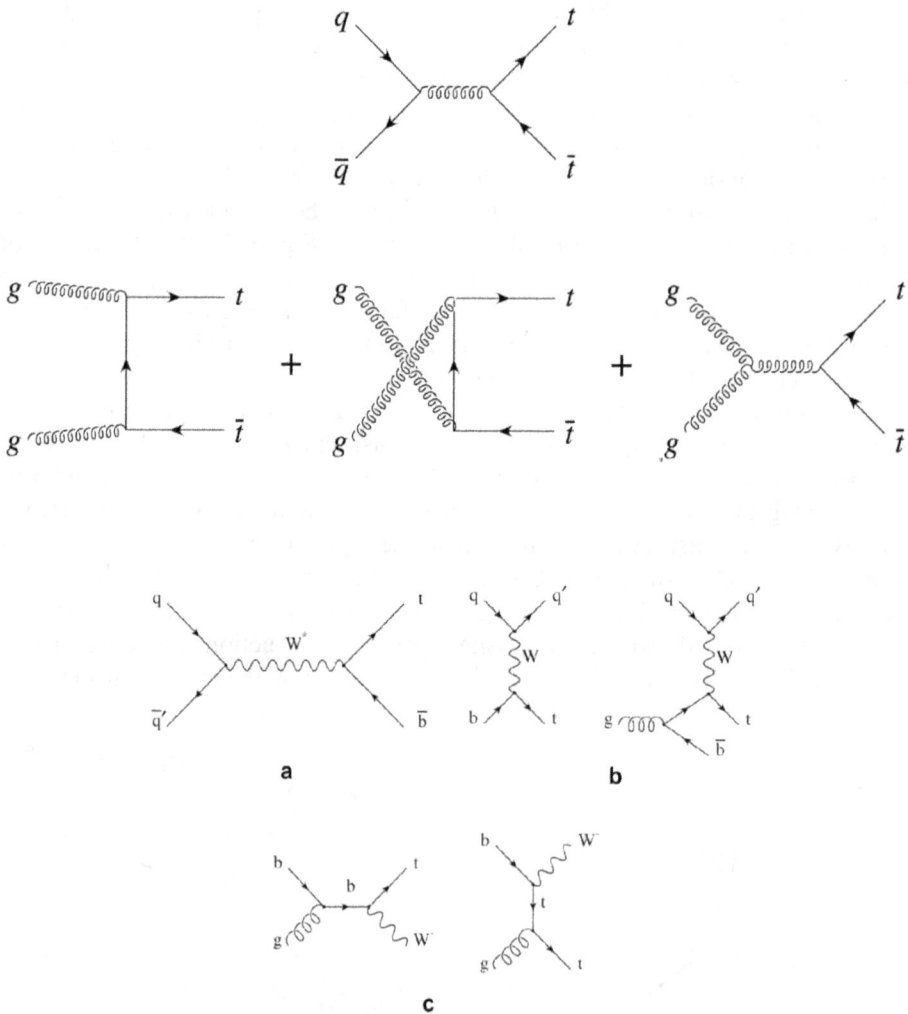

Figure 11.3. (a) Leading-order Feynman diagrams for QCD top-quark pair-production. (b) Leading-order Feynman diagrams for single-top-quark production: a) s-channel, b) t-channel, and c) Wt modes.

semileptonic $t\bar{t} \to b\bar{b}jj\ell\nu$) is the most commonly used top-pair process for precision measurements.

The nice thing about single-leptonic top reconstruction is that we can largely recycle the principles used in the previous section for leptonic W reconstruction—after all, the SM $t \to b\ell^+\nu_\ell$ decay *is* a leptonic W^+ decay with an additional b quark. This additional quark is not trivial—as can be seen from Figure 11.3, the leptonic top will either be in a $t\bar{t}$ event, in which case there will certainly be another b-jet (from the other, hadronic, top-decay), or will be in a single-top event where a b-quark is almost certainly co-produced with the top due to the dominant charged-current production vertex. Even in the diagrams without an explicit b-quark in the final state, the initial-state b has to be produced from a QCD $g \to b\bar{b}$ splitting, with the additional b often entering the analysis phase-space rather than just being 'absorbed' into the PDF. The mix of implicit versus explicit $g \to b\bar{b}$ splitting in such processes is the subject of active theory discussion about the best mixture of calculational four- and five-flavour schemes. The extra b-jet, via either production mechanism, leads to a combinatoric ambiguity which again requires an extra reconstruction heuristic.

As for W reconstruction, if the top mass is not being *measured* by the analysis, the invariant mass $m(b, \ell, \nu) = m_t \sim 172.5$ GeV can be used as a constraint in addition to the W mass standard-candle. In principle this gives a perfect solution up to now four-fold quadratic ambiguities, and we could stop the description here. However, this more complex system is a good place to discuss the more practical realities of reconstruction with unknown p_z^{miss}. These are:

- experimental resolutions;
- truly off-shell tops and W's;
- unresolvable neutrino pairs from leptonic tau-decay feed-down (electrons and muons from direct-tau decays are necessarily treated as if themselves direct).

These effects spoil the perfect numerics of solving the quadratic systems, and hence realistic top reconstruction attempts to take them into account. The simplest approach is rather lazy: perform the W and top reconstruction sequentially, first fixing the neutrino p_z via the W mass, then using the (mostly known) top mass for ambiguity resolution. The first step will not in general produce real-valued solutions for p_z^{miss}, in which case some heuristic is needed: one common approach is to permit directional variations to the $\boldsymbol{p}_{\text{T}}^{\text{miss}}$ vector, setting $p_z^{\text{miss}} = 0$ and setting the transverse mass m_{T} equal to the W mass. An alternative common heuristic is to obtain the complex solutions and just set the imaginary parts to zero—with a maximum tolerance beyond which the event is declared un-reconstructible. Both are simple, but improveable.

A more holistic approach is that of kinematic likelihood fitting. In its simplest form this defines a simple, χ^2-like metric of reconstruction quality based on how closely the W and top components of the system match their pole-mass values:

$$\chi^2 \sim \frac{\left(\tilde{m}_{\ell\nu b} - m_t^{\text{pole}}\right)^2}{\sigma(\tilde{m}_{\ell\nu b})^2} + \frac{\left(\tilde{m}_{\ell\nu} - m_W^{\text{pole}}\right)^2}{\sigma(\tilde{m}_{\ell\nu})^2}, \tag{11.11}$$

where $\tilde{m}_{\ell\nu}$ and $\tilde{m}_{\ell\nu b}$ are the leptonic W and top-quark mass estimators from the combination of the charged lepton ℓ, and the neutrino candidate ν (with free/reconstructed z momentum), and b-jet b. The $\sigma(\tilde{m}_i)$ terms encode Gaussian uncertainties around the pole masses, combining the intrinsic $\Gamma_t \sim 1.3$ GeV and $\Gamma_W \sim 2.1$ GeV widths with typically larger reconstruction resolutions $\Delta\tilde{m}_i$, combined according to e.g. $\sigma(\tilde{m}_i)^2 = \Gamma_i^2 + \Delta\tilde{m}_i^2$. It is then a question of numerical optimisation and exploration to identify the best-fit configuration, classify whether it is acceptably good, and to perhaps propagate some measure of the resulting uncertainty. An example application of this technique, at truth-particle level for high-p_{T} 'boosted' $t\bar{t}$ events, is shown in Figure 11.4.

Obviously this approach can be refined into a more careful (log-)likelihood metric in which the Lorentzian resonance is not approximated as Gaussian, but rather convolved into a Voigt profile (see Section 5.1.6). Other extensions include putting the hadronic top reconstruction into the fit (in the case of semi-leptonic $t\bar{t}$, and with a correspondingly larger experimental resolution term), and potentially even accounting for systematic correlations between the W and top-quark masses, given their common constituents.

A further recent refinement in semi-leptonic $t\bar{t}$, in exchange for some clarity about the mechanism of operation, is to use modern neural network technology in addition to this leptonic reconstruction to assist with reducing or resolving the light-jet combinatoric backgrounds for the accompanying hadronic top: these can be avoided by requiring that there be only two b-jets and two light-jets in the event, but doing so greatly reduces the number of available events, particularly for high-$m_{t\bar{t}}$ events where there is a large phase-space for significant extra QCD radiation. Deep neural network classifiers can help to improve the efficiency and purity of semi-leptonic $t\bar{t}$ reconstruction in the presence of multiple extra jets, but there is no free lunch: if using such methods, validations and systematic error propagation will likely need to be performed to capture any model-dependences implicit in the neural network training.

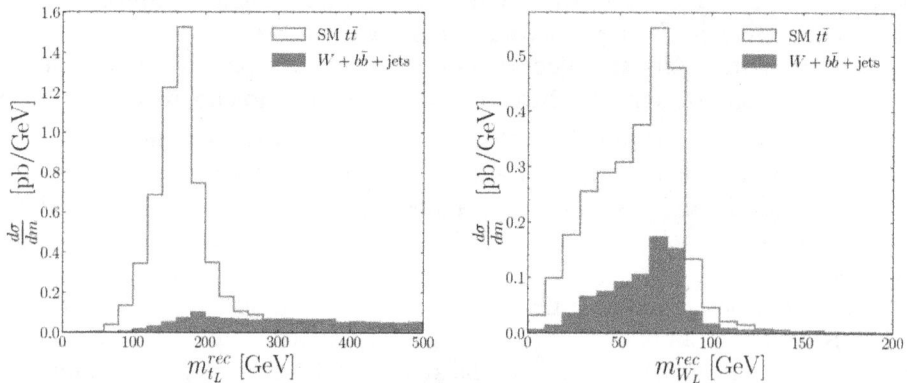

Figure 11.4. Illustration of invariant-mass kinematic fitting reconstruction of high-p_{T} semileptonic $t\bar{t}$ events, and their main W + jets background process. Both the leptonic W and the leptonic top-quark reconstructed mass spectra are shown, with the extra suppression of the background clear in the latter case. Credit: MC analysis and plots by Jack Araz.

11.2.3 Dileptonic top-pair

Events containing a dileptonic top pair are a further iteration in difficulty, due to the p_T^{miss} vector being only the *sum* of the two neutrino momenta (or those of other invisible species, beyond the assumption of SM top-decays), with the information about the individual neutrino momentum components now ambiguous.

The reason to spend effort grappling with this difficult reconstruction is precision: in particular, measurements of asymmetries and spin-correlations between the top and anti-top require an angular precision on the top-decay orientation which cannot be equalled by events in which one of the pair decays hadronically. At present it is in the most complex class of composite-object reconstructions attempted for precision analyses, although greater numbers of invisible particles are encountered with less bespoke reconstructions in BSM search analyses, e.g. in SUSY-partner decay cascades.

As a matter of principle, we of course need to know that the system *can* be solved, by comparing the number of independent kinematic variables to the number of constraints available:

Variables: four momentum components for each of the physics objects: two charged leptons, two b-jets, and two neutrinos (or invisibles in general). A total of **24 variables**.

Constraints: three measured momentum components for each of the leptons, four measured energy-momentum components for the b-jets, two measured momentum components total for the neutrinos. Plus assumptions: two $m_\ell \sim 0$ approximations for the charged leptons, two $m_\nu = 0$ assumptions for the invisibles, and finally two $m_W \sim m_{\ell\nu}$ and two $m_t \sim m_{b\ell\nu}$ composite-state constraints. A total of **24 constraints**.

So the system can be solved—but only just. This counting assumes many things:

1. Perfect charged-lepton, b-jet, and p_T^{miss} identification and resolution;
2. Absence of any extra QCD radiation (the degree of inaccuracy of this assumption effectively further degrading momentum and mass resolutions);
3. Correct assignment of b-jet/lepton pairings (in practice the b-quark charge cannot be identified from the reconstructed b-jet);
4. SM decay kinematics via the intermediate W;
5. Perfectly on-shell W and top-(anti)quark resonances, with perfectly known pole-mass values.

Even with these assumptions, the solution can only be obtained up to discrete polynomial root-finding ambiguities, which in practice are multiplied by the assignment ambiguities between leptons and b-jets. To minimise the impact of Z + jets backgrounds in further confusing the resolution of these ambiguities, it is common for experimental studies of the dileptonic $t\bar{t}$ system to focus on the mixed lepton-flavour $e\mu$ channel, rather than the same-flavour ee and $\mu\mu$ ones, or to use explicit p_T^{miss} and $m_{\ell\ell}$ cuts to reduce the impact of $Z \to \ell\ell$ backgrounds.

Several approaches have been developed to solve this reconstruction problem, from analytic solution of the system of kinematic constraints in the main unknowns (the three three-momentum components of each neutrino), to numerical minimisation and weighting approaches. While the detail of each is beyond the scope of this book, a review of their core concepts gives a useful insight into their pros and cons.

Generic m_{T2}-based reconstruction
Our first method is as discussed in Chapter 10, for BSM decay chains resulting in two invisible particles: use the m_{T2} variable introduced in Section 8.4.2. This is a generalisation of the transverse mass m_T to the case of pair-produced particles where each member of the pair decays to a set of visible particles and one invisible one, denoted χ.

The benefit of m_{T2} reconstruction for $t\bar{t}$ dileptonic studies is its reduction of combinatoric backgrounds, and minimal SM assumptions—the W and t pole-mass resonances are not used, and even the assumed invisible-particle momentum, m_χ can be scanned over rather than assuming $m_\chi = m_\nu \approx 0$. These advantages are offset by the resulting lack of precision: the $t\bar{t}$ system can only be reconstructed in the transverse plane, which is not sufficient for spin-correlation and forward-backward asymmetries which need some handle on longitudinal momentum components. Nevertheless, the potential role of m_{T2} as a degeneracy-breaker for explicit full-reconstruction approaches is significant, or e.g. as a complementary input to machine-learning approaches to dileptonic $t\bar{t}$ reconstruction.

Analytic reconstruction
Given the equal numbers of variables and constraints on the system, analytic solution is an obvious approach to take: the first such approach in 1992 was solved numerically, and subsequent attempts reduced the sensitivity to singularities in the constraint equations. The current *de facto* analytic approach is the 'Sonnenschein method', a 2006 refinement to minimise the number of solution steps[3].

This method separately considers the top-quark and antiquark systems, entangled by the ambiguity between their (anti)neutrino decay products, assuming that the b-jet/lepton pairing can be made correctly. By the usual arguments, the neutrino energy on either side of the top-pair decay can be expressed via both the W and top mass-constraints, as

$$E_\nu = \frac{m_W^2 - m_\ell^2 - m_\nu^2 + 2\boldsymbol{p}_\ell \cdot \boldsymbol{p}_\nu}{2E_\ell} \tag{11.12}$$

$$= \frac{m_t^2 - m_b^2 - m_\ell^2 - m_\nu^2 - 2E_bE_\ell + 2\boldsymbol{p}_b \cdot \boldsymbol{p}_\ell + 2(\boldsymbol{p}_b + \boldsymbol{p}_\ell) \cdot \boldsymbol{p}_\nu}{2(E_b + E_\ell)}, \tag{11.13}$$

---

[3] See arXiv:hep-ph/0603011, but note there is an important sign-error in the original formulae! Using a debugged pre-written implementation of the method is recommended.

whereupon the two right-hand sides can be equated to eliminate the dependence on E_ν. Doing so for each of the top and anti-top sides gives rise to linear equations in the (anti)neutrino three-momentum components,

$$a_1 + a_2\, p_{\nu,x} + a_3\, p_{\nu,y} + a_4\, p_{\nu,z} = 0 \tag{11.14}$$

$$b_1 + b_2\, p_{\bar\nu,x} + b_3\, p_{\bar\nu,y} + b_4\, p_{\bar\nu,z} = 0, \tag{11.15}$$

where the a_n and b_n coefficients are functions of the other kinematic variables for that decay. These can be used to express the neutrino z-components in terms of their x and y counterparts, and together with the neutrino dispersion relation $E_\nu^2 = m_\nu^2 + p_{\nu,x}^2 + p_{\nu,y}^2 + p_{\nu,z}^2$ and the complementarity of the neutrino and anti-neutrino x and y momenta in producing the total $\boldsymbol{p}_{\mathrm{T}}^{\mathrm{miss}}$, can be used to obtain a pair of two-variable quadratic equations,

$$c_{22} + c_{21}\, p_{\nu,x} + c_{11}\, p_{\nu,y} + c_{20}\, p_{\nu,x}^2 + c_{10}\, p_{\nu,x} p_{\nu,y} + c_{00}\, p_{\nu,y}^2 = 0 \tag{11.16}$$

$$d_{22} + d_{21}\, p_{\nu,x} + d_{11}\, p_{\nu,y} + d_{20}\, p_{\nu,x}^2 + d_{10}\, p_{\nu,x} p_{\nu,y} + d_{00}\, p_{\nu,y}^2 = 0. \tag{11.17}$$

As these two quadratics are expressed in the same variables, solutions for $p_{\nu,x}$ and $p_{\nu,y}$ can be found where the two surfaces intersect. This is done by computing the resultant with respect to $p_{\nu,y}$, which is a quartic polynomial in $p_{\nu,x}$ and hence in general has four solutions. In fact, with perfect resolution and particle-pair assignment only two would usually be distinct, but in practical applications all four need to be considered. For each $p_{\nu,x}$ solution, a single corresponding $p_{\nu,y}$ can be computed with good numerical stability.

The critical issue with the Sonnenschein method is its assumptions of perfectly on-shell decays, perfect experimental resolutions, and absence of assignment ambiguities. The latter of these is not such a problem, merely ensuring that the fourfold solution ambiguity is always relevant; the former are bigger issues, especially given significant jet and $p_{\mathrm{T}}^{\mathrm{miss}}$ systematic uncertainties, and may lead to no real solutions being found. A similarly analytic, but distinct 'ellipses method' has been developed in which solutions are identified as the intersections of ellipses along which valid solutions of the dilepton-$t\bar{t}$ system's kinematic constraints can be found. The advantage of this more geometric approach is that it provides a clear heuristic for recovering a solution in the case of 'near-miss' events with no intersections, instead using the points of closest approach between the two ellipses.

Kinematic likelihoods
Given the relative algebraic complexity of analytic reconstruction, and especially since it eventually runs aground on issues of finite resolution and resonance widths, a more popular method in LHC dileptonic $t\bar{t}$ studies has been to obtain 'best fit' estimates of the $t\bar{t}$ system kinematics by an explicit **kinematic likelihood fit**, bypassing the analytic solution in favour of a numerical optimiser in the space of neutrino momentum parameters—perhaps with analytically informed starting values. The W and top mass

constraints then naturally enter the target loss function not as delta-functions as assumed in the Sonnenschein approach, but as broad pdfs incorporating knowledge of both the Breit–Wigner resonance shapes and known widths, and experimental uncertainties encoded to in-principle arbitrary levels of detail. Priors derived from SM MC simulations can potentially also be included—via the usual dance between Bayesian and frequentist semantics, and concerns about inadvertent model-dependence. This overall approach is simply the higher-dimensional equivalent of kinematic likelihood fits for semileptonic $t\bar{t}$, and can either be used to map the whole pdf (on an event-by-event basis) or to find the single best-fit point within it.

Neutrino weighting

A similar approach goes by the name of **neutrino weighting**, which is most used in top-mass measurements with dileptonic $t\bar{t}$: in this situation, the m_t and $m_{\bar{t}}$ constraints cannot be used, leaving the system underconstrained even if the typical $m_t = m_{\bar{t}}$ assumption is made. The neutrino weighting method extracts an m_t measurement from a set of events by *assuming* top mass values in a set of special MC simulations propagated through a comprehensive detector simulation. In general, the predictions from each hypothesised top mass can be compared to a whole set of observables $\{\mathcal{O}\}$, and the likelihood or posterior density computed accordingly. The best-fit probability score in each event can be used as an event weight, and the distributions of these weights—typically their first two moments—used to place a confidence interval on m_t.

So much for the general concept. However, a reduced version of this has become usual to the extent of being synonymous with 'neutrino weighting', in which $\{\mathcal{O}\} = \{\boldsymbol{p}_{\mathrm{T}}^{\mathrm{miss}}\}$, i.e. the weight is computed purely based on the missing-momentum components. To do this, of course, the missing momentum cannot be used as a constraint, otherwise there would be a circular-logic problem. The approach taken is to replace those two constraints (in p_x^{miss} and p_y^{miss}) with a scan over hypothesised values for the unmeasureable neutrino and antineutrino (pseudo)rapidities η_ν and $\eta_{\bar{\nu}}$; each $(\eta_\nu, \eta_{\bar{\nu}})$ combination is then translated into a *prediction* for $\boldsymbol{p}_{\mathrm{T}}^{\mathrm{miss}}$ via the other constraints, which can be compared to the measured value to compute a likelihood weight,

$$w_i(\eta_\nu, \eta_{\bar{\nu}}; m_t) = \exp\left(-\frac{(p_x^{\mathrm{miss}} - p_x^\nu - p_x^{\bar{\nu}})^2}{2\sigma_{\mathcal{X}}^2}\right) \cdot \exp\left(-\frac{(p_y^{\mathrm{miss}} - p_y^\nu - p_y^{\bar{\nu}})^2}{2\sigma_{\mathcal{X}}^2}\right), \quad (11.18)$$

where $\sigma_{\mathcal{X}}$ is an effective Gaussian resolution on each of p_x^{miss} and p_y^{miss}. Smearing of other resolutions and resonance widths is typically introduced in the propagation of $(\eta_\nu, \eta_{\bar{\nu}})$ to $\boldsymbol{p}_{\mathrm{T}}^{\mathrm{miss}}$, to ensure that those resolution effects are incorporated in the weight computation, albeit stochastically rather than by explicit likelihood fitting or scanning/propagation of pdfs. The huge event datasets available at the closure of LHC Run 2 and beyond, mean that CPU budget can be a limiting factor in analyses, and the relatively simplicity of $\boldsymbol{p}_{\mathrm{T}}^{\mathrm{miss}}$-based neutrino weighting (as compared to a full kinematic likelihood fit per event) means that it is comparable with simplified kinematic fitting as a viable method for dileptonic $t\bar{t}$ analyses.

11.3 Detector corrections and unfolding

A central component of the 'fiducial' picture of measurement is that whatever is measured by the imperfect detector can be mapped—with some necessary inflation of uncertainties—to a detector-independent view of the physics present in nature. A good motivation for this process is that experimentalists alone have a detailed knowledge of their detector and reconstruction algorithms, so it is naturally their responsibility to publish the information needed to connect the detector-specific measurement to detector-independent model predictions.

This mapping could be provided in either a forward or a backward direction: as a set of analysis-specific transfer functions capturing the effects of detector biases on simulated particle-level events, or by making corrections to the biased, detector-specific reco-level measurements. In practice the latter has been the preferred route, as the forward 'folding' transfer functions would in general require publishing (and expensive running) of the vast and complicated experimental simulation and reconstruction software chain, including systematic variations. By contrast, publishing 'unfolded' particle-level measurement data—the experiment's best possible estimate of what a perfect, unbiased experimental apparatus would have measured—is far more compact and more amenable to re-use. Unfolding does, however, require extra data processing steps and validation within the experimental analysis: unfolded analyses typically take significantly longer than reconstruction-level ones to publish, although the extent to which this is directly due to unfolding issues as opposed to the general focus of such measurements on 'precision' issues is not always clear.

In practice there is no single way to obtain the detector-correction map: detector effects are often probabilistic and a multitude of particle-level values may be consistent with a single reconstruction-level observation with varying degrees of probability. In this section we will explore several different approaches to detector-bias corrections, with differing levels of sophistication and probabilistic rigour: which is correct for you will depend on the intrinsic precision of your measurement, and the downstream use-cases you intend for your published results.

11.3.1 Efficiency and acceptance corrections

Detector correction, as we shall see, is a one-to-many mapping largely dependent on comparisons of particle-level and reconstruction-level observables on MC-simulated events—so-called **unfolding**. These 'before and after' event samples may be intuitively understood, but the usually non-parametric processes by which a space of mappings between them is obtained tend to be rather a computational 'black box'. Most physicists reasonably feel some discomfort at relying on opaque machinery to make large corrections, and so it is desirable to first apply *known* detector corrections as much as possible before engaging the unfolding algorithms. This step is sometimes called **pre-unfolding**.

The main inputs used in pre-unfolding are data-driven efficiency factors, as used in physics-object calibration. For example, it is standard for tracking performance groups to estimate tracking efficiency maps for charged particles as binned functions in $(p_T, |\eta|)$, and similar maps are common for isolated charged lepton and photon reconstruction, and for the performance of jet flavour taggers. Additional key variables may be included, such as the number of tracks associated to a vertex for vertexing or tagging performance: it is important that such efficiency tabulations do include all key variables rather than assuming that $(p_T, |\eta|)$ is always sufficient, and that the calibration distributions in these variables are similar enough to those in the signal phase-space[4].

While these tabulated efficiency numbers are only averages—per-event sampling effects will be considered in the coming sections—their reciprocals can be treated as per-object weights to offset some average biases,

$$w_i^{\text{eff}} = 1/\epsilon_i \qquad (11.19)$$

for object i with efficiency ϵ_i. By construction, these weights are always >1.

The effect is that e.g. a low-p_T or high-$|\eta|$ reconstructed track (with typically lower reconstruction efficiency) will receive a higher weight than one with higher-efficiency kinematics, and observable contributions containing such objects will be increased in importance. Note that this is a per-object, per-event correction to be derived in an event-analysis loop, not something that can be performed as a *post hoc* correction on the observable distributions alone.

Mis-identifications, e.g. accidental labelling of an electron as a prompt photon, or contributions from objects outside the fiducial acceptance, may be handled similarly, but now with correction factor

$$w_i^{\text{mis}} = (1 - f_i), \qquad (11.20)$$

where f_i is now the expected mistag rate for an object i's kinematics, i.e. the **fake fraction** of the reconstructed objects. By construction, this weight is always <1. A good example of such corrections in real-world use is the ATLAS track-based minimum-bias analysis from early LHC operation.

It is important that such efficiency weights not be misunderstood as up-scalings of momenta or similar: an efficiency tells us nothing about per-object mismatches between particle and reco levels, and often this correction is made under the assumption that they coincide, with shortcomings in that assumption to be offset by the black-box unfolding step to follow. Efficiency weights rather indicate that if one *a priori* low-efficiency object is present at reconstruction level, there were probably more, or more events containing such objects, at particle level. The exact propagation of these weights into observables depends on how the observable is constructed from the inefficiently constructed objects: of course, inefficiencies or impurities in objects irrelevant to the analysis should not affect the results at all.

We now proceed to the more complicated issue of handling the residual biases, known as unfolding. Most such methods are based on MC-driven comparisons between particle-level and reco-level observables, but in the interests of continuity we note here that one—the 'HBOM' method—is a natural extension of the (in principle) MC-independent pre-unfolding corrections described here. Those interested in following this logic immediately should turn to the HBOM description later in this chapter.

---

[4] An excellent, forensic, and very instructive examination of errors caused by subtleties in efficiency-correction estimation for LHCb D^0 meson differential cross-section measurements may be found in Chapter 8 of CERN-THESIS-2018-089.

11.3.2 Unfolding

In common use, 'unfolding' is both an umbrella term used to refer to the entire detector-correction process, and a more specific reference to the 'magic' bit of that inference process. We will use it in the second form.

But first, to understand why correction is in general a difficult problem to solve, and hence why complicated numerical machinery needs to be involved at all, it is useful to take a step back and consider the inverse process: the **forward folding** of particle-level observables (cf. the fiducial definitions discussed earlier) into reconstruction-level ones as would be seen by an analyser accessing LHC-experiment data files. This process includes the effects of finite detector resolutions, systematic calibration offsets (both additive and multiplicative), and efficiency and mistag rates as invoked in the previous section. Other than the systematic offsets, these processes are stochastic: repeated passes of the exact same particle-level objects through the detector would not produce exactly the same reconstructed observables. Folding is hence a broadening of particle-level delta functions into finitely wide reco-level probability distributions, which are then convolved with the particle-level input distributions. We should note immediately, that as the original folding process was stochastic, the unfolding logically cannot be deterministic: whether it is explicitly stated or not, unfolding is a probabilistic inference process, and has to work in terms of indefinite probability distributions.

On a per-event basis, smearing is not a deterministic process, but aggregated over a statistically large number of events, the expected particle-to-reco mapping is well defined. The problem is that this map may be many-to-one: more than one particle-level distribution may produce the same reco-level one, and hence—even in the infinite-statistics limit—a simple inversion of the 'folding function' is not in general possible. Technically, this makes unfolding an *ill-posed problem* (as are many inverse problems). As we shall see, such issues are typically dealt with by *regularizing* the inversion—modifying the algebra such that a stable and unique solution can be obtained with minimal bias. In the real world, we have not just this algebraic singularity to deal with, but the additional complications of finite statistics, and dependences of our detector-behaviour estimates on unknown nuisance parameters. Before we consider these difficulties, it is useful to establish some common concepts and terminology with which to frame the problem and its attempted solutions.

Unfolding formalism

Unfolding is always an attempt to find a map, or more generally a probabilistic family of maps, between two data spaces: the reco observable space R, corresponding to the measured data we want to correct, and the particle-level 'truth' space T we are targetting in order to make a fiducial measurement. Barring some unforeseen development, the values in each space are real, so $R \in \mathbb{R}^N$ and $T \in \mathbb{R}^M$ where N and M are the number of measurements, i.e. number of distribution bins, in each space. Folding corresponds to $T \to R$ maps, and unfolding to $R \to T$. This relationship is illustrated in Figure 11.5, with the 'detector effects' map from $T \to R$ labelled $\mathcal{D}$, and therefore its inverse $\mathcal{D}^{-1}$ providing the ideal map from $R \to T$. These ideal maps not being available, we instead typically progress with an approximation $\widetilde{\mathcal{D}}$ and its

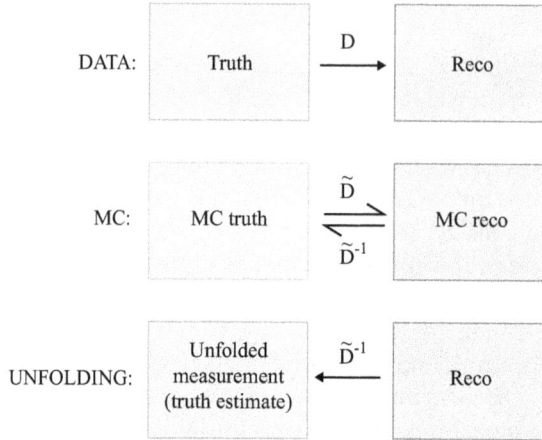

Figure 11.5. Relationship between 'truth' and 'reconstruction' views of collision processes, in both real and simulated events. MC simulations may be used to obtain an estimate $\widetilde{\mathcal{D}}$ of the real detector-effect transform $\mathcal{D}$, and effectively (not always literally) invert it to find an estimate of the true physics parameters.

schematic inversion, obtained from e.g.~MC simulation. Note that in this setup we are not making any requirement that the 'physical meaning' of the values be the same in both spaces: we will return to this, as it provides a powerful framing with which to explore the limits of what can and cannot be unfolded.

The natural way to encode such a map is through linear algebra, that is by expression of forward folding as the transformation of a tuple of truth-space bin values t_j to a tuple of reco-space ones r_i:

$$r_i = \sum_j \mathcal{R}_{ij} t_j, \tag{11.21}$$

where $\mathcal{R}_{ij}$ is the **response matrix**. It is good practice for the r_i and t_i tuples to include bins for migrations from and into regions outside the analysis acceptance, so we have an algebraic handle on such migrations and can include what information we have about them when inverting the folding map.

While written deterministically here, we can view this response equation as the infinite-statistics limit (and maximum-likelihood estimator) of the stochastic folding process, with $\mathcal{R}$ encoding the fractional contributions that each truth-level bin will make on average to each reco-level one. Expressed as a conditional probability,

$$\mathcal{R}_{ij} = P(i \mid j), \tag{11.22}$$

and hence it is normalised as $\sum_i \mathcal{R}_{ij} = 1 \, \forall j$. The intention is, of course, that the response matrix should have a localised structure, ideally with each truth-level bin mapping to one and only one reco-level one; in practice there will always be some width to the distribution, with each truth bin populating several reco bins to different finite degrees.

Note that we have avoided saying that $\mathcal{R}$ should be *diagonal*: while this is often seen, the concept introduces an assumption of equal bin numbers at truth and reco levels, $N = M$, and also the idea that bins with adjacent i or j indices are necessarily adjacent or at least close in the observable space. There is no such requirement for unfolding to work: as framed here, there is no longer a concept of the observables themselves, just the discrete bin indices that they induce. This is an important realisation for e.g. unfolding observables with more than one binning dimension, since it is impossible to create a linear index that captures the multidimensional nature of proximity.

It is freeing to realise that proximity is overrated in this situation: the dimensionality of histograms is a presentational detail, and our interest is purely in the bin indices. As long as bins in any dimensionality can be ordered—in *any* deterministic way—and hence assigned unique indices, the mapping between their indices in two distinct spaces can be defined and assigned a probabilistic structure. The response matrix for such observables will certainly not have a 'diagonal' response matrix as usually understood, but this is no indication that they cannot be unfolded. Even in the one-dimensional case, diagonality is too strict a condition: as an extreme example, any response matrix with exactly one entry per row (i.e. for each reco bin) can be exactly unfolded, even though the layout may appear far from diagonal. The logical conclusion of this thinking is that there is no need for the truth and reco observables to even be computed using related definitions: an unfolding's power is determined purely by the degree of correlation or mutual information as expressed through the response matrix.

So the response matrix is a key quantity for assessing the degree to which detector effects *migrate* observable contributions between bins. How can we obtain an estimate of it? $\mathcal{R}_{ij}$ can be estimated within an analysis by use of MC simulation event samples, as these provide us with event-by-event pairs of (i, j), i.e. the mappings between bin-fills at truth and reco levels. Running over an MC event sample, we can assemble a 2D histogram in (i, j) known as the **migration matrix**:

$$M_{ij} = \sum_{n \in \text{events}} w_n \, \Theta(r^{(n)} \in i) \, \Theta(t^{(n)} \in j), \tag{11.23}$$

where the Θ functions indicate whether the truth and reco observable values for event n fall into the jth and ith bins in their respective space, and w_n is the event weight (possibly including pre-unfolding efficiency-correction factors). The migration matrix is proportional to the joint probability $p(i, j)$ of populating reco-bin i *and* truth-bin j, and indeed we will assume from here that it has been normalised from its raw form such that $\sum_{i,j} M_{ij} = 1$. A crucial feature is that M encodes not just the way in which detector effects split a given truth-level contribution into multiple reco bins, but also the truth-level distribution across the bins. We cannot allow this assumption, dependent on the event generator physics, to bias fiducial observable extractions from data, which is why we instead work in terms of the response matrix, calculated via

$$\mathcal{R}_{ij} = M_{ij} \Big/ \sum_i M_{ij}. \tag{11.24}$$

Unfolding is conceptually an inversion of $\mathcal{R}_{ij}$, i.e. an attempt to perform the map

$$t_j \approx \tilde{t}_j = \sum_i \mathcal{R}_{ji}^{-1} r_i. \tag{11.25}$$

This should not be taken too literally as an algebraic statement. In particular, our insistence that the numbers of bins N and M do not necessarily need to match precludes a literal matrix inversion, although there is a simple extension to deal with this situation. However, it is reasonable to note from an information perspective that we cannot reliably 'invert' from a small number of reco bins to a large number of truth-level ones—at least not without introducing a significant model dependence to interpolate the fine-resolution details, which would be a betrayal of our no-extrapolation fiducial principles! While rarely followed up in practice, one could potentially unfold from one observable—possibly multi-dimensional—at reco level to a different but correlated observable at truth level provided there is sufficient information in the reco-level input.

In the following sections we explore, in roughly increasing complexity, how various real-world unfolding schemes attempt to generalise this inversion concept into a form able to cope better with numerical instabilities and to acknowledge explicitly the statistical limitations and probabilistic nature of the inference.

Matrix inversion and bin-by-bin unfolding

The simplest approach to unfolding is to assume the infinite-statistics limit, in which the relative widths of the distributions whose means are given by the response matrix go to zero, and the linear algebra can be taken literally.

Differing reco and truth bin numbers, $N < M$, can be handled by replacement of the matrix inversion with the Moore–Penrose **pseudoinverse**, based on the singular-value decomposition (SVD) of an $m \times n$ matrix M:

$$M = U\Sigma V^*, \tag{11.26}$$

where U and V are $m \times m$ and $n \times n$ unitary square matrices, respectively, and Σ a real, rectangular diagonal matrix. The pseudoinverse of this is $M^+ = V\Sigma^+ U^*$, with Σ^+ computed by simply taking the reciprocal of each non-zero element on the diagonal, and transposing.

Unfortunately, this literal algebraic inversion is usually too naïve, due to numerical ill-conditioning of the inversion. (Technically, it is typical for the response matrix's condition number—the maximum absolute ratio of output change to an infinitessimal input change—to be large, precluding a stable matrix inversion.) Relatively overbinning the reco observable *may* help, but the risk of taking the reciprocals of infinitesimally non-zero values in the final step means that the pseudoinverse is intrinsically no less numerically fragile than the standard square inverse.

Confronted with this problem, we can either try our best to ignore it, or we can attempt to **regularise** the inversion. 'Ignoring' here means giving up on unfolding the structure of migrations between bins, requiring $N = M$, and assuming the response matrix to be diagonal, estimated from MC event samples as

$$\tilde{\mathcal{R}}_i = \mathrm{diag}(r_i/t_i). \tag{11.27}$$

This **bin-by-bin unfolding** simply corresponds to making 1D histograms of the observables at truth and reco levels, and obtaining per-bin correction factors by dividing one by the other. The inversion then becomes a matter of multiplying each data value by the reciprocal of its corresponding bin-response factor,

$$\tilde{f}_i = d_i/\tilde{\mathcal{R}}_i. \tag{11.28}$$

This approximation is clearly simplistic, but is also undeniably easy to perform and debug, and as such has frequently been used in practice for 'quick and dirty' unfoldings where the aim is less a precision result than to publish *some* indication of what is going on in a certain important phase-space in lieu of a longer-timescale detailed study: two examples are the time-sensitive first measurements of minimum-bias QCD distributions at LHC start-up and in particular from the first pp runs above the Tevatron's 1960 GeV energy frontier, and the relatively new development of roughly unfolded distributions in control regions as a valuable side-effect of reco-level BSM search analyses. The main requirement for using bin-by-bin unfolding is that bin-migrations are minimal, or at least below the statistical and systematic uncertainties of the measurement: this tends to mean that regrettably large bins have to be used, but this is better than no measurement at all.

The other approach is to attempt to numerically regularise the matrix inversion. Many methods exist for matrix-inverse regularisation in general, of which the most famous is **Tikhonov regularisation**, adding positive numbers to the matrix diagonal to reduce its condition number. In HEP unfolding, the most well-known regularisation scheme is the **SVD unfolding** method, again making use of the singular-value decomposition, but in a different way. The approach used is rather involved, but its key elements are

- the numerical instability in algebraic inversion of the response matrix can be traced directly to the appearance of large off-diagonal response terms (or their $n \neq m$ equivalents);
- the folding equation (11.21) can be rewritten as a least-squares problem, in which the quantity to be minimised (by finding values $\tilde{t}_j$) is $y = \sum_i (\sum_j \mathcal{R}_{ij}\tilde{t}_j - r_i)^2/\Delta r_i$, where Δr_i is the uncertainty on r_i;
- the numerical inversion instability can be regularised by adding a constraint term $y \to y + \tau \Delta y$ with $\Delta y \sim \tau \sum_i (\sum_j C_{ij}\tilde{t}_j)^2$. Here C is a matrix,

$$C = \begin{pmatrix} -1 & 1 & 0 & 0 & \cdots \\ 1 & -2 & 1 & 0 & \cdots \\ 0 & 1 & -2 & 1 & \cdots \\ & \vdots & \vdots & & \ddots \end{pmatrix} \tag{11.29}$$

which ensures that Δy is the sum of squares of numerical second-derivatives across the unfolded $\tilde{t}_i$ spectrum: its inclusion in the minimisation term y hence suppresses high-curvature oscillating solutions $\tilde{t}_i$, corresponding to a physical requirement of smoothness in the extracted spectrum;
- minimisation of this least-squares system requires calculation of C^{-1}: this can itself be regularised by the Tikhonov method;
- finally, the size of the regularisation term is governed by the τ parameter, and the optimal value of this can be obtained by study of the singular values

(the diagonal terms in the SVD Σ matrix), which approach zero to suppress the high-frequency solutions: for a rank-k system, in which only the first k contributions are significant, the optimal $\tau = \Sigma_{kk}^2$.

This method also permits full propagation of the effective covariance matrix due to the unfolding process. The SVD unfolding has, for reasons that are not entirely clear, been largely eclipsed by the 'iterative Bayes' method to be described in the next section; however, it is more explicitly about numerical well-behavedness, as compared to the other's more probabilistic motivation, and it is worth being aware of its existence and the approaches used within it, should other methods prove numerically troublesome in practice.

Iterative Bayes

An alternative approach to the regularisation problem is that of **iterative Bayes unfolding** (IBU), for some time the most popular approach to detector-effect correction at particle colliders. The core motivation for this method is complementary to that of SVD unfolding: its concern is less the numerical instabilities and more that the linear algebra approach to building the map from the space of reco-level observables to truth-level observables focuses on the asymptotic statistical limit in which the mean bin-migration effects encoded in the response matrix deterministically connect truth-level fill values to their reco-level counterparts.

The IBU picture—unsurprisingly—uses Bayes' theorem to express the probability of the fiducial value vector t as a function of the reco values r and the response matrix $\mathcal{R}$,

$$P(t|r, \mathcal{R}) = \frac{P(r|t, \mathcal{R}) \cdot P(t)}{\sum_t P(r|t, \mathcal{R}) \cdot P(t)}. \tag{11.30}$$

The left-hand side here is not quite what we want, as it holds a dependence on the response matrix, which in general is not perfectly known—a conceptual refinement of some importance from here on. To eliminate this dependence, much as the space of t configurations is eliminated in the denominator of equation (11.30), this result must be integrated over the space of response matrix configurations with its own prior probability density,

$$P(t|r) = \int_{\mathcal{R}} P(t|r, \mathcal{R}) p(\mathcal{R}) \mathrm{d}\mathcal{R}. \tag{11.31}$$

Limiting to the case of a binned differential distribution, and unit 'fills' of its bins from the events being processed (including out-of-range overflow bins), the truth-to-reco probabilistic map for a given truth bin containing t_j fills is not a continuous Poisson but a discrete multinomial distribution, with $j \to i$ index-mapping probabilities $\mathcal{R}_{ij}$,

$$P(r|t_j, \mathcal{R}) = \frac{t_j!}{\prod_i r_i!} \prod_i \mathcal{R}_{ij}^{r_i}. \tag{11.32}$$

The desired likelihood term (the RHS numerator in equation (11.30)) is the sum of such multinomial distributions over all truth bins j. Were we in the statistical limit where smooth Poisson distributions could be used, the sum of multiple Poissons would result in yet another Poisson, amenable to further analytic simplification, but this is not the case for the discrete multinomial and so analytic attempts to define the unfolding for finite fill-statistics run aground. The currently used version of the IBU algorithm (as opposed to the first version) uses sampling from these multinomial distributions via a Dirichlet distribution (their unbiased Jeffreys prior) to go beyond the infinite-statistics assumptions of previous methods.

The other defining feature of the IBU method is of course its introduction of the prior over truth spectrum values. Setting this term to unity produces the maximum-likelihood estimators of the fiducial values, but implicitly assumes a flat spectrum—a bias which leaves some residual effect in the final results. The iterative algorithm essentially implements an intuitive procedure to reduce this bias: if one were to make a likelihood-based inversion by hand, with a flat prior on one set of data, the resulting $\tilde{t}$ would be the new best estimate of the real spectrum. One could then (ideally on a new and independent dataset, to avoid reinforcing statistical fluctuations), use that estimated $\tilde{t}$ as the prior in a second unfolding, and so on. In practice, the algorithm recycles the same data several times to iterate its prior estimate, which leads to instabilities: it is important in IBU to check the dependence of the result on the number of iterations used, a rather unsatisfying feature and one of many cross-checks required by the method (see Section 11.3.2).

The final step in IBU is the integral over response matrix uncertainties, which is performed by sampling. Systematic uncertainties are fed in, *incoherently* with respect to the response matrix variations which are fed by the same sources, by explicit use of '1σ variation' r inputs.

IBU is a very popular method, in large part due to its readily available implementation in software packages such as RooUnfold and pyunfold, and has been used for the majority of LHC unfolded analyses in Runs 1 and 2. But it is not beyond criticism: as the iteration process is statistically *ad hoc*, it is more 'Bayes-inspired' than fully Bayesian, and its regularising effect is semi-incidental by contrast with the explicit treatment in the SVD method. (The original code applied a further *ad hoc* regularisation via polynomial-fit smoothing in the prior iteration, but this is not enabled by default in the more commonly used RooUnfold implementation.) There are also instabilities inherent to recycling the input data through multiple iterations, hence in practice it is not uncommon to see very small numbers of iterations, N_{iter}, in use, and without a clear, probabilistic way to treat the *a priori* N_{iter} as a nuisance parameter with its own probability mass function. Finally, current implementations propagate systematics point-wise rather than holistically mapping the full likelihood function—a limitation for re-interpretations as analysis data becomes more precise, and correlations become limiting factors. In the following section we move to unfolding methods which embrace modern computing power to more completely and probabilistically map the full, high-dimensional pdf of fiducial cross-sections and their systematic uncertainties.

Likelihood-based and fully Bayesian methods

We move now to the final, and arguably most complex form of simulation-based unfolding: mapping probability density functions in the space T of truth-level unfolded bin values, given the observed data and our knowledge of both the nominal detector response matrix and its sensitivity to systematic uncertainties. The resulting object is a full functional form in at least M parameters (the number of truth-level bins) plus any nuisance parameters, as opposed to other methods' list of M bin-values in nominal and systematic-variation modes.

While this seems complicated, the method itself is conceptually rather simple. If we had unlimited computing resources, for every point in the parameter space $\theta = \{\phi, \nu\}$ (divided into both parameters of interest (POIS) ϕ from T, and the nuisance parameters ν) we would evaluate a function proportional to the likelihood $\mathcal{L}(\theta; D) = p(D|\theta)$ or the posterior pdf. From this we can extract a nominal bin value (as the maximum likelihood estimator, posterior mode, or similar parameters indicating a region of high probability density), error bars (from the confidence interval (CoI), credible interval (CrI) etc), and higher-order approximations to the full pdf or likelihood, such as the covariance matrix and its nuisance projections (see Section 5.2.3). This is unfolding viewed as parameter estimation, with all the accompanying statistical machinery.

A key simplification of this method relative to others is that the calculation is principally based on the forward map $\tilde{t} \to r$, as opposed to the inversion $r \to \tilde{t}$ attempted by our coverage thus far. Unlike those inversions, likelihood or pdf determination is not intrinsically plagued by numerical instabilities from highly fluctuating potential solutions, and so one technical problem is avoided, or at least greatly reduced in importance.

The key objection, of course, is that we do not have unlimited computing resources: these methods are significantly more computationally intensive than the others, hence were for a long time not tractable. In particular, a super-naïve strategy based on repeated forward propagation of the response matrix on random points in the high-dimensional θ space will *never* find the typical set. It is essential that the likelihood function is smooth enough to guide the exploration of truth-values toward high-probability configurations. For this to work, some probabilistic 'width' is needed in the forward map: if the map $\tilde{t} \to r$ were one-to-one, i.e. $p(r|t)$ as a delta function in R, the chance of the sampler finding any configuration with non-zero probability would be negligible. Fortunately, the Poisson or multinomial distributions governing event distribution, and reasonable priors/penalty terms to be applied to nuisance parameters have broader support over the parameter space, and their gradients can be exploited as usual by optimisation and sampling algorithms. The unfolding evaluation time is nevertheless likely to be measured in hours rather than seconds or minutes as for other approaches, but this is not usually a show-stopper and may be worth the effort for the automatically richer output[5].

---

[5] In particular, more detailed likelihood information permits more coherent combination of many analyses into a composite likelihood for *post hoc* 're-interpretation' studies—see Chapter 12.

We have been perhaps conspicuously coy about the exact function to be estimated. This is, of course, because of the usual frequentist/Bayesian dichotomy on the one true way to interpret observations. Just as BSM search efforts are split into those preferring frequentist or Bayesian methods for hypothesis testing (cf. Sections 5.4 and 5.5), these very closely related unfolding approaches have both likelihood-oriented and posterior-oriented implementations. Canonical examples of these are, respectively, TRexFitter and PyFBU, following **likelihood-based unfolding** and '**fully Bayesian unfolding**' (FBU), respectively. We will here present the idea through the FBU formalism, as in practice the difference is less in the form of the pdf and more in its interpretation and how parameter estimates are obtained from it.

The FBU posterior probability for each observable is constructed as the product of Poisson probabilities over all reconstruction-level bins i as a function of the model parameters θ,

$$p(\theta|D) \propto \mathcal{L}(\theta; D) \cdot \pi(\theta), \tag{11.33}$$

where π is the prior probability density. The FBU and other frequentist models are based (as are the SVD and IBU techniques) on the discrete event/fill counts in the observable bins, but rather than attempt to invert the map to find the most probable true number of fills (cf. IBU's use of the multinominal distribution for bin-migration of a fixed number of truth-level events), the θ are identified with the *mean expected* event yields in each bin, $\lambda(\theta)$, and so a per-bin Poisson distribution of reco-level events, $r_i \sim \mathcal{P}(\lambda_i(\theta))$. The total likelihood can hence be expressed as a product of Poisson likelihoods over the bins,

$$\mathcal{L}(\theta; D) = \prod_{i \in \text{bins}} \mathcal{P}(D_i; \lambda_i(\theta)). \tag{11.34}$$

Now we introduce the $t \to r$ map, using as before the response matrix $\mathcal{R}$ specific to the analysis phase-space and the observable's truth and reco binnings. We are only bothered about making this map for the signal process, however, so can separate contributions to $\lambda(\theta)$ into the integrated luminosity (and its systematic uncertainty) $L(\nu)$ and the reconstruction-level background and signal cross-sections b_i and s_i, and further deconstruct the signal portion into the set of signal-level bin cross-sections $\sigma \in \phi$ and the response matrix:

$$\lambda_i(\theta) = L(\nu) \cdot (b_i(\nu) + s_i(\theta)) \tag{11.35}$$

$$= L(\nu) \cdot \left(b_i(\nu) + \sum_j \mathcal{R}_{ij}(\nu) \cdot \sigma_j \right). \tag{11.36}$$

An important feature of this encoding is that the response matrix is made dependent on the nuisance parameters ν, i.e. the uncertainty in the response structure can be smoothly encoded as a function of the nuisances, usually by use of interpolated MC templates as for the background modelling $b(\nu)$. In some circumstances, this may be overkill and the nominal response matrix can be used instead, with just a 'diagonal' modification of signal responses to nuisance-parameter variations,

$$\lambda_i(\boldsymbol{\theta}) \approx L(\boldsymbol{\nu}) \cdot \left(b_i(\boldsymbol{\nu}) + \left[1 + \sum_{\nu \in \boldsymbol{\nu}} (\nu_i - \nu_{0,i}) \delta_i \right] \left[\sum_j \mathcal{R}_{ij}(\boldsymbol{\nu_0}) \cdot \sigma_j \right] \right), \qquad (11.37)$$

where ν_0 are the nominal nuisance parameter values (usually encoded as $\boldsymbol{\nu} = 0$), and the δ_i are the (MC-estimated) reco-level multiplicative modifications to bin values induced by the usual 1σ systematic variations corresponding to $\nu_i = \pm 1$.

Now we have a posterior function, and for fixed data $\boldsymbol{D}$ this expression can be evaluated for a given parameter set $\boldsymbol{\theta}$. The prior $\pi(\boldsymbol{\theta})$ (and its penalty-term equivalent in likelihood-based unfolding) is most usually set to a collection of unit Gaussians in the nuisance parameters $\boldsymbol{\nu}$,[6] and a uniform or Jeffreys prior[7] in the case of the POIs $\boldsymbol{\phi}$, where biasing is to be avoided.

You may be wondering about the difference between the FBU approach, and the frequentist likelihood-based unfolding approach. In the frequentist approach too, standard Gaussian penalty terms are added to the likelihood to stop the nuisances moving too far from their nominal values. As with their equivalents in profile fits for Higgs and BSM searches, these occupy a vague space between frequentist and Bayesian approaches; the remaining distinction between FBU and likelihood-fit unfolding is hence largely in whether the nuisances are to be marginalised over (FBU) or profiled out (likelihood), and whether CoIs or CrIs are used to construct the resulting unfolded nominal bin-values and error bands. These treatments are standard and follow the procedures described in Chapter 5. The practical distinction, for many purposes, is the convergence time of the fit or scan, which tends to be much faster for the profiling approach: it is easier, especially given a reasonable sampling proposal model, to find a global maximum and map its functional vicinity, than to comprehensively integrate over the entire typical set of the high-dimensional pdf.

The HBOM method

The common thread running through all of the methods described so far in this section has been their reliance on MC modelling of the fundamental physics process and of the detector and reconstruction response to fiducial observables. It is the event-by-event association of truth-level to reco-level values enabled by MC simulation that underpins estimates of $\mathcal{R}$ and its systematic variations, from which the inverse map from reco-level observations to fiducial values (or probability structure) is attempted. But what if the MC modelling, either from the generator or the detector, is unreliable? A fundamental bias— in the response and in the IBU iterated prior—would be imposed on the measurement.

The HBOM method (a backronym for 'Hit Backspace Once More') is a neat alternative to MC-driven unfolding, relying instead on repeated application of

---

[6] This is a major assumption about the statistical role of the template variations from object-calibration prescriptions, particularly for ad hoc two-point uncertainties such as alternative MC generator choices, but a common one nonetheless.

[7] The maximally uninformative prior for a Poisson statistical model is $p(\lambda) \sim 1/\sqrt{\lambda}$.

known, calibrated detector uncertainties. In this sense it is like a non-linear extension of the pre-unfolding ideas introduced in Section 11.3.1, especially in that it is applied at the level of physics objects, rather than resulting observable values (or their corresponding bin indices).

The main building-block of HBOM is that every physics object measured at reco-level can be related to its truth-level equivalent by a detector function $\mathcal{D}$, which performs both efficiency losses (i.e. an object like a track or a flavour-tag which is present at particle-level but has disappeared in the reconstruction view), and kinematic distortions. The mean rates of such losses are given by the efficiency and mistag numbers introduced earlier, parametrised or tabulated as a function of object proper-ties—most notably $(p_T, |\eta|)$. The smearing effects can be described by pdf functions, again dependent on object properties. A single 'history' of random detector effects on a particle-level event can hence be obtained by sampling one set of losses, mistags, and distortions from these distributions for each affected object in the event, and returning the resulting modified event. One can of course construct in parallel many such random histories, and hence map the effects of the detector response with some statistical weight, but one would need generated MC input events for that, and the point of HBOM is that it is for use where we do not trust the MC.

The key insight from HBOM is to use the already-reconstructed set of observed events as the target for the $\mathcal{D}$ function. A good example is track-jets, which we interpret at reco level as the result of jet-finding on the set of primary charged particles, with some lossiness due to tracking efficiencies and momentum misrecon-struction. To perform HBOM, we retrieve the reco-level event before application of the jet-finding to the set of tracks $\{t_i^{(0)}\}$ from which track-jets $\{j_i^{(0)}\}$ were formed, and apply the tracking detector-function $\mathcal{D}_t$ *again*, on every track $t_i^{(0)} = (p_{T,i}, \eta_i, \phi_i)$

$$\tilde{t}_i^{(1)} = \mathcal{D}_t \cdot t_i^{(0)}, \tag{11.38}$$

where the '(n)' index indicates how many times the HBOM $\mathcal{D}$ function has been applied[8]. The track-jet finding is then run, giving a new set of track-jets $\{j_i^{(1)}\}$; we can view this as a collective—and non-linear—track-jet detector function $\mathcal{D}_j$. From here, a new set of 'reco-squared' binned observables can be computed, $r_i^{(1)}$ to complement the standard reco-level $r_i^{(0)}$. If we wanted we could linearly extrapolate this effect backward to the $r_i^{(-1)}$ configuration, i.e. before the *real* detector effect and hence an estimate of the fiducial values:

$$r_i^{(-1)} \approx r_i^{(0)} - \left(r_i^{(1)} - r_i^{(0)} \right) \tag{11.39}$$

$$= 2r_i^{(0)} - r_i^{(1)}. \tag{11.40}$$

But the non-linearity gives pause for thought, and so we apply the $\mathcal{D}$ functions again on the whole event to give the second, third, nth etc. HBOM iterations, denoted by the (n) index:

---

[8] $\tilde{t}^{(1)}$ can be 'null', due to the sampling deciding that track does not get reconstructed this time: this is more easily handled in source code than in mathematical formalism.

$$j_i^{(n)} = \left(\mathcal{D}_j\right)^n \cdot j_i^{(0)}. \tag{11.41}$$

In practice this should be done with full statistical independence for each $j^{(n)}$, rather than naïve iteration $j^{(m+1)} = \mathcal{D}_j \cdot j^{(m)}$. The resulting set of $r_i^{(n)}$, for every bin i, is then a non-linear series of iterated detector effects on the observable, which can easily be fit with a polynomial and extrapolated back to the $n = -1$ iteration. This is the core of the HBOM method, which has been applied with success and minimal simulation-dependence in several LHC analyses, especially for non-perturbative and low-p_T physics where the MC modelling is not expected to be robust, such as in the minimum-bias particle-correlation analyses shown in Figure 11.6.

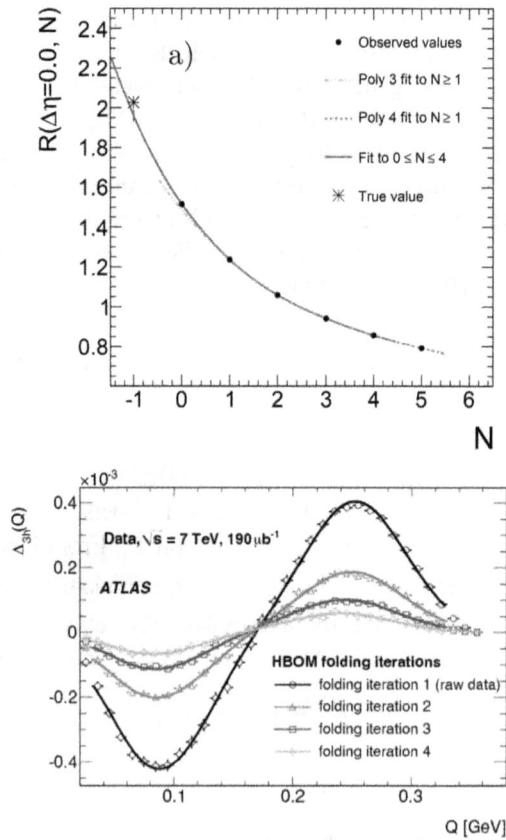

Figure 11.6. Use of the HBOM method for unfolding of 2-particle correlation data (the original HBOM application, top), and azimuthal ordering of hadron chains (bottom). Both analyses were studies of phenomena unimplemented or incompletely implemented in MC event generators, mandating a data-driven unfolding method. Reproduced with permission from Monk J and Oropeza-Barrera C *Nucl. Instrum. Meth.* A **701** 17–24 (2017), arXiv:1111.4896; and ATLAS Collaboration *Phys. Rev.* D **96** 092008 (2017), arXiv:1709.07384.

Unfolding uncertainties and cross-checks

As alluded to several times, unfolding is an inference from imperfect information regardless of which approach is used to tackle the map from detector-level to fiducial observables. As such, it is inevitable that the uncertainties on the particle-level results must be greater than those on the reconstruction quantities: most strikingly, a perfectly error-free integer number of observed events in a bin has to correspond to an uncertain differential cross-section.

It is also worth stressing here that the propagation of these uncertainties assumes that the observables being unfolded have a free normalisation, rather than 'shape' distributions with fixed integral. By construction, the latter incur crossing points or 'pinch points' of zero or reduced width in their uncertainty bands, since increases in one part of the distribution must be compensated by decreases elsewhere. This complicates the migration structure, and hence unfolded fixed-normalisation observables should be obtained by performing the correction and unfolding steps on the floating-normalisation observables, then performing the normalisation post hoc on the resulting distributions. As usual these uncertainties can be classed as statistical and systematic, with the latter category including not only the 'usual' theoretical modelling and experimental calibration uncertainties, but also imperfections in the unfolding mechanism itself.

Statistical uncertainties from the observed data are treated most completely in the IBU and pdf-based unfolding approaches. In both, the explicit use of a Poisson distribution (whose variance, recall, is related to the mean rate λ as $\sigma^2 \sim \lambda$) for the event yield automatically encodes statistical fluctuations in data numbers into the resulting error-bars and probability distributions. In the IBU case, a second source of width is encoded via the multinomial distribution used for the bin-migrations of finite numbers of events. IBU implementations automatically propagate this uncertainty; pdf methods require it to be extracted instead as the 'remaining' uncertainty, once systematic uncertainty projections have been subtracted from the total CoI or CrI in (assumed) quadrature. In more basic methods like bin-to-bin or SVD unfolding, or in the HBOM method, which do not explicitly acknowledge the probabilistic nature of the physics process or bin-migration process, a more manual approach will be needed: for example, creation of an ensemble of unfolded results via Poisson bootstrapping of their inputs, for extraction both of statistical errors on the result values and estimates of their statistical correlation.

Systematic uncertainties are also treated manually in most unfolding treatments, requiring explicit re-runs of the unfolding process with discrete alternative versions of the reco-level bin counts, background estimates, and migration matrices obtained through '1σ' variations of event generator parameters, calibration working points, and so on. The likelihood and fully-Bayesian methods that compute a full likelihood or posterior function are the outliers here, the richness of their outputs justifying to some extent their greater runtime: the likelihood or posterior function from a single run explicitly includes dependences on the systematic nuisance parameters ν on an equivalent footing to the parameters of (physics) interest, ϕ. The marginalising or profiling of nuisance parameters directly contributes to the broadening of the data-compatible region of ϕ, and hence systematic contributions to the total CoI or CrI can be extracted again via covariance projection (cf. Section 5.2.3). For systematic correlations, most methods have to extract this by the 'toy MC' method of sampling from the nuisance parameters (e.g. again using ad hoc Gaussian distributions) and using a template-interpolation between nominal and 1σ unfolding results. For the pdf methods, this is again 'built in' to the resulting distribution, with the covariance obtainable directly from the pdf samples: this can be cut down to the ϕ covariance sub-matrix to represent overall bin-to-bin correlations.

Unfolding uncertainties are the remaining issue to deal with. These are of course dependent on the method, with more complex methods providing more explicit handles for evaluating uncertainty, e.g.

SVD unfolding: dependence on the number of singular values considered statistically significant when choosing the regularisation scale;

Iterative Bayes: dependence on the number of iterations used to obtain the final prior, MC inputs to the prior determination, and dependence on the regularisation technique;

FBU: dependence on choices of priors, particularly on the parameters of interest;

HBOM: ambiguities in both the maximum number of detector-function iterations, and the fitting method used for extrapolation. Use of ensembles of HBOM 'histories' can help to assess the stability and uncertainty of the result extraction;

All but HBOM: uncertainties from MC statistics, particularly in stably populating all bins of the migration matrix—this is likely to be impossible in the lowest-probability bins, but fortunately these by construction contribute to very subleading migration modes: the most important limitations are in the leading migrations.

These require detailed checking in an analysis, which can be a difficult and time-consuming process, especially as the number of cross-checks requested in both internal and external review can be large. In addition, it is normal to perform further standard MC-based 'stress tests' and 'closure tests' to verify that the method *should* be performing adequately for the chosen observable. These typically include:

- showing that particle-level predictions from MC model A will be reobtained to within statistical precision when unfolding reco-level A with MC prior A;
- showing that closure remains when unfolding reco-level A using prior model B;
- reweighting the prior model such that its forward-folding approximately matches the reco-level data.

Even in perfect closure-test conditions, some degree of numerical non-closure may be observed due to issues such as the residual impact of discreteness from e.g. Poisson-distribution sampling. If significant, the degree of non-closure is typically added in quadrature to affected bins in the final results.

Developments in unfolding

Unfolding techniques are an ever-evolving area of technical development, spurred by the open-endedness of the topic: the incompleteness of the information means there are myriad approaches that can be taken; observables can always be made higher-dimensional or otherwise more difficult to handle; and increasing computational power makes possible methods that had to be dismissed as unworkable by earlier generations.

At present most analyses are content with the 'standard' solution of iterative Bayes unfolding, but its shortcomings in the completeness of the returned pdf in the new era of LHC precision physics may stimulate either refinements and extensions of the technique and software, or a shift to use and publish the more explicitly pdf-oriented methods: this is linked to many important issues of maximising analysis-data reusability, to be discussed in Chapter 12.

As with everything else in HEP, the front line of development includes initiatives based on leveraging machine-learning technologies. A notable instance of this is the development of Omnifold method, built on the **particle-flow networks** framework, which applies neural networks with a 'deep set' structure to characterise the dominant momentum configurations within the high-dimensional 'full phase-space' of collider events: these can be leveraged to perform the iterative-Bayes method's reweighting iterations into output particle-level observables without needing to bin or reduce the dimensionality of the reco-level events. In practice, lower-dimensional inputs are currently necessary, but the concept of being able to perform very differential and generic unfolding opens new possibilities that are likely to grow further with computing firepower.

Further reading

- Truth definitions for fiducial analysis at the LHC are publicly discussed in ATL-PHYS-PUB-2015-013, https://cds.cern.ch/record/2022743/.
- Not many public resources exist for review of top-quark reconstruction methods. One of the few is Kvita J, *Nucl. Instrum. Meth.* A **900** 84–100 (2018), arXiv:1806.05463, which presents a comparison of optimisation targets in kinematic fits.
- The SVD unfolding method is described in Hoecker A and Kartvelishvili V, SVD approach to data unfolding, *Nucl. Instrum. Meth.* A **372** 469–81 (1996), arXiv:9509307. The updated form of the IBU unfolding method is described and discussed in D'Agostini G, Improved iterative Bayesian unfolding, arXiv:1010.0632 (2010), and Choudalakis G, Fully Bayesian unfolding, arXiv:1201.4612 (2012).

Exercises

11.1 Which of the following need to be included in the list of fiducial cuts for an SM analysis? a) primary vertex position, b) displaced vertex cuts, if any, c) b-hadron secondary-vertex positions in flavour-tagging, d) identification of jet constituents, e) requirements on electron hits on inner-tracker silicon, f) photon promptness, g) which visible objects are used to define the MET vector.

11.2 Let's imagine a BSM physics effect manifests in top-quark events, modifying the kinematics such that reconstruction methods tuned for the SM top are less efficient than expected. If placing statistical limits on the BSM model via measurements of partonic top-quark observables, would you expect the effective exclusion range to be reduced (conservative) or enhanced (aggressive) compared to the correct value? What is the equivalent effect for fiducial analyses?

11.3 Using the same approach as for the lepton p_T distribution in equation (11.6), derive the equivalent distribution for m_T. Comparing to the lepton p_T plot in Figure 11.1, what effect would you expect a non-zero W p_T to have on the m_T distribution?

11.4 Show the missing steps in the derivation of equation (11.10).

11.5 Given the ambiguities in leptonic W and top-quark reconstruction, why are precision W and top mass measurements not made via the hadronic channels?

11.6 If detector and reconstruction effects smear an observable uniformly across its monotonically falling spectrum, a) in what direction will the net migration of events occur from a single bin? b) in what direction will the whole spectrum distort? How can these be reconciled?

11.7 This is a programming task about detector resolutions, distortions, and unfolding, using toy sampling from a single variable.

    (a) Using a standard numerical computing library, sample $\mathcal{O}(10000)$ 'truth event' p_T from a $d\sigma/dp_T \propto p_T^{-4}$ distribution, for $p_T > 20$ GeV.

    (b) For each event, model the reco-jet p_T's by applying a Gaussian-noise smearing with width σ_{p_T} equal to 10% at 20 GeV, and decreasing linearly to a fixed 5% above 200 GeV.

    (c) Now make 10-bin truth- and reco-level p_T histograms, and corresponding migration and response matrices from your distribution of truth and reco events. How diagonal are the matrices?

    (d) Try a naïve unfolding by matrix inversion applied to the reco distribution: how stable and close to the true distribution are the results?

    (e) Use the same distributions and matrices as inputs to the pyunfold or PyFBU toolkits.

11.8 Certain state-of-the-art fixed-order and resummed calculations of high-energy b-quark production require use of a special 'flavour k_T' jet algorithm to cancel IR divergences. This algorithm behaves similarly to the k_T algorithm, but requires parton-level information about its clustering constituents. Putting together what you now know about QFT and MC calculations, fiducial definitions, and folding/unfolding procedures, what issues of theory/experiment comparison arise, and what possible strategies could be employed to solve them? Which seems the best compromise to you?

IOP Publishing

Practical Collider Physics

Andy Buckley, Christopher White and Martin White

Chapter 12

Analysis preservation and reinterpretation

We have now completed our review of the searches and measurements that are performed at hadron colliders, along with the complicated theoretical, experimental and computational knowledge that underpins the Large Hadron Collider (LHC) programme. We have seen that precision measurements of Standard Model (SM) quantities allow us to search indirectly for physics beyond the Standard Model (BSM), whilst resonance and semi-invisible searches allow us to search directly for new particles and interactions.

What we have not yet considered is how to take the *results* of these various LHC analyses, and use them to learn something about the theories we introduced in Chapter 4. To take a concrete example, we have not yet uncovered any evidence for the existence of supersymmetry at the LHC. This must therefore tell us something about which particular combinations of masses and couplings of the superpartners are now disfavoured. To take a second example, we have also not seen any new resonances beyond the Standard Model Higgs boson, which must place constraints on any theory that involves new resonances within a suitable mass range, such as composite Higgs theories, or two Higgs doublet models. For a third example, consider the fact that we have also not made any measurements of SM quantities that differ significantly from their predicted values. This imposes non-trivial constraints on the parameters of BSM theories that are able to provide sizable loop-corrections to SM processes.

In all of these cases, we need to understand how to take LHC data that was developed in a very specific context, such as a search optimised on a particular assumed theory, and *reinterpret* it in a generic theory of interest. Note that *all* LHC results might in principle be relevant to a theory we are interested in, and we should take care to consider any results that we think might be relevant. In particular, this means that precision measurements can be as useful as searches for new particles for constraining the parameter spaces of new theories.

Accurate reinterpretation is a crucial topic for both theorists and experimentalists. The former must learn how to rigorously make use of published LHC data,

doi:10.1088/978-0-7503-2444-1ch12

whilst the latter must learn how to present their data in a way that is most useful for theorists, and how to design future experimental searches that use a full knowledge of which areas of a theory are no longer viable.

12.1 Why reinterpretation is difficult

A simple example will suffice to demonstrate the typical challenges encountered in the reinterpretation of LHC data. Let us say that we are interested in the Minimal Supersymmetric Standard Model, for which 24 parameters are typically required to describe the possible LHC phenomenology[1] We wish to know which regions of that 24-dimensional parameter space remain viable after a decade of null search results at the LHC, including literally hundreds of searches in a huge range of final states. To simplify matters, we will not even consider the precision measurements that might be relevant—it will turn out that the problem is already difficult enough!

The typical presentation of a supersymmetric search result by the ATLAS and CMS collaborations involves the simplified models that we first introduced in Chapter 10. Recall that a simplified model assumes that only particular super-partners are produced, that they decay in a specified way, and that the masses are fixed such that there are only two free mass parameters. Every given simplified model can therefore be viewed as a particular 2D plane of the more general supersymmetric parameter space, and a null search result can be used to define an exclusion limit at the 95% confidence level (CL) following the statistical techniques presented in Chapter 5. Each of the simplified model assumptions may be revisited to provide a further simplified model, including variations of the assumed masses of intermediate sparticles in decay chains, the branching ratios to specific final states, or the types of sparticle produced. For example, Figure 12.1 shows 95% exclusion limits from the CMS experiment for several simplified models of electroweakino production. To produce the left panel, it is assumed that a lightest chargino and second-lightest neutralino are produced, with the chargino decaying to a W boson and a lightest neutralino, and the second-lightest neutralino decaying to a lightest neutralino and either a Higgs boson or a Z boson. The lightest chargino and second-lightest neutralino are assumed to have the same mass. To produce the right-panel, the production and mass assumptions are retained, but it is assumed that the decays now involve slepton production and decay. Three different scenarios are considered, being 1) a model with the same branching fractions to all slepton flavours, 2) a model with 100% branching fraction of the chargino to a $\tilde{\tau}$ whilst the neutralino decays to all three flavours and 3) a model with both the chargino and neutralino decays proceeding exclusively via a $\tilde{\tau}$. The x-values in the legend indicate the ratios of the mass differences $(m_{\tilde{l}} - m_{\tilde{\chi}_1^0})/(m_{\tilde{\chi}_2^0} - m_{\tilde{\chi}_1^0})$, and thus the plot summarises the results of a bunch of different searches in a bunch of different simplified models, all shown in the same mass plane.

---

[1] Other parameters exist, but will not typically affect measurement and search results at the LHC, so we can forget about them.

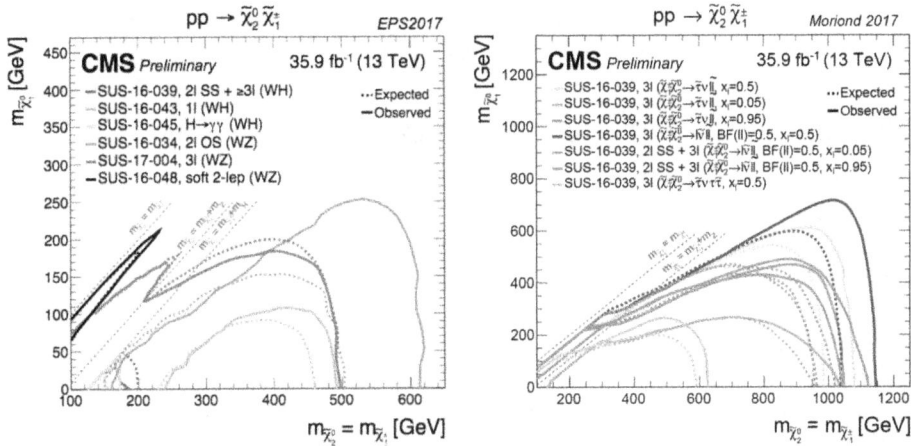

Figure 12.1. Two examples of simplified model presentations of null results in supersymmetry searches. CMS Collaboration, https://twiki.cern.ch/twiki/bin/view/CMSPublic/SUSYSummary2017, Copyright CERN, reused with permission.

Figure 12.1 is instructive in a number of ways. First, it shows us something of the variety of simplified models, even if the models shown are a small fraction of the total number that are used for the presentation of LHC results. Second, it shows us that the strength of the exclusion limit in the mass plane clearly depends on the details of the assumed simplified model, with the exclusion limit getting much weaker for the model that assumes all decays proceed via a $\tilde{\tau}$. A realistic supersymmetric model would complicate things further by having a number of competing production and decay processes, with branching ratios, sparticle couplings and sparticle masses that differ substantially from each of the simplified assumptions of Figure 12.1. Clearly we cannot assume that we know anything about the viability of such a general model based on looking at a set of simplified model figures, even if there might be useful information contained in them that one can extract with further effort. If we want to reinterpret the search within the framework of a non-supersymmetric model, the situation seems totally hopeless.

In the rest of this chapter, we will briefly review some common approaches to the reinterpretation of LHC data. From the very start of the LHC programme, theorists have been reinterpreting LHC results in the context of their favourite model. Extensive conversations between theorists and ATLAS and CMS experimentalists have led to an ever increasing amount of public data for each analysis, which has then further improved the scope and rigour of theoretical analyses, and the number of approaches for making use of the public information. Typically, the information published by an analysis will be similar within each category of semi-invisible particle searches (e.g. supersymmetry searches), resonance searches and precision measurements, with different information preferred across the categories. In all cases, public results usually consist of a minimum of a detailed analysis paper or note, along with electronic data published online in a dedicated database called

HepData. As we shall see, creative use of this information makes it possible to perform very sophisticated reinterpretations of LHC data.

12.2 Fundamentals of reinterpretation: global statistical fits

Before we get into the details of how to reinterpret specific LHC results, let us consider the fundamental principles that apply to all reinterpretation efforts. Assume that we have some physics model with a set of parameters θ. We then have a massive range of LHC data that might possibly constrain the parameters of that model, both from direct particle searches and from precision measurements. How do we determine which parameters of the model are now consistent with LHC data? In the case of null results for BSM physics, we should expect to exclude certain parameter combinations, leaving some region of the space as viable (unless the entire model is now excluded). In the case of positive discoveries, we should instead determine which values of the parameters are consistent with the observed signal.

In the case of exclusion limits, there are a variety of popular approaches for reinterpretation in the theory literature that will give incorrect results. For example, many theorists simply take the published 95% CL exclusion curves for a relevant parameter space, and overlay them in a 2D plane of interest after fixing most of the other parameters in a model. The obvious deficiency of this approach is that the exclusion obtained in the 2D plane of interest depends strongly on the fixed choices of the other parameters. The less obvious deficiency of this approach is that a simple overlaying of exclusion curves is not statistically sound, since it neglects the possibility that a slightly worse fit to one set of observations can be compensated for by a much better fit to another set of observables, leading to good viability overall. Another popular approach is to perform a random sampling of the parameters of a model with many parameters, and to perform a naïve combination of exclusion curves to accept or reject models based on whether they fail any single 95% CL exclusion result. This retains the deficiency of the naïve combination of exclusion limits and adds a further problem: random sampling will not properly explore a parameter space of large dimension due to the problems discussed in Chapter 5, and the results of such a procedure tell us nothing rigorous about the parameter space at all.

Fortunately, the *correct* way to reinterpret LHC data is not only well-known, but there exist open-source public software codes for implementing it. The procedure is valid for both null BSM search results and positive discoveries, and is known as a **global statistical fit**. The starting point for any reinterpretation should be the likelihood of the observed data given a particular set of the model parameters, since this is the correct statistical quantity for determining the viability of that set of parameters. Likelihoods for different results can be combined in a rigorous way to form a *composite likelihood* which determines the viability of a set of model parameters with all relevant data. Although we will confine ourselves to LHC data only in this book, it is important to note that one could easily add likelihood terms for other relevant experimental data, such as astrophysical or cosmological measurements.

As an example, assume that we are at a point $\boldsymbol{\theta}$ in the parameter space of a model, and that we are trying to reinterpret a search for semi-invisibly decaying particles that uses a single signal region in which n events were observed. Assume also that we can predict the number of signal events $s(\boldsymbol{\theta})$ that will be produced at the model point $\boldsymbol{\theta}$, with a systematic uncertainty of σ_s. Meanwhile, assume that the ATLAS or CMS paper for the analysis states that the background prediction was b events, with a systematic uncertainty of σ_b. A common approximation to the likelihood in this case is[2]

$$\mathcal{L}(n|s(\boldsymbol{\theta}), b) = \int_0^\infty \frac{[\xi(s(\boldsymbol{\theta}) + b)]^n e^{-\xi(b+s(\boldsymbol{\theta}))}}{n!} P(\xi)\mathrm{d}\xi. \tag{12.1}$$

This is a Poisson distribution which is marginalised over the probability distribution of a rescaling parameter ξ, to account for the systematic uncertainties on s and b. The ξ probability distribution is peaked at $\xi = 1$, with a width characterised by $\sigma_\xi^2 = (\sigma_s^2 + \sigma_b^2)/(s + b)^2$, and it is common to assume either a Gaussian form,

$$P(\xi|\sigma_\xi) = \frac{1}{\sqrt{2\pi}\,\sigma_\xi} \exp\left[-\frac{1}{2}\left(\frac{1-\xi}{\sigma_\xi}\right)^2\right], \tag{12.2}$$

or a log-normal form,

$$P(\xi|\sigma_\xi) = \frac{1}{\sqrt{2\pi}\,\sigma_\xi}\frac{1}{\xi} \exp\left[-\frac{1}{2}\left(\frac{\ln \xi}{\sigma_\xi}\right)^2\right]. \tag{12.3}$$

If the search has a number of signal regions rather than a single region, one can either define the likelihood using the signal region with the best expected exclusion at each point or, if covariance information is available, one can define a likelihood which correctly accounts for signal region correlations. Alternatively, one can use the likelihood formation supplied by the ATLAS and CMS experiments where available. In cases where signal regions are obviously not overlapping (and one can neglect correlated systematic uncertainties), one can combine the likelihoods for each signal region simply by multiplying them together. This then gives the composite likelihood that will be our measure of the model viability at each parameter point. Although we have discussed semi-invisible particle searches in the above example, broadly similar principles apply in the case of resonance searches and precision measurements. One must simply add the relevant likelihood terms to the composite likelihood.

Given this likelihood for a single model point, we must now consider how to determine which regions of the parameter space as a whole remain viable. The short

---

[2] It is approximate because it does not contain all of the nuisance parameters of the experiment, nor does it contain the details of the data-driven background estimates over the various control regions. It also uses a single rescaling parameter to account for the signal and background systematics, rather than performing a 2D integration.

answer is that we must use a valid statistical framework to find the regions of the parameter space that lie within a specific confidence region. For frequentists, this means that we must find the parameters that maximise our composite likelihood, and define confidence intervals around that maximum. If we want to plot the shapes of these intervals for each single parameter, or for 2D parameter planes, we can make use of the profile likelihood construction. Bayesians must instead define a suitable prior on our model parameters, then use the likelihood to define a posterior that is mapped via sampling. The set of posterior samples can then be used to plot 1D and 2D marginalised posterior distributions. Both of these approaches were covered in Chapter 5.

12.3 Reinterpreting particle searches

For particle searches, the key to obtaining the likelihood of equation (12.1) is the prediction of the number of signal events expected for each parameter point, or the prediction of the signal distribution that is fitted to the data in the case of a bump hunt or a shape fit. The most rigorous way of obtaining this prediction is to act like an ATLAS and CMS experimentalist, and perform a dedicated Monte Carlo simulation of the LHC proton-proton collisions for a given parameter point, followed by a suitable detector simulation, and a reproduction of the ATLAS or CMS analysis selections. Whilst it will never be possible to *exactly* reproduce the proprietary procedures of ATLAS and CMS, one can get surprisingly close thanks to the generosity of the LHC collaborations in the scope of their published information.

This information starts with a description of the signal regions used, and what their kinematic selections were. This description must also contain a careful definition of all relevant kinematic variables, particularly when they are highly non-trivial. It is now common to also provide code snippets that implement the analysis logic, which may be published in HepData or released on the relevant experiment website. For resonance searches, the analysis description is usually simpler, consisting of a description of the variable that one performs the bump-hunt in (usually an invariant mass of the expected resonance decay products), plus the kinematic selections that were applied to the objects used in the definition of the kinematic variable. In the case that the resonance produces a species that itself promptly decays (e.g. top quarks or weak gauge bosons), the analysis description must also include the details of how those objects were reconstructed. Finally, no analysis description is complete without a detailed description of the statistical treatment used to derive any exclusion limits presented in the analysis paper.

The main 'result' of an LHC semi-invisible particle search is the number of events in each signal region found in the relevant experimental dataset (e.g. ATLAS or CMS). Along with simple event counts, it is common to show a number of kinematic distributions in the signal region, e.g. for important signal region variables that are expected to distinguish the background from the signal of interest. It may also be the case that a final exclusion limit was derived by performing a binned likelihood fit with a particular variable distribution, in which case that distribution must be provided. This applies both to resonance searches, and to some semi-invisible searches that rely on fitting the shape of particular kinematic variables. All of this **primary event data** is usually given in both the analysis paper and in HepData in electronic format.

Along with the data yields, it is essential to provide the expected background yields or distribution. For semi-invisible searches, this typically consists of the number of background events in each signal region, which is usually obtained via a data-driven method such as those outlined in Chapter 10. These numbers are practically impossible to reproduce outside of the ATLAS and CMS experiments, making their publication essential. For binned likelihood fits, the equivalent information is the number of expected background events in each bin of the relevant kinematic variable. For resonance searches, we have seen in Chapter 9 that it is instead common to fit the background with a smooth function, in which case one should aim to publish the functional form used, the range over which it was fitted, the fit procedure, and the best fit values of the function parameters along with their associated uncertainties. For analyses with multiple signal regions and/or multiple bins of a kinematic distribution, a complete reproduction of the experimental statistical procedure is only possible if the correlations between signal regions (or bins) are published. These are often published as part of simplified likelihood treatment, such as that detailed in Chapter 5. Where they are not, theorists will be forced to rely on approximations in their reinterpretation efforts.

The above is a perfectly workable procedure, but can go wrong for a variety of reasons. For example, how do we ensure that our Monte Carlo generation of the signal process is a faithful reproduction of what we expect to see at the LHC? The detector simulations of the ATLAS and CMS experiments are not available to theorists, and would be too slow to use in practice in any case. Assuming that we are using a fast simulation of the LHC detectors, how can we ensure that it is up to the job? To aid with this, it is routine to publish a variety of extra information for each LHC analysis that either aids debugging, or directly assists in the accurate reproduction of LHC detector effects. This includes:

- Files that describe the physics models used for the results in the analysis paper. These are usually released in a standard format called SLHA, and are made available through the HepData database. The advantage of releasing SLHA files is that theorists can debug their Monte Carlo simulations on the precise models utilised by the experiments; any attempt at describing the models in the analysis papers is typically sufficiently ambiguous to leave open questions.

- Tables of the number of simulated signal events passing each successive kinematic selection of the analysis, derived using the actual ATLAS and CMS generation, simulation and reconstruction chain. These **cutflow tables** allow theorists to check that they get similar numbers using their own simulations, and to hunt for bugs until they see closer agreement. Typically, a theorist can expect to see an agreement of the final signal region yields at the level of 20% or so.

- Smearing functions and efficiencies, so that theorists can ensure that the parameters of their detector simulation are up-to-date. Although generic trigger and reconstruction efficiencies can be found for both ATLAS and CMS in a variety of detector performance papers, they are often superceded by the time an analysis comes out, and the original references are rarely available in electronic form. Useful public information currently includes analysis-specific object reconstruction efficiencies as a function of p_T and η, official configuration files for public detector-simulation packages such as Delphes, and parametrisations of resolution function parameters for the invariant-mass variables used in resonance searches. Many long-lived particle searches also provide analysis-specific efficiencies, without which external reinterpretation would be very difficult indeed.

12.3.1 Reinterpretation using simplified models

The Monte Carlo simulation approach is arguably the most rigorous possible way of reinterpreting LHC analyses, but unfortunately it is also the most computationally expensive. Statistically-convergent global fits of parameter spaces with even a handful of parameters often require millions to billions of likelihood evaluations, making Monte-Carlo simulation tricky unless one has sufficient CPU hours to speed up the simulation via parallelisation. For this reason, it is often useful to make approximations that, whilst not fully rigorous, give a reasonable idea of the viability of each set of model parameters.

We have previously considered simplified models in the context of supersymmetry, and it remains true that 95% CL exclusion contours in 2D parameter planes in these models are *not* useful for reinterpretation. However, simplified models can be made very useful through the publication of additional information, and one can also generalise the concept beyond supersymmetry to other physical theories of interest. For example, limits from resonance searches are commonly presented as bounds on the production cross-section times branching ratio to the final state of interest, as a function of the resonance mass, for different choices of the width and spin of the resonance. Such limits count as simplified models, since they do not require a Lagrangian of any particular BSM physics model to be specified. They will typically remain valid provided that interference or threshold effects are absent in the theory of interest.

One way to make simplified models useful for supersymmetry searches is to borrow this idea from resonance searches and publish 95% CL upper limits on the signal production cross-section in each simplified model plane. For each parameter point in these planes, the upper limit results from the fact that a null result cannot remain compatible with an arbitrarily high signal production cross-section. Some searches also publish signal acceptances and efficiencies in the same 2D model planes, which allow contributions from several simplified models to be combined for arbitrary supersymmetry models that depart from the simplified assumptions. Note, however, that such combinations are based simply on working out whether any of the 95% exclusions apply to the arbitrary model, and they thus are not technically rigorous. More rigorous are the rare cases where experiments publish a correlation matrix alongside the simplified model results, which allows theorists to calculate an approximate likelihood.

12.4 Reinterpreting fiducial measurements

Fiducial measurements, in the form of total or differential cross-section measurements, are routinely used to tune the non-perturbative parameters of Monte Carlo event generators, to extract the parameters of parton distribution functions and the SM and to constain BSM physics. BSM tests can be carried out within effective field theory frameworks, or explicitly within the context of a specific BSM model. These measurements have been mostly targeted at regions of phase-space dominated by single SM processes, and with high enough statistical rates to permit a stable detector unfolding to cross-section estimates, by contrast with BSM-search analyses which are typically background-dominated and/or have low event rates due to tight selection cuts.

Differential fiducial cross-sections are usually presented in ATLAS and CMS measurement papers following an unfolding procedure such as those described in Chapter 11. It is also useful to publish additional theoretical correction factors such as those arising from electroweak final-state radiation and hadronisation, so that one can perform comparisons to theoretical predictions at different levels. Backgrounds to measurements are generally subtracted prior to, or during, the unfolding procedure, and the associated uncertainties are included in the measurement uncertainties. Whilst this does not affect reinterpretation for instrumental backgrounds, it is problematic in the case of irreducible backgrounds that contribute to the fiducial phase space, since the subtraction procedure may involve assuming that the SM is correct. It is therefore desirable, where possible, to define measured cross-sections in terms of the final state only, with no background subtraction. Distributions are usually published in electronic form in HepData, along with breakdowns of their systematic uncertainties. In order to make full use of measurement data, it is necessary to also publish covariance matrices or even more complete likelihood characterisations.

Accurate reinterpretation of measurements requires that the final state particles are described in detail, along with the fiducial kinematic region used. This information is usually somewhat simpler than that required in the description of particle searches, but the higher precision involved means that theorists will strongly benefit from being given extra information such as the precise version of the Monte Carlo generators used in the analysis, and the values of their relevant parameters. Comparison of the data with the SM expectation requires a detailed knowledge of what the SM prediction actually is. Typically, this will be evaluated directly by theorists, or produced by the ATLAS and CMS experiments by running open-source codes provided by theorists, and is increasingly also recorded in detail in HepData.

Interpretation of fiducial measurements is predominantly performed using Rivet, an MC event-analysis framework which implements standard definitions of many event properties and physics objects. Indeed it was the primary platform for developing many of the fiducial analysis ideas discussed in Section 11.1. Rivet also includes code routines corresponding to a significant fraction of suitable LHC experimental analyses, mostly unfolded but also several making use of smearing-based fast-simulation. These routines are largely provided by the LHC experiments themselves, each of which operates a Rivet-based analysis preservation programme for fiducial measurements: if you work on a measurement analysis, you should expect to write a Rivet analysis code to accompany the submission of the numerical results to HepData. The Rivet platform is further used in applications for MC generator tuning and development with the Professor and other optimisation toolkits, and in BSM model interpretations using a package called Contur.

12.4.1 Reinterpretation of Higgs searches

Measurements of the Higgs boson production cross-sections and branching ratios are particularly sensitive to BSM physics, since many popular BSM theories are expected to extend or perturb the Higgs sector of the SM. For this reason, a great deal of attention has been given to how best to present Higgs results for reinterpretation. Typically, it is assumed that the signals observed in various Higgs search channels originate from a *single* narrow resonance with a mass near 125 GeV, and that the particle is a CP-even scalar as expected in the SM. If ones further assumes that the Higgs has a negligible width (true for the SM, and thus true for theories that do not show a strong deviation from SM-like behaviour), the production and decay of the Higgs factorise. The simplest approach for presenting Higgs measurements is then to present them in terms of **signal strength** parameters. If we assume that the Higgs is produced and decays via some process $i \rightarrow H \rightarrow f$, the signal strength μ is the observed rate normalised by the SM prediction:

$$\mu_{if} \equiv \frac{\sigma_i \times B_f}{(\sigma_i \times B_f)^{\mathrm{SM}}} = \mu_i \times \mu_f, \qquad (12.4)$$

where σ_i is the Higgs production cross-section starting from the state i, B_f is the branching ratio to the final state f, and the final two quantities have been defined as the ratio of production cross-sections and ratio of branching ratios, respectively. Disentangling the production (μ_i) and decays (μ_f) can be performed if one supplies the additional assumption of SM branching ratios in order to extract μ_i, or of SM production cross-sections in order to extract μ_f.

The kappa framework
Whilst useful, this approach has clearly hidden a lot of information. The signal strength in any particular series of channels might change because the couplings of the Higgs boson to *specific particles* have changed, and it would be nice to parameterise these deviations directly. This is the basis of the so-called **kappa framework**, which dresses the predicted SM Higgs cross sections and partial decay widths with scale factors κ_i. The κ_i are defined so that the cross sections σ_i or the partial decay widths Γ_i associated with the SM particle i scale with the factor κ_i^2 when

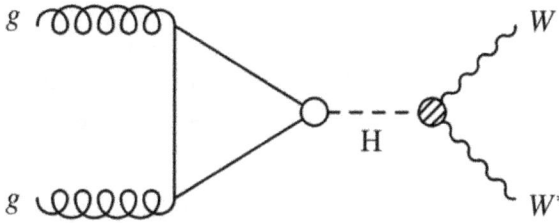

Figure 12.2. Feynman diagram for Higgs boson production via gluon–gluon fusion, with subsequent decay to W bosons (one of which must be off-shell).

compared with the corresponding SM prediction. For the process $i \to H \to f$, we can write the signal strength in terms of these κ_i factors as:

$$\mu_{if} = \frac{\kappa_i^2 \times \kappa_f^2}{\kappa_H^2} \qquad (12.5)$$

where we have defined $\kappa_H^2 = \Gamma_H/\Gamma_H^{SM}$. In the absence of new Higgs decays (to non-SM particles), κ_H can be expressed in terms of the individual κ_i values as

$$\kappa_H^2 \equiv \sum_j \frac{\kappa_j^2 \Gamma_j^{SM}}{\Gamma_H^{SM}}, \qquad (12.6)$$

which is nothing more than the statement that the total Higgs width has changed because the partial widths have changed. Why have we dressed the SM cross-sections and partial widths, rather than simply parameterising the couplings directly? The answer is that it allows us to retain the current best knowledge for the Higgs cross-sections, which include higher-order quantum chromodynamics and electroweak (EW) corrections. The κ_i correspond to additional leading-order degrees of freedom whose values tell us whether the observed Higgs couplings are compatible with SM predictions or not and, if not, which couplings show hints of BSM physics. Such a result would then motivate tests of explicit models of BSM physics in order to find which best matches the observed data.

To get a better idea of what the κ_i represent, consider the Feynman diagram for Higgs boson production via gluon–gluon fusion, with subsequent decay to W bosons, shown in Figure 12.2. Since the total mass of two W bosons exceeds 125 GeV, one of these W bosons must be off-shell. For the decay vertex, we are assuming that the vertex factor gets modified from the SM value by a factor κ_W, such that the branching fraction scales as κ_W^2. For the production, things are more complicated since there is no direct Higgs-gluon vertex. The lowest-order diagram instead involves a loop, and the ability of different particles to run in this loop means that a variety of κ_i factors are in principle relevant to the modification of the Higgs production cross-section in this case. We have two choices for how to proceed. The first is to simply replace the loop with an effective vertex, and treat the gluon fusion production cross-section as if it is modified by a single effective factor κ_g. The second is to use the known combination of diagrams in the SM alongside the κ_i factors for each diagram to calculate a **resolved scaling-factor**. For the present case, this can be shown to be $\kappa_g \approx 1.06 \cdot \kappa_t^2 + 0.01 \cdot \kappa_b^2 - 0.07 \cdot \kappa_t \kappa_b$. Whether to use an effective or resolved scaling factor depends on how many parameters a physicist wants to test against the Higgs data. It might be worth dropping multiple κ_i factors if they are poorly constrained by the data, since a lower-dimensional space requires less computational effort to explore. A full list of the resolved scaling factors for different production and decay modes is given in table 12.1.

Given a choice of which κ_i factors to include, experimental results can then be presented via a global fit of the set of κ_i parameters.

Simplified template cross-sections
The kappa framework also has its limitations, though. While it is of benefit for the immediate interpretation of Higgs measurements by an experiment, the reliance of the framework on the *current* precision of SM calculations makes signal-strength or kappa interpretations a secondary rather than primary research output. To be able to incorporate measurements into new fits and comparisons to improved theory in future, a simpler and more direct primary output is needed.

The obvious candidate, as we have argued earlier, is fiducial cross-sections—but they require high-enough bin populations to be stable in an unfolding process. For phenomena in an early stage of exploration, as the Higgs boson is at the time of writing, and as other particles may be in future, few single $i \to H \to f$ production+decay

Table 12.1. List of resolved scaling factors in the kappa framework.

Production/Decay mode	Resolved scaling factor
$\sigma(ggH)$	$1.04\kappa_t^2 + 0.002\kappa_b^2 - 0.038\kappa_t\kappa_b$
$\sigma(\text{VBF})$	$0.73\kappa_W^2 + 0.27\kappa_Z^2$
$\sigma(WH)$	κ_W^2
$\sigma(qq/qg \to ZH)$	κ_Z^2
$\sigma(gg \to ZH)$	$2.46\kappa_Z^2 + 0.47\kappa_t^2 - 1.94\kappa_Z\kappa_t$
$\sigma(ttH)$	κ_t^2
$\sigma(gb \to WtH)$	$2.91\kappa_t^2 + 2.31\kappa_W^2 - 4.22\kappa_t\kappa_W$
$\sigma(qb \to tHq)$	$2.63\kappa_t^2 + 3.58\kappa_W^2 - 5.21\kappa_t\kappa_W$
$\sigma(bbH)$	κ_b^2
Γ^{ZZ}	κ_Z^2
Γ^{WW}	κ_W^2
$\Gamma^{\gamma\gamma}$	$1.59\kappa_W^2 + 0.07\kappa_t^2 - 0.67\kappa_W\kappa_t$
$\Gamma^{\tau\tau}$	κ_τ^2
Γ^{bb}	κ_b^2
$\Gamma^{\mu\mu}$	κ_μ^2
Γ^H	$0.58\kappa_b^2 + 0.22\kappa_W^2 + 0.08\kappa_g^2$
	$+ 0.06\kappa_\tau^2 + 0.026\kappa_Z^2 + 0.029\kappa_c^2$
	$+ 0.002\,3\kappa_\gamma^2 + 0.001\,5\kappa_{Z\gamma}^2 + 0.000\,025\kappa_s^2 + 0.000\,22\kappa_\mu^2$

channels have enough available observed events to compute a fiducial cross-section, let alone a differential one. In addition, such rare-process analyses use every trick available to maximise signal-sensitivity, and the multivariate analysis (MVA)-based machinery ubiquitous to such searches is not yet well-explored for fiducial measurements.

For these reasons, the **simplified template cross-section** (STXS) framework has been developed, occupying a middle ground between signal-strength and kappa-parameter interpretations and true fiducial cross-sections. The key features of STXS measurements are:

1. they are defined inclusively in Higgs decay channels and kinematics (up to an overall rapidity-acceptance cut on the Higgs itself); and
2. they are specific to the Higgs production mode, and to the topology and kinematics of the associated physics-objects.

The first of these properties enables combination of multiple Higgs decay channels, giving the necessary statistical stability to unfold to the common set of 'STXS bins' in the production mode and event topology introduced by the second property. The simplifications of STXS—in integrating over all decay modes and in unfolding to parton-level, only indirectly observable physics objects—permit the use of MVA techniques such that the STXS-bin cross-sections can be mapped via a likelihood fit or similar techniques.

The STXS bins are standardised across the HEP community by the LHC Higgs Cross-section Working Group, such that different experiments have a common set of 'semi-fiducial' phase-spaces in which measurements can be compared and combined. The schematic design of the STXS bin-set is shown in Figure 12.3, where the full set of production+decay channels is decomposed into first the ggF, VBF, and VH production modes; then into sub-bins in jet-multiplicity, $p_T(V)$ etc; and as events accumulate and finer sub-divisions become possible, further nested bins. The initial set of STXS bins was denoted 'Stage 1', followed by a 'Stage 1.1' refinement, with anticipation of later 'Stage 2', etc—eventually, it will become possible to drop the simplification and perform even higher-precision Higgs interpretations via a holistic set of fiducial cross-section measurements.

12.5 Analysis preservation and open data

So far, we have largely assumed that reinterpretations are based on an external reconstruction of the analysis performed by the ATLAS and CMS experiments, which will necessarily not be a perfectly accurate recreation. In recent years, there has been a substantial effort invested in making the actual experimental analysis toolchain available for re-use by theorists, via a mechanism that also preserves the analysis code for re-use by the experimentalists themselves. The **Reana/Recast** system provides a fully functional environment that preserves the experimental software platform on which an analysis was originally performed, along with an implementation of the analysis code and data flow. Theorists outside of the ATLAS and CMS experiments are able to submit requests for new models to be analysed with existing data, in order to determine their viability. Although this

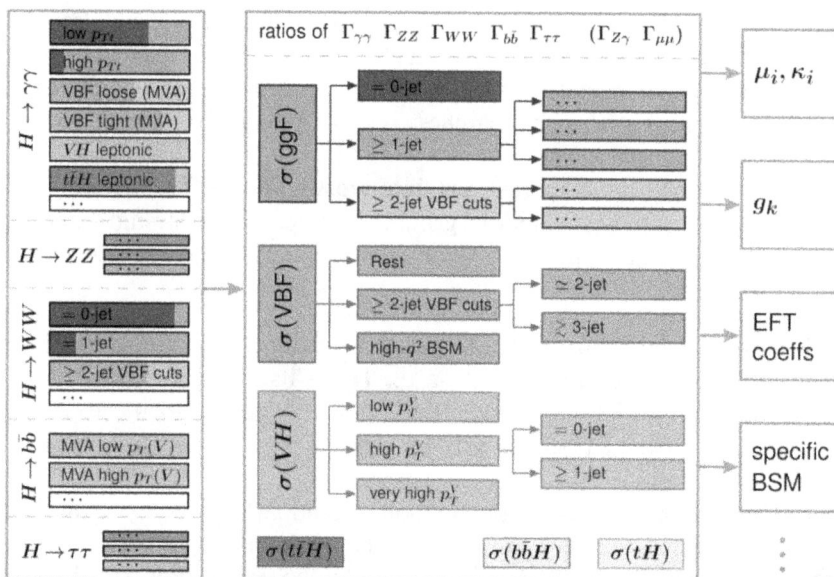

Figure 12.3. Schematic operation of the STXS framework for Higgs-boson production interpretations. The combination of the many different Higgs-boson decay modes is shown on the left-hand side, feeding into a set of standardised bins in production topology and kinematics, which become more deeply nested as data statistics permit.

currently comes with a substantial CPU and person-hour cost, it may be possible in future to use such systems for complex phenomenological studies. In the meantime, the system offers great utility for members of the ATLAS and CMS collaborations who wish to add value to their analyses by studying new theoretical models.

An even more extreme way to make ATLAS and CMS results fully interpretable by theorists is to release the actual data itself, and provide theorists with tools for analysing it. This is occurring through the CERN Open Data Portal, and it is important to note that it is just as important to release the ATLAS and CMS simulations of SM background events as it is to release their recorded proton–proton collision data. At the time of writing, the CMS experiment is leading the charge, having promised to make 100% of data available within 10 years of it first being recorded. ATLAS has made data available primarily for outreach and education projects, but it can be reasonably expected that more serious efforts will follow within the next few years.

One should never be naïve when starting an Open Data project. Extensive documentation is required to understand the content of the data, and how to reconstruct and use physics objects. This is much more natural for an experimentalist than a theorist, and there is currently a significant effort to create working analysis examples and simplified data formats for those not trained in the requisite techniques. Even if a physicist has the technical skills to process the data, the sheer size of the datasets means that a realistic analysis will require serious computing power. With cloud computing becoming ever more affordable, however, performing physics analyses on real LHC data is becoming a realistic option for phenomenologists, mirroring developments in other fields such as gamma ray astronomy.

Further reading

- The HepData database can be found at https://www.hepdata.net/.
- Various parts of this chapter are based on the excellent report 'Reinterpretation of LHC Results for New Physics: Status and Recommendations after Run 2', published in *SciPost Phys.* **9** 022 (2020). This reference contains links to many public software tools for the reinterpretation of LHC results.
- Public codes for reinterpreting LHC analyses in the context of new theories include GAMBIT (https://gambit.hepforge.org), Rivet (https://rivet.hepforge.org) and MadAnalysis (https://launchpad.net/madanalysis5).
- The κ framework for parameterising Higgs production and decay modes is described in arXiv:1307.1347.
- The STXS framework was originally introduced in the proceedings of the Les Houches Physics at the Terascale 2015 workshop (arXiv:1605.04692), and a report from the LHC Higgs Cross-section Working Group (arXiv:1610.07922). The Stage 1.1 STXS report (arXiv:1906.02754) also serves as a useful introduction, with the perspective of several years of Stage 1 operation.

Exercises

12.1 Prove equation (12.5).

12.2 Explain why the exclusion limits in the right-hand plot of Figure 12.1 change with the assumed branching ratios of the sparticles.

12.3 Write a more general form of equation (12.1) that is suitable for the case of two signal regions, with a 2×2 covariance matrix.

12.4 Explain how a cutflow table could be used to debug a generation and simulation software chain, assuming that one had the SLHA file for the physics model of interest.

12.5 If correlation information between analyses is not explicitly provided, which of these sets of analyses can be assumed safe to combine as if independent?

    (a) Any measurements/searches at different experiments,

    (b) any measurements/searches at different $\sqrt{s}$ energies,

    (c) same experiment and energy analyses with orthogonal cuts,

    (d) same experiment and energy analyses with overlapping cuts, one much tighter than the other?

12.6 SMEFT model spaces include all possible admixtures of BSM operators, including many configurations which are extremely hard to create through explicit BSM models. With this in mind, are marginalised/profiled EFT constraints always better than single-parameter fits? Can you think of a Bayesian way to better reflect the mixture of top-down and bottom-up constraints?

12.7 The Contur package for interpretation of fiducial measurements makes a default assumption that the measured data points represent the SM, with BSM effects treated additively. What does this imply about use of the method for BSM-model

    (a) discovery,

    (b) exclusion?

IOP Publishing

Practical Collider Physics

Andy Buckley, Christopher White and Martin White

Chapter 13

Outlook

In this book we have presented a sometimes intimidating array of information: from the theoretical foundations of the Standard Model and its extensions to the simulation of particle collisions in such models, from detector designs to reconstruction methods and techniques, and from abstract statistical concepts to their application in various types of collider data analysis.

But these ideas, while powerful both individually and in concert, are simply building blocks from which phenomenological or experimental data analyses can be constructed. Exactly how to assemble and connect them is for you to decide: what we have provided here is a set of components, not an end-to-end blueprint. And like any building project, it never runs precisely to plan: finding novel solutions to bridge the gaps between the well-defined techniques requires invention, creativity, and problem-solving. These live at the heart of collider physics research, which goes some way to explaining why it can remain so compelling even after many decades of practice. They are where science becomes, if not an art, certainly a craft.

Before concluding this book, we will give a brief summary of what to look out for in the near future of collider physics, which is set to remain an exciting field for many years to come.

13.1 The future of the LHC

Despite having been active for over a decade, we are currently very early in the life of the LHC, in terms of the amount of data that will eventually be taken. It is currently assumed that the LHC will run up to 2039, following the schedule shown in Figure 13.1. This alternates physics runs with long-shutdown periods during which accelerator and detector upgrade work can be performed.

A third run from 2022 to 2024 will operate at a luminosity of $2.0 \times 10^{34} \text{cm}^{-2} \text{ s}^{-1}$, with a centre-of-mass energy that finally reaches the original design value of 14 TeV. Following this, a much more major upgrade will occur that will keep the LHC energy fixed but dramatically increase the luminosity, which will reach $5.0 \times 10^{34} \text{cm}^{-2} \text{ s}^{-1}$ in a

doi:10.1088/978-0-7503-2444-1ch13

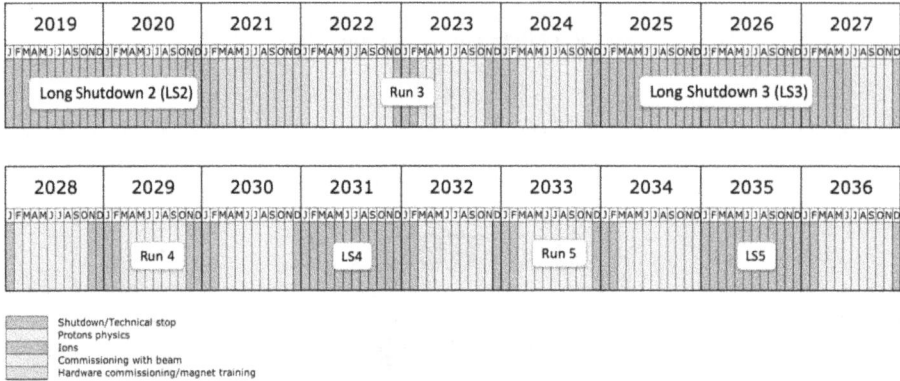

Figure 13.1. The LHC operations schedule up to 2036. Credit: https://lhc-commissioning.web.cern.ch/schedule/images/LHC-longterm-schedule-sep20.png.

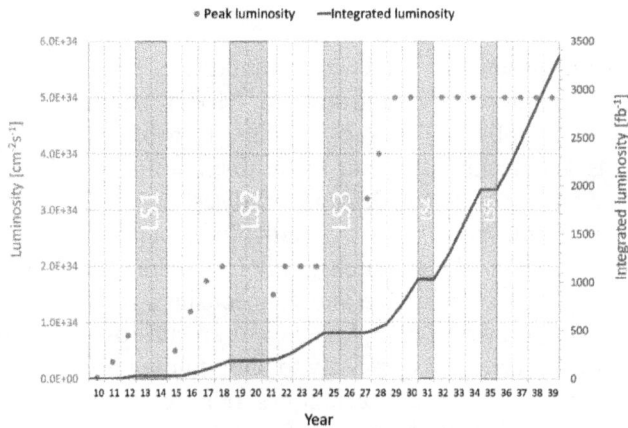

Figure 13.2. The expected LHC luminosity shown as a function of year. Red dots show the peak luminosity (which reflect the past or anticipated operating conditions of the LHC), whilst the blue solid line shows the past or expected integrated luminosity. Credit: https://lhc-commissioning.web.cern.ch/schedule/images/LHC-nom-lumi-proj-with-ions.png.

run period starting in 2027. This is called the **high-luminosity LHC**, or **HL-LHC**. These machine conditions will persist for the remaining LHC run periods scheduled up to 2039, with the final aim of collecting approximately 50 times more data than has currently been obtained. The accumulation of integrated luminosity out to 2039 is shown in Figure 13.2, which makes the impact of the HL-LHC very clear. Making the most of this deluge of data, however, will require not just development of new detector elements with the necessary resolutions, readout rates and radiation hardness, but also major reworkings of the computational data processing chain, both for collision data and simulation. The scale of this challenge is clear from the comparison of computing resources to requirements in Figure 13.3.

Note, however, that the accumulation of integrated luminosity is relatively modest before then, which is often a cause of pessimism in the high-energy physics

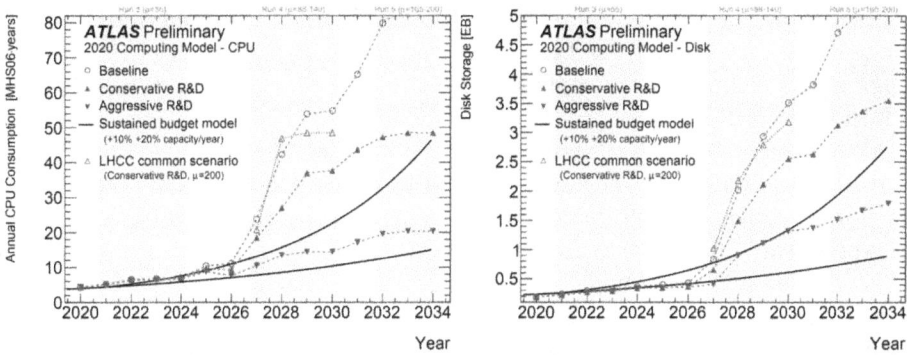

Figure 13.3. The ATLAS computing model's comparison of CPU and disk capacity to expected demand, looking forward from Run 3 to the HL-LHC runs. The need for intense technical development through the simulation and data-processing chain is starkly clear. Credit: ATLAS HL-LHC Computing Conceptual Design Report, CERN-LHCC-2020-015, https://cds.cern.ch/record/2729668.

community. It is often assumed, for example, that the lack of evidence for beyond-the Standard Model (BSM) physics in the LHC Run 1 and Run 2 datasets implies that the mass of new particles is high enough to guarantee a relatively small cross-section for their production. One therefore needs to accumulate a lot more data in order to find evidence for these particles, and the relatively slow increase in the amount of data in the next few years is unlikely to produce a dramatic discovery. Against this, however, is the fact that it is known that many searches for new particles contain substantial gaps in their reach *even within the Run 1 and Run 2 data* (see, for example, Section 10.1.2), and that we need cleverer, and more model-independent analysis techniques to close these gaps. We therefore remain confident that it is all still to play for at the LHC, and that commencing graduate students can expect exciting discoveries at any time provided they retain a commitment to innovation.

One obvious tool—or rather a whole suite of tools—for such innovation is statistical machine learning. Techniques under this umbrella, such as deep neural networks, are huge subjects in their own right, and are evolving so rapidly that it is not currently possible to cover them in this volume without both inflating its size and ensuring rapid obsolescence. But the reader will undoubtedly be aware of the potential of such methods for addressing thorny problems, from quantum chromodynamics (QCD) calculations to detector simulation, to smoothing and interpolation of background estimates, and directly in statistical inference from measurements. They are no 'silver bullet' with guaranteed easy wins, but in combination with solid physical foundations it is likely that the future of collider physics will heavily rest on novel applications of such techniques.

13.2 Beyond the LHC

Running in parallel with the remaining data taking of the LHC is a considerable international effort to design the particle accelerators that will ultimately supersede it. The recent history of the field has seen the field alternate between using a hadron collider for discovery (e.g. the Super Proton Synchrotron at CERN) and a lepton collider for precision (e.g. the Large Electron Positron collider at CERN), before swinging back again to hadron colliders (e.g. the Tevatron at Fermilab and the Large Hadron Collider at CERN).

This would suggest that the next wave of collider proposals would be biased towards lepton colliders, and indeed this is currently the case. The argument for precision machines stems ultimately from the discovery of the Higgs boson. Much as the experiments of the Large Electron Positron collider made precision measurements of the W and Z bosons after their discovery by the Super Proton Synchrotron experiments, future lepton colliders could make precise measurements of the Higgs boson. By the end of the HL-LHC run, the ATLAS and CMS experiments will be able to study the main Higgs boson properties with a precision of approximately 5%–10%, limited by the large QCD background and the expected rates of pile-up. A proposed circular collider in China—the Circular Electron Positron Collider (CEPC)—would collide electrons and positrons at a centre-of-mass energy of 240 GeV, which represents the optimum energy for the production of a Z boson with a Higgs boson. The same collider could run at 160 GeV to produce pairs of W bosons and 91 GeV to produce Z boson events, and the tunnel could be used for a later hadron collider. The experiment could be operating by 2030, which is the most aggressive timescale of the current proposals, overlapping substantially with the running of the HL-LHC.

A competing European proposal, the Future Circular Collider at CERN, would involve the construction of a new 100 km circumference tunnel at CERN, that would first host an electron–positron collider running at energies up to 240 GeV, then on to 350 GeV to 365 GeV by 2045 or so. Again, the tunnel can be used for a later hadron collider, which in this case is proposed to have a centre-of-mass energy of 100 TeV. The catch is that, even if funded, the hadron collider would not realistically start taking data until around 2050 at the earliest.

Competing with these circular collider proposals are linear colliders, which, by virtue of not suffering from the synchrotron-radiation problem that affects circular colliders, are able to target higher centre-of-mass energies. Current proposals include the International Linear Collider (ILC)—that would commence at a centre-of-mass energy of 250 GeV shortly after 2030, before running at 500 GeV and 350 GeV from 2045—and the Compact Linear Collider (CLIC) that would start at a centre-of-mass energy of 380 GeV before targeting energies up to 3 TeV in the decades thereafter. These higher energies would permit an extensive top quark physics programme.

Whatever the future holds (and there are many more proposals, e.g. for lepton–hadron colliders, muon colliders, or entirely new ways of accelerating particles), collider physics is sure to remain an active field with tens of thousands of international participants. With its blend of topics at the frontier of both theoretical and experimental physics, it remains hard to imagine a more exciting field in which to commence study.

Part III

Appendix

Appendix A

Useful relativistic formulae

A.1 Summary of useful relativistic quantities

In Section 1.2, we stated various formulae that are useful in relativistic kinematics, including the relativistic energy and three-momentum, given, respectively, by

$$E = \gamma mc^2, \quad \boldsymbol{p} = \gamma m\boldsymbol{v}, \tag{A.1}$$

where the γ factor is given in terms of the velocity v by

$$\gamma = \frac{1}{\sqrt{1 - \beta^2}} = \frac{1}{\sqrt{(1 - \beta)(1 + \beta)}}. \tag{A.2}$$

As this is a little awkward to work with, it is usual to define the variable $\beta \equiv v/c$, giving the slightly clearer

$$\gamma \equiv \frac{1}{\sqrt{1 - \beta^2}} = \frac{1}{\sqrt{(1 - \beta)(1 + \beta)}}. \tag{A.3}$$

We can invert this equation to obtain

$$\beta = \sqrt{1 - \frac{1}{\gamma^2}}. \tag{A.4}$$

Note that we can also write the useful formulae $\gamma = E/m$ and $\beta = p/E$.

A.2 Changes of variables in four-vector representations

After reconstructing the energy and momentum of each particle at the LHC, there are a variety of ways to store them in a four-vector. Assuming that we are using natural units, one may, for example, store and manipulate the (E, p_x, p_y, p_z) vector for each particle, and calculate quantities such as the rapidity and invariant mass directly from that. Note, however, that when $|\boldsymbol{p}|$ is much greater than the invariant

doi:10.1088/978-0-7503-2444-1ch14

mass of the particle m, calculating m relies on the subtraction of two large numbers (through $E^2 - |\boldsymbol{p}|^2$). This in turn can generate an apparently negative mass through numerical instability, which will break any code that relies on a positive mass. It is therefore preferable to store four-vector components in an alternative form such as (m, p_x, p_y, p_z) from which one can calculate the energy safely via $E^2 = |\boldsymbol{p}|^2 + m^2$. Since analyses typically rely on using η and p_T, it can also be convenient to use the choice of variables (p_T, η, ϕ, m), or the true-rapidity version (p_T, y, ϕ, m). The polar angle θ with respect to the positive beam direction may also be useful, particularly for cosmic-ray or forward-proton physics where other parametrisations are numerically unreliable for expressing small deviations from the z axis.

Below, we provide a shopping list of formulae that will assist in converting between different choices of variables. Typically, these will be implemented inside an analysis package, such as the ubiquitous ROOT.

$$\eta = -\ln \tan(\theta/2) = \text{sign}(z) \ln\big[(|\,p\,| + |z|)/p_T\big]$$

$$y = \frac{1}{2} \ln\left[\left(E + p_z\right)\Big/\left(E - p_z\right)\right]$$

$$\theta = 2 \tan^{-1}(\exp(-\eta))$$

$$p = p_T / \sin(\theta)$$

$$p_T = \sqrt{(p_x^2 + p_y^2)}$$

$$E = \sqrt{p^2 + m^2} = \left(\sqrt{p_T^2 + m^2}\right)\cosh y$$

$$p_x = p_T \cos(\phi)$$

$$p_y = p_T \sin(\phi)$$

$$p_z = \left(\sqrt{p_T^2 + m^2}\right)\sinh y = p \cos(\theta).$$

www.ingramcontent.com/pod-product-compliance
Lightning Source LLC
Chambersburg PA
CBHW082120210326
41599CB00031B/5826